P9-CKX-313

DESIGNING LEARNING ENVIRONMENTS FOR DEVELOPING UNDERSTANDING OF GEOMETRY AND SPACE

STUDIES IN MATHEMATICAL THINKING AND LEARNING

Alan H. Schoenfeld, Series Editor

Carpenter/Fennema/Romberg (Eds.) • *Rational Numbers: An Integration of Research*

Cobb/Bauersfeld (Eds.) • *The Emergence of Mathematical Meaning: Interaction in Classroom Cultures*

English (Ed.) • *Mathematical Reasoning: Analogies, Metaphors, and Images*

Fennema/Nelson (Eds.) • *Mathematics Teachers in Transition*

Fennema/Romberg (Eds.) • *Mathematics Classrooms That Promote Understanding*

Lajoie • *Reflections on Statistics: Learning, Teaching, and Assessment in Grades K–12*

Lehrer/Chazan (Eds.) • *Designing Learning Environments for Developing Understanding of Geometry and Space*

Ma • *Elementary Teachers' Mathematical Knowledge and Its Relationship to Teaching Competence: A United States–China Comparison*

Reed • *Word Problems: Research and Curriculum Reform*

Romberg/Fennema/Carpenter (Eds.) • *Integrating Research on the Graphical Representation of Functions*

Schoenfeld (Ed.) • *Mathematical Thinking and Problem Solving*

Sternberg/Ben-Zeev (Eds.) • *The Nature of Mathematical Thinking*

Wilcox/Lanier (Eds.) • *Using Assessment to Reshape Mathematics Teachers: A Casebook for Teachers and Teacher Educators, Curriculum, and Reform*

DESIGNING LEARNING ENVIRONMENTS FOR DEVELOPING UNDERSTANDING OF GEOMETRY AND SPACE

Edited by

Richard Lehrer
University of Wisconsin–Madison

Daniel Chazan
Michigan State University

LEA LAWRENCE ERLBAUM ASSOCIATES, PUBLISHERS
1998 Mahwah, New Jersey London

Lawrence Erlbaum Associates, Inc., Publishers
10 Industrial Avenue
Mahwah, NJ 07430

Cover design by Kathryn Houghtaling Lacey

Library of Congress Cataloging-in-Publication Data

Lehrer, Richard.
Designing Learning Environments for developing understanding of geometry and space / Richard Lehrer, Daniel Chazan
p. cm.
Includes bibliographical references and indexes.
ISBN 0-8058-1948-7 (alk. paper). —ISBN 0-8058-1949-5 (pbk. : alk. paper)
1. Geometry—Study and teaching. I. Chazan, Daniel. II. Title.
QA461.L45 1998
516'.0071—dc21 97-48323
CIP

Books published by Lawrence Erlbaum Associates are printed on acid-free paper, and their bindings are chosen for strength and durability.

The final camera copy for this work was prepared by the author, and therefore the publisher takes no responsibility for consistency or correctness of typographical style.

Printed in the United States of America
10 9 8 7 6 5 4 3 2 1

Contents

Preface

Despite a long intellectual history dating back to the origins of civilization, and a recent resurgence as cutting-edge mathematics, geometry and spatial visualization in school are often compressed into a caricature of Greek geometry, generally reserved for the second year of high school. The resulting impoverished view of the mathematics of space rebounds throughout schooling generally to diminish student (and adult) understanding of mathematics.

Among mathematicians and mathematics educators, there is increasing consensus, however, that geometry and spatial visualization deserve a more prominent role in school mathematics. Formalist views of mathematics as a "game" in which abstract symbols are manipulated (views ascendant in the second half of the 19th and the early part of the 20th century) are now challenged by views emphasizing the role of "empirical" methods in mathematics. Contemporary mathematicians studying chaos, fractals, and nonlinear dynamics rely on computer-generated visual representations to perform and display the results of experiments. Moreover, not only do new computer technologies make mathematical experimentation possible and plausible, but these technologies have been widely adopted in a range of cultural practices. At the same time, the mathematics education reform movement accords a central role to mathematical exploration and sense-making and supports the use of technology and visual representations. This shift implies the need to reexamine the nature of the school mathematics curriculum, the goals and aims of teaching, and the design of instruction.

Rather than looking to high-school geometry as the locus (and all too often, the apex) of geometric reasoning, the authors of this volume, many of whom were active in the National Center for Research in Mathematical Sciences Education (NCRMSE), suggest that reasoning about space can and should be successfully integrated with other forms of mathematics, starting at the elementary level and continuing through high school. Reintegrating spatial reasoning into the mathematical mainstream (indeed, placing it at the core of K–12 mathematics environments that promote learning with understanding) will mean increased attention to problems in modeling, structure, and design and reinvigoration of traditional topics like measure, dimension, and form: Geometry education should include contributions to the mathematics of space that were developed after those of the Greeks.

This volume reflects our appreciation of the interactive roles of subject matter, teachers, students, and technologies in designing classrooms that promote understanding of geometry and space. Although these elements of geometry education are mutually constituted, the volume is organized to highlight our vision of a general geometry education, the development of student thinking in everyday and classroom contexts, and the role of technologies. Every contributor addresses all of these themes, albeit in different proportions.

Part I provides readers with a broad view of geometry education, embracing both tradition and innovation in the nature of the mathematics of space. In chapter 1, Goldenberg, Cuoco, and Mark illustrate the central role of spatial reasoning not only in all strands of mathematics—not solely those traditionally deemed "geometry"—but also in the development of reasoning (habits of mind), conjecturing, and argumentation, a theme Koedinger returns to in Part II. In chapter 2, Gravemeijer suggests that everyday experiences (and the informal knowledge that even very young children construct from them) serve as the foundation of a mathematics of space: What children first understand intuitively and informally about space can be progressively "mathematized" (redescribed and elaborated mathematically) through guided reinvention. This approach, he argues, together with the use of real-world contexts, not only provides a solid base for geometry education but also supports students' ability to reflect and reason outside the realm of mathematics. In the third chapter, Chazan and Yerushalmy decant the old wine of Euclidean geometry into the crystal of classroom argument and electronic technologies. They remind us that for the Greeks, proof was a form of argument, not a ritual, and they suggest that electronic tools make new forms of argument in the classroom both possible and fruitful. Devaney concludes this first section of the volume by extending the reach of geometry education to include the contemporary topic of chaos. He describes how new computer technologies for envisioning and playing chaos games in the classroom can be used to extend the mathematical reach and grasp of high-school students.

Part II of the volume consists of a series of studies in geometry education. The studies span the K–12 years and encompass a variety of methodologies, ranging from longitudinal studies of classroom learning (and broad-scale developmental trends) to case studies of individual reasoning about space (and transitions in individual strategies). What students explore in these studies is diverse, ranging from contexts in physics (e.g., floating and sinking), design (e.g., quilts), and architecture (e.g., structure and loads), to traditional problems involving shape, measure, and form.

The first two chapters in Part II give broad-scale perspectives on the growth and development of reasoning about space in traditional contexts. In the first chapter, Pegg and Davey suggest a synthesis of the van Hiele framework and SOLO taxonomy as a way of looking at levels of thinking, cycles of learning, and modes of reasoning. The second chapter, by Lehrer, Jenkins, and Osana, takes a closer, longitudinal look at continuity and change in student reasoning in geometry in the elementary grades.

The next three chapters in the second section make a transition from traditional contexts to those designed explicitly to promote the development of understanding of space. Lehrer, Jacobson, and colleagues describe primary-grade classrooms where children develop understanding about geometric ideas (e.g., transformations, area measure) through the interplay of tasks and contexts like wayfinding and quilt design and through participation in a classroom culture that emphasizes mathematical justification and argument. Clements, Battista, and Sarama report studies of transitions in children's thinking about length and angle as children explore linear-measure ideas and turns in computer (a modified form of Logo) and noncomputer activities. In a related vein, Battista and Clements explore students' understanding of three-dimensional cube arrays and the enumeration strategies students use as they begin to coordinate and integrate side, top, and front views to develop concepts of spatial structure.

The remaining four chapters in Part II describe conceptual change in older (middle- and high-school) students. Emphasizing that geometry need not involve only traditional topics, Middleton and Corbett examine middle-school students' ideas about rigidity and deformation of structure in contexts of architecture and engineering design. Also looking at middle-school students, Raghavan, Sartoris, and Glaser examine the power of physics contexts involving floating and sinking as opportunities to foster conceptual change about area and volume measure. Chapters by Dennis and Confrey and by Koedinger use case-study methodologies to examine spatial reasoning at the high-school level. Dennis and Confrey examine the development of one student's "voice" and reasoning as he works with geometric curve-drawing devices (resurrected by the authors' analysis of the history of the development of function) to (re)discover algebraic equations about the curves, or functions that he (physically) creates and explores. In the last chapter of Part II, Koedinger analyzes students' conjecturing skills (examining and creating two-dimensional kites), develops a cognitive model, and suggests (from a base of deductive proofs) appropriate tasks and software to enhance or develop students' conjecturing and argumentation skills, noting that these skills are important to adult life as well as to high-school geometry.

The third and final section of the volume explores the potential of electronic technologies in geometry education. The interactivity of these tools makes even abstract qualities of space tangible, and so affords new opportunities for invention and exploration of models of space. Even when these technologies produce screen objects with properties not found in the world around us (for example, when the user can "drag" points on a shape to deform it), those screen objects often still retain a concreteness and tangibility usually associated with physical objects. The first three chapters in Part III explore the implications of dynamic-geometry tools for the development of student thinking. Goldenberg and Cuoco take a close look at dynamic geometry, specifically at the feature known as "dragging" and the potential consequences of this feature for student perception and conception of geometric figures. De Villiers then criticizes the traditional use of proof in geometry and suggests the use of proof (in the environment of dynamic geometry) as a means of exploration, verification, and most importantly, explanation. Olive, in the following chapter, surveys the variety of ways in which young and old students explore shape and form using dynamic software, suggesting that one implication of the use of dynamic tools is the eventual fusion of algebra and geometry.

The final three chapters of Part III take a different tack. Watt examines fifth-grade students' mapping of a large-scale space (their classroom) with the aid of a computer-assisted design tool—a KidCAD. Here the technology facilitates a transition between experience and geometric representation. Further highlighting the diverse uses of technology, Zech and her colleagues at the Cognition and Technology Group at Vanderbilt University discuss and illustrate their work in developing visual (computer-based) toolkits to help middle-school students carry out real-world activities such as wayfinding and the designing of playgrounds. In the last chapter, Renninger, Weimar, and Klotz indicate that if we recast the form, substance, and tools of a geometry education, then it also becomes imperative to reconsider teachers' professional development. They discuss the development of a national electronic forum for geometry teachers and their students that encourages learning, discussion, and problem solving, as well as providing a resource for the field.

As Hershkowitz suggests in the epilogue, the contributors to this volume advance several related agendas for mathematics education. First, the authors help us better understand the wide range and influence of spatial reasoning and geometry in mathematics. The research presented here suggests that instead of the current arrangement of years of arithmetic with occasional small helpings of geometry, geometry and spatial reasoning can and should be incorporated as a central feature of a general mathematics education: geometry for all. Second, the contributors emphasize the diversity and range of student thinking encompassed by spatial reasoning and geometry. Not only are existing theories called into question, but several fruitful avenues for new theoretical development in mathematics education are suggested. Third, contributors explore how the development of spatial thinking is tied to tools, ranging from modest (but powerful) ones like Polydrons™ to mechanical curve-drawing devices and the new notational forms made possible by computer-based technologies. Taken together, the research suggests renewed curricular focus on geometry and space: Geometry is not only central to reform in mathematics curricula and the instructional focus on learning with understanding, but with its inherent (and in these studies, enhanced) emphasis on conjecture, argumentation, deductive proof, and reflection, is also central to a solid general education and, as many in this volume note, to good habits of mind.

ACKNOWLEDGEMENTS

Preparation of this volume was supported by the National Center for Research in Mathematical Sciences Education (NCRMSE), which in turn was supported by a grant from the Office of Educational Research and Improvement, U.S. Department of Education (grant No. R117G1002) and by the Wisconsin Center for Education Research, School of Education, University of Wisconsin–Madison. Any opinions, findings, or conclusions are those of the authors and do not necessarily reflect the views of the supporting agencies.

The work reported in this volume had a wide sponsorship, including the NCRMSE and the Geometry Working Group at the 1992 International Congress of Mathematics Education. We all thank Fae Dremock, our technical editor, for shepherding us through the many phases of creation and production involved in such a large scope of work. Thanks also to Lynn Levy for technical assistance (and moral support), and to our colleagues, especially the series editor, Alan Schoenfeld, all of whom helped us craft a collective vision of new spaces for teaching and learning mathematics.

Part I

Why Teach Geometry?

Over the past half-century in the United States, geometry has taken less and less of a central role in mathematics curricula. Many have even questioned its relevance in this high-tech, computer-dependent world. Among mathematics educators, researchers, mathematicians, and scientists, however, criticism has focused on the lack of emphasis in mathematics education on reasoning about geometry and space, on the traditional (Euclidean) emphasis on proof, and on the approaches taken to teaching geometry and space. These criticisms and the mathematics reform movement associated with the NCTM *Standards* (1989, 1991, 1995) form the backdrop to the first section of this book. Why, indeed, should students study geometry? How and when do we begin to teach geometry? What should the study of geometry entail? How do we generalize geometry to children's overall education?

Without linking their discussion to particular grade levels or courses, Goldenberg, Cuoco, and Mark make a strong plea for regular attention to geometry throughout the curriculum. Their examples—from the dissection of shapes to the use of area models for multiplication, from proofs without words to dynamically linked graphs—suggest a wide range of activities possible in a geometry curriculum. Throughout their chapter, they examine the central role of visualization in mathematics—noting along the way the importance of developing habits of mind that create a fertile environment for reflection, argument, and thought.

Gravemeijer, in the following chapter, takes a close look at how we teach geometry, suggesting (from 20 years of experience in The Netherlands with Realistic Mathematics Education) that geometry instruction build on children's informal knowledge of their environment. Drawing on Dutch curricular materials and examples taken from an American reform curriculum, he suggests ways of supporting children as they build their own models of geometric concepts and of guiding them through the reinvention (and abstraction) of the intended mathematics.

In the context of secondary mathematics reform, Chazan and Yerushalmy suggest that it is more important to focus on students' classroom activity than on the particular type of geometry taught. Although they appreciate the argument for introducing students to new mathematical advances, they argue that there still are benefits to be derived from courses in Euclidean geometry, provided that the way such courses are taught changes. Noting Schoenfeld's (1988) scathing critique of well-taught traditional geometry courses, they argue that dynamic-geometry programs support the creation of radically different Euclidean-geometry courses. In such courses, students' conjectures play a central role, and deductive reasoning, rather than being merely a two-column ritual, becomes a method of justification and communication.

In the last chapter in this section, Devaney suggests bringing the fields of contemporary mathematical research into middle- and high-school geometry classrooms. Noting the availability of new technologies, he argues that access to such topics as chaos and fractals—when introduced appropriately (e.g., through "The Chaos Game" he describes)—can not only introduce students to a wide range of concepts, but also stimulate, perhaps even excite their interest in mathematics as a whole.

Together, these four chapters attempt to redirect our thinking on geometry education, pointing us in rich directions for enhancing and reforming the teaching and learning of geometry and space.

REFERENCE

National Council of Teachers of Mathematics. (1989). *Curriculum and evaluation standards for school mathematics.* Reston, VA: Author.

National Council of Teachers of Mathematics. (1991). *Professional standards for teaching mathematics.* Reston, VA: Author.

National Council of Teachers of Mathematics. (1995). *Assessment standards for school mathematics.* Reston, VA: Author.

Schoenfeld, A. (1988). When good teaching leads to bad results: The disasters of "well taught" mathematics courses. In T. P. Carpenter & P. L. Peterson (Eds.), *Learning from instruction: The study of students' thinking during instruction in mathematics* [Special issue]. *Educational Psychologist, 23*(2), 145–166.

1

A Role for Geometry in General Education

E. Paul Goldenberg, Albert A. Cuoco, and June Mark
Education Development Center, Inc.

Readers of a book with this title certainly don't need to be sold on the virtues of geometry. But virtues are not enough to justify elevating any one body of knowledge or ideas—whether as a discrete course, or as a thread interwoven into other courses—to a status that takes significant time away from other virtuous bodies of knowledge or ideas. There is a great deal worth learning, and it cannot all be taught in school. So, why geometry?

Geometry is not merely an attractive side dish in a balanced mathematical diet, but an essential part of the entrée. We make two claims: Geometry, broadly conceived, can help students connect with mathematics, and geometry can be an ideal vehicle for building what we call a "habits-of-mind perspective." After a brief outline of the argument, the majority of this chapter is devoted to giving meaning and support, with examples: the notion of "geometry, broadly conceived," the proposition that it helps students *connect*, the further claim that the connection is *with mathematics,* and the notion of a habits-of-mind perspective and its importance.

Claim 1: Geometry Can Help Students Connect with Mathematics

- Students connect well with properly selected geometric studies. The many hooks include no less than art, physical science, imagination, biology, curiosity, mechanical design, and play.
- Geometry also connects richly with the rest of mathematics. Assuming a kind of transitive property, geometry's connections, both to students and to the rest of mathematics, might help us build good bridges, attracting more students—and a more diverse group of students—to mathematics in general.

As we explain briefly next (for greater detail, see Goldenberg, 1996), the needs of both the general education student and those students who will follow on in mathematics-intensive fields are best served by adding a new perspective to the mathematics curriculum—a perspective that roughly parallels the role that "the scientific method" plays in science classes, providing a counterpoint to a facts-centered curriculum.

To pursue any field (craft, trade, profession, or discipline) we need, of course, to know its specialized facts and tools, but we also need to know its way of thinking. All the more so in mathematics, the very facts and tools of which are, or are about, objects of the imagination, and whose entire substance is thoughts and ideas.

Now, it makes little sense to try to teach ways of thinking in isolation from the relevant facts, contexts, and so on. Still, just as ingredients such as love, hate, greed, and power can be rearranged to tell stories with many different messages, so also can basic mathematical ingredients be rearranged and emphasized (in a course) to tell a variety of mathematical stories. A current trend in mathematical storytelling focuses on applications—how a body of facts and ideas is used. Occasionally in college (but rarely before), the story brings out the historical development of mathematical ideas. Perhaps the most prevalent curricular story is still a tale of logic. (In actual implementation, this tale is often so abridged that the original logic is apparent only to the teacher, not to the student.) In this story, mathematical facts and ideas are presented in a sequence, with each rung of the ladder building directly from the previous one. But mathematical facts are seldom discovered in such an orderly fashion, nor does this logical sequence accord well with what is known about how students learn mathematics. (How common it is to read a problem solution and feel, "What a clever trick to use here! Why, it makes things so simple, even *I* can understand it, but how on earth did anyone ever think of it in the first place?")

Along with whatever other story a mathematics course tells, we believe it must also tell a story about thinking—powerful, mathematical thinking. Such a story organizes its facts not around themselves but around how people find the facts—ways of thinking that are demonstrably essential in just about any endeavor that requires serious reasoning or investigation.

For a general education, such a "habits-of-mind" perspective is important precisely because these mathematical ways of thinking have valuable application outside of mathematics as well as within it. But this approach also serves those students who will someday pursue interests that require advanced mathematical study. For students to understand mathematics, they must learn how to think from a mathematical point of view. For students to pursue advanced mathematical study, they must spend some of their time learning to "think like the professionals."

Claim 2: Geometry Can Be an Ideal Vehicle for Building the "Habits-of-Mind" Perspective

- Within mathematics, geometry is particularly well placed for helping people develop these ways of thinking. It is an ideal intellectual territory within which to perform experiments, develop

visually based reasoning styles, learn to search for invariants, and use these and other reasoning styles to spawn constructive arguments.
- Geometry, broadly conceived, is also ideally placed for helping to expand a student's conception of mathematics.

To achieve the goal of attracting a wide variety of students to a richer mathematics in this way, we must create curricula that empirically do connect geometry with students, do connect it with the rest of mathematics, and do highlight important mathematical ways of thinking—including the interplay of experimentation (both thought and concrete), informal and formal description, and theoretical analysis. And we must also not exaggerate the message: Mathematical ways of thinking are not mathematical without mathematics, so the geometrical (and other mathematical) content must be clear and rich.

In this chapter, we discuss all these features of curriculum—the hooks or connections with the student, the rich interconnections with other mathematics, and the thinking styles or habits of mind that geometry can foster—but the examples we give do not fall neatly into any one of these categories, for wherever it is reasonable it must also be deemed optimal that each activity contain all three features.

Before entering this main argument, however, we need to say more about visualization and visual thinking, which are at the heart of what makes geometry a special case within mathematics.

VISUALIZATION AND VISUAL THINKING

In the most common curricula, both in and out of the United States, geometry represents the only visually oriented mathematics that students are offered. Curricula tend to present an otherwise visually impoverished, nearly totally linguistically mediated mathematics, a mathematics that does not use, train, or even appeal to "the metaphorical right-brain" (see Tall, 1991, for an argument that visual thinking can play a key role in developing students' understanding).

There is a huge cost in this state of affairs: Some students who would like a visually rich mathematics never find out that there is one because they've already dropped out before they've had the chance to encounter any of the more visual elements.[1] We lose not only potential geometers

---

[1]Only about 50% of U.S. students ever take high-school geometry. Even within geometry, we often see more emphasis on verbal than on visually based reasoning. Part of the difficulty students have with proof as a mathematical method may indeed stem from its apparent redundancy: Proof is often first emphasized in geometry, a subject in which many (although certainly not all) of the proofs seem unnecessary because the visual, itself, makes the result so obvious. Healy (1993), a Los Angeles teacher, related the story of his class in which groups of students present their own experimental results and reasoning to the class and vote on whether or not to include these in a cumulative volume representing the class's

and topologists in this way, but all students who might enter mathematics through its visually richer domains and then discover other worlds, not as intrinsically visual, to which they can apply their visual abilities and inclinations.

For some students, a visual approach may be absolutely essential (Krutetskii, 1976; Mayer, 1989). For such students, including many who consider themselves to be poor at mathematics, visual approaches are *access*. Thus, for many students, visualization and visual thinking serve not only as a potential hook, but also as the first opportunity to participate.

Deliberately enriching the visual education in the curriculum may also contribute to reducing an old equity problem. Certain visualization skills correlate well with, and plausibly contribute to, better mathematical achievement in the upper grades and college. In turn, these skills are themselves plausibly honed through informal experiences at home, like building models, manipulating structured visual materials like blocks or Legos, and taking things apart and putting them back together. To the extent, then, that out-of-school experiences with such materials may help build the foundations for mathematical achievement, their differential availability to children—more boys are encouraged to build models and play with Legos than girls, and more affluent than nonaffluent children have these toys—might further encourage us to provide equalizing opportunities in school.

But visualization and visual thinking are far more than access, preparation, and motivation: They are worthy content themselves. Visual representations help people gain insights into calculations in arithmetic and algebra. Properties of mathematical processes are often discovered by studying the geometric properties of their visual representations. In fact, geometry has historically played a highly generative role in the development of mathematics, and geometric techniques and visual images remain essential tools and sources of inspiration for mathematicians. By ignoring visualization, curricula not only fail to engage a powerful part of students' minds in service of their mathematical thinking, but also fail to develop students' skills at visual exploration and argument. These, according to Eisenberg and Dreyfus in their article "On the Reluctance to Visualize in Mathematics" (1991), are skills that most mathematicians use in their daily work, despite their reluctance to accept visual approaches as finished work.

---

work over the year. In one class, he recalls, students were adamant about not including a proof that the perpendicular bisector of a line segment was equidistant from the endpoints of the segment. They considered the result trivial and unworthy of the text they were writing. Although we could argue that by regarding the result as self-evident they were ignoring proof as an important element of mathematical thinking, we cannot reasonably expect beginning students to become embroiled in the subtleties of mathematical proof before they have experienced the kind of refutation of their conjecture that makes them feel a need for further analysis. Fourteen-year-olds are not, after all, likely to behave like Lakatos's (1976) "advanced class." Proof, especially for beginners, might need to be motivated by the uncertainties that remain without the proof, or by a need for an explanation of *why* a phenomenon occurs. Proof of the too obvious would likely feel ritualistic and empty.

On purely pragmatic grounds, there seems more reason to hone students' perception of shape and space than to teach them the contents of the traditional first-year algebra course. In building a house or bookcase, painting a picture, designing a blouse or a plastic part, or rearranging the furniture, the ability to "see" the end product in the mind's eye is, at the least, an invaluable aid, and often an absolute prerequisite. There are almost certainly more people who use such visualization than who need (or remember how) to set up and solve a pair of simultaneous equations. The curriculum at all grade levels should help develop such visual skills.[2] But precisely what are the skills we're talking about?

First of all, they are mental skills. Geometry involves finding patterns (invariants) in the relationships among and between elements of some figure. Prerequisite to that is the ability to take a figure apart in the mind, see the individual elements, and make sufficiently good conjectures about their relationships to guide the choice of further experimental and analytic tools.

Now, how students learn the skills may involve a rich (and fun) combination of drawing, manipulating, imagining, paper folding, building, and using computerized enhancements or simulations of any or all of these. The activities may be physical, but the essential skills are mental. For example, students can learn to attend to the size and relationship of parts by measuring, creating a layout for, and then sketching a face. The geometer, like the artist, must learn to see clearly what most of us see in a schematic way: The portrait artist sees that eyes are not at the top of the head, but in the middle (see Fig. 1.1).

In a similar way, most people find it very hard to draw "impossible figures" like the one in Fig. 1.2a without first dissecting it in their minds somehow, perhaps as in Figs. 1.2b and 1.2c.

In addition to seeing details very clearly, the mathematician, like the artist, must also learn to see beyond assumed details and make mental images that highlight only the essential features. This is part of what Poincaré had in mind when he considered it "no joke" that "geometry is the art of applying good reasoning to bad drawings."

---

[2]This is quite different from saying that the development of visual skills is the job of a mathematics education. Courses in the design of clothing or stage sets or furniture could do that as well. Nor does mathematics need visualization to justify it. But in thinking through the role of mathematics in general education, we must look for principles that may help select from among the many, many worthy and accessible topics and methods within mathematics. Finally, classical geometry is not the only candidate for a visual mathematics. We'd like to see graph theory and visual topology join the curricular options. Graphs are stylized diagrams that can be used to represent certain relationships among members of sets. As such, they can be quite useful representations of information in a wide variety of contexts and make a useful addition to a student's repertoire. Some elements of graph theory provide a mathematically rich opportunity to deal with the broader-than-graph-theory issue of helping students acquire skills at diagramming relationships. There are also ways of introducing far more visualization into the so-called nonvisual realms of mathematics. For some good examples, see the MAA volumes on *Visualization in Teaching and Learning Mathematics* (Zimmermann & Cunningham, 1991) and *Proofs Without Words* (Nelsen, 1992).

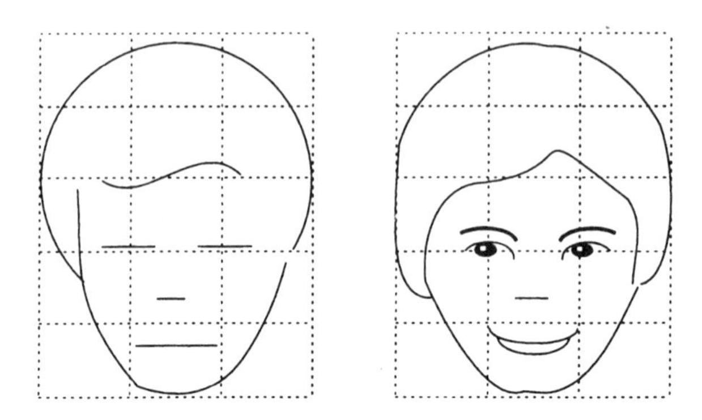

FIG. 1.1 Learning to see clearly: The eyes are in the middle of the head.

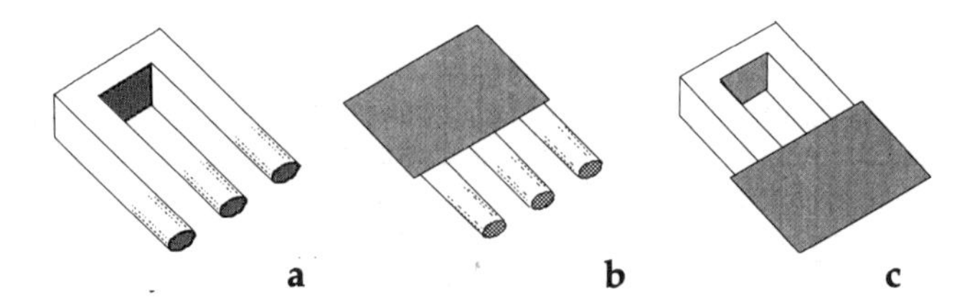

FIG. 1.2 Learning to see clearly: drawing the "impossible figure." (a) The figure. (b,c) Dissecting the figure to "see" it.

Senechal (1991) distinguished visualization and visual thinking, using the term *visualization* to mean bringing inherently visible things to mind (spatial visualization) and the term *visual thinking* to refer more broadly to a visual rendering of ideas that are not inherently spatial (e.g., visual thinking about multiplication). Where the distinction matters, we follow her convention, but we often find ourselves needing to include both senses in our arguments.

Spatial Visualization: Recreating the Visible

Senechal's *visualization* involves doing things in our heads that, in the right situation, can be done with our eyes. For example, we might approach the question "How many windows are there in your home?" by constructing and manipulating a mental picture. In a similar way, figuring out how many shelves there are in some familiar (but not visible) bookcase, how tall a book can fit on the lowest shelf, and what part of the body is the same height as the top shelf is perhaps best accomplished by examining a mental picture.

With the goal of helping children to build, hold, and manipulate mental pictures, it seems reasonable to work from objects of greatest interest to the children. The world of the youngest children is three-dimensional, and so it makes sense to take three-dimensional geometry very seriously early. Its object should be to give young children a sense of "what that shape can do for me." At this stage, as we see it, structure and design are paramount whereas measurement and nomenclature are of little importance. Can I identify the object by its feel? By its various shadows? Can I predict those shadows? Can I draw what I see?[3] Can I draw what *you* see when you look at it from *your* angle? What kinds of structures can I build with it? What will it look like if I slice through it? What shape will it stamp on paper if I ink it? How can I wrap it smoothly? *Can* I wrap it smoothly? How can I build it from other shapes? How can I build it from toothpicks and gumdrops? Can I build it from a photograph of it? Can I identify some rotated view of it?

Now, the *it* in the previous questions may be a simple "geometric solid," a more bizarre object like the teething ring shown in Fig. 1.3, something quite strange like the "EGB" sculpture pictured on the cover of Hofstadter's (1979) book, or even a moderately complicated construction made of blocks or Legos. Whether we favor "pure geometric" figures or not should be based on the nature and purpose of the task. It is plausible that the figures that will interest students most will frequently not be the spartan shapes of classical geometry, but figures of greater complexity, including fractals, mechanical devices, and natural forms.

Mechanical devices suggest a kind of three-dimensional "geometry" sufficiently far from the usual fare to warrant special mention. Rube Goldberg gadgets require a kind of visual analysis that is exactly the kind we want to develop, and they are amusing and appealing. Simpler devices—such as Lego Technic constructions, pinball constructions, and so on—can make good introductions.

What follows is a variety of thought puzzles that challenge us to create, manipulate, "read," and analyze mental pictures. Try each in your head, or using only a rough sketch as an aid. Then, if you like, try more concrete manipulatives.

Shapes Constructed by Dissecting Other Shapes. All three of these puzzles help develop or use ideas of symmetry and congruence. The third, with related activities, can also lead to ideas of area and the logical development of area formulas.

FIG. 1.3 Example of geometric solid: the teething ring.

[3]Children are particularly fascinated by learning to draw the third dimension. When they get an idea of how to do it, they will rehearse that idea over and over, embellishing their name with a sense of depth (ARIEL), making endless copies of the canonical three-dimensional representation of a cube, and so on.

- Which letters of the alphabet can be cut into two identical parts? Are there any that can be cut into three? Four? One youngster might picture a **T** cut into two parts by slitting the vertical down its center (⅂Γ), making use of the symmetry of the letter. Another might solve the problem in a very different way, dissecting the horizontal from the vertical element (T).
- How many different ways can you fold a square (rectangle, equilateral triangle, regular pentagon, etc.) so that one part lies entirely on the other? What shape(s) do you create when you do this?
- How many ways can you place a single straight cut through a rectangular sheet of paper so that the resulting two pieces can be rearranged and taped together to form a perfect triangle?

Shapes Constructed by Combining or Moving Other Shapes. The first of these involve images that calculus students are often asked to work with, but have never had practice making in their heads.

- Rotate a square (triangle, pentagon, and so on) around one of its edges. What shape does it sweep out? What shape(s) can the head of a lollipop sweep out if the lollipop is rotated in various ways around the very tip of its stick?
- Cut straight from the middle of one side of a square (rectangle) to an opposing corner. What shapes can you assemble, from the two parts you have just created, by taping them together at two congruent edges?

Shapes Created Using Specified Perspectives. When artists create pictures from their imagination, they must create a convincing look—three-dimensionality, shadows, and so on—without the benefit of a model. Designers must also make parts that fit into or against each other, often without making a model first. Geometers study the rules by which these artists and designers must work.

- Make a straight slice through a pipe in various ways. What does the cross section look like? Do the same with a cube. An egg. A donut. A grapefruit. What do the cross sections look like?
- Using the sun or other light source at some distance, what shape(s) can a soda can's shadow be? A cube's shadow? Is it possible to know an object's shape if you have enough shadows of it? Can the shadow of a square on a flat floor or wall ever not be a square? Or not be a rectangle? Or not be a parallelogram? What happens if you create these shadows using a nearby light source?
- Fold each figure along the dotted line (see Fig. 1.4). How many sides does the resulting shape have?

Visualization of Motion or Change in a Static Picture. Shown a picture of some simple Lego construction with gears and a crank on one gear, describe how some other gear will turn. Or, shown a simple construction with a lever or some pulleys, describe how point x will move as point y is

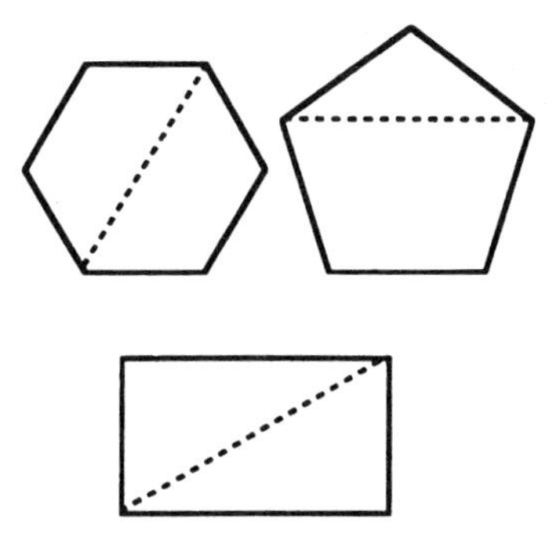

FIG. 1.4 Creating shapes by folding: How many sides does the resulting shape have?

moved. Or, shown a Rube Goldberg gadget, what happens when the cat finishes lapping up the milk?

Conceptual Visual Thinking: Seeing the Invisible

Senechal's *visual thinking* involves constructing visual analogues to ideas or processes that are first encountered in nonvisual realms. This includes, for example, using an area model (see Fig 1.5) to visualize multiplication and to help understand the distributive law and the multiplication algorithm. Various manipulatives have been designed for this exact purpose.

In early grades, students might also learn to sketch pictures like the one shown in Fig. 1.5 as a kind of intermediate stage between manipulatives and numbers alone. If they've learned to structure the computation for themselves, they may later use it to help themselves recall the logic and keep track of the steps as they dare to perform "$3\frac{1}{2}$ times $8\frac{1}{3}$" mentally.

Of course, it is something of an oversimplification to suppose that such pictures are pure help to young children. One reason why pictures such as these are useful in mathematics (and why diagrams, in general, are useful) is that they are compact, nonsequential representations of a great deal of information. That also makes them nontrivial to interpret

	20	6
10	10 · 20	10 · 6
7	7 · 20	7 · 6

```
   26
 × 17
   42
  140
   60
+ 200
  442
```

FIG. 1.5 Using an area model to visualize multiplication.

(see Eisenberg & Dreyfus, 1991). When curricula do make any use of such diagrams, they are often tossed in as if students are born knowing how to interpret them, with little or no effort put into developing interpretive skill, even though such skills take time and effort to develop.

In later grades, the task is often made more difficult. Students might be asked to explain how pictures like the ones shown in Fig. 1.6 relate to the symbolic expressions next to them. Other examples of conceptual visualization include visualizations of things too small, too large, or too diverse to be seen; visualizations of relationships rather than of objects themselves; and so on. As with spatial visualization, there are many categories here, but the first is just unfamiliar enough to require a disproportionately long explanation.

Seeing Change. In 1989, a group at the Education Development Center (EDC) began exploring the concept of Dynagraphs (Goldenberg, Lewis, & O'Keefe, 1992), a kind of dynamic visualization tool that allows students to "drive" the variable of some very broadly defined function and watch the resulting behavior of the function. Consider a simple example in which two number lines appear on the screen, one above the other, with a range of -100 to $+100$ represented on each line. The student enters a simple function, such as $f(x) = 2x$. Then, using a mouse, the student moves a pointer on the bottom number line, thereby manipulating the value of x. As the student does so, a second pointer moves on the top number line to show the value of $f(x)$.

If $f(x) = x$, the two pointers move together, in the same direction at the same speed, and always point to the same number on their respective number lines. If $f(x) = x - 1$, the two pointers move in the same direction at

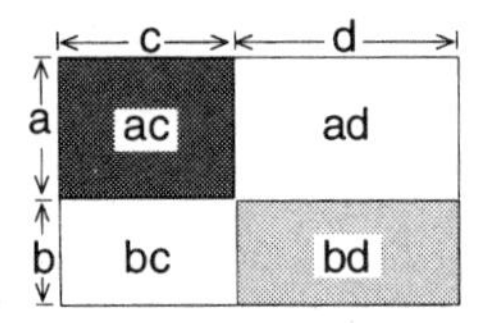

$$(a+b)(c+d) = ac + ad + bc + bd$$

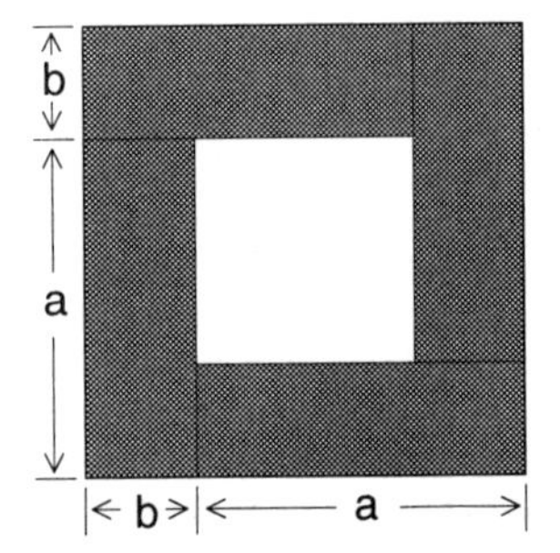

$$(a+b)^2 - 4ab = (a-b)^2$$

FIG. 1.6 Relating pictures to symbolic expressions.

the same speed, but the $f(x)$ pointer always stays one unit to the left of the x pointer (see Fig. 1.7).

When $f(x) = -x$, the pointers move at the same speed, but in opposite directions. When $f(x) = 3x$ the pointers move in the same direction, but the top one moves three times as fast as the bottom one. The reverse happens when $f(x) = x/3$. A remarkably different kind of behavior is seen when the student tries $f(x) = 3/x$.

Now consider what happens when $f(x) = x^2$. As the student moves the bottom pointer away from 0 in the positive direction, the top pointer also moves in the positive direction, but its speed increases as the student moves farther away from 0.

The pointers on the two number lines line up directly above each other only at $x = 0$ and $x = 1$. When the student moves x away from 0 in the negative direction, the $f(x)$ pointer still moves in the positive direction (see Fig. 1.8).

Pause to picture what students see when they explore the behaviors of trigonometric functions, step functions, and functions with minima or maxima, asymptotes, or discontinuities.

What is salient in these visualizations is not so much the values that the functions assume, but the ways in which the visualizations change with small changes in their variables. When number is further suppressed by showing no tick marks on the number lines, students from pre-algebra through college mathematics tend to describe what they see in ways that reflect geometric and topological ideas. Even before students have formally encountered the concepts and sophisticated vocabulary, they nevertheless notice, and find ways to express, distinctions between functions that *translate* the number line (e.g., $x \mapsto x + c$), ones that *dilate* (e.g., $x \mapsto ax + c, a > 0$), ones that *reflect* (e.g., $x \mapsto ax + c, a < 0$), ones that *fold*

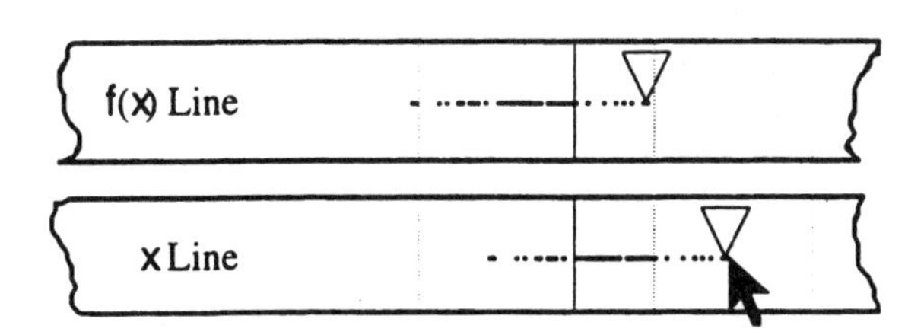

FIG. 1.7 "Watching" the behavior of a function: $f(x) = x - 1$.

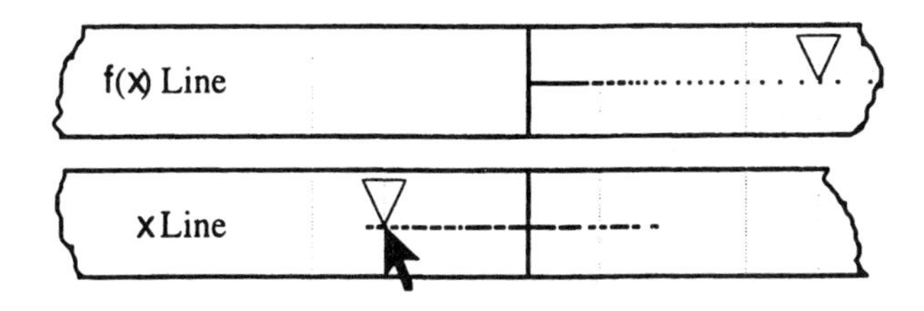

FIG. 1.8 "Watching" the behavior of a function. $f(x) = x^2, x < 0$.

(e.g., $x \mapsto |ax + c|$), ones that *distort* in various ways (e.g., $x \mapsto x^a$), ones that *tear* (e.g., $x \mapsto 1/x$), and so on.

The Dynagraph we built extended beyond functions from **R** to **R**. For example, instead of having two number lines on the screen, it could put up two planes. In such a case, a student could define a function on points in the plane, use the mouse to control the two-dimensional movement of a point on one coordinate plane, and observe the effect of the movement on a point in the other plane. Here, the connections with transformation geometry and the geometry of complex numbers are even more apparent.

Equally, the Dynagraph might put up a number line and a plane; the student controls a point in one and watches the result in the other. If the student controls the point on the number line, then the point on the plane might trace out, among other possibilities, the familiar Cartesian graph of the function. In this context, parametric equations feel like a minor extension of the same idea.

With a focus on functions, not on geometry, we envisioned extending the Dynagraph to manipulate functions defined through geometric rather than algebraic processes, but did not build the tools. The various "dynamic geometry" tools that showed up soon afterward—Cabri II, Geometer's Sketchpad, Geometry Inventor, and SuperSupposer, to name four—serve this purpose well (examples are given later). The essential feature of the Dynagraph idea that is common to all of these is that the independent variable is literally in the hands of the student and dynamically controllable, so that the student feels in direct control of the function's value.

Seeing Processes. Students should also think in terms of machines. Some machines—linkages, for example—are explicitly geometrical. But visual metaphors for other kinds of processes are also important. Students should use many visual representations for input–output pairings (functions), including meat-grinders, specialized calculators, Dynagraphs, and (if the process being considered happens to be a function from **R** to **R**) ordinary Cartesian graphs because these various visualizations highlight different aspects of a process. For example, icons for functions (see Cuoco, 1993; Feurzeig, 1994) highlight composition of functions, but show little about function behavior. By contrast, dynamic representations (e.g., Dynagraphs) show behavior well, but draw no attention to the structure of the algorithm that computes the function.

Seeing Quantity. Visual components to "mental arithmetic" and estimation—some blend of seeing process and seeing quantity—are too often ignored. For example, picture a bar divided into fifths. Mentally shade in two and a half of those fifths (see Fig. 1.9). How much of the bar is now

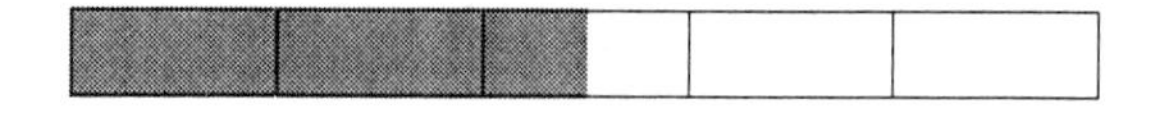

FIG. 1.9 "Seeing" quantity: two and a half fifths.

shaded? How many fifths are in a half? The mental image of two and a half fifths is one good scaffolding for understanding why $^1/_2$ divided by $^1/_5$ is a "big" number.

In a similar way, one of the authors often uses mental images to support his unreliable arithmetic facts: When an automatic answer for 8 + 5 is, for whatever reason, unavailable to him, he *sees* the 3-in-excess-of-10 in the form of a spillover on a vague 10-high column.

Seeing Structure. Visual thinking can also play an important role in exploring conjectures and devising proofs. Consider, for example, the potential role of visual thinking in a problem like the following:

Pick any two numbers in column A in Fig. 1.10, and add them. Which column contains the answer? Try that for a different pair of numbers from column A. What pattern do you see in your results? How might you explain that pattern?

Now do the same experiment on column B. Pick any two numbers in column B and add them. Which column contains the answer? Explain why.

What is column C's pattern for addition?

What is the pattern of answers when the addends come from two different columns? Again, explain.

Good algebraic and geometric explanations of this pattern can be given, but they are not equally accessible to all students. In particular, young students can show (see Fig. 1.11) that all the numbers in column A can be constructed with light green Cuisenaire rods only; that the numbers in column B require one additional white rod; and that the numbers in C require one additional red rod. Once the student really *sees* the number in this way, the pattern of the sums of any two numbers (one from column B and one from column C [Fig. 1.10]) then becomes "obvious" (see Fig. 1.11), as we mathematicians like to say. To such young students, the algebraic argument from that same pattern hardly seems as obvious.

Column A	Column B	Column C
0	1	2
3	4	5
6	7	8
9	10	11
12	13	14
15	16	17
18	19	20
21	22	23
24	25	26
27	28	29
30	31	32
...	...	...

FIG. 1.10 "Seeing" patterns: Two numbers are selected from a single column. In which column is their sum?

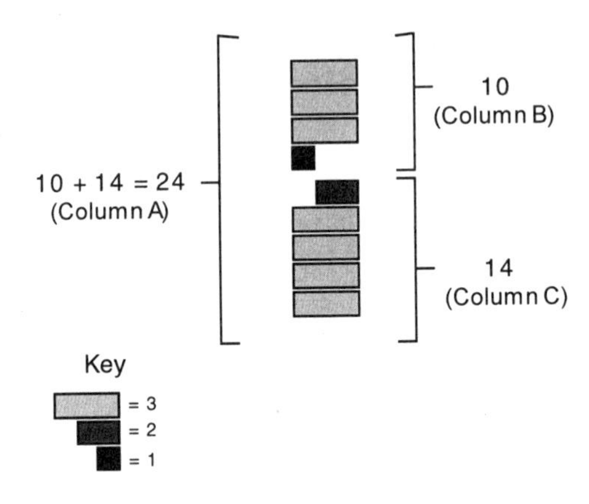

FIG. 1.11 "Seeing" patterns: A visual model suggesting a general proof: why the sum of a number from Column B and a number from Column C must be a number in Column A.

The Physics of Mathematics. We are accustomed to talking about the mathematics of physics, but there is a kind of "physics of mathematics" that students may benefit from by experiencing. As mathematical objects take on familiarity, they often appear to "behave" as if governed by some natural, physical law. An elementary example arises from the parallel number line Dynagraph with which a student can feel arbitrarily constructed functional relationships as if their behavior were dictated by mechanical devices. Other examples follow.

Physical Behavior in Mathematical Contexts. Arcavi and Nachmias (1990) developed a piece of software that visually represents **R** to **R** functions by connecting points in the domain, visualized as a number line, with their images on a separate, parallel number line. Most functions generate a tangle of wires, but linear functions are special cases.

In functions that involve no rescaling of the number line (functions $f(x) = mx + b$ where $m = 1$), the connectors between domain points and their images are all parallel; all such functions merely translate the number line (see Fig. 1.12).

When $m \neq 1$, the connectors (extended, if necessary) converge at a single point. If the slope is negative, this point is between the two lines. On a Dynagraph, such functions "feel" like first-class levers controlled by the students. (Arcavi's software shows the "lever," but does not allow the dynamic control.)

If the slope is, let us say, -0.5, the image moves more slowly than the domain point—exactly half its speed—and in the opposite direction (see Fig. 1.13). As a result, the "fulcrum" will be closer to the range line than to the domain.

A slope of 0.5 resembles a second-class lever—the image moves half the speed of the domain, in the same direction. With $m > 1$, the function feels like a third-class lever, with the image moving faster than the preimage and in the same direction (see Fig. 1.14).

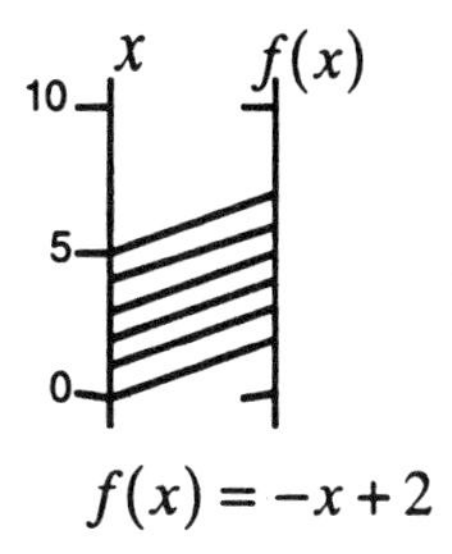

FIG. 1.12 "Seeing" physical behavior of a function: functions that involve no rescaling of the number line. Note that connectors are parallel.

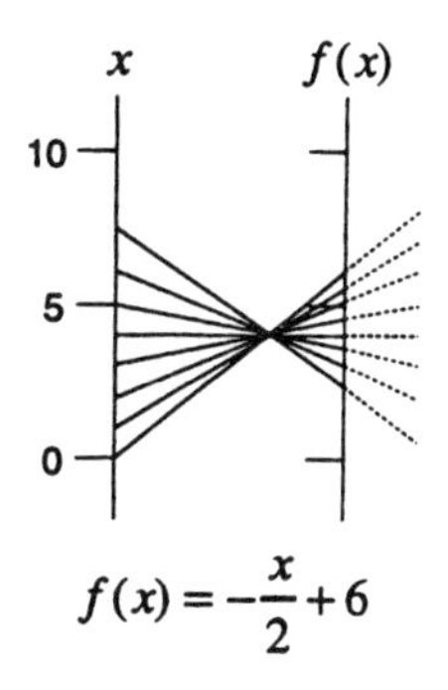

FIG. 1.13 "Seeing" physical behavior of a linear function: when $m = -0.5$.

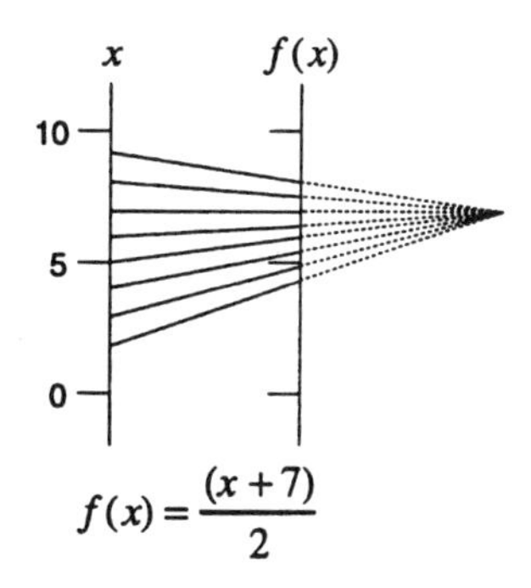

FIG. 1.14 "Seeing" physical behavior of a function: when $m > 1$.

Using dynamic tools, this effect can be shown geometrically. When a student moves the point on the domain, he or she sees its image move in a coordinated way. The effect so much resembles the behavior of a lever that all of the student's experience with levers immediately is brought to bear on the mathematical abstraction. The physical behavior helps to give concrete meaning to the mathematical formalisms, and the mathematics helps to quantify the observed and already familiar behavior.

Classifying the Behavior of Mathematical Systems. Wittgenstein (1983), in his *Remarks on the Foundations of Mathematics*, presented his readers with the following thought experiment.[4]

A rigid rod slips loosely through a pivoted bearing that is fixed in position (see Fig. 1.15). One end of the rod is driven around clockwise by a rotating wheel. What path does the other end of the rod trace?

In what way(s) does that path depend on the length of the rod and placement of the pivoted bearing? In Fig. 1.16, for example, the rod is

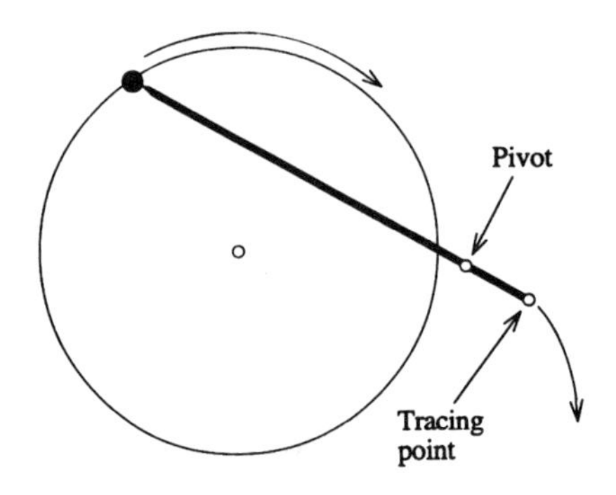

FIG. 1.15 Wittgenstein's thought experiment: What path does the end of the rod trace?

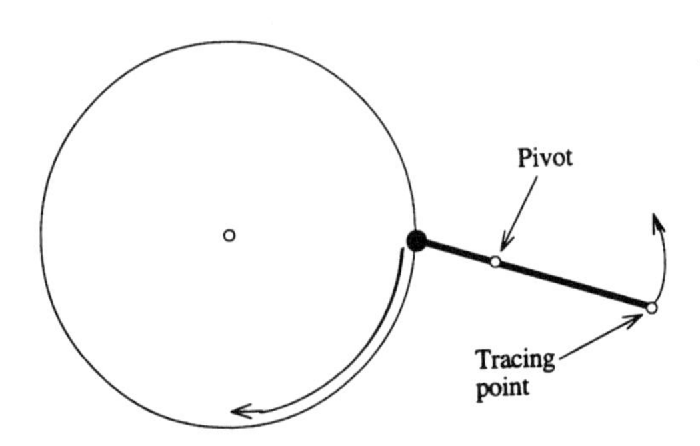

FIG. 1.16 Wittgenstein's thought experiment: How does the locus of the tracing point depend on the length of the rod and the placement of the pivot?

[4]When his book was first published in 1956, Wittgenstein could not easily perform this experiment except in his mind. Today it is easy to experiment quite concretely with the parameters, collect the pictures, and use them to illustrate a written analysis of the experiment. The pictures shown here were generated on Geometer's Sketchpad.

shorter, and the pivot farther from the circle. Clearly, if the rod is short enough, or the bearing far enough away, the rod may at times slip out of the bearing entirely. In such cases, it is deemed to aim in such a way that its extension continues to pass through the bearing. The images are curious and varied. Some are shaped like figure eights.

In Fig. 1.17, we see the mechanism in two stages of drawing a figure eight. In Fig. 1.17a, the tracing point is moving up around the outer lobe of the figure eight. In Fig. 1.17b, the rod, which had slipped out of the pivoted bearing roughly half a rotation earlier, continues to aim toward it and is about to slip back into the bearing.

With a longer rod, and a pivot near enough to the circle (see Fig. 1.18a), the rod never slips out of its bearing, and the locus of its endpoint is quite different. With a short rod and distant pivot (see Fig. 1.18b), the rod never quite reaches the pivot: Again the picture is qualitatively different. With the pivot inside the circumference of the wheel (see Fig. 1.18c), still other pictures are possible.

Just classifying the pictures by their qualitative differences and stating the conditions required to achieve those characteristic shapes is a very respectable mathematical investigation. No formulas and no advanced mathematical techniques are required, just clear thinking and systematic investigation.

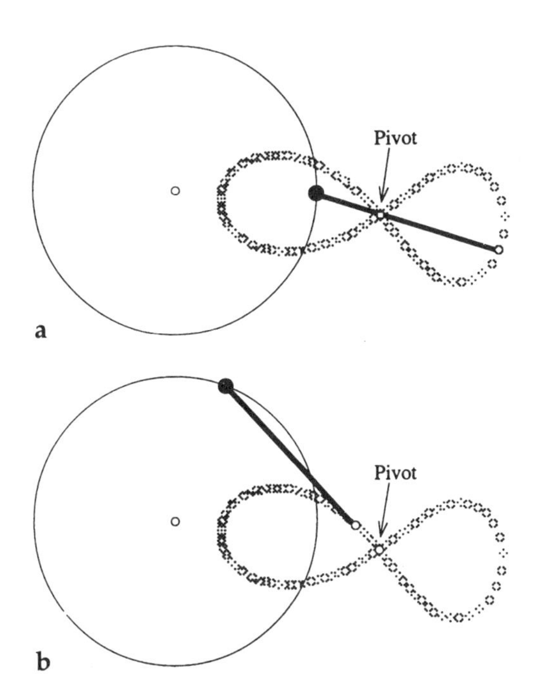

FIG. 1.17 Wittgenstein's thought experiment: two stages in a drawing in which the rod "slips out of the pivot."

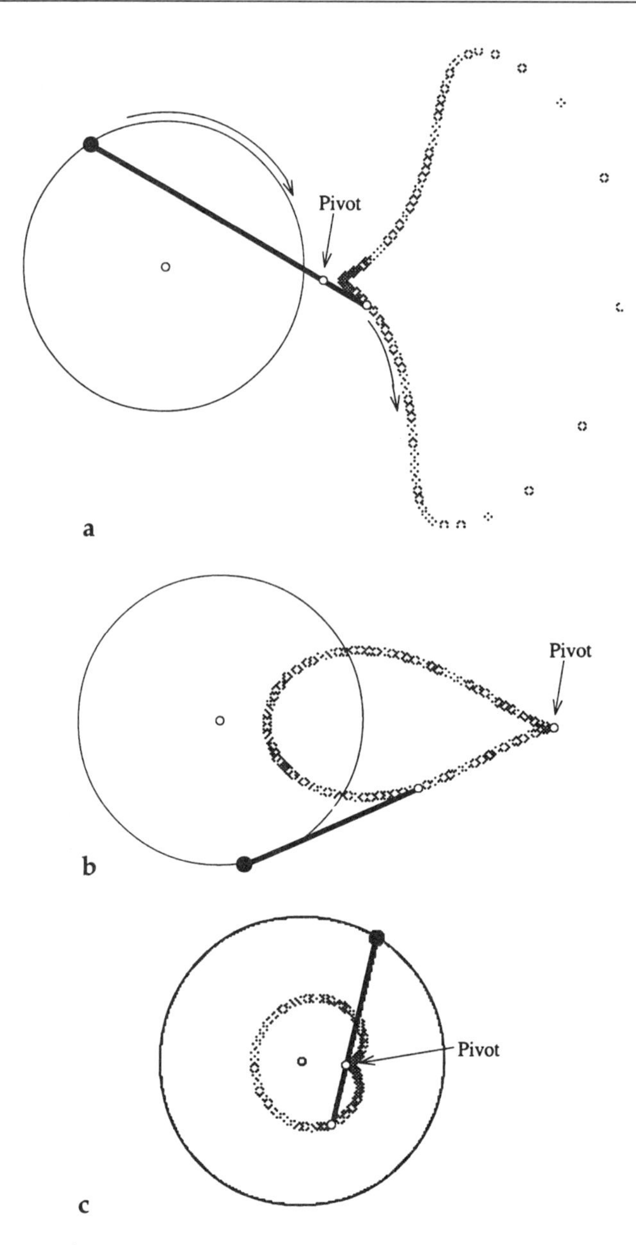

FIG. 1.18 Varying the parameters in Wittgenstein's thought experiment. (a) A long rod and a pivot near the circle. (b) A short rod and distant pivot. (c) The pivot inside the circumference of the wheel.

HOOKING STUDENT INTEREST: AN EXAMPLE FROM *THE SHAPE OF SPACE* (WEEKS, 1985)

Geometry has a dual character of being "both abstract and concrete." Although geometry deals in abstractions as much as any other branch of mathematics—points, lines, and planes are just as much things of the imagination as are polynomials—it is natural to relate *visually* to geometric objects, "approaching geometry somewhat as a branch of physics [as if] geometric objects really exist and we want to learn more about them" (Klee, 1991). Whatever way we choose to manipulate the objects of geometry—with the words and style of synthetic methods or with the symbols of analytic geometry—the currency with which geometry deals is inherently visual.

Weeks (1985) used this dual character to advantage in *The Shape of Space*. He began with a fantasy based on Abbott's (1980) *Flatland*. In Weeks's story, everyone assumes the world is flat, except for one character—A Square, by name—who has the theory that the world is a "hypercircle" (i.e., a sphere). Like Abbott's creatures, Weeks's characters are two-dimensional. They live *in* their surface world, not atop it, and are unable to escape into the third dimension to view it from afar. How could such creatures check out A Square's theory?

A Square thinks that perhaps a journey might settle the issue:

> [He] reasoned that if he were willing to spend a month tromping eastward through the woods, he might just have a shot at coming back from the west.
>
> He was delighted when two friends volunteered to go with him. The friends . . . didn't believe any of A Square's theories—they just wanted to keep him out of trouble. To this end they insisted that A Square buy up all the red thread he could find in Flatsburgh. The idea was that they would lay out a trail of red thread behind them, so that after they had traveled for a month and given up, they could then find their way back to Flatsburgh.
>
> As it turned out, the thread was unnecessary. Much to A Square's delight—and [his friends'] relief—they returned from the west after three weeks of travel. Not that this convinced anyone of anything. Even [his friends] thought that they must have veered slightly to one side or the other, bending their route into a giant circle in the plane of Flatland. (Weeks, 1985, pp. 5, 6)

Undaunted, A Square sets out on another expedition, this time to the north, laying a trail of blue thread:

> Sure enough, [he] returned two weeks later from the south. Again everyone assumed that he had simply veered in a circle, and counted him lucky for getting back at all.
>
> A Square was mystified that his journey was so much shorter this time, but something else bothered him even more: he had never come across the red thread they laid out on the first journey. (pp. 6, 7)

In just this fashion, Weeks invited his readers to visualize various manifolds—in this case a two-manifold embedded in three-space, but with the aim of later visualizing three-manifolds in four-space—distinguishing geometric and topological, local and global, and intrinsic and extrinsic features. Can A Square's detractors still claim their world is a plane and that he has again veered in a smaller circle? More generally, what experiments would help the two-dimensional inhabitants of various surface worlds—for example, a sphere, a cube, a cylinder of finite radius but infinite length, a doughnut, or the teething ring—determine their world's shape?

Is Weeks's story a good hook? For most people, yes, and it is worth thinking about why. It is certainly not for the pragmatic value: Neither the content nor the context are readily "applied." Pragmatism has been vastly overrated as a motivation in mathematics. The pragmatic value of the mathematics generally taught is also vastly exaggerated. High-school students can hardly fail to notice that virtually everyone they meet who seems to be doing just fine will eventually announce, even proudly, that they were never any good at the subject.[5]

What works about this story is not the applicability of the mathematics—although there *are* applications to science topics of current public interest—but the fact that it captures (most) people's imagination and curiosity: (Most) people do not ask what they're learning or why they need it. (Of course, no context works for everyone, so the texture and style of a curriculum must be varied.)

But personal appeal is not enough. Even if it is sufficient reason for a student to engage in an activity, it is insufficient for a curriculum writer to include the activity. What else is there in Weeks's story that is of mathematical interest?

Weeks's thought experiment captures many of the faces of mathematics. Weeks felt (and encouraged his readers to feel) perfectly free to play with the rules and change things about in a game of mathematics *lishmah*,[6] which is characteristic not only of mathematics games but also of pure mathematics. Yet this pure mathematics is not meaningless or inaccessible. It is, at the same time, both abstract and concrete: The student has never experienced *these* worlds—they are pure abstractions—but the student can easily enter the fantasy and have very "concrete" impressions of the worlds Weeks described. In that way, the mathematics is (in the real world of the imagination) "applied math"—geometry in its original sense of world-measuring. It is (in *that* real world) "practical" but never mundane. If only all students could experience mathematics from this perspective.

Weeks's own creative brilliance in introducing his topic in this fashion is not to be underestimated, but neither should the character of the subject matter be considered irrelevant to his success.

---

[5]If this is a peculiarly U.S. phenomenon, as we've occasionally heard, let us hope we do not export this piece of "culture."

[6]*lish • mah.'* Hebrew, "for its own sake."

CONNECTIONS

Geometry can capture and hold students' imagination and interest through connections with ideas outside of mathematics, but that alone, without rich connections within mathematics, would probably not justify much attention in a general education. In fact, geometry *does* connect richly with the rest of mathematics, including such topics and themes in discrete and continuous mathematics as combinatorics, algorithmic thinking, geometric series, optimization, functions, limits, trigonometry, and more. Geometric approaches can be a route into new mathematical ideas and can provide new insights into mathematics that the student already knows.

Perhaps the most obvious place to look for an example is an idea taught in algebra or precalculus but named after geometry—geometric series. Dennis Meyer, a Minnesota teacher, showed us a powerfully convincing geometric derivation of one familiar formula for the sum:

> Take a piece of paper representing "a whole." Rip it roughly into thirds, and place one third to your right and one third to your left, retaining the last third in your hand. Rip *that* piece into thirds, and place one piece (a ninth of the whole) on the pile to your right, one piece on the left-hand pile. Continue in this manner, ripping the piece in your hands into thirds of itself, and placing one of the three pieces on the pile to the right and one on the pile to the left, keeping the last to rip again into thirds.
>
> The two accumulating piles contain the same amount of paper, and what remains in your hands gradually diminishes to nothing. In other words, this process is a gradual division of all of the paper equally into two piles. The piles contain one third of the original, one ninth, one twenty-seventh, and so on, demonstrating that their sum is $1/3 + 1/3^2 + 1/3^3 + \cdots = 1/2$.

Another experiment shows that if you rip the original "whole" into five equal parts, create four piles, and repeat the process with the remaining piece, each pile will gradually accumulate one fourth of the whole and will contain the sum of powers of fifths of the whole. By reflecting on the process, we see its generalization: $1/n + 1/n^2 + 1/n^3 + \cdots = 1/(n-1)$, at least for integer values of n.

Mathematics Magazine regularly publishes "proofs without words," which employ geometry to help explain classical mathematical results. The example by Klein and Bivens (1988) used two similar right triangles (see Fig. 1.19)—one whose longer and shorter legs are $1 + x + x^2 + x^3 + x^4 + \cdots$ and 1, respectively, and the other whose longer and shorter legs are 1 and $1 - x$—to derive (using the ratio of the legs) a related formula for the sum of a converging geometric series.

This demonstration takes more work to understand but, unlike the paper-ripping example, does not depend on the discreteness of integer values. Between the two, these methods connect ideas of limits, continuity, slope, series, similarity, self-similarity, and more.

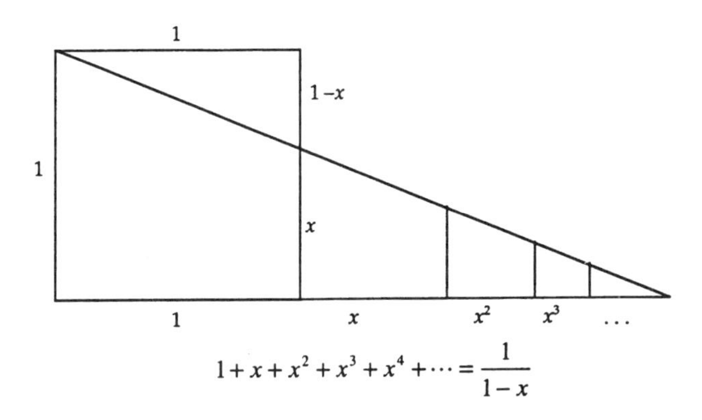

$$1+x+x^2+x^3+x^4+\cdots=\frac{1}{1-x}$$

FIG. 1.19 Using the ratio of legs in two similar right triangles to derive a formula for the sum of converging geometric series.

Banchoff (1991), in his chapter on dimension in *On the Shoulders of Giants*, remarked that Froebel, the inventor of the kindergarten, gave students three-dimensional geometrical "gifts," not as objects whose names were to be memorized, but as stimuli to their imagination. To the extent that these gifts are still in the schools, they tend now to *be* the curriculum, not the stimulus to it. Banchoff's presentation shows a rich interplay between geometrical and algebraic reasoning and gives a convincing picture of how such an interplay might be translated into a form accessible to students throughout the grades. Among other connections he made, he showed how the geometric notion of dimensionality can inform and enrich ideas in number—two- or three-dimensional quantities.[7]

The following sections give more examples of the interplay among geometry, the rest of mathematics, and the human experience.

Billiards, Optics, and the Remarkable Art of Michael Moschen

The physics of light beams bouncing off mirrors is different from the physics of billiard balls bouncing off the billiard table sides, but the geometry is the same. And it is a geometry from which we can reach a great deal of mathematics. In his classic *Mathematics: A Human Endeavor*, Jacobs (1970) got to number theory from geometry with an activity in which students play a peculiarly constrained billiard game that leads to the concept of relatively prime numbers. The constraints of the game are (a) rectangular tables with integral side lengths and pockets only at the corners and

[7]Cuoco, Goldenberg, and Mark (1995) illustrated a range of mathematical connections with a sequence of activities for high-school geometry students drawn from the *Connected Geometry* series (EDC, in press). In these, classical geometry blends naturally with ideas from elementary algebra (including inequalities), reasoning about algorithms, concepts of optimization (maxima and minima), notions of functions from the plane to the real line, and reasoning by continuity (understanding the properties of continuous variation).

(b) that the ball must be shot from a corner at a 45° angle (see Fig. 1.20). On an $m \times n$ table, $(m,n) = a$, a ball that starts at a corner will bounce $(m/a) + (n/a) - 2$ times before it reaches a different corner. Closed paths are possible only when $a \neq 1$ and the ball does not start at a corner.

Michael Moschen, a choreographer/dancer/juggler with a mathematical bent, choreographed a dance within a vertically oriented triangular "billiard table" (Blumberg, 1991). In the piece, he is both dancer and musical accompanist. His co-dancers are three rubber balls, and his musical instrument is the triangular frame. The result is visually, aurally, and intellectually fascinating. In the piece, he dances within the frame, bouncing balls off its walls. The rhythm tapped out by the balls as they hit the frame depends on the time between bounces and, therefore, on the lengths of free flight between bounces. To get that rhythm to make musical sense, the times must add up properly—each must be a small integer multiple of some unit time (e.g., eighth notes or sixteenth notes), and their total must be a full measure.

The path angles and the lengths of free flight between bounces are (almost) all geometry. With the right tools, we can explore this problem on horizontal billiard tables, investigating the bounce angles, the path lengths, the conditions under which the path closes, and so on. Michael Moschen's task is, in fact, more complex than the triangular billiard table because the vertical orientation of his frame means that the balls' paths between bounces are not segments of straight lines, but segments of parabolas. But, observing that the *local* geometry of the bounces is unchanged in the vertical orientation, we can even begin to explore this more complex real task. And what if the table were elliptical instead of polygonal? Or perhaps the shape of a parabola with a chord across it?

This context has "hooks" aplenty for all tastes: art, dance, music, physics, and, of course, billiards. Watching Moschen in motion is dazzling.

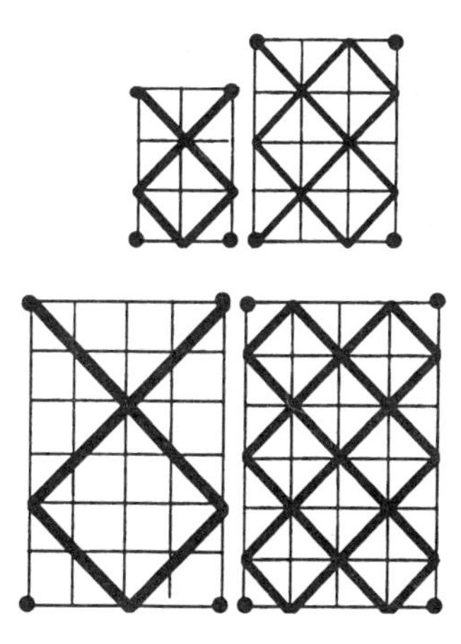

FIG. 1.20 Connecting geometry with number theory: a billiard game (Jacobs, 1970).

Sequences, Series, Mathematical Induction, and More Number Theory from Iterated Geometrical Constructions[8]

The Construction. Imagine that a triangle (see Fig. 1.21) has three children, each of which is similar but exactly half the linear size (one quarter the area) of its parent. The children grow from their parent's vertices and are oriented as their parent is. The parent is shown in Fig. 1.21a. Each succeeding frame shows one more generation of children. The following are three problems based on this construction.

The Nature of the Triangular Boundary. Fig. 1.21d looks neatly bounded by a triangle. What is the shape of that boundary triangle?

There are many approaches we might take. One way is to notice that the vertices of this new triangle appear to be along extensions of the sides of the original triangle. A geometric series based on the construction tells how far out along these extensions the new vertices lie. It is possible to do the rest of the work with algebra alone, but unless the ancestral parent is a right triangle, a trigonometric approach is probably more straightforward than a strictly algebraic one.

Amy Nelson, a high-school junior, took a different approach (see Fig. 1.22). She chose to measure one side of the limiting triangle and defined "pseudosides" (the dashed line in Fig. 1.22), which approximated this goal ever more closely with each succeeding generation of children. Amy first used *measurement* and noticed that the ratios of pseudosides appeared to be rational. The pseudoside in Fig. 1.22c seemed to be $(3/2)s_1$ and pseudoside s_3 seemed to be $(7/6)s_2$. She then used *geometric reasoning* to show that the first two ratios were, indeed, what her measurement

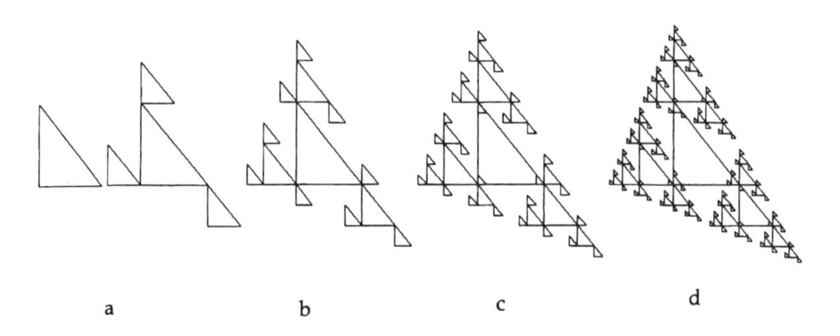

FIG. 1.21 An example of iterated geometrical constructions: successive generations of "children."

[8]This example is drawn from Cuoco and Goldenberg (1992), which discussed mathematical induction in a visual context and showed how the geometric concept of self-similarity can be used to support and enhance an understanding and use of the "self-similar logic" entailed in reasoning by mathematical induction.

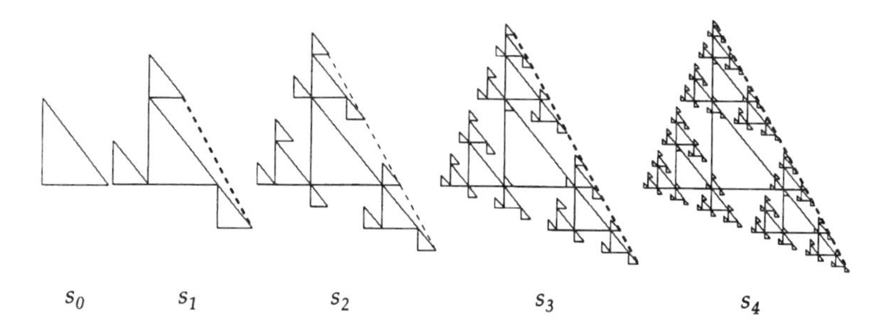

FIG. 1.22 Using pseudosides to determine the shape of the boundary triangle of an iterated construction.

seemed to show. Her argument was based on noticing that pseudoside s_1 did not touch the construction in any way except at its endpoints, but that pseudoside s_2 touched the construction twice between its endpoints, distinguishing three smaller segments. She showed that these segments were all $s_1/2$.

With help, she then used *mathematical induction* with her geometric argument to generate a series of other ratios (which she checked out by measuring). She showed two things:

- The *sizes* of the subsegments in the pseudoside at any given generation were half the size of the previous subsegments: $S_{g+1} = S_g/2$.
- The *count* of the subsegments in any given generation was one greater than twice that of the previous generation: $C_{g+1} = (2 \cdot C_g) + 1$.

She also expressed all of the pseudosides in terms of s_1, allowing her to come up with an expression for s_n in terms of s_1:

$$s_2 = \frac{3}{2} \cdot s_1$$

$$s_3 = \frac{7}{6} \cdot \frac{3}{2} \cdot s_1$$

$$s_4 = \frac{15}{14} \cdot \frac{7}{6} \cdot \frac{3}{2} \cdot s_1$$

$$s_5 = \frac{31}{30} \cdot \frac{15}{14} \cdot \frac{7}{6} \cdot \frac{3}{2} \cdot s_1$$

Amy saw how much canceling she could do and moved easily from a computation like this one:

$$s_5 = \frac{31}{\cancel{30}_2} \cdot \frac{\cancel{15}^1}{\cancel{14}_2} \cdot \frac{\cancel{7}^1}{\cancel{6}_2} \cdot \frac{\cancel{3}^1}{2} \cdot s_1$$

$$= \frac{31}{2} \cdot \frac{1}{2} \cdot \frac{1}{2} \cdot \frac{1}{2} \cdot s_1$$

$$= \frac{2^5 - 1}{2^4} \cdot s^1$$

$$= \left(2 - \frac{1}{2^4}\right) \cdot s_1$$

to its generalization:

$$s_n = \left(2 - \frac{1}{2^{n-1}}\right) \cdot s_1$$

and a statement about the limit, namely, that $s_n \mapsto 2s_1$ as $n \mapsto \infty$. She then consulted a friend, who showed her how to use the law of cosines to find s_2. What Amy did here was to construct and evaluate the infinite product:

$$s_1 \prod_{n=2}^{\infty} \frac{2^n - 1}{2^n - 2}$$

Of course, the notation (and some of the notions about limit and convergence) were clearly new ideas for her, and she was not comfortable with them. But the interplay between the geometric constructions and arithmetic calculations at each level was quite concrete. What was more important is that Amy was able to capture an inductive pattern in the geometry that allowed her to move from the length of any pseudoside to the length of the next one and that worked at every level.

Children in a Row. Amy's construction calls attention to a row of triangles along her pseudoside and measures and counts the spaces in between these triangles. In any one generation, how many triangles are in the *other* rows (see Fig. 1.23) parallel to that right-side row?

Each generation produces its own sequence of numbers corresponding to the number of triangles in parallel rows. In Fig. 1.23d, for example, there are eight rows of tiny triangles, containing 1, 2, 2, 4, 2, 4, 4, and 8 triangles, respectively. What characterizes these sequences? What, for example, is the sequence of numbers we would see in Fig. 1.23f, the next generation? Or how many 1s, 2s, 4s, 8s, and so on, are in any particular sequence?

Reasoning about this problem combines geometric arguments and mathematical induction and connects nicely to number theory. (In the sequence: 0, 1, 10, 11, 100, 101, 110, 111,..., there are 0, 1, 1, 2, 1, 2, 2, 3, ...

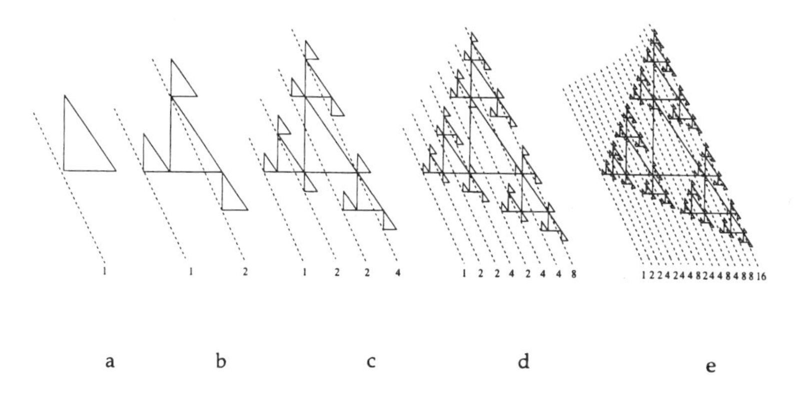

FIG. 1.23 Deriving number sequences from iterated constructions.

ones, respectively. These exponents of 2 generate the sequence in Fig. 1.23d. Explain the connection.)

Not Taking Things for Granted. We observed earlier (see again Fig. 1.21) that "the rightmost frame looks neatly bounded by a triangle," but looks can be deceiving. The vertices appear to be on what Amy called a pseudo-side, but are they indeed all collinear? This particular problem is a wonderful opportunity to develop or apply a collection of elementary facts of Euclidean geometry and to exercise mathematical induction to complete the proof for all generations. Do the spaces between them ever "close up"? That is, does the growth originating at one vertex of the primordial ancestor ever reach, or perhaps even overreach, the growth from another vertex? Remarkably, this, too, uses only elementary geometry and an understanding of the sum $1/2 + 1/4 + 1/8 + 1/16 + \cdots$.

Opportunities for Problem Posing. Triangles oriented 180° with respect to their parent raise different but related questions, as do nontriangles (e.g., squares or regular pentagons). Richer conjectures and theorems come from examining this "bigger picture."

Converting Facts into Functions

Seeing a Theorem as a Function. One heuristic for inventing experiments is to recast static statements of fact as dependencies on some variable element, often as a result of trying to find a suitable visual representation of the fact. The following two examples illustrate how such recasting can lead from an apparently isolated fact to a host of interrelated mathematical ideas.

Here is a theorem as it is often expressed in high-school geometry texts (example adapted from Goldenberg, Lewis, & O'Keefe, 1992):

> Theorem 1: The midpoint of the hypotenuse of a right triangle is equidistant from the three vertices of the triangle.

Presented verbally, the theorem just sits there. An attempt to present it visually reminds us that it is a statement about *all* right triangles: A single picture does not suffice.

Many theorems, like this one, have a distinctly dynamic sense to them, and we must learn to look for the implicit *for all* in wordings like "a right triangle." The *for all* identifies a potential variable and suggests a functional rewriting of the theorem: in this case, a mapping over all right triangles. But what might characterize this mapping, and how can we organize the set of all right triangles in order to "see" this function?

One way is to put all the triangles at the same scale by fixing the length of the hypotenuse, and then arranging them in order by the size of one angle. So we can imagine—and, with dynamic geometry tools, actually create—a dynamic construction in which we can track the midpoint of a fixed-hypotenuse right triangle while varying the other features of the triangle by some parameter such as leg length.

On paper, we can do little more than create separate "frames" of this animation to show the set under discussion, but it is hard to make much visual sense out of the display: In Fig. 1.24, eight right triangles with congruent hypotenuses (with marked midpoints) are lined up to show variation in the non-right-angles, but little or nothing of the relationship of the midpoint to the right-angle vertex is apparent.

Using dynamic geometry tools, we can trace the locus of the hypotenuse by dragging the highlighted point along the base of the triangle (see Fig. 1.25a). The shading shows locations the hypotenuse has occupied. If the point is moved to the left of the vertical leg (reversing the sense of the horizontal leg), we still see the pattern of locations of the hypotenuse (see Fig. 1.25b). Already, new conjectures suggest themselves from the pictures created for the purpose of rendering the theorem visually.

It is worth considering how this picture and the feeling of moving the highlighted point may affect what Vinner (Vinner, 1983; Vinner & Dreyfus, 1989) called students' *concept image* of function. In particular, although this function might conceivably be rendered as a mapping from **R** to **R** (e.g., from the horizontal position of the highlighted vertex to the slope of the hypotenuse), that is almost certainly not how the student perceives it. The variable is the location of a point on the plane, and the func-

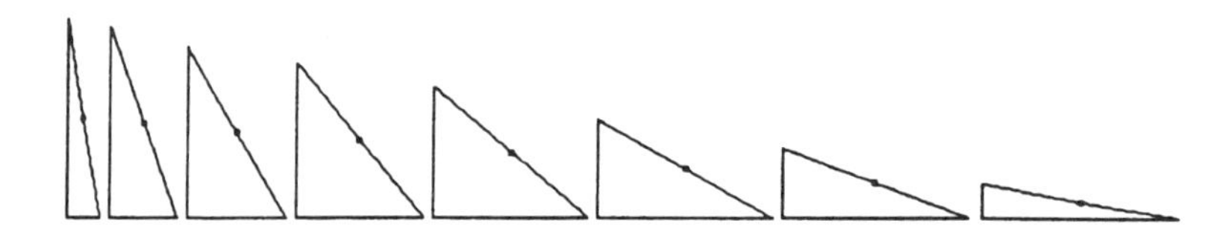

FIG. 1.24 The relationship of the midpoint of the hypotenuse of a right triangle to the three vertices of that triangle seen through "separate frames" that vary the non-right angles.

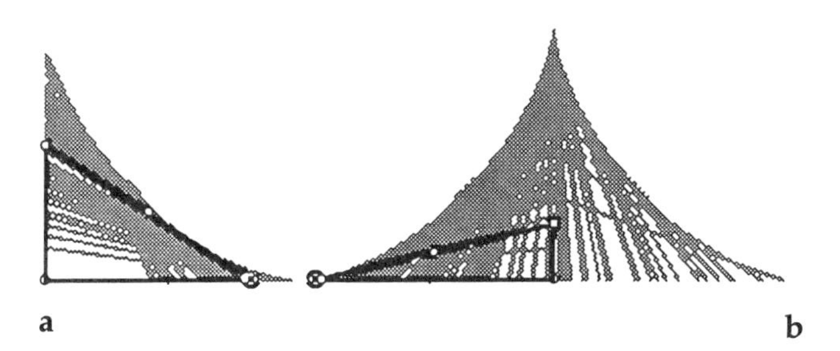

FIG. 1.25 Using dynamic geometry to show the relationship of the midpoint of the hypotenuse of a right triangle to the three vertices of that triangle.

tion value is the position of a line segment. Furthermore, the definition of the function involves only geometric acts: This is certainly not the stereotype of "function" that is defined by a collection of letters arranged according to algebraic syntax.

Now let's return to the purpose of the experiment. The theorem's claim is about the midpoint of the hypotenuse, so let's trace only that point. If the path is, as it seems to be, a circle around the right angle vertex (see Fig. 1.26), then it is some fixed distance R from that vertex. As the hypotenuse approaches the horizontal leg, the vertical leg evaporates into the center of the circle, and we can reason by continuity that R is half the hypotenuse. So the theorem is now supported (if that path is a circle), but with new insights.

The heuristic of converting static facts into dynamic visual representations aids experimentation in other ways as well. Verbal statements rarely invite experimentation. For example, this claim is simply wrong:

> Claim: A *non*-midpoint of the hypotenuse of a right triangle is equidistant from the three vertices of the triangle.

But such a change in the *construction of a function* is quite reasonable to perform, and such constructions almost beg students to tinker.

Experimentation focuses our attention on process and algorithm, but such a focus also aids experimentation. If we generate examples of some theorem by algorithm, it becomes quite easy to change the algorithm and ask what new theorems would arise. For example, what do we see if we choose to trace a point 20% or (80%) of the way along the hypotenuse (see Fig. 1.27) instead of the midpoint?

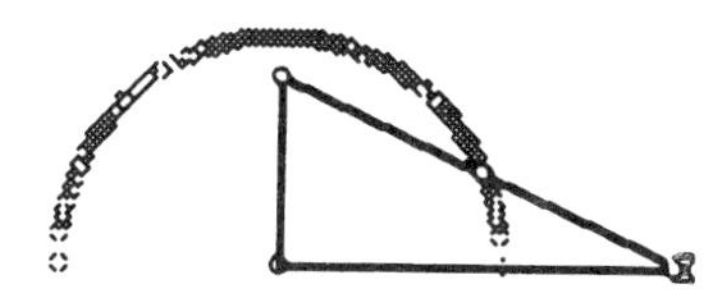

FIG. 1.26 Tracing the path of the midpoint of the hypotenuse of a right triangle.

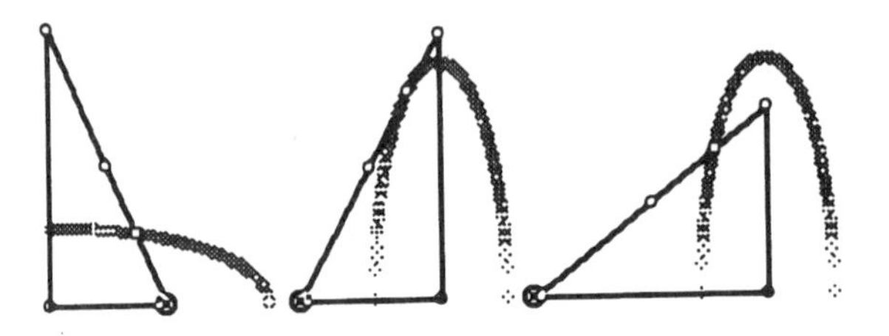

FIG. 1.27 Tracing the path of non-midpoints on the hypotenuse of a right triangle.

We gain insight into this experiment by shifting perspective (see Fig. 1.28). As the experiment is performed, we act as if we are observers sitting on the right angle vertex "planet," watching the satellite on the hypotenuse rotate around us. But observers on the satellite will see a different picture. They will feel as if *they* are the stationary point. The hypotenuse will pivot around them, and the legs, as before, will remain in a fixed orientation (although their lengths and positions will vary).

Now, if we focus on the vertices instead of the sides, we see a new dynamic picture (see Fig. 1.29). The dynamic picture shows what is difficult to convey in any static image: The two endpoints of the hypotenuse trace out circles of different radii, and the right angle vertex—the satellite we are tracing—follows horizontal projections from one circle and vertical projections from the other. This plants the seeds for a new idea that might eventually be expressed as $x = \alpha \cos \phi$ and $y = \beta \sin \phi$.

But students can use, explore, and come to understand the algorithm—building triangles and displaying them a certain way—before they know the algebra of parametric equations and the particular trigonometric functions involved.

The dynamic approach and the heuristic of tinkering are powerful problem-posing methods. What if the fixed angle between the two projections of the ends of the "hypotenuse" is not 90° but, say, 60°? Again, the satellite (see Fig. 1.30) appears to trace out an ellipse, but one whose axes are tilted. Is it still an ellipse? If so, how does the tilt angle relate to the projection angle?

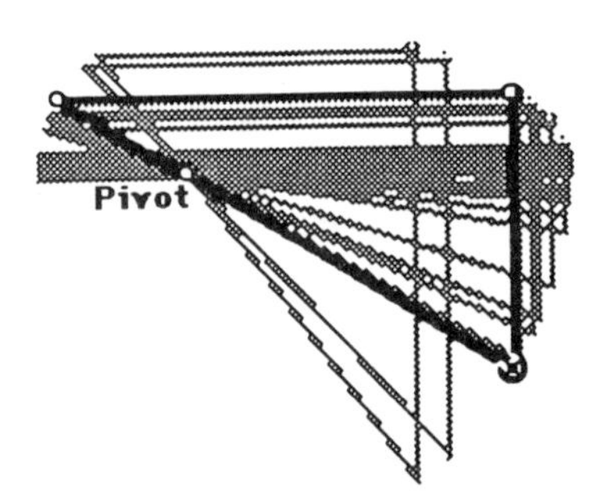

FIG. 1.28 Shifting perspective: "Sitting as a point on the hypotenuse," and watching the right-hand vertex move.

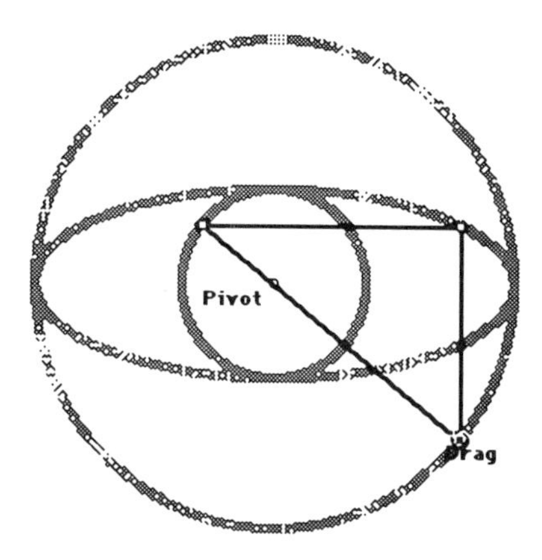

FIG. 1.29 Seeing the paths of the vertices of a right triangle using dynamic geometry tools.

A Classical Theorem Grows from an Experiment in Optimization. The heuristic of approaching geometric truths as functions of the conditions over which they are defined is such a powerful one that it deserves a second example. Here, as before, we can see a function arise from a classical Euclidean theorem, but in this case we will investigate the reverse route as well, seeing how the theorem might arise from a function in a somewhat surprising way. (For a term-length curricular sequence in which this problem and the generalizations, extensions, and modes of analysis described here form a centerpiece, see EDC, 1996c.)

Several years ago, a student, "Rich," was faced with a problem on a standardized test in which he was given an equilateral triangle of side 10 and a point D in the triangle's interior (see Fig. 1.31). He was asked to find the sum of the distances from D to the sides of the triangle.

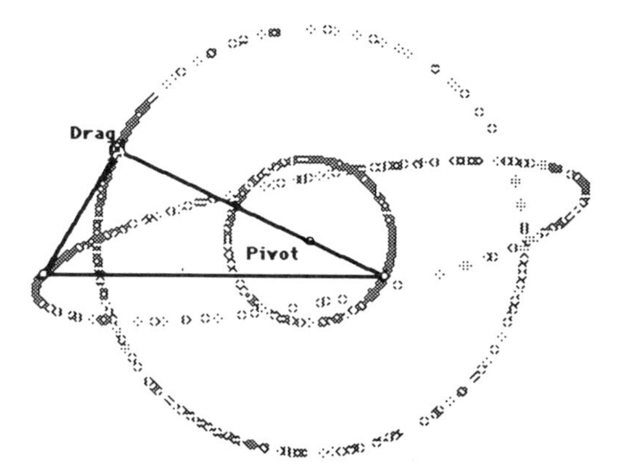

FIG. 1.30 Seeing the paths of the vertices of a non-right triangle using dynamic geometry tools.

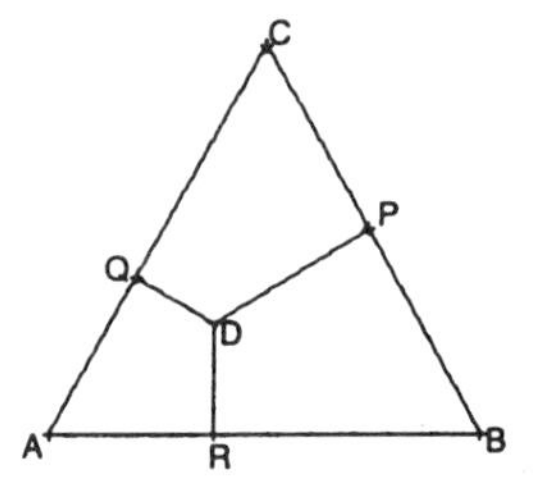

FIG. 1.31 Rich's function: Given an equilateral triangle with point D in the interior, find the sum of the distances from D to the sides of the triangle.

The question is based on a classical theorem:

> Theorem 2: The sum of the distances from an interior point to the three sides of an equilateral triangle is equal to the altitude of the triangle.

Rich, in many respects not an extraordinary mathematics student, did not know the theorem. But Rich was a thinker.

Not knowing any special facts, Rich reasoned that, because the problem didn't say anything special about D, he could move it anywhere he liked. So Rich moved D very close to vertex C where two of the distances were almost 0, and the other was almost an altitude of the triangle. He reasoned that the answer to the question must therefore be the length of the triangle's altitude, which he could easily find.

Rich assumed that the function defined on the triangle and its interior, given by $D \mapsto DP + DR + DQ$, is continuous and constant. His assumption that the function is constant came from the artificial environment of a standardized test, but his assumption that it is continuous came from his ability to do the thought experiment of "moving the point around" inside the triangle, seeing that small perturbations to D produce only small changes in the sum of the distances. It's not surprising that most students don't realize that the function is constant, but our experience with students tells us that many of them also have a hard time visualizing the dynamics that Rich carried out in his head.

Rich's ability and inclination to perform such experiments in his head grew without apparent prior (curricular) experiences doing similar experiments with his hands and eyes. That is clearly unusual. But we believe that many students could learn to perform such mental experiments if they first had appropriate experiences performing them concretely. Using appropriate software, students can design their own experiments, get a "feel" and "vision" for the dynamics, and develop the kind of reasoning by continuity that Rich used.

By relaxing the restriction that the triangle be equilateral, this rather specialized experiment becomes part of a broader class of optimization questions: At what point is the total distance to the three sides of a given triangle minimized? Students might set up the experiment with an arbitrary triangle (see Fig. 1.32), move D throughout the triangle's interior,

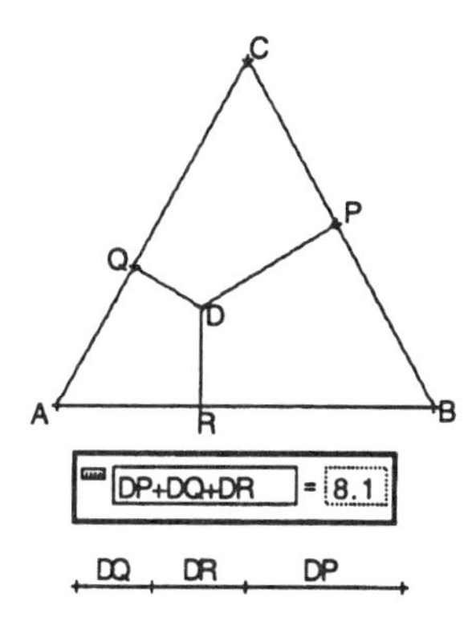

FIG. 1.32 The contour lines for Rich's function. Here D is free to be anywhere in the plane of the triangle.

and observe the effect that D's position has on the sum of the distances and on the individual contributions of DQ, DR, and DP to the sum.

The continuity of this function becomes apparent as soon as they start dragging D around. With scalene triangles, students will find that the minimum point is at the vertex opposite the longest side. But clearly that raises a question about what happens with isosceles triangles. Are there *two* best locations? What happens between them? And in the special case of the equilateral triangle, the surprise is that although the individual contributions of DQ, DR, and DP change as D moves, their sum seems to stay constant over the entire triangle.

Many conjectures may emerge from these experiments. But one direction particularly well illustrates the interplay between different mathematical perspectives and tools. (This point of view—organizing mathematics [and other] curricula around those ideas that support both specialized mathematical [or other] and general education needs—is the focus of Goldenberg, 1996.)

Even in the equilateral case, in which the sum of distances to the sides appears to be constant over all internal locations for D, the sum must surely increase if we move D far enough outside the triangle. So, even in the equilateral case, there are values for this function other than the minimum that we might examine. We might also wonder about the loci of points that produce one of these other values. One way to visualize these is to look at the function's *contour lines* or *level curves*.

In Fig. 1.33, we see an equilateral triangle with a side of 10 on the Cartesian plane. The gentle curves and wild wobbles in the graph result from inaccuracies in the graphing program, but even this picture is sufficient to help us conjecture that, for an equilateral triangle, the level curves are hexagons whose vertices lie on the extensions of the triangles' sides and whose sides are parallel to those of the triangle. After having come to this conjecture experimentally, it proves easy enough to verify through geometric reasoning. The threefold symmetry of the contour lines, for example, is immediately explained by the symmetry of the triangle itself.

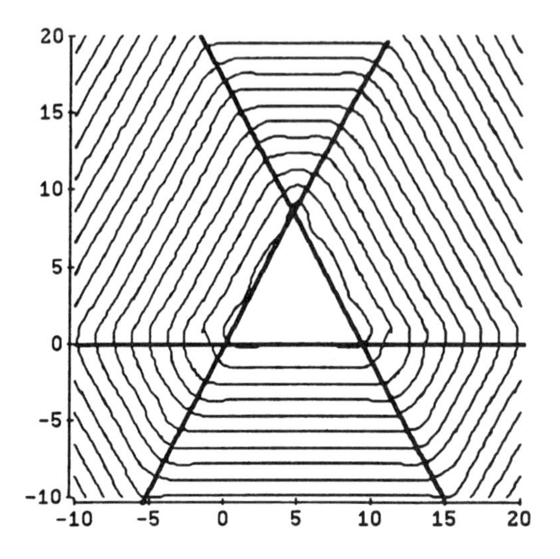

FIG. 1.33 Looking at a function's contour lines: Rich's function, given an equilateral triangle and point D outside the triangle.

But what happens in the case of a nonequilateral triangle, such as the one in Fig. 1.34 with vertices (0,0), (15,0), and (4,3)?

Experiments in a dynamic geometry environment convince us that the minimum value is at the vertex opposite the longest side, and the maximum value is at the vertex opposite the shortest side. The level curves ap-

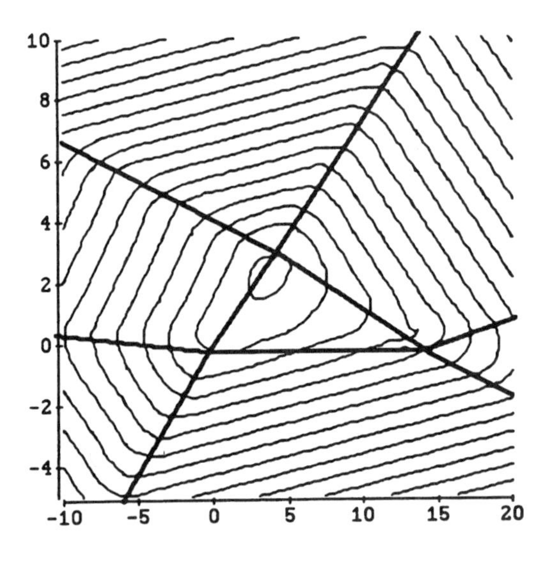

FIG. 1.34 The contour lines for Rich's function using a nonequilateral triangle. Again D is free to be anywhere in the plane of the triangle.

pear to confirm that observation, but in all other ways they are quite mysterious. The contour lines again seem to be hexagons, at least when any order is discernible from the mess of wiggles in the graph, but the shape and orientation of these hexagons seem, at first, almost totally unrelated to the triangle that spawned them.

Another kind of graph not typically encountered in geometry courses helps to explain the mystery and also returns us to thinking about the geometry. This time, we imagine our triangle sitting on the x-y plane (see Fig. 1.35). At each point (x,y), we evaluate the function and plot the value at the appropriate height z above the point.

As we might expect, the equilateral case looks regular. The surface is a basin with planar walls. The region of points directly above the triangle are all at the same height because the function value is invariant in that region. The level curves may be interpreted as watermarks along the walls. Thinking about why the walls are *planes* instead of curved surfaces gives some insight into the original function.

We can use the same visualization technique to investigate the general case (see Fig. 1.36). Again the sides of the basin are planes, but the bottom of the basin—the interior of the triangle—isn't horizontal any more because the values of the function inside the triangle are not constant. This visualization helps us see why the watermarks will not be parallel to the sides of the triangle—and we can see even more. Observing that the minimum value is at the lowest point in this basin, a vertex of the triangle, we see that the watermarks immediately surrounding it must be quadrilaterals, not hexagons, because only four planes intersect at the minimum point.

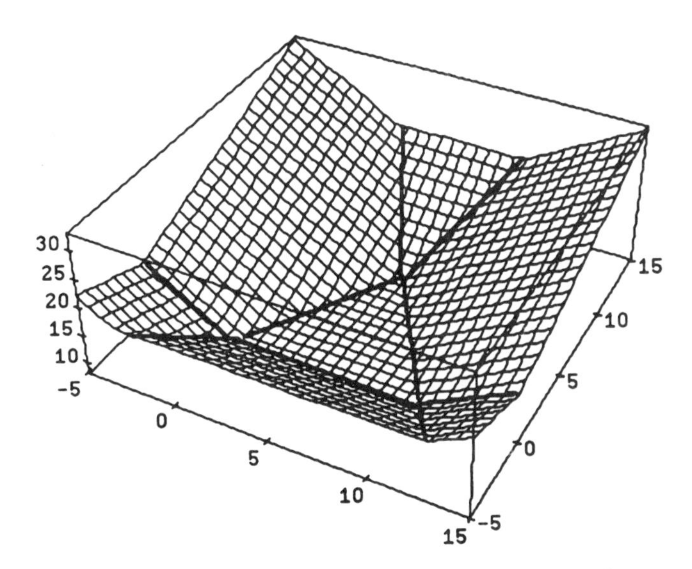

FIG. 1.35 Using a three-dimensional graph to look at a Rich's function as a surface. Here, the triangle is equilateral. Its image lies parallel to, but above the x-y plane.

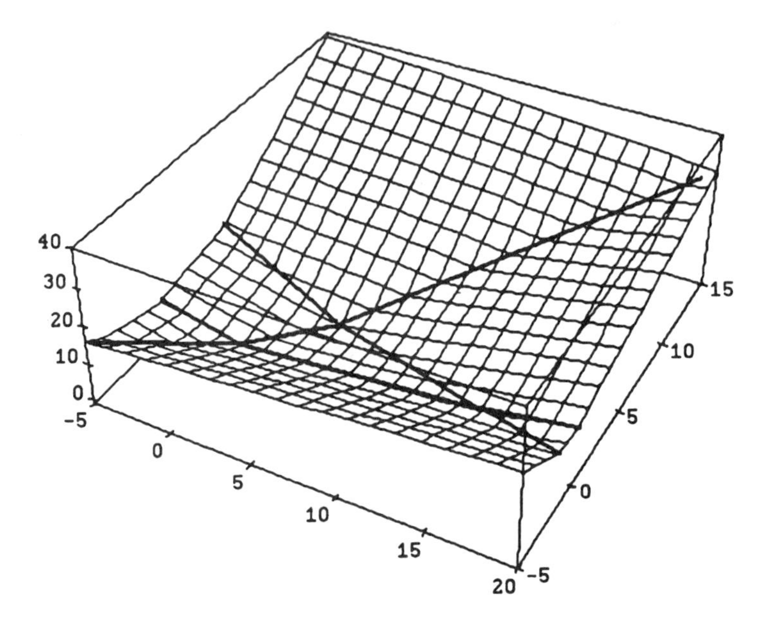

FIG. 1.36 A surface plot of Rich's function using a nonequilateral triangle. In this case, the triangle's image is not parallel to the *x-y* plane.

HABITS OF MIND

The smorgasbord of ideas we've just given does not lend itself well to the traditional curricular organization schemes. In particular, it is hard to incorporate such an eclectic collection of problems and, at the same time, present a logical development of topics. If topics cannot be the organizing principle, then something else must bring coherence to a course. On many grounds, it seems particularly successful to organize a course around the mathematical ways of thinking, or "habits of mind," that we use in solving problems (see Cuoco, Goldenberg, & Mark, 1996; Goldenberg, 1996).

Mathematical Power

We're greedy. It's not enough to draw more students into more mathematics: We would like to see this larger and more diverse group of students have more mathematical knowledge *and* more mathematical power. What makes someone a "power user" of mathematics is not the ability to recite formulas or to perform algorithms. Neither is it the knowledge of the "facts" of modern mathematics. These *are* assets—important assets—in the hands of a true power user, but by themselves they are neither necessary nor sufficient. Everyone has known people who have studied a great deal of mathematics yet who have little intuition about how to attack a problem; conversely, we all know people with little mathematical training who are capable of pulling unorthodox and elegant problem solutions seemingly out of thin air.

Mathematical power is best described by a set of *habits of mind*. People with mathematical power perform thought experiments; tinker with real and imagined machines; invent things; look for invariants (patterns); make reasonable conjectures; describe things both casually and formally (and play other language games); think about methods, strategies, algorithms, and processes; visualize things (even when the "things" are not inherently visual); seek to explain *why* things are as they see them; and argue passionately about intellectual phenomena.

If we listen to a mathematician describe the way mathematics is done, or look at the blackboard of a room that has just hosted a mathematical discussion, we will hear and see very little about numbers, accuracy, and precision. Instead, we will find visual images, and we will hear about the shapes and rhythms of calculations, about motion and machines, about rearrangements of pieces. Mathematics is a way of *looking* at things.

It is possible to build curricula that embody this reality. In both *Seeing and Thinking Mathematically* and *Connected Geometry*, National Science Foundation–funded curricula developed at EDC for middle and high school, respectively, the development of mathematical habits of mind is taken as a primary goal. The activities are explicitly designed to help students learn to seek and find invariants; to make and refine conjectures; to verify their conjectures to the community of young mathematicians that make up their class; to treat visual imagery as a central ingredient in mathematical discovery, invention, and explanation; and to seek and construct logically and coherently reasoned arguments.

When the mathematics curriculum is viewed from the habits-of-mind perspective, the criteria for choosing content and context are quite different from the typical benchmarks used to construct courses and programs for schools. Looking at the high-school mathematics curriculum from this perspective, we seek out topics and activities that allow students to connect up the various parts of mathematics, to employ central ideas like function and iteration in a wide variety of contexts, to use mathematical symbolism as a medium for expression, to revisit important themes in seemingly different situations, and, perhaps most importantly, to develop the mathematician's knack for visualizing phenomena.

The examples presented in this chapter do these things. Thought experiments that ask students to rearrange geometric objects in their heads lead naturally to activities that ask them to think about the "look and feel" of functions using tools like Dynagraph, Function Machines, and the dynamic geometry tools. Thought experiments like Wittgenstein's ask students to *picture* the workings of a machine; software makes it easier to gather up the pictures of the results and look for theoretical underpinnings for what those computer-generated pictures show. The classic activity of the billiard table, and its extensions to dance and motion, are similar invitations to students to experiment, conjecture, and argue. In the course of conjecturing and arguing, students bump into number theory, geometry, physics, and analysis.

Iterated geometric constructions provide students with an arena in which they can obtain theoretical results by reasoning about the algorithms that generate the images; Amy's proof amounts to a careful analysis of the genesis of her pictures—what Wittgenstein called a "picture of an experiment."

In turning the theorem about right triangles into a dynamic experiment, students not only develop a deeper insight into the result and its explanation; they also tinker with the experiment in ways that suggest new results and new explanations. At the same time, they develop a sense for continuous functions that makes intuitive use of the delicate properties of real numbers in ways that would have been quite familiar and pleasing to Euler, but that have become lost in present-day curricula, which concentrate on algebraic formalisms. So, too, does Rich's function add life and generality to a static geometric result. Students see how the result changes if the original constraints are relaxed, giving them a continuum of *theorems* to think about. Activities built around Rich's function dramatically show how the different ways of visualizing phenomena can give new insights and clear up old mysteries.

Practicality in Teaching

This is good mathematics, but is this geometry? Not in the traditional school sense, but more broadly conceived, yes. These activities require students to develop a sense for the properties of space and measure, and they encourage the development of many of the classical themes of Euclidean geometry. These activities also encourage students to visualize things in more than one way, to employ tools from the entire spectrum of mathematics, and to move beyond data-driven discovery toward a way of explaining what they see in a way that makes essential use of their experiments.

The emphasis on ways of thinking also means that the most important facts, methods, and procedures tend to recur often: They are important precisely because they are useful. This kind of recurrence affords practice without repetitiveness, and teachers and students have reported appreciating that greatly. And facts that are less central need not be slighted in the process: They, too, tend to be learned more solidly (apparently because they are connected to other mathematical ideas) rather than being learned as individual topics.

Developing a Mathematical Method

In mathematics, as in most disciplines, it takes years of study and specialization before a student develops a background sufficient to understand issues that are at the cutting edge of work in the field. But long before students reach the stage where they can understand the topics of current interest to mathematicians, they can and should be using the *methods* fundamental to the field. Something much like this is attempted in science courses in schools: Science educators do try to help students develop a sense for the way scientists work. Although not accounting for serendipity and other important aspects of true scientific inquiry (and despite an occasional tendency to ossify it into The Three-Step Method For All Things Scientific), the official school-taught "scientific method" is quite a reasonable first approximation of the real thing and a salutary alternative to an exclusive focus on facts and results.

This kind of dual focus has, for the most part, never been available in mathematics curricula. There is no mention of a "mathematical method" for approaching things, and the methods that *are* developed are concerned with low-level techniques for carrying out calculations. For a single example, conventional curricula may (or may not) emphasize that mathematicians "prove things," but rarely make clear to students *why* precise justification—the thing that separates mathematics from almost every other discipline—is so central to mathematical thinking. In fact, the lack of an apparent rationale—when proof is included at all, it tends to be treated as a kind of post hoc ritual—is one source of the popular calls to deemphasize or totally remove proof from the curriculum.

This trend is perhaps most evident in geometry courses, which have a tradition of asking students to work on nonroutine problems and, hence, have the potential for opening up a discussion about basic mathematical methods. But over the years, geometry courses have tended to fall into recognizable categories:

- Attempts at faithful replicas of Euclid. These tend to be dogmatic expositions of established mathematics, using the axiomatic method; they are the definition, theorem, two-column-proof, corollary courses.
- Euclid without proof. These courses follow essentially the same route as their more formal cousins, but the main results of geometry are usually stated instead of derived, and the emphasis is on "applications"—often clever problems that require students to apply various area formulas and the Pythagorean theorem.
- "Inductive" geometry. We put the word inductive in quotes because, as used here, it has nothing to do with mathematical induction, but refers only to reasoning from the specific to the general, to making conclusions on the basis of experiment, to the "discovery method."

What these courses have in common is the sharp distinction between how a result is obtained and how that result is established. Each type of course has its own approach, but all three take the point of view that justification of a result is a separate matter from its formulation.

In fact, that point of view does not represent the best mathematical thinking: Proof is not merely to support conviction, nor to respond to a distrustful nature or self-doubt, nor to be done as part of an obsessive ritual. Proof serves to provide *explanation*, and therefore is a central technique in research. Looking for a working justification for emerging and half-formulated facts that arise in the midst of experiments helps us fine-tune conjectures and design new experiments. In real life, both in and out of mathematics, the distinction between empirical and theoretical investigations breaks down: We must move back and forth between "doing stuff" and understanding what we have done. Seen in this way, proof (or at least the precursors to proof) is a natural step in satisfying curiosity to *understand* what we have observed. Proof's centrality and unique role in mathematics are preserved and clarified, but it no longer stands as a peculiarity of mathematics. Instead, it becomes a specialization of a way of thinking that we would hope students extend into other interests and disciplines. Although mathematics cannot live without proof, proof itself is a larger idea that is part of our broader intellectual heritage. The notion of proof, the inclination to prove, and even many of the attributes of mathematical proof are habits of mind we can see in many disciplines—something suited both to the special needs of those who specialize in mathematics and to the general needs of a general education.

New Roles, New Demands, New Payoffs

Such a habits-of-mind point of view entails looking at both geometry and pedagogy in new ways. Teachers need to see how results in geometry fit into a more general mathematical framework so that, for example, the results springing from generalizations of Theorem 2 earlier in this chapter are seen as attempts to locate minimum values of functions. Teachers also need to adopt the point of view that such results evolve slowly as students develop experiment-based conjectures, attempt to justify them, develop new experiments on the basis of what they've learned, and then refine their conjectures. What is most important (and most difficult) is that teachers need to view the development of a *style of work* and a *style of thinking* in their students as the primary goal of mathematics class. This new role, one of senior research partner, creates a new set of expectations of students.

The habits-of-mind point of view even casts technology in a somewhat new role. Technology allows the design of activities that put experimental power into the hands of students and teachers. The focus can be on the facts (theorems) discovered or, as we are proposing, on the ways of thinking that lead us to discover and understand the facts, and to *know* that they are facts. The direction that classroom experimentation takes is determined in part by curricular materials but also by interpretations imposed on those curricular materials by teachers. Hence, materials must convey to teachers a sense of mathematical purpose and must help them "buy into" the habits-of-mind point of view.

For a general education and equally for the education of the mathematical specialist, a geometry curriculum of this sort must therefore straddle a fine line. To be accepted in mathematical classrooms as they currently exist, geometry curricula must trade in the accepted currency—the objects, facts, and methods of geometry as they are generally known in high schools now—but these curricula must also, in every activity, make mathematical methods and style very apparent and salient. Students and teachers (and parents and curriculum administrators) must be able to start out feeling that what they do, they do for the sake of geometry. But they must be able to end up feeling that what they have done, they have done for the sake of mathematics, and, even more, for the sake of good thinking.

ACKNOWLEDGMENTS

This chapter is adapted from the paper "Making Connections With Geometry," originally prepared for the keynote address to the Geometry Working Group at the International Congress of Mathematics Education, ICME 7, Université Laval, Québec, August 1992, and first published in 1994 in Czech translation in *Pokroky matematiky, fyziky a astronomie* [Advances in Mathematics, Physics and Astronomy], Association of Czech Mathematicians and Physicists, 39, 275–304. This work was supported in part by the Science Foundation, grant numbers MDR-8954647, MDR-9054677, and MDR-9252952, with additional support from Apple Computer, Inc. The views represented here are not necessarily shared by any of the funders. We are very grateful for the substantial intellectual and editorial contributions made by Glenn Kleiman at various stages of the preparation of this document and its many ancestors.

REFERENCES

Abbott, E. A. (1980). *Flatland*. San Francisco, CA: Arion.

Arcavi, A., & Nachmias, R. (1990). Desperately looking for the focus. *Mathematics in School, 19*(2), 19–23.

Banchoff, T. (1991). Dimension. In L. Steen (Ed.), *On the shoulders of giants* (pp. 11–59). Washington, DC: National Academy Press.

Blumberg, S. (1991). *Michael Moschen: In motion*. Great Performances Series, Dance in America. New York: In Motion Productions with WNET.

Cuoco, A. A. (1993). Action to process: Constructing functions from algebra word problems. *Intelligent Tutoring Media, 4* (3/4), 117–127.

Cuoco, A. A., & Goldenberg, E. P. (1992). Mathematical induction in a visual context. *Interactive Learning Environments, 2* (3/4), 181–203.

Cuoco, A. A., Goldenberg, E. P., & Mark, J. (1995). Connecting geometry with the rest of mathematics. In P. A. House & A. F. Coxford (Eds.), *Connecting mathematics across the curriculum; NCTM 1995 yearbook* (pp. 183–197). Reston, VA: National Council of Teachers of Mathematics.

Cuoco, A. A., Goldenberg, E. P., & Mark, J. (1996). Habits of mind: An organizing principle for mathematics curriculum. *Journal of Mathematical Behavior, 15*(4), 375–402. [http:\\www.edc.org\LTT\ConnGeo\HOM.html]

Education Development Center, Inc. (in press). *Connected geometry*. Chicago: Everyday Learning Corporation.

Eisenberg, T., & Dreyfus, T. (1991). On the reluctance to visualize in mathematics. In W. Zimmermann & S. Cunningham (Eds.), *Visualization in teaching and learning mathematics* (pp. 25–37). Washington, DC: Mathematical Association of America.

Feurzeig, W. (1994). Explaining function machines. *Intelligent Tutoring Media, 4* (3/4), 97–108.

Goldenberg, E. P. (1996). "Habits of mind" as an organizer for the curriculum. *Journal of Education, 178*, (1), 13–34.

Goldenberg, E. P., Lewis, P. G., & O'Keefe, J. (1992). Dynamic representation and the development of an understanding of functions. In G. Harel & E. Dubinsky (Eds.), *The concept of function: Aspects of epistemology and pedagogy* (MAA Notes, Vol. 25, pp. 235–260). Washington, DC: Mathematical Association of America.

Healy, C. C. (1993). *Build-a-book geometry: A story of student discovery*. Berkeley, CA: Key Curriculum Press.

Hofstadter, D. (1979). *Gödel, Escher, Bach*. New York: Basic Books.

Jacobs, H. R. (1970). *Mathematics: A human endeavor*. San Francisco: Freeman.

Klee, V. (1991). Convex geometry in undergraduate instruction. In J. Malkevitch (Ed.), *Geometry's future* (pp. 33–36). Arlington, MA: Community Map Analysis Project.

Klein, B. G., & Bivens, I. C. (1988). In proof without words. *Mathematics Magazine*, 61(4), 219.

Krutetskii, V. A. (1976). *The psychology of mathematical abilities in schoolchildren*. Chicago: University of Chicago Press.

Lakatos, I. (1976). *Proofs and refutations*. Cambridge, England: Cambridge University Press.

Mayer, R. E. (1989). Models for understanding. *Review of Educational Research, 59*(1), 43–64.

Nelsen, B. R. (1992). *Proofs without words*. Washington, DC: Mathematical Association of America.

Senechal, M. (1991). Visualization and visual thinking. In J. Malkevitch (Ed.), *Geometry's future*, (pp. 15–21). Arlington, MA: Community Map Analysis Project.

Tall, D. (1991). Intuition and rigour: The role of visualization in the calculus. In W. Zimmermann & S. Cunningham (Eds.), *Visualization in teaching and learning mathematics* (pp. 105–119). Washington, DC: Mathematical Association of America.

Vinner, S. (1983). Concept definition, concept image and the notion of function. *International Journal of Mathematical Education in Science and Technology 14*, 293–305.

Vinner, S., & Dreyfus, T. (1989). Images and definitions for the concept of function. *Journal of Research in Mathematics Education 20*(4), 356–366.

Weeks, J. (1985). *The shape of space*. New York: Marcel Dekker.

Wittgenstein, L. (1983). *Remarks on the foundations of mathematics*. Cambridge, MA: Massachusetts Institute of Technology Press.

Zimmermann, W., & Cunningham, S. (1991). *Visualization in teaching and learning mathematics*. Washington, DC: Mathematical Association of America.

2

From a Different Perspective: Building on Students' Informal Knowledge

Koeno P. Gravemeijer
Freudenthal Institute

This chapter describes the Dutch approach to geometry education, an approach that builds on students' informal knowledge about geometric aspects of everyday life situations. Most students have a great deal of informal geometrical knowledge at their disposal, and even young children can model situations with gestures and imagine events and objects from different perspectives. In this chapter, we argue that this informal knowledge can be explicated and built on in geometry education within the framework of the domain-specific instruction theory of Realistic Mathematics Education (RME), an approach developed in the Netherlands. The key characteristics—reinvention through progressive mathematization, didactical phenomenological analysis, and use of emergent models—function as the scaffolding for an RME curriculum. Such an approach enables students to construct their own mathematical knowledge, fosters a reflective attitude toward the world, and fits in with the overall philosophy that views mathematics as a human activity. As such, this approach also accommodates the critique by Ehrenfest-Afanassjewa (1931), van Hiele-Geldof (1957), Freudenthal (1971), and van Hiele (1973) of traditional instruction of Euclidean geometry in the Netherlands.

Although other countries endorse a similar approach (see, e.g., National Council of Teachers of Mathematics [NCTM], 1989), the presentation in this chapter is limited to the Dutch curriculum. The first third of this chapter looks at ways in which the Dutch curriculum builds on informal knowledge. The next section looks at the theoretical background for Dutch mathematical education. The final third of the chapter elaborates on the theory of RME as applied to geometry education.

FROM A DIFFERENT PERSPECTIVE

Every day, whether modeling a physical activity for someone else (e.g., teaching a child to tie a shoe), or mirroring the movements of another (e.g., following an aerobics instructor's movements), we adapt frequently to literal variations of our perspective. In the Netherlands, geometry education is commonly based on such everyday reasoning about spatial relations, and, in Dutch mathematics reform, "looking at the world from a geometrical perspective" is indispensable. As an example of what this can mean, let's consider the following problem (created by G. S. Monk, 1992):

> Imagine that a man walks away from a light pole at a very slow but constant speed, or imagine that the man moves away from the light pole in equal jumps. In either case, describe the pattern by which the top of the shadow moves.

This problem makes use of a physical device that shows how the situation changes if the man moves (see Fig. 2.1) and was set up to generate student thinking about (the rate of) change in general terms as a preparation for calculus. Students had major problems figuring out how and at what rate the shadow moves.

But let's examine this problem from a geometric point of view. In this situation there are three points characterizing the actual state of affairs: the light, the man's head, and the tip of the shadow. One point is fixed, and the other points move along essentially parallel lines. One of these parallel lines is the path of the tip of the man's shadow as it moves along the ground. The second line is the path of the top of the man's head as he crosses that ground. These two parallel lines are situated in the triangle formed by the lamppost and the light ray projecting the man's head as a shadow on the ground (see Fig. 2.2).

The picture of the triangle with the intersecting line segment parallel to the base of the triangle (the path of the man's head) calls to mind the equal ratios in such a triangle. These ratios relate to the speed of the shadow. There is a fixed relation between the length of the line segment representing the distance traveled by the man's head, the line segment defined by the lamppost, and the tip of the man's shadow. This fact implies a constant ratio between the speed of the man (his head) and the tip of his shadow. This can be seen if we segment the dynamics of the situation into

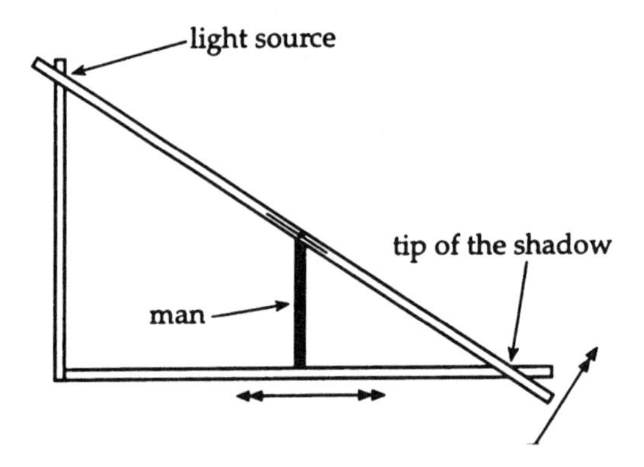

FIG. 2.1 Rate of change: the shadow of a man in motion.

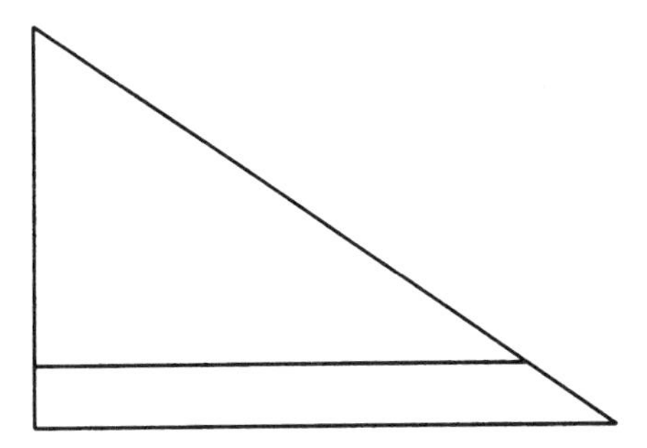

FIG. 2.2 Triangle with intersecting line segment: geometric model of the moving shadow.

equal time intervals (e.g., as if the man is jumping). For each interval, a small triangle may be discerned in which the segment h_i has a fixed ratio to its base s_i, which is defined by the ratio a over b (see Fig. 2.3). So every time the man moves an interval Δh, the tip of his shadow moves:

$$\Delta s = (a/b)\ \Delta h$$

Perhaps we cannot expect students to see this solution easily, but if they have had experience with shadows before—perhaps having reflected on the variation of the length of their shadows at different times on a summer day—then they will have had an opportunity to develop this type of solution. Reflecting on phenomena like this can produce the basic underpinning for the concepts of vision lines and angles. Of course, other variations of the situation—the differences between situations involving lamplight and sunlight—should be investigated (see Fig. 2.4). For example, when a man walks in a straight line past a streetlight, what path does the tip of the man's shadow describe?

These problem situations and others can support the elaboration of a wide variety of geometric models. In Monk's problem, however, we have to be aware of an abstraction from reality, not only in the way the situation is represented, but more importantly because of the outside point of view. We are not looking at the situation through the eyes of the man in

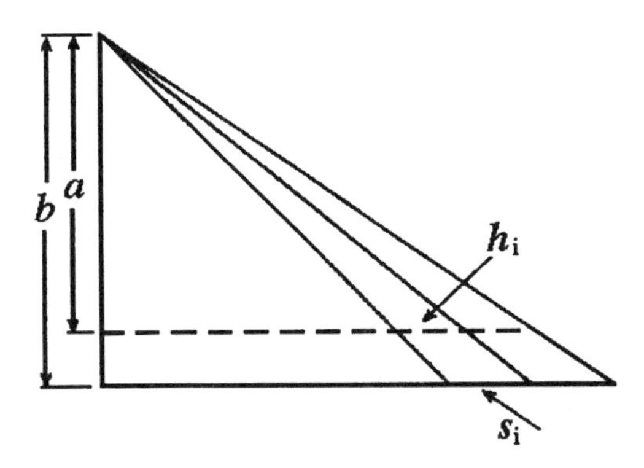

FIG. 2.3 Ratio of speeds: geometric model of the moving shadow.

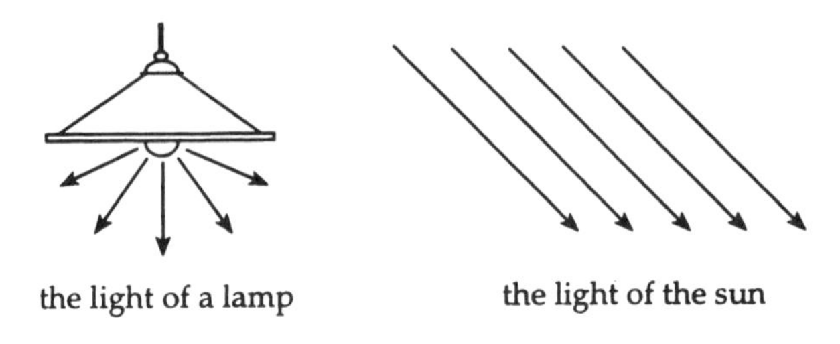

FIG. 2.4 Effect of light source on projected shadows: lamplight versus sunlight.

the problem. If we did limit ourselves to the man's point of view, it would be remarkably difficult to solve the problem. In stepping outside the problem, we construct a model of the situation. In this example, two issues are at stake: the role of a knowledge base in reflecting on everyday situations (like the length of a shadow on a sunny day or the differences between lamplight and sunlight), and the ability to look at a situation from a point of view outside that situation.

GEOMETRICAL LITERACY

Reflecting on geometrical aspects of everyday life situations is a salient characteristic of a mathematical attitude (Freudenthal, 1971). This attitude is expressed in mathematizing everyday situations (like those exemplified previously) and other subject matter spontaneously and in building notions about when such mathematizing is appropriate and when it is not. Such a mathematical attitude also expresses itself in the so-called mathematical literacy that is most often described for number sense and statistics (Paulos, 1988; Treffers, 1991). Number sense implies a relational framework constituted by reflecting on number relationips, the role of numbers, and their values in everyday life (McIntosh, Reys, & Reys, 1992). In a similar way, geometric sense implies a relational framework for, among other things, vision lines, shadow lines, mental images, side views, top views, and maps.

The questions used to develop geometrical literacy should be closely related to applied problems in everyday situations. Determining a position on a map is a useful skill when we are in an unfamiliar place. Such problems can be brought into the classroom with the help of photographs. One of the main goals of the instruction, however, should be to foster the inclination of students to reflect upon their own tacit knowledge. That focus may demand the implementation and/or development of specific instructional activities.

Generally we do not recognize our own tacit knowledge. However, as a quick investigation of Fig. 2.5 (adapted from Kennedy, 1974) shows, we can infer ample information—but generally without explicit awareness—from lines that intersect, lines that break, and lines that bend.

In Fig. 2.5, we "know" that the house is in front of the dunes, that there is a sea, and that the sea is behind the dunes. A fence partially en-

FIG. 2.5 Tacit or informed knowledge: three-dimensional clues in a two-dimensional picture.

closes the house (the inside angle and an outside angle of the fence). We infer the presence of the roof, and there are windows, doors, and a chimney. The position of the photographer is above sea level and probably slightly above the level of the house.

Geometrical knowledge is applied in the art of drawing either tacitly or consciously. There are, for instance, several ways to suggest depth. In Fig. 2.6, we can see the use of variation in size (a) and height (b) to show distance between objects. Perspective lines (c) are used to show distance from foreground to point of intersection. Occlusion, the fixed example (d) in Fig. 2.6, can be used to show distance between objects that are quite close to each other. All of these methods involve both tacit and conscious geometric knowledge.

Model and Perspective

Look closely at Fig. 2.7. Why does the figure in the background look out of place? Many of us can recognize that something is wrong in Fig. 2.7; some may even understand that the figure in the background is either a mistake or a giant. Few of us, however, will consciously attribute this understanding to the tacit application of geometric knowledge.

Techniques used in drawing can also be used as the starting point for lessons in geometry (see Goddijn, 1980). Even simple tricks, such as using a pencil as a measuring stick can raise discussion-provoking questions. In Fig. 2.8, a pencil can "measure" a church (to compare it to other buildings in the drawing)—and can conceal it. Why does this method of measurement work? And how is it possible to hide an entire church behind a pencil?

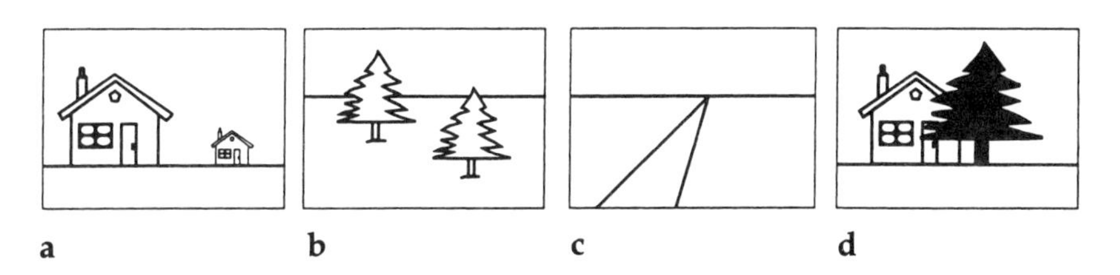

FIG. 2.6 Application of tacit and conscious geometric knowledge concepts: depth in illustration: (a) variation in size, (b) variation in height, (c) perspectives, and (d) occlusion.

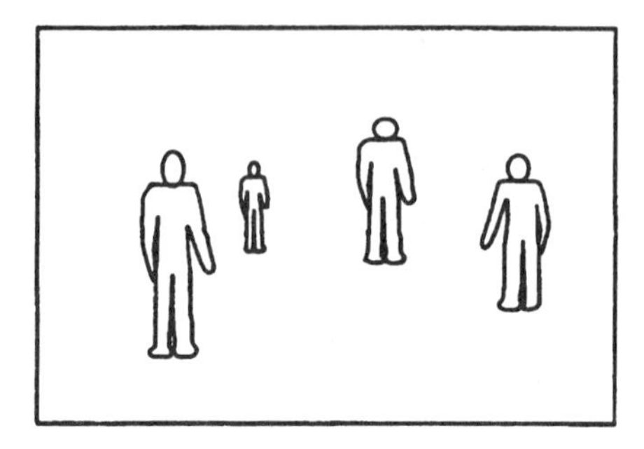

FIG. 2.7 Tacit application of knowledge: the giant in the background.

Modeling a real-life geometrical problem often involves looking at the situation from a different perspective: an outside point of view. In Fig. 2.9, both (a) and (b) depict the same buildings, but in (a) the church appears taller than the building in the foreground. In (b), the church, in comparison to the other building, appears smaller.

This apparent discrepancy can be unmasked if we look at a side view (see Fig. 2.10). From a distance, a larger portion of the church tower is visible (a). If we come closer (b), the tower appears to sink behind the building in the front. It is the same effect we experience when we approach a large city with skyscrapers. From a distance, the skyscrapers tower above the city, but as we approach the city, the skyscrapers "sink" behind other buildings (see Fig. 2.11) even though, logically, they should "grow" as we approach.

Another difference between these two pictures involves the apparent width of the building in the foreground. The building appears broader at a closer distance. This can be understood intuitively by again referring to Fig. 2.8, in which the church "hides" behind a pencil. The more formal explanation is that the scale on which the object appears depends on its distance. When both objects are far away, they will appear to be more or less on the same scale because both are more or less equally far away. As we approach, the relative difference between the two distances becomes greater, and so does the difference in scale. The effect of the distance on the relative scale can be more easily seen from the top view (see Fig. 2.12). The change in perspective as we approach "widens" the building in the foreground.

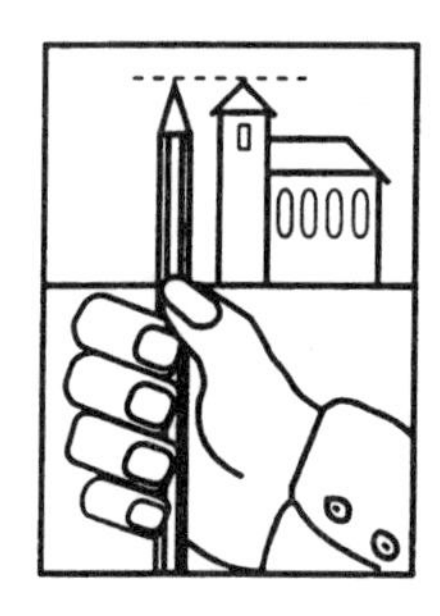

FIG. 2.8 Perspective: How big is a church?

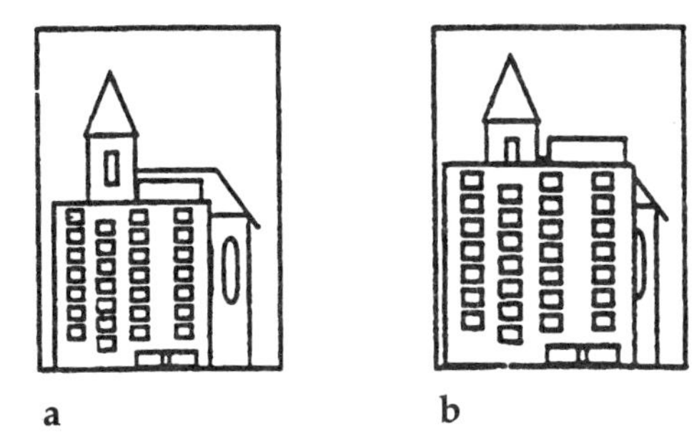

FIG. 2.9 Perspective: the shrinking church.

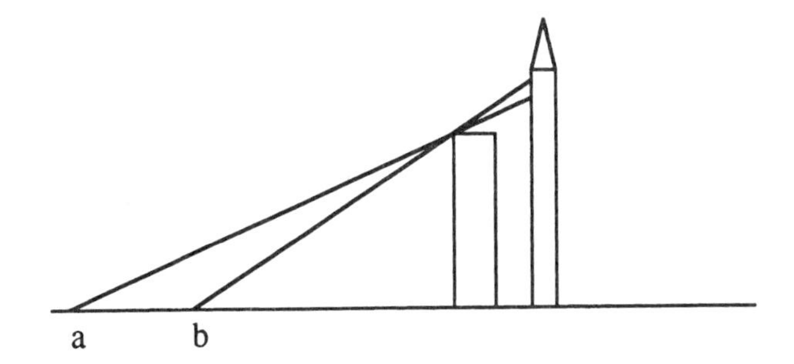

FIG. 2.10 Perspective: the effect of distance on height.

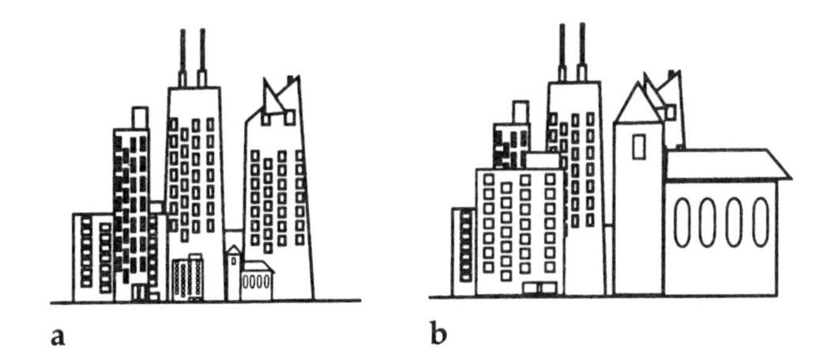

FIG. 2.11 Perspective: the sinking skyscrapers.

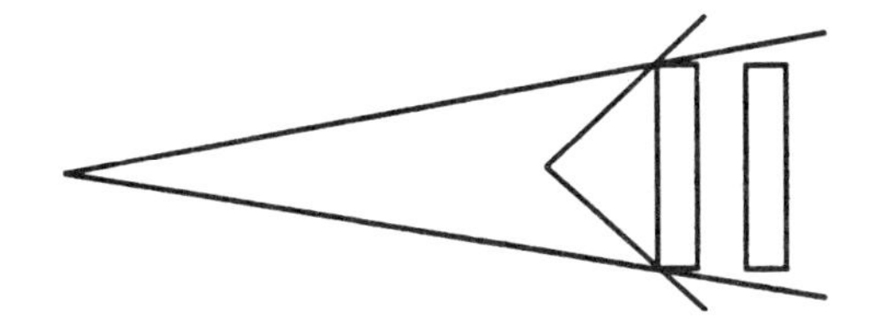

FIG. 2.12 Perspective: the effect of distance on width.

Modeling

As seen in the last section, geometrical models can prove very useful in explaining visual perception in everyday situations and in transposing those situations into the classroom. More important, such situations can be used to help students develop their own models of perspective. Young children can and often do develop their own geometrical models. Let's look at the following anecdotes:

> Alexli, about six years old, sees a freighter on the river, carrying the captain's car on its deck (see Fig. 2.13a). The difference in size between the car and the boat helps Alexli to understand the size of the boat. With her gestures, Alexli creates a simple linear model to describe the ratio between the length of the car and the length of the boat (see Fig. 2.13b), saying, "If the car is this big, the boat is this big."
>
> Onno, eight years old, is taking a walk with his dad. Approaching a bridge they had passed earlier, Onno remarks, "We walked a circle. Oh, no, it's more the shape of a water drop" (see Fig. 2.14a.) At the end of the walk, Onno adequately describes the shape of the route with his finger (see Fig. 2.14b).

In our view, this ability of children spontaneously to create geometrical models should be exploited in primary school: Young students should be given the opportunity to develop, generalize, and formalize these models, instead of postponing these concepts until secondary school

FIG. 2.13 Modeling ratios: How big is the boat?

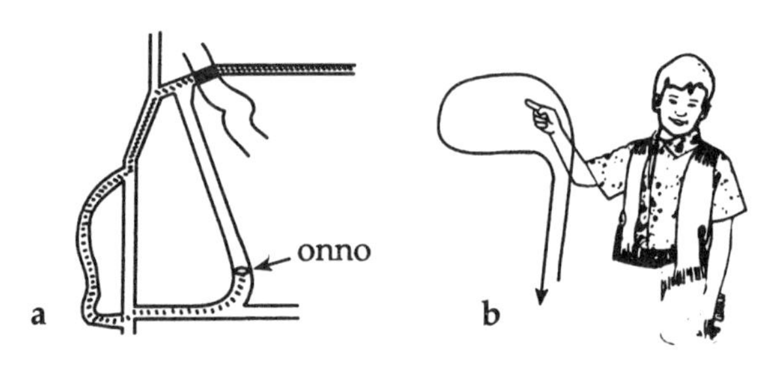

FIG. 2.14 Modeling a route: Where did you walk?

and confronting the students at that point—and with no preparation—with prefabricated abstract models (see Clements, Battista, & Sarama, chap. 8, and Lehrer, Jacobson, et al., chap. 7, this volume).

DUTCH GEOMETRY EDUCATION

This long discussion on "looking at the world from a geometrical perspective" is meant as an extensive argument for a geometry education that is rooted in real-life experiences and experientially real situations. The contemporary approach to mathematics education in the Netherlands has its roots in an innovation that started about 25 years ago. At the end of the 1970s, the National Institute for the Development of Mathematics Education (IOWO) was given the task of developing a new approach to mathematics education in primary and secondary schools. The point of departure was Freudenthal's (1971) philosophy of "mathematics as a human activity":

> It is an activity of solving problems, of looking for problems, but it is also an activity of organizing a subject matter. This can be a matter from reality, which has to be organized according to mathematical patterns, if problems from reality have to be solved. It can also be a mathematical matter, new or old results, of your own or (of) others, which have to be organized according to new ideas, to be better understood, in a broader context, or by an axiomatic approach. (Freudenthal, 1971, pp. 413–414)

It is important to note that this organizing activity—"mathematizing"—applies both to mathematical matter and to real-world subject matter. According to Freudenthal (1973), mathematics education for young children should start with the mathematization of everyday reality. Besides the mathematization of problems that are real to students, there also has to be room for the mathematization of concepts, notations, and problem-solving procedures. Treffers (1987) made a distinction between horizontal and vertical forms of mathematization. The former involves converting a contextual problem into a mathematical problem; the latter involves taking mathematical matter onto a higher plane. Vertical mathematization can be induced by setting problems that admit of solutions at different mathematical levels.

Although Freudenthal espoused mathematics as an activity, he did not lose sight of mathematics as a product. In his view the process and the product should be connected: The combination of horizontal and vertical mathematizing should enable the students to reinvent mathematical insights, knowledge, and procedures. Freudenthal (1973, 1991) connected the idea of mathematization with the principle of guided reinvention. According to the reinvention principle, a learning route has to be mapped out, along which the student can find the result by him- or herself. Of course, the students cannot be expected to reinvent, in a short period of time, mathematics that took outstanding mathematicians centuries to invent. Therefore, the developer has to develop a set of instructional activities that give the students the opportunity to experience such a reinvention process in a condensed form. The emphasis is on the character of the learning process rather than on inventing as such. The process has to allow learners to regard the knowledge they acquire as their own private knowledge, knowledge for which they themselves are responsible. Applied to teaching, this idea means that students should be given the opportunity to build their own mathematical knowledge store. To design instructional activities that offer such an opportunity, we can look to the history of mathematics to inform the design process. History might tell the developer, on the one hand, what dead alleys to avoid, and on the other, what short cuts to take. The developer will also look for crucial breakthroughs, paradigmatic problems, useful tools, and so on. With this historical knowledge as background, the guiding question is, how could someone (ideally) have invented this?

In geometry, the reinvention approach has meant a radical break with the traditional Euclidean geometry curriculum in secondary school. Such a break was endorsed by people like Freudenthal (1971) and Ehrenfest-Afanassjewa (1931; see de Moor, 1993). Critiques on the shortcomings of traditional Euclidean geometry, as it was taught in Dutch schools, also formed the basis for the work of van Hiele (1973, 1985) and van Hiele-Geldof (1957).

Critique of the Traditional Euclidean Curriculum

The starting point for analysis concerns the communication problems between teachers and students. Van Hiele (1973, 1985) illustrated this with a discussion of the concept *rhombus*. In his opinion, the expression "This is a rhombus" will have different meanings for a student and a teacher (see Pegg & Davey, chap. 5, this volume). We can expect the student to recognize the shape and connect the shape with the label *rhombus*. To students who recognize a rhombus in this way, it may not be self-evident that a square is a rhombus as well, unless the square is tilted (see Fig. 2.15).

For a mathematician as well as for a math teacher, the concept *rhombus* has an entirely different meaning. To such a person, the name rhombus signifies a set of properties and relations. The figure is a quadrangle. All sides are equally long. The figure is a parallelogram. The figure has

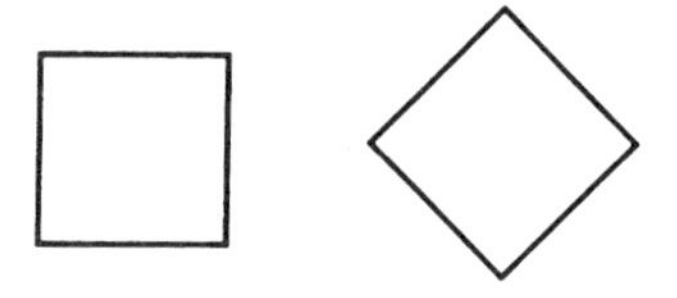

FIG. 2.15 Student versus teacher interpretation of geometric meaning: square or rhombus?

pairs of parallel sides. The diagonals are perpendicular at intersection. Consequently, a teacher will consider a square a rhombus because of its properties. He or she will also accept a rough sketch as a rhombus, if it is agreed that the sides are equally long and opposite sides are parallel.

The difference between the teacher and the student concerns, in essence, a difference in referential framework, and that difference hinders communication between them. Both the teacher and the student use the same words, but these words do not have them same meaning for them both.

The only way to overcome this problem, van Hiele argued, is to have the students construct a referential framework. To do so, they should start at the so-called "ground level." Van Hiele wrote not of building a referential framework, but of building a relational framework. The availability and the quality of the relational framework define three levels of thinking (see Pegg & Davey, chap. 5, this volume):

- *Ground level.* A relational framework has not yet been formed.
- *First level.* The students have at their disposal a relational framework.
- *Second level.* The relationships among relationships have become the subject of analysis.

Treffers (1987) characterized these three levels as concrete, descriptive, and concerned with subject-matter systematics. The crux of the matter is that these levels are subject-matter dependent. In this sense, too, the word *concrete* must be defined. Concrete in this context does not mean tangible, but experientially real—familiar and meaningful. As a consequence, what is concrete for someone will change over time: What is at the second level in one topic at one time will be at the ground level in another topic at a later time.

According to van Hiele, the traditional Euclidean geometry curriculum started at too high a level. The concrete level, and often the descriptive level as well, were omitted. Students started at the level of the subject matter systematics, with all the problematic consequences we might expect. Again, according to van Hiele, geometry education should start at a concrete level.

REALISTIC MATHEMATICS EDUCATION

Realistic Mathematics Education (RME), developed in part in response to critiques of traditional mathematics curriculum, is a domain-specific instruction theory that proceeded from the work done at IOWO in the Netherlands. In the following sections, the core principles of RME—guided reinvention, didactical phenomenological analysis, and the concept of emergent models (Gravemeijer, 1994)—are used to describe the RME approach to geometry.

Guided Reinvention

Embedded in the instructional activities developed by van Hiele-Geldof (1957; see also Freudenthal, 1971) we find the principle of guided reinvention, one of the cornerstones of the theory of RME. Van Hiele-Geldof gave students cardboard triangles, squares, and the like to use. The cardboard tiles were not so perfectly made that everything fit precisely, but that was no problem for the students. They perceived them to be perfectly regular, and they created patterns as if the pieces were. Irregularities were ignored, and interesting things happened when the students had to make drawings of the patterns. In describing a pattern made with congruent triangles, almost all students ended up drawing parallel lines (see Fig. 2.16).

Initially, the students recognized only one or two sets of parallel lines. Eventually, however, they realized that there were three sets of parallel lines. The students figured out that the parallel lines related to the way the triangles fit in the pattern. At each intersection of three triangles, three different angles (α, β, ϕ) combine to create a straight line (see Fig. 2.17). The total of the angles of a triangle adds up to the measure of a straight line, 180°. Because these angles are the same as those in any of these congruent triangles, the sum of angles in any of the triangles will always be 180°. The students, as a result of these activities, "rediscovered" certain mathematical knowledge (here, the sum of angles in a triangle). The students also found other structures, such as parallelograms, and regularities that had corresponding angles in a variety of patterns. Figure 2.18 shows one such pattern of corresponding angles created by the intersection of straight lines with a set of parallel lines.

According to the reinvention principle, the students should be given the opportunity to reinvent the intended mathematics. For this to happen, the curriculum developer has to act as a trailblazer (Freudenthal, 1991),

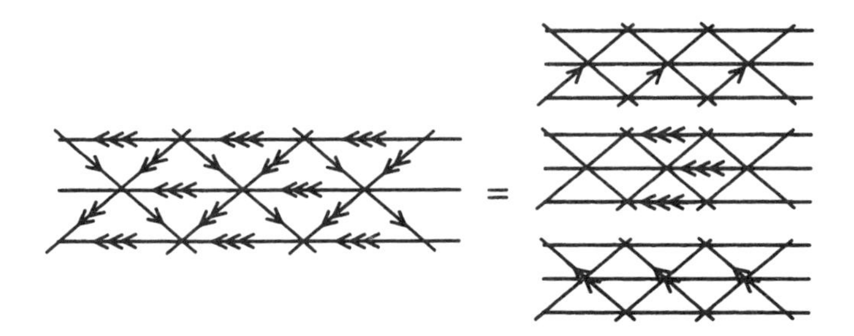

FIG. 2.16 Tesselating triangular tiles: How many sets of parallel lines?

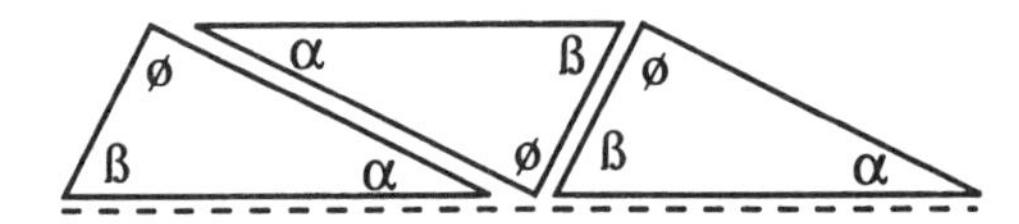

FIG. 2.17 The sum of angles in a triangle: combining congruent triangles to form a straight line.

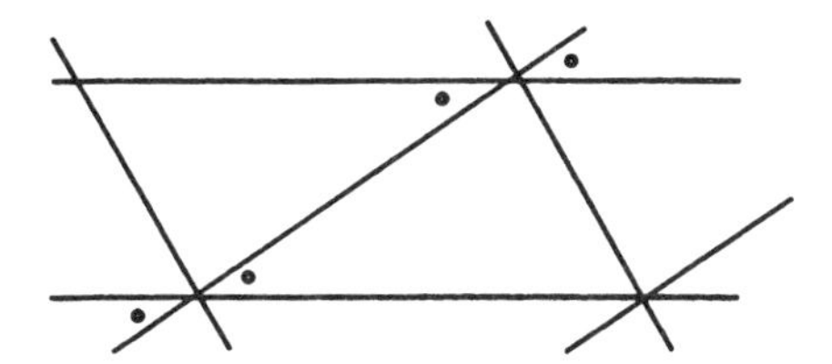

FIG. 2.18 Corresponding angles: straight line intersection of a set of parallel lines.

setting out a path of instructional tasks along which the reinvention process can proceed. Knowledge of the history of mathematics can be used as a heuristic device: Knowing how certain knowledge developed may help the developer find a curtailed (re)invention route. The developer can also be inspired by informal solution procedures: Students' informal strategies often anticipate more formal procedures. To facilitate a reinvention process, we need to find or develop contextual problems that allow for a wide variety of solution procedures, preferably those that, when considered together, indicate a possible learning route of progressive mathematization. As an example of how such a sequence might appear, let's look at one such sequence involving the Pythagorean theorem taken from *Mathematics in Context* (National Center for Research in Mathematical Sciences Education & Freudenthal Institute, in press), a National Science Foundation (NSF)–funded middle-school curriculum.

Exploring the Pythagorean Theorem (Adapted from Kindt, 1980; see also Abels, de Lange, Brendefur, & Halevi, 1998). As a first step, students may be asked to figure out the length of a side of a square with a given area. In the *Mathematics in Context* unit "Building Formulas," the area of a giant fungus (154,000 m^2) is used as the context (Weijers, Roodhardt, Burrill, & Cole, 1998). The area of a square is subsequently used to define diagonal distances on a square grid (see Fig. 2.19), which in turn can be used to define square root. These diagonal distances can be explored further by systematically varying their properties, (e.g., looking at the areas of squares, the sides of which are hypotenuses of right triangles of varying heights but with fixed bases; see Fig. 2.20).

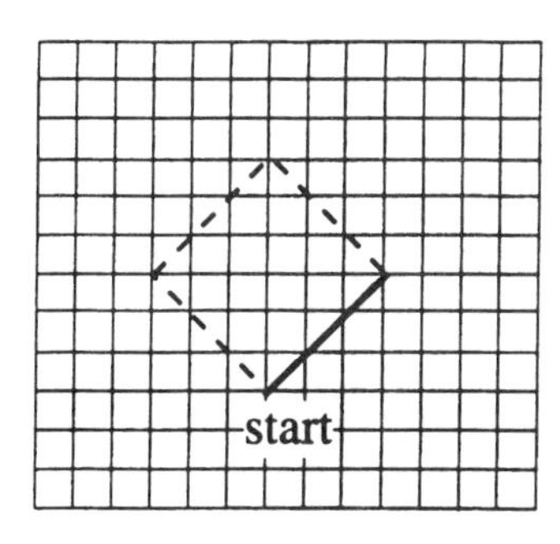

FIG. 2.19 Defining diagonal distances: using the area of a square.

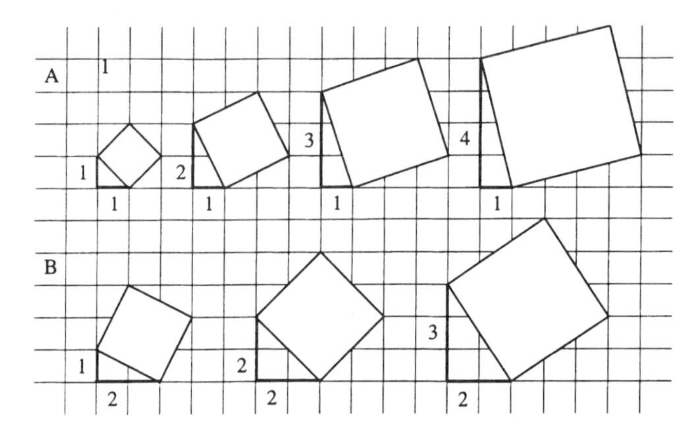

FIG. 2.20 Defining diagonal distance: systematic analysis of hypotenuses.

Solving these problems will produce the following rows:

series (a): 2,5,10,17, . . .
series (b): 5,8,13,20, . . .

Students familiar with square numbers (1, 4, 9, and so on) may recognize a pattern in these series of squares of the hypotenuses of these triangles:

series (a): 1 + 1, 4 + 1, 9 + 1, 16 + 1, . . . (a square plus 1)
series (b): 1 + 4, 4 + 4, 9 + 4, 16 + 4, . . . (a square plus 4)

The investigation of this pattern can be supported by questions such as "What would be the area of square 10 in this row?" and "What would this pattern look like for right triangles with a base of 3?"

We might remark that such an investigation does not really match the idea of real-life geometry, although we could imagine some context where measuring distances on a square grid fits a practical function. Looking for such patterns as the ones just shown, however, would seem to exceed direct practical application, but realistic mathematics education is not limited to real-life situations. In our conception of realistic mathematics education, realistic refers to what is experientially real to the students. The notion of "what is experientially real" can be related to Freudenthal's conception of reality. According to Freudenthal (1991), "real" is to be understood "commonsensically," in the meaning intended by someone who uses the term unreflectingly. Reality is thought of as a mixture of interpretation and of sensual experience. It is not bound to the space–time world and includes mental objects and mental activities. What is "experientially real" is not static but grows under the influence of the learning process of the person in question. This implies that mathematics, too, can become part of the reality of the students. For these students, the "context" of a context problem can consist of "experientially real" mathematics. Once the students have mastered some mathematics, mathematics itself can become a "realistic" context. Nevertheless, in RME we try to make connections to real-life situations as much as possible—whether in authentic or reinvented situations. Let's look at an example of the latter.

The ancient Egyptians allegedly used "knotted ropes" to measure the area of the fields of fertile soil along the Nile. Each year these fields were flooded. When the river returned to its summer norm, the Egyptians pegged out the fields again using a variety of methods. One of the most common involved the use of a 12-knot rope (see Fig. 2.21).

In this activity, students are first asked to find the triangles that can be made, with the stipulation that each angle corresponds to a knot. In this process, triangles with sides 4, 4, 4, and 3, 4, 5, work as special cases. The 3-4-5 triangle is interesting because of the right angle. Students investigate whether other knotted ropes would produce right triangles. In fact, there are no ropes with fewer than 12 knots that produce right triangles. Several possibilities for larger numbers—24 (sides: 6, 8, 10), 30 (sides: 5, 12, 13), 36 (sides: 9, 12, 15), and so forth—produce such triangles. Through the question "What is the secret behind these numbers?" a connection can be made with the square-pattern activity, and the Pythagorean theorem can be formulated in its standard form: $a^2 + b^2 = c^2$.

All examples here seem to corroborate the theorem, but would this theorem be true for every right triangle? As a closure to this activity, students are asked to do a "reallotment proof" (see Fig. 2.22), which can be

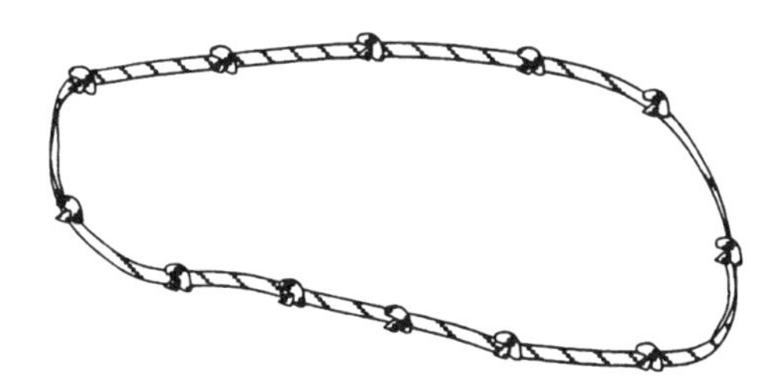

FIG. 2.21 Measuring fields: a 12-knot rope.

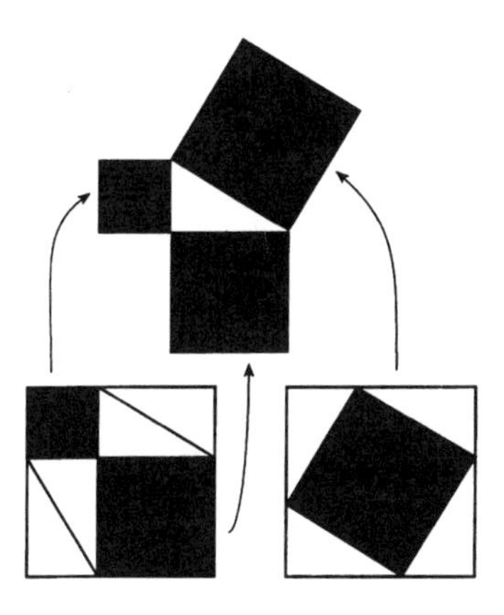

FIG. 2.22 Pythagorean theorem: a reallotment proof.

extended through several variants (Kindt, 1980). Here, as in the earlier activity, students reinvented the intended mathematics.

Didactical Phenomenological Analysis

According to didactical phenomenology (Freudenthal, 1983), situations where a given mathematical topic is applied are to be investigated for two reasons: first, to reveal the kind of applications that have to be anticipated in instruction and, second, to consider the suitability of such applications as points of impact for progressive mathematization. If we view mathematics as evolving historically from the process of solving practical problems, it is reasonable to expect to find problems that give rise to mathematical concepts. We can then imagine that formal mathematics came into being in a process of generalizing and formalizing situation-specific problem-solving procedures and concepts about a variety of situations. The goal of our didactical phenomenological analysis then becomes to find problem situations for which situation-specific approaches can be generalized and to find situations that can evoke paradigmatic solution procedures, which can be taken as the basis for vertical mathematization. In the following section, the concept *angle* is used in an illustration of didactical phenomenological analysis.

Angle. Looking at angles from a phenomenological perspective reveals many characteristics: Angles can be tangible (e.g., the right angles of a rectangular table), or imaginary (e.g., the angle between a light ray and the ground it strikes). Angles can also be static or dynamic (e.g., a 90° turn or 90° angle). Angles can also be used to indicate direction, as on a compass card, or to indicate a position in a coordinate system, such as in the system of latitude and longitude.

This short list shows how complex the concept of angle is—from a phenomenological perspective (see Lehrer, Jenkins, & Osana, chap. 6, this volume). This complexity, in turn, argues for a broad phenomenological exploration and also suggests contextual problems that can be used in an instructional sequence. For instance, the notion of the "four winds" can be built on with contextual problems in which more precise descriptions are used. Such problems may lead naturally toward the use of angle as descriptor. In another context, making turns to change direction, measurement of angles comes up naturally. Other situations that can be exploited involve vision lines and light rays, where angles are "invisible" or "imagined." More informally, the concept of angle can be dealt with implicitly, as in fitting together pieces of a tangram or when comparing the similarity of triangles.

A core element of such an exploration, however, is the relationship between direction and turn (Clements & Battista, 1989; Hoyles, 1992; see also Clements, Battista, & Sarama, chap. 8, and Lehrer, Jenkins, & Osana, chap. 6, this volume). That at least is the stance taken both by Kraemer (1990) and de Lange, Feijs, van Reeuwijk, and Middleton (1997). Most students are familiar with a compass and the four wind directions, but students can "know" north, east, south, and west on a map and still not understand what those directions actually mean. If we ask middle-school students to point to the north, we are not surprised if they point to the ceiling or to the map that hangs in front of the classroom (Feijs & de Lange, 1995). From the four wind directions, de Lange et al. (in press) also raise the issue that when the whole class is pointing north, the arms are almost parallel, but if students started walking north from different places on earth, they would all end up at the North Pole. Further exploration of the concept is therefore necessary, and de Lange et al. take great care to make students realize that north and south are relative to a given position (e.g., North Carolina is north of South Carolina, but south of Wisconsin).

From the four wind directions, a further refinement is possible: northeast in between north and east, and north-north-east, right between north and northeast. (See Fig. 2.23 for a compass card created by folding a square of paper.)

We can next refine this system by introducing a description of directions that makes use of degrees. In Fig. 2.24 (de Lange et al., 1997), we see the use of the polar coordinate system, with an air traffic control system as the context.

From here, the step to dynamic angles can easily be made. In the context of flying, the notion of a turn as a change in direction comes up naturally. This situation, however, is more complicated than it might look like at first sight, for the angle we see in the route of the plane is the supplement of the turn that the plane makes (see Fig. 2.25).

Exploration of such contexts can be followed by measurements of, and computations with, angles. Later, the air-traffic polar coordinate system can be elaborated on with the introduction of three-dimensional polar coordinates, of which the global coordinate system is a special case. The analysis presented here can similarly be done on other concepts, generating problem situations that eventually lead to vertical (and horizontal) mathematization.

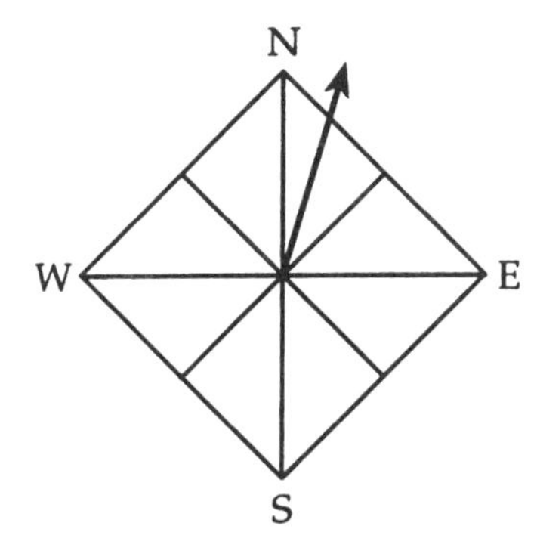

FIG. 2.23 Refinement of wind directions: using a compass card.

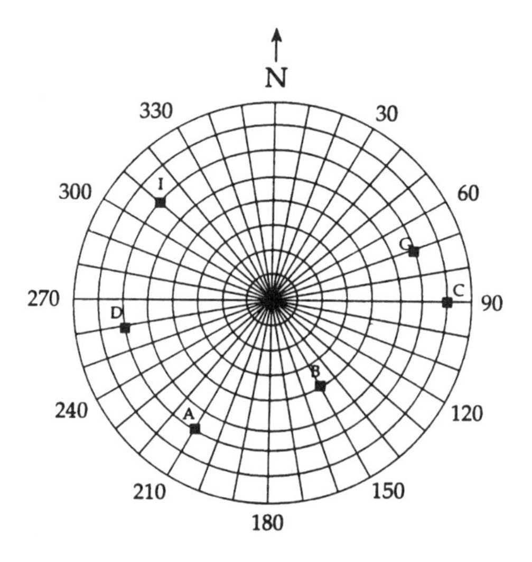

FIG. 2.24 Polar coordinate system: an air traffic control system.

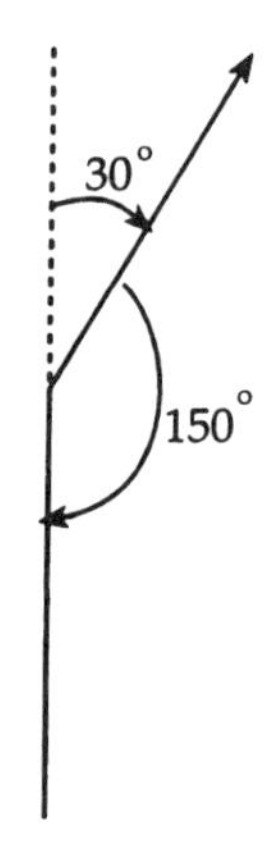

FIG. 2.25 Supplementary angles: alteration in flight path caused by 30° turn.

Emergent Models

The third principle in RME concerns the role that emergent models play in bridging the gap between students' informal knowledge and formal mathematics. In product-oriented mathematics education, a tool is presented as a preexisting model, whereas in realistic mathematics education a model emerges from the activities of the students themselves. Note that the term *model* should not be taken too literally. A model can be a situation, a scheme, a description, or a way of notation. Initially, models refer to concrete or paradigmatic situations that are experientially real for the students. At this level, the "model of" allows informal strategies that correspond with solution strategies situated within the contextual problem. From that point on, the role of the model changes. As the student gathers more experience with similar problems, the model becomes more general in character. At the same time, a process of reification takes place, by which the "model of" becomes an entity in itself, a "model for" mathematical reasoning (Gravemeijer, 1994; Streefland, 1985; Treffers, 1991). At that point, the model becomes more important as a referential base for mathematical reasoning than as a way to represent a contextual problem. This transition follows the theoretical reconstruction of the genesis of subjective mathematical knowledge: "What is proposed is that by a vertical process of abstraction or concept formation a collection of objects or constructions at lower, preexisting levels of a personal concept hierarchy become 'reified' into a object-like concept, or noun-like term" (Ernest, 1991, p. 78). In the following section, we use a shadow model to look more closely at this transition from "model of" to "model for."

Shadow Model. Students who are familiar with light rays and vision lines can be asked to infer the length of the shadow of one pole from the length of the shadow of a second pole (see Fig. 2.26). The shadow of the second pole can be found by drawing parallel lines, but we can also reason that, in sunlight, a rod twice as small as another produces a shadow twice as small as that of the larger rod. The combination of these two strategies forms an intuitive base for understanding the relationship between the ratios of the lengths of the sides and the angles of a right triangle.

Students can further explore these relationships in the context of lamplight and the shadows it creates. Other analogous situations involve determining the position of a ship (Gravemeijer, 1990) or predicting the position where a hang glider will land (de Lange et al., 1997). At this point, the shadow model has evolved into a more general triangle model in which no references to context are needed in order to reason about fixed ratios between the lengths of the sides in a right triangle with given angles. The relationship between the (trigonometric) ratios and the measures of the angles in a right triangle becomes the formal properties of a triangle. The shadow model—a "model of"—has become a triangle model—a "model for" mathematical reasoning. With this more abstract model in mind, students can explain why the ratio between the lengths of line segment EF (parallel to side AB) and line segment AB is equal to the ratio of the lengths of segments CE and CA (see Fig. 2.27).

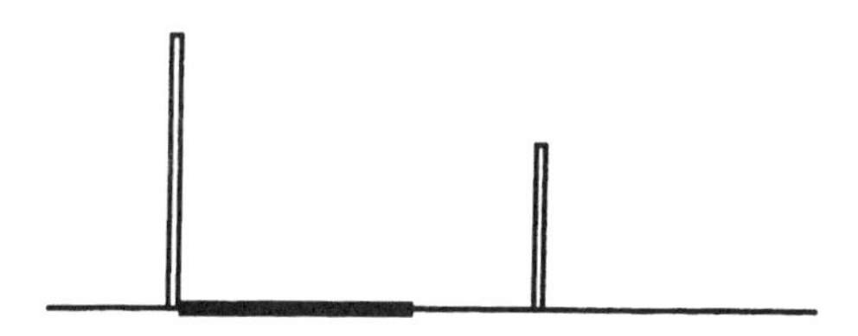

FIG. 2.26 Ratios: a shadow model.

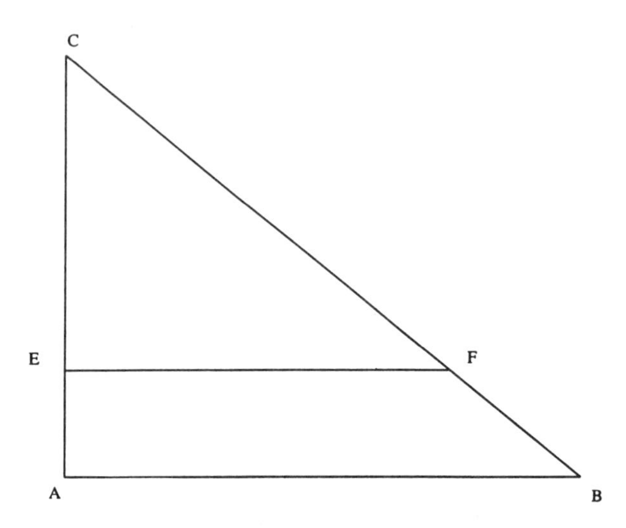

FIG. 2.27 Ratios between sides of right triangles: evolution of context-specific shadow model into abstract triangle model.

Armed with this knowledge, the student will be able to explain why the ratio between h_i and s_i is constant in the problem with which we started this chapter (see again Fig. 2.3). This process of abstraction through an emergent model uses the student's informal knowledge as a basis for later mathematical reasoning.

CONCLUSION

In this chapter, through a look at RME as it is applied in the Dutch curriculum, we have argued that student's informal knowledge can be used as a basis for geometry education and spatial reasoning. In the Netherlands, the point of departure for "realistic" geometry education is the world itself, seen from a geometric perspective. Real-life problem situations that can evoke problem-solving paradigms, which can serve as a basis for progressive mathematization, are construed through phenomenological analysis of mathematical concepts. Students are given activities that guide them to "reinvent" the intended mathematics and from which models emerge that eventually serve as a referential base for formal mathematics. Throughout this process, students' informal knowledge is made mathematically explicit and elaborated on: Their informal knowledge serves as a launching point into formal mathematics. We argue that this approach to geometry education not only supports the view of mathematics as as a human activity, but more importantly allows students to construct their own knowledge and reinforces their ability to reflect. The key characteristics of RME—reinvention through progressive mathematization, didactical phenomenological analysis, and use of emergent models—erect a framework of geometry education consistent with a view of mathematics as a socially constituted activity.

REFERENCES

Abels, A., de Lange, J., Brendefur, J., & Halevi, T. (1998). Going the distance In National Center for Research in Mathematical Sciences Education & Freudenthal Institute (Eds.), *Mathematics in context: A connected curriculum*. Chicago: Encyclopaedia Britannica.

Clements, D. H., & Battista, M. T. (1989). Learning geometric concepts in a Logo environment. *Journal for Research in Mathematics Education, 20*, 450–467.

de Lange, J., Feijs, E., van Reeuwijk, M., & Middleton, J. (1997). Figuring all the angles. In National Center for Research in Mathematical Sciences Education & Freudenthal Institute (Eds.), *Mathematics in context: A connected curriculum*. Chicago: Encyclopaedia Britannica.

de Moor, E. (1993). Het gelijk van Tatiana Ehrenfest-Afanassjewa [Tatiana Ehrenfest-Afanassjewa was right]. *Niewe Wiskrant, 12*(4), 15–24.

Ehrenfest-Afanassjewa, T. (1931). *Übungssammlung zu einer geometrischen Propädeuse* [Set of exercises for an introduction to geometry]. The Hague, the Netherlands: Martinus Nÿhoff.

Ernest, P. (1991). *The philosophy of mathematics education*. Hampshire, England: Falmer.

Feijs, E., & de Lange, J. (1995). *From a different angle*. Manuscript in preparation.

Freudenthal, H. (1971). Geometry between the devil and the deep sea. *Educational Studies in Mathematics, 3*, 413–435.

Freudenthal, H. (1973). *Mathematics as an educational task*. Dordrecht, the Netherlands: Reidel.

Freudenthal, H. (1983). *Didactical phenomenology of mathematical structures*. Dordrecht, the Netherlands: Reidel.

Freudenthal, H. (1991). *Revisiting mathematics education*. Dordrecht, the Netherlands: Kluwer.

Goddijn, A. (1980). *Shadow and depth*. Utrecht, the Netherlands: OW&OC.

Gravemeijer, K. P. E. (1990). Realistic geometry instruction. In K. Gravemeijer, M. van den Heuvel, & L. Streefland (Eds.), *Contexts, free productions, tests and geometry in realistic mathematics education* (pp. 79–91). Utrecht, the Netherlands: OW&OC.

Gravemeijer, K. P. E. (1994). *Developing realistic mathematics education*. Utrecht, the Netherlands: Cd-ß Press.

Hoyles, C. (1992). *Logo mathematics in the classroom*. London: Routledge.

Kennedy, J. M. (1974). *Icons and information*. In D. R. Olson (Ed.), *Media and symbols: The forms of expression, communication, and education* (pp. 211–230). National Society for the Study of Education. Chicago: University of Chicago Press.

Kindt, M. (1980). *Pythagoras*. Utrecht, the Netherlands: IOWO.

Kraemer, J. M. (1990). *Meetkunde op de basisschool* [Geometry in primary education]. Rotterdam, the Netherlands: Project Onderwijs en Sociaal Milieu.

McIntosh, A., Reys, B. J., & Reys, R.E. (1992). A proposed framework for examining basic number sense. *For the Learning of Mathematics*, 12(3), 2–8.

Monk, G. S. (1992, April). *A study of calculus students' constructions of functional situations: The case of the shadow problem*. Paper presented at the Annual Meeting of the American Educational Research Association, San Francisco, CA.

National Center for Research in Mathematical Sciences Education & Freudenthal Institute. (in press). *Mathematics in context*. Chicago: Encyclopaedia Britannica.

National Council of Teachers of Mathematics. (1989). *Curriculum and evaluation standards*. Reston, VA: Author.

Paulos, J. A. (1988). *Innumeracy: Mathematical illiteracy and its consequences*. New York: Hill & Wang.

Streefland, L. (1985). Wiskunde als activiteit en de realiteit als bron [Mathematics as an activity and reality as a source]. *Tijdschrift voor Nederlands Wiskundeonderwijs Nieuwe Wiskrant*, *5*(1), 60–67.

Treffers, A. (1987). *Three dimensions: A model of goal and theory description in mathematics education: The Wiskobas project.* Dordrecht, the Netherlands: Kluwer.

Treffers, A. (1991). Innumeracy at primary school. *Educational Studies in Mathematics, 22*(4), 309–332.

van Hiele, P.M. (1973). *Begrip en inzicht* [Understanding and insight]. Purmerend, the Netherlands: Muusses.

van Hiele, P.M. (1985). *Structure and insight: A theory of mathematics education.* Orlando, FL: Academic Press.

van Hiele-Geldof, D. (1957). *De didacttiek van de Meetkunde in de eerste klasse van het VHMO* [The didactics of geometry education in Grade 7]. Unpublished doctoral dissertation, University of Utrecht, Utrecht, the Netherlands.

Weijers, M., Roodhardt, A., Burrill, G., & Cole, B. (1998). Building formulas. In National Center for Research in Mathematical Sciences Education & Freudenthal Institute (Eds.), *Mathematics in context.* Chicago: Encyclopaedia Britannica.

3

Charting a Course for Secondary Geometry

Daniel Chazan
Michigan State University

Michal Yerushalmy
Haifa University

> Fortunately, I see mathematics as a very big house, and it offers teachers a rich choice of topics to study and transmit to students. *The serious problem is: how to choose.* (Mandelbrot, 1994, p. 80, emphasis added)

Traditionally, in the United States, geometry instruction is concentrated at the secondary level; high-school students study Euclidean geometry in a concentrated, year-long course that emphasizes deductive proof. Even though educational policy is a local matter theoretically decided school district by school district, teacher by teacher, there are nonetheless strong commonalities in the mathematics education offered to many students; their opportunities to study geometry are limited.

In the past, many arguments were advanced to support the traditional Euclidean geometry course. For example, Moise (1975) touted the traditional Euclidean geometry course as "the only mathematical subject that young students can understand and work with in approximately the same way as a mathematician" (p. 477). For Moise, mathematicians work deductively; studying Euclidean geometry gives students an opportunity to experience the deductive development of an axiomatic system. More recently, others (e.g., Malkevitch, n.d.) objected to the study of Euclid. They claimed that limiting students to the study of Euclid misrepresents modern geometry. Is the traditional geometry course a defensible and effective "rudimentary version" of geometry well suited to a wide range of secondary school students, or is it an anachronism? As we near the millennium and as technological tools provide new types of geometrical representations, should Euclid be replaced in the secondary mathematics curriculum, or should the traditional course be maintained or modified?

These questions interest us because we do not see curriculum as fixed; in our view there are curricular choices to be made. In making such choices, we take seriously Bruner's suggestion "that children should encounter 'rudimentary versions' of the subject matter that can be refined as they move through school" and Schwab's notion of a curriculum "in which there is, from the start, a representation of the discipline" (quoted in Ball, 1993, p. 376). But it is a complicated matter to determine what an "authentic" representation of a discipline is. Geometry may be "what geometers do" (Usiskin, 1987), but the work of geometers can be described differently. Some, like Moise, may emphasize geometers' deductions based on undefined terms within well-defined systems, and others may highlight the role of computer-based experiments (Gleick, 1987). Similarly, people conceive of the relationship between the objects of study in Euclidean geometry and the world of sensory experience in different ways. To reiterate, there are choices to be made in deciding how to represent geometry authentically.

Like Ball in her teaching of elementary school mathematics, we wonder, "What constitutes a defensible and effective 'rudimentary version'? And what distinguishes intellectually honest 'fragments of the narrative of inquiry' from distortions of the subject matter?" (Ball, 1993, p. 376). Although we believe that there are no indisputable answers to such questions, our experiences, values, and perhaps aesthetics lead us to argue for the kinds of revisions in the traditional course that geometry construction programs can support. By geometry construction programs, we mean programs like Cabri Geometry II (J.-M. Laborde & Bellemain, 1994), the Geometer's Sketchpad 3.0 (Jackiw, 1995), the Geometric SuperSupposer (Schwartz & Yerushalmy, 1993), and the Geometry Inventor (Logal Software, 1994), which allow students to carry out geometrical constructions on a computer screen. Although these revisions may not seem radical and may be criticized for not representing modern geometry authentically, we value the ways in which they change the traditional course's representation of how mathematics is done.

In order to help clarify the kinds of changes that such programs support, before turning to an analysis of the opportunities for changing the traditional course that these programs create, we begin by presenting four aspects—axioms, objects of study, proof methods, and how mathematics is done—for describing how a curriculum represents geometry. To illustrate and explicate these four aspects, we use them to describe our view of the traditional course and to present a range of suggestions people have made for changing the traditional course and its representation of geometry.

Having outlined different aspects of geometry curricula, we focus our discussion on computer-based innovations, which may best be used to change the traditional (high-school) Euclidean geometry course's representation of how mathematics is done. Teachers can use geometry construction programs to support the inclusion of experimentation and the development of conjectures, in addition to the traditional focus on geometric proof and justification. We exemplify how geometry construction programs can be used to change the nature of two problem types traditionally central to high-school Euclidean geometry (construction problems and problems to prove) and thereby address concerns raised by researchers (e.g., Schoenfeld, 1988) about inadvertent lessons that students might learn from traditional courses. We also examine difficulties with the implementation of such uses of geometry construction programs and then conclude by arguing for the continuing value of the study of Euclidean geometry, *if* students are actively engaged in creating their own ideas.

TRADITIONAL HIGH-SCHOOL GEOMETRY AND PROPOSED CHANGES

Geometry is often described as the study of shape and space. To situate the work with geometry construction programs with respect to the traditional high-school curriculum and the wide range of options that could be chosen for revising and altering the high-school curriculum, we now use the four aspects listed earlier to contrast the traditional curriculum and proposed changes.

Axioms

The traditional course chooses to focus on geometrical axioms, like the ones developed by Euclid, which assume that—given a line and a point not on the line—there is exactly one line through the given point and parallel to the given line. These courses also use a Pythagorean metric (the distance between two points is the square root of the sum of the squares of the legs of the right triangle, the hypotenuse of which is the line segment between the two points). These choices are introduced without investigating alternatives that have different assumptions about parallelism or that use different metrics.

Others suggest that students in high school should be aware that Euclidean geometry is only one geometrical axiom system. Proponents of such changes argue that changes might help students understand geometry more abstractly. For example, van den Brink (1995) and Lenart (1993) argued for engaging students in comparison of spherical, non-Euclidean geometry and planar, Euclidean geometry to help students come to a better understanding of what "geometries" are.

Proof Methods

In the traditional course, the preferred proof methods are synthetic rather than analytic. Also, in such proofs, appeal is made only to the intrinsic properties of the shape, not to the placement of the shape with respect to a coordinate system.

Coxford and Usiskin (1971) chose a transformational approach to geometry. They argued that their approach keeps the same axioms, but changes the proof methods (Usiskin, 1987). Their proofs use coordinates for describing shapes extrinsically and also use algebraic methods for describing such coordinates. Similarly, others have proposed that proofs in high-school geometry could use a variety of tools as appropriate—synthetic techniques as well as analytic geometry and vectorial methods (e.g., Education Development Center, 1995).

Objects of Study

The objects of study in the traditional plane geometry course are points, lines, and arcs. Although students in a traditional course are assumed to be familiar with these objects from their day-to-day life and although these objects are always represented by diagrams in the traditional way, these objects are introduced as undefined terms; as a result, diagrams are aids for intuition and are not the objects of study themselves. The rest of the objects in plane geometry, such as ray, triangle, and polygon, are defined with reference to the undefined terms. In the traditional course, these objects are viewed from outside the plane as complete entities. In part as a legacy of the Greek concerns about motion and the infinite, they are not viewed as the traces of motion or as the result of operations carried out infinitely many times.

Others have suggested that the approach to the objects of study in geometry should more explicitly build on students' experience of the world around them. For example, Aleksandrov (1956/1963) argued that "Geometry has as its object the spatial forms and relations of actual bodies, removed from their other properties and considered from the purely abstract point of view" (p. 22); a geometric object is the shape of an actual object abstracted from its other properties such as weight, color, and dimension. Unlike the objects studied in the physical sciences, in this view geometrical objects are also abstracted from their position and removed from any particular frame of reference. Such a conception of the nature of the objects of study in geometry changes the role of the diagram and is embraced by those using computers. For example, computers allow users to treat screen representations of Euclidean objects as tangible and manipulatable objects (Pimm, 1995). They can even support viewing more complex objects, triangles and quadrilaterals, as the primary objects of study in geometry, rather than points and lines (e.g., the Geometric SuperSupposer [Schwartz & Yerushalmy, 1993] has users start with triangles and quadrilaterals, in addition to points, lines, and circles).

Goldenberg and Cuoco (chap. 14, this volume) show how dynamic geometry's stretchable segments introduce motion into geometry. They suggest that changing our characterization of a segment from a fixed-measure abstract object into a dynamically stretchable computer-based object changes the properties of the segment. Such changes to the objects of study in geometry have implications for connections between the geometry curriculum and other aspects of the mathematics curriculum. They argue that dynamic geometry develops "reasoning by continuity" and suggest that the inclusion of "reasoning by continuity" in the geometry curriculum enhances opportunities to make connections with analysis (Cuoco, Goldenberg, & Mark, 1995).

Abelson and diSessa (1981, pp. 13–14) also used motion in geometry and argued for a local rather than global perspective on geometrical objects—a perspective from inside the plane, not outside the plane. They suggested that in Logo the emphasis is on the tracing out of a curve; as in differential geometry, the user is aware of what is happening in the immediate vicinity of the traveling point, the turtle. They claimed that, as a result, Logo's circle is a different object than the Euclidean object even though when completed it may seem to be the same physical diagram.

Mandelbrot (1994) and Devaney (chap. 4, this volume) suggested that we should allow operations in geometry to be iterated infinitely many times, thus creating objects, fractals, that more accurately portray the nature of shapes in the world than do the shapes of Euclidean geometry. In their view, the shapes of Euclidean geometry often cannot accurately describe the complex shapes in the world around us (e.g., clouds). In their view, fractals, like Euclidean shapes, are abstractions of the shape of objects in the world around us. Also, fractals are like Euclidean shapes in that they are not tied by definition to a particular coordinate system. But, fractals, unlike Euclidean shapes, "characterize objects with structures on values of scale, large as well as small, and thus reflect the hierarchical principle of organization" (Peitgen & Richter, 1986, p. vi). Because of their self-similarity property, fractals, unlike Euclidean shapes, "do not change significantly when viewed under a microscope" (Peitgen & Richter, 1986, p. vi).

How Mathematics Is Done

The traditional definition of geometrical objects as undefined terms carries with it ramifications for how mathematics is done. In this view, geometrical objects are not present in our environment and cannot be the subjects of an experiment; we do not measure to see if a proof holds (see Scheffler, 1965, p. 3). This conception of the object of study leads to an emphasis on deductive reasoning. Indeed, for many years, Euclidean geometry served as an exemplar of the certainty that deductive proof provides for mathematical knowledge and as an ideal of mathematical practice for emulation in other fields of mathematics (Moise, 1975).

Computer-based "experimental" approaches to geometry and earlier construction-based approaches (e.g., Schoenfeld, 1983, pp. 42–50, for an illustration of Polya's approach) emphasize other aspects of the ways mathematics is done. Such approaches are based on the work of mathematicians and philosophers of mathematics (see Tymoczo, 1986) who have sought to highlight the importance of "plausible reasoning" and "conjecturing" in the work of mathematicians. Recently, there have been new developments in the practice of mathematics and in philosophical conceptualizations of the practice of mathematics that support such views. Emerging computer technologies have supported "experimental" approaches to mathematics. In the study of chaos, researchers have "run" their models to develop and test conjectures (Gleick, 1987). Thus, when reformers now view geometry as a field of study where students can appreciate how mathematicians work, their focus is less on the deductive proof and axiomatics, and more on conjecturing and local deductive justification.

Geometry Construction Programs: Changing the Curricular Representation of How Mathematics Is Done

In the rest of this volume, several authors describe the excitement felt by teachers and researchers interested in using computers in many levels of schooling in order to change the emphasis of geometry instruction away from the focus of the traditional Euclidean course. At the high-school level, many teachers in the United States are interested in geometry construction programs, because of the mixture of change and stability that they represent (e.g., Healy, 1993; Houde, 1993; Lampert, 1993, 1995). Although each of these programs is different and was designed for different reasons, secondary teachers view all these programs as tools for revitalizing their teaching of the traditional Euclidean geometry course. Much of the excitement about these programs is generated by the possibility of changing teachers' and students' roles in the classroom, thus changing students' views of what it means to do mathematics and of their ability to do mathematics. It is hoped that these software programs will provide students with feedback about what is and is not true in particular cases, which will scaffold their ability to make general statements or construction procedures. Thus, geometry construction programs have the potential to support teachers interested in creating experimental environments where collaborative learning and student exploration are encouraged.

These tools have the potential to help students assume a role as mathematical authorities in the classroom and to help teachers create classroom mathematical discourse communities, in which the truth of a statement is determined by evaluating the arguments put forth in its behalf instead of by appealing to the teacher's authority. The hope is that such a community will evolve norms for mathematical argumentation and use them to assess the strength of arguments advanced for statements (for two interpretations of these ideas, see Chazan, 1990b, and Healy, 1993). These desired changes in what students are expected to do in class also have an impact on how the role of the teacher is conceptualized; they change the role of the teacher from expositor to guide and partner (Chazan & Houde, 1989; Lampert, 1993, 1995; Wiske & Houde, 1993).

We now analyze how geometry construction programs can support teaching that is committed to changing the nature of students' activity in the high-school geometry course. In such approaches, much stays the same; the postulates are the postulates of Euclidean geometry, the objects of study are the abstractions of the shape of objects in the world around us, as described by Aleksandrov (1956/1963), and proof methods usually remain synthetic. But there is a strong emphasis on plausible reasoning and the relationships between inductive and deductive reasoning. We next analyze how the capabilities of geometry construction programs support changes in two types of problems traditionally used in Euclidean geometry instruction, and we review the available research about such changes.

CONNECTING CONSTRUCTION AND DEDUCTION WITH CONSTRUCTION PROBLEMS

In "When Good Teaching Leads to Bad Results," Schoenfeld (1988) described the inadvertent lessons about mathematics that he believed students learned in a "well taught" high-school geometry class. He illustrated his contention that "students learn to separate the worlds of deductive and constructive geometry" (p. 155) by describing the way in which "construction problems" were taught. In a construction problem, a description was given of the desired object that students were to construct. For example, students might be given an angle and asked to construct an angle congruent to the given angle.

To solve such a problem, students must create a procedure that can be shown to produce the desired result. Yet, in traditional instruction, these problems are often thought to be difficult for students to do on their own. Students are often asked to reproduce a construction introduced by the teacher (e.g., Schoenfeld, 1988, p. 156). When approached this way, construction problems become a matter of memorizing steps in a procedure. In Schoenfeld's study, "the idea that practice is essential, and that the students should have the constructions committed to memory, was repeated throughout the unit" (p. 157). Procedures tend to get short shrift. Instead of writing out procedures, teachers often ask that the final product be reasonably accurate and that students leave a trail of arcs that indicate their construction procedure (p. 159). For students taught this way, the process of determining, on their own, whether or not a construction is correct becomes an empirical matter. Without a written description of the procedures used by students, it becomes difficult for teachers to identify whether students have correctly carried out a procedure.

Such paper-and-pencil practice conflates the diagram, which results from a procedure, with the procedure itself. Unless the teacher insists on a separate documentation of the procedure, the only way to judge a procedure is by its product. Here geometry construction programs can help. The program separates the particular diagram, which results from a construction procedure, from the procedure itself. Each of the programs has a different way of capturing and representing the procedure. Each program also has a different way of allowing students to see their procedure applied to other shapes. Building on the ease and speed with which computers can draw diagrams and their capacity to capture and store procedures, geometry construction programs provides users with the ability to

- Draw starting shapes (points, lines, segments, polygons).
- Make compass and straight-edge constructions (perpendicular and parallel lines, circles centered on one point and through another, etc.).
- Measure aspects of those drawings and constructions (lengths, areas, angles, etc.) and display those measurements (in charts and/or graphs).
- View a representation of the procedure they have created.
- Generate other diagrams that satisfy the same construction (by repeating the construction on a new starting shape or by dragging one of the vertices of the starting shape[s]).

Some of the reasons for excitement about the potential of geometry construction programs are that these tools (a) provide more precise visual and numerical feedback about a construction, (b) capture and allow for inspection of a user's construction procedure, and (c) emphasize the difference between constructing and drawing freehand. For example, when a Sketchpad user attempts the construction problem outlined earlier (construct an angle from a given angle), after creating (or drawing) an angle with the freehand tools in the toolbox (see Fig. 3.1), the user can then use the construction menu's geometric language to construct a congruent angle out of the intersection of the appropriate arcs and use the measurement menu to display the measures of the original and constructed angles (see Fig. 3.2).

Then the original angle can be dragged by the user to see if the constructed angle changes as the original angle changes. The user has both visual and numerical feedback to decide whether or not a given construction works. Both teachers and students can immediately ascertain whether a student's construction is correct or not. Meanwhile, the procedure used to make the construction can also be captured, inspected, and saved as a script (see Fig. 3.3), and students can then be asked to explain why the construction does indeed work.

By way of contrast, if the second angle is drawn with the freehand tools in the toolbox instead of constructed through the construction menu, it will remain unchanged as the original angle is dragged (see Fig. 3.4).

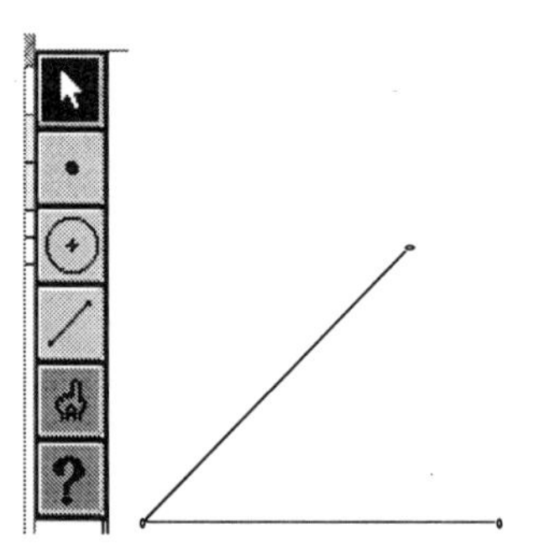

FIG. 3.1 An angle created using the freehand tools of Geometer's Sketchpad.

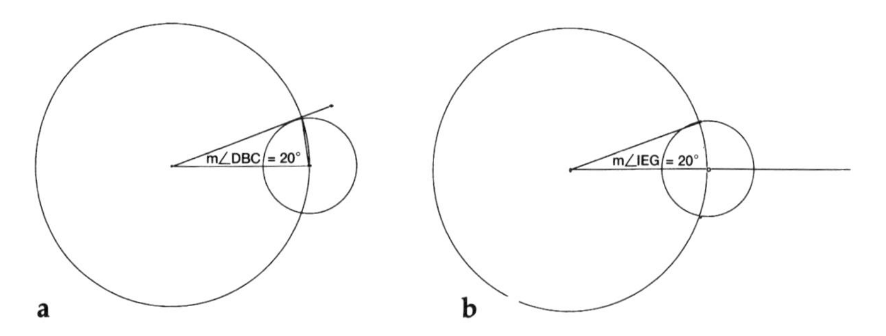

FIG. 3.2 Constructing an angle congruent to a given angle (Geometer's Sketchpad). (a) Original angle. (b) Constructed angle.

Given:
1. Point B
2. Point A
3. Point C
4. Point E
5. Straight Object 1

Steps:
1. Let [j] = Segment between Point B and Point A.
2. Let [k] = Segment between Point C and Point B.
3. Let [c1] = Circle with center at Point B passing through Point C.
4. Let [A] = Intersection of Circle [c1] and Segment [j].
5. Let [c2] = Circle with center at Point C passing through Point [A].
6. Let [c3] = Circle centered at Point E with radius length [k].
7. Let [B] = Intersection of Circle [c3] and Straight Object 1.
8. Let [1] = Segment between Point C and Point [A].
9. Let [c4] = Circle centered at Point [B] with radius length [l].
10. Let [C] = Intersection of Circle [c3] and Circle [c4].
11. Let [D] = Intersection of Circle [c3] and Circle [c4].
12. Let [m] = Segment between Point [D] and Point E.
13. Let Measure [A] = Angle ([A]–B–C).
14. Let Measure [B] = Angle ([D]–E–[B]).

FIG. 3.3 Script of the procedure used to construct an angle congruent to a given angle (Geometer's Sketchpad).

This difference between drawing and constructing an image is also available in Cabri Geometry II. To distinguish these two uses, the designers suggest that teachers talk with students about what makes a Cabri square—which remains a square no matter how it is dragged—as opposed to a square drawn on the screen with the freehand tools (for descriptions of an artificially intelligent environment that stresses this distinction, see Balacheff, 1993; J.-M. Laborde & Strasser, 1990). Research with Cabri Geometry (see, e.g., Balacheff, 1993; Capponi & Strasser, 1991; C. Laborde, 1992, 1993; C. Laborde & Capponi, 1994) suggests that the presence both of diagrams that preserve desired properties under the trans-

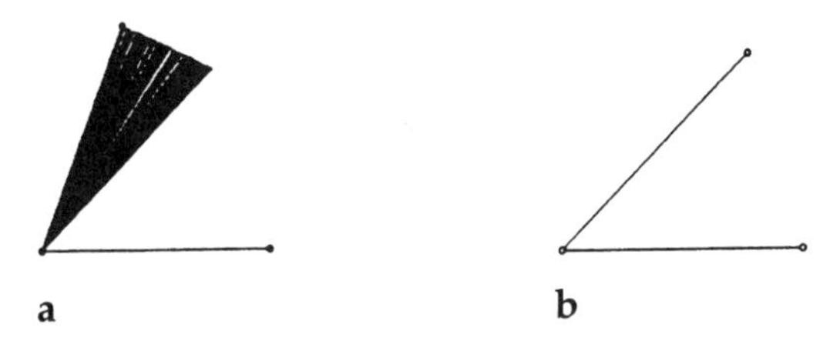

FIG. 3.4 Paired (congruent) angles created separately using the freehand tools of Geometer's Sketchpad. Dragging one side of the original freehand angle (a) leaves traces of the movement. Note that the second freehand angle (b) does not change with changes in the first angle.

formation of dragging and those that do not breaks down the separation between deduction and construction described by Schoenfeld (1988).

But teachers do not usually pose the "copy an angle" type of traditional construction problem with the Geometer's Sketchpad (Jackiw, 1995). A tool like Sketchpad allows teachers to embed such traditional problems inside more complex tasks. For example, in the context of exploring the properties of quadrilaterals, the problem of constructing an angle with the same measure as another angle can be embedded inside the task of constructing an isosceles trapezoid. (Such a task comes up naturally in Sketchpad because Sketchpad does not provide the user with the built-in capability of drawing an isosceles trapezoid.) At the same time, exploration of this construction problem can also lead to a reexamination of the definition of a trapezoid.

Fig. 3.5 summarizes three of the four strategies we have seen for constructing isosceles trapezoids. The first, "freehand drawing," does not create a "Cabri" isosceles trapezoid; dragging point D creates figures that are

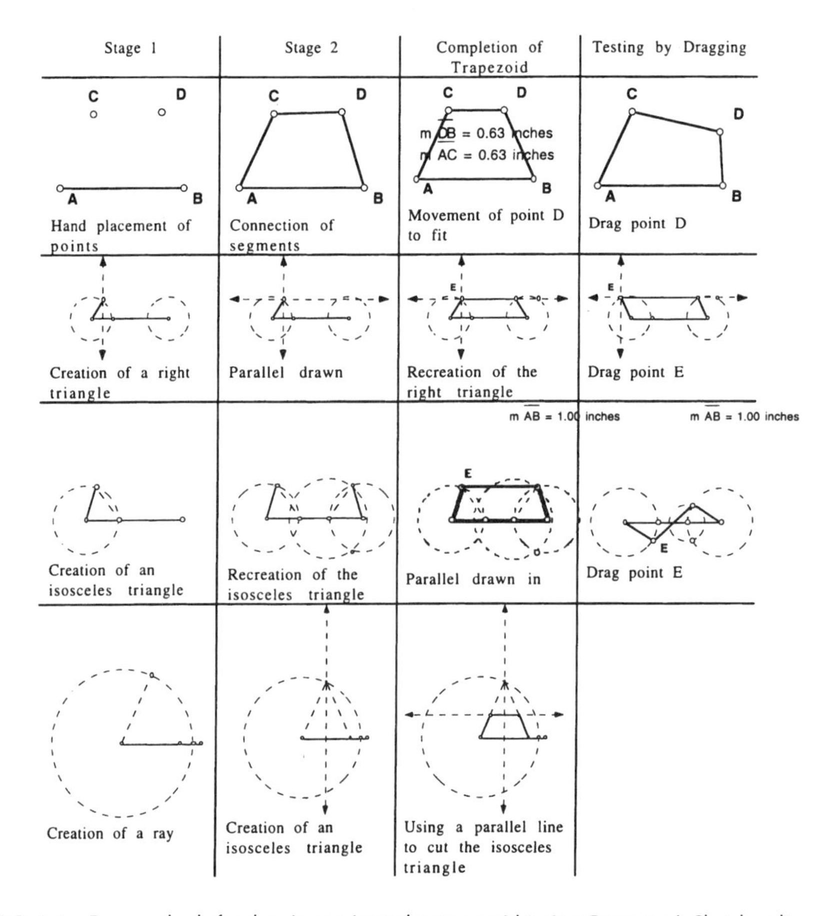

FIG. 3.5 Four methods for drawing an isosceles trapezoid (using Geometer's Sketchpad).

not isosceles trapezoids. The second strategy, which involves the copying of an angle, creates an isosceles trapezoid, but when tested can also create parallelograms. This construction strategy then raises the question of whether a parallelogram is also an isosceles trapezoid (in the way that a square is also a rectangle). The third strategy also involves the copying of an angle, but can lead to a crossed shape. Once again, the question is whether such a shape is a trapezoid or not. The final strategy, which uses reflection to copy the angle, always creates an isosceles trapezoid, but not a parallelogram or a crossed figure. Depending on the user's definition and goals, this last may not be viewed as an adequate procedure.

INTRODUCING CONJECTURING AND PROOFS OF CONJECTURES

Schoenfeld (1988) reported two other findings from his study of the high-school geometry class: "Students view themselves as passive consumers of others' mathematics" and "Students perceive that the form of a mathematical answer is what counts." Both of these findings are reflections on the way in which mathematical proof and "problems to prove" were presented in the class that he studied: "There was little sense of exploration, of the possibility that the students could make sense of the mathematics for themselves" (p. 18).

In the type of class that he studied, students were asked to produce short deductive proofs for statements provided in the textbook (e.g., "Show that the midpoint of the hypotenuse of a right triangle is equidistant from the three vertices"). Often in traditional texts, these assignments are given with a diagram and a description of the "given," that which the student is to assume to be true from the start, and the "to prove," that which the student must show to be true (see Fig. 3.6). Students know that these statements have been "proven" year after year by others sweating at their desks. Teachers, texts, and students assume that the truth value of these statements has been established. (See Lampert, 1993, pp. 145–147, for another description of typical use of problems to prove.)

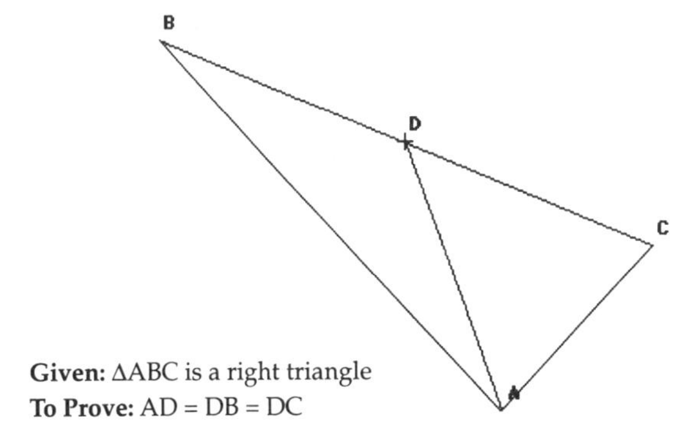

FIG. 3.6 A traditional deductive proof problem.

With the aid of geometry construction programs, this type of problem is also dramatically changed. Instead of having students write deductive proofs for statements from the text, constructions ("the givens"), usually complex ones, are described for students. Rather than tell students what to prove, students are asked to use the visual feedback and the measurement capabilities of the program to develop conjectures about relationships in the diagrams that result from the construction (which includes the "givens"). Compared to traditional textbook-based courses, diagrams proliferate. Students' observations that withstand testing in a number of cases become conjectures that students then attempt to explain by creating a deductive argument using their prior knowledge (Koedinger, chap. 13, this volume; National Council of Teachers of Mathematics [NCTM], 1989, p. 158).

For example, instead of the problem to prove described earlier, teachers using the Geometric Supposers might have their students use the new shape menu (instead of the Sketchpad's freehand toolbox) to select a right triangle and the "Draw" menu to construct the median from the right angle vertex to the hypotenuse and then medians from the midpoint of the hypotenuse of ΔABC in each of the two newly created triangles (see Fig. 3.7).

Students might examine a diagram created by this construction and notice that four new triangles (ADE, DEB, ADF, and DFC) are created inside the original triangle. Measurements might begin to suggest that all four of these triangles are congruent and that, when taken in pairs, they create what seem to be isosceles triangles of equal area. In Fig. 3.7, for example, triangles ADF and FDC taken together create ΔADC, and triangles AED and DEB taken together create ΔADB.

Students could then test these observations by repeating the construction on other right triangles (rather than dragging) and examining

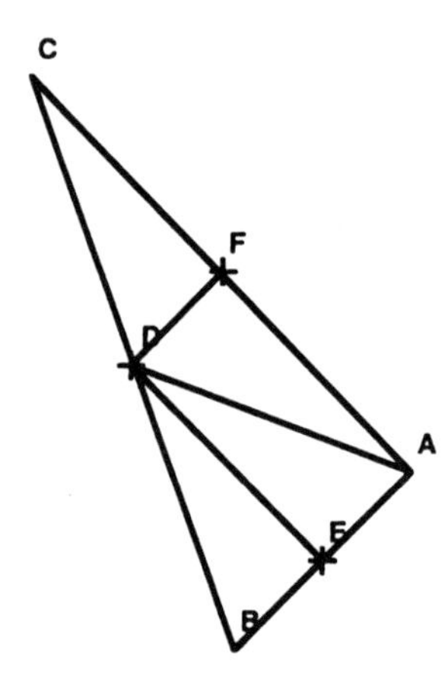

FIG. 3.7 A diagram created using the Geometric Supposer.

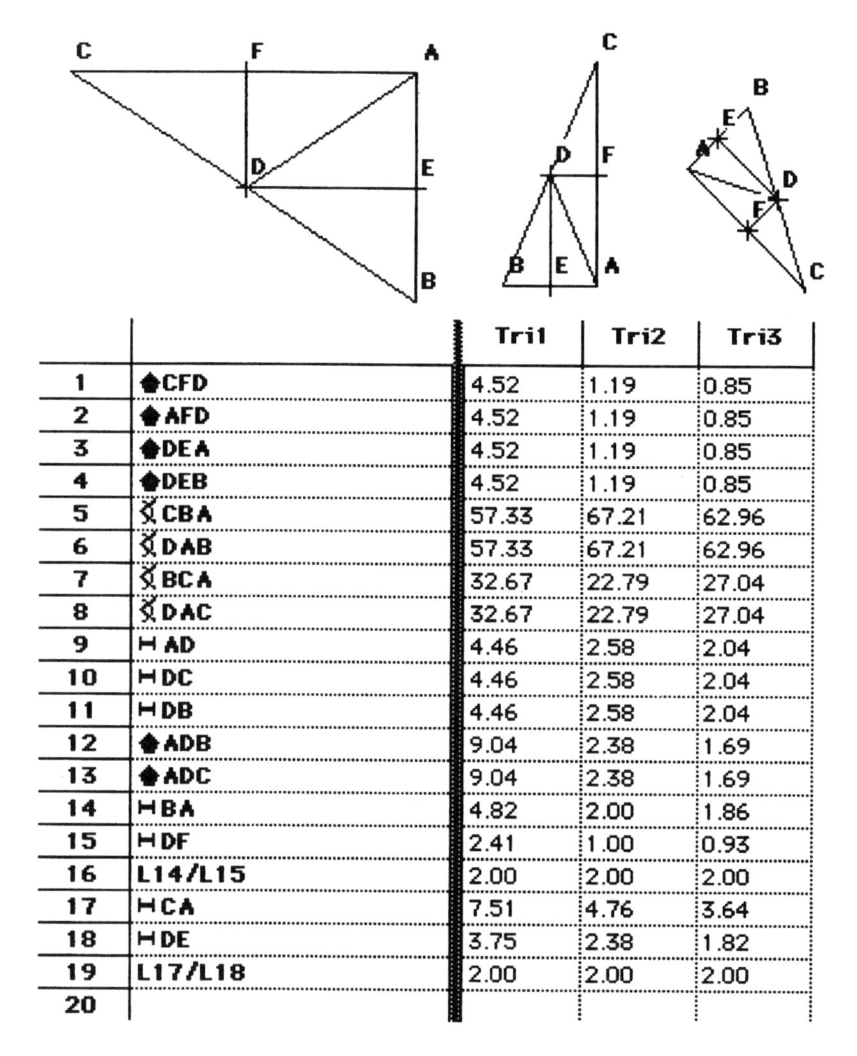

		Tri1	Tri2	Tri3
1	◆CFD	4.52	1.19	0.85
2	◆AFD	4.52	1.19	0.85
3	◆DEA	4.52	1.19	0.85
4	◆DEB	4.52	1.19	0.85
5	∢CBA	57.33	67.21	62.96
6	∢DAB	57.33	67.21	62.96
7	∢BCA	32.67	22.79	27.04
8	∢DAC	32.67	22.79	27.04
9	⊢AD	4.46	2.58	2.04
10	⊢DC	4.46	2.58	2.04
11	⊢DB	4.46	2.58	2.04
12	◆ADB	9.04	2.38	1.69
13	◆ADC	9.04	2.38	1.69
14	⊢BA	4.82	2.00	1.86
15	⊢DF	2.41	1.00	0.93
16	L14/L15	2.00	2.00	2.00
17	⊢CA	7.51	4.76	3.64
18	⊢DE	3.75	2.38	1.82
19	L17/L18	2.00	2.00	2.00
20				

FIG. 3.8 Visual and numerical feedback from the Geometric Supposer.

the measurement data, which appear in the table provided by the program (see Fig. 3.8).[1]

Examination of non-right-triangle cases can help students identify what is special about right triangles. For example, in non-right triangles all four small triangles (ADE, DEB, ADF, and DFC) are not congruent, and triangles ADC and ADB are not isosceles although they continue to have equal area. Furthermore, the relationships between the midsegments (DE and DF) and the sides (AC and AB, respectively) remain the same (see Fig. 3.9).

[1]As other modes of feedback, the Geometry Inventor would also allow one to create a graph from such a table of measurement data. Some versions of Cabri have an artificially intelligent oracle that informs the student whether a given relationship will always hold.

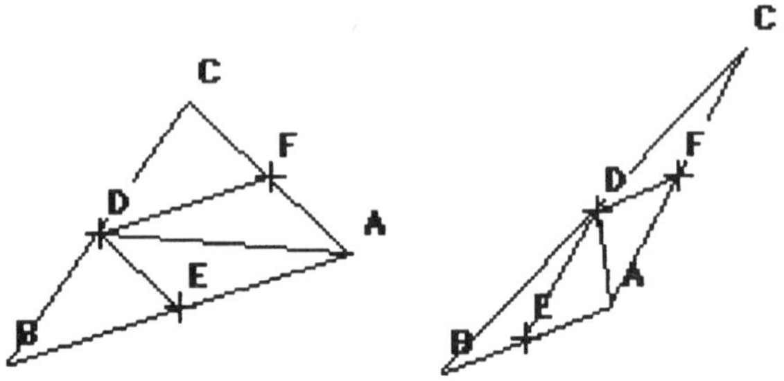

		Tri1	Tri2
1	⬟CFD	1.78	0.86
2	⬟AFD	1.78	0.86
3	⬟DEA	1.78	0.86
4	⬟DEB	1.78	0.86
5	∡CBA	37.20	24.22
6	∡DAB	22.32	102.95
7	∡BCA	82.01	14.97
8	∡DAC	38.47	37.86
9	⊢AD	3.63	1.35
10	⊢DC	2.28	3.21
11	⊢DB	2.28	3.21
12	⬟ADB	3.56	1.73
13	⬟ADC	3.56	1.73
14	⊢BA	5.17	2.62
15	⊢DF	2.58	1.31
16	L14/L15	2.00	2.00
17	⊢CA	3.16	4.17
18	⊢DE	1.58	2.08
19	L17/L18	2.00	2.00
20			

FIG. 3.9 Non-right-triangle examples.

When students seek to justify and explain their conjectures, they can use prior knowledge about parallel lines to argue that, for right triangles, all four small triangles are congruent, that triangles ADB and ADC are thus isosceles, and therefore that AD, DC, and DB are segments of equal length. When justifying conjectures developed for the general case, students may develop an alternative to the base-and-altitude argument for the equal areas of the triangles created by drawing a single median in a triangle (see Chazan, 1990a, for a more elaborate description of conjectures and arguments developed for the general case).

Again, the research carried out so far seems promising. Students experience a different type of mathematical activity (Chazan, 1990b): They learn to make conjectures, to distinguish and appreciate both inductive and deductive reasoning, and to have deeper appreciation for the role of deductive proof within the mathematics community. Chazan (1993) indicated that students using the Geometric Supposers distinguished between arguments based on the measurement of examples and those based on synthetic deductive proofs. In particular, students appreciated the explanatory aspect of deductive proofs, which empirical argument based on the measurement of examples lacks. Yerushalmy (1993) suggested:

> Use of the Supposer focuses students' attention on the nature of the sample (from which a generalization is derived): the quantity of needed information, the variability of the information, and the objectivity of the information (it is generated by the computer and not created by the individual). They learn to appreciate the ease of getting many examples but become aware of the need to form a strategy of collection in order not to get lost. They learn to use extreme cases, negative examples, and non-stereotypic evidence to back up their conjectures. (p. 82)

REFLECTIONS ON THE USE OF GEOMETRY CONSTRUCTION PROGRAMS

The changes in "problems to prove" and "construction problems" outlined earlier share three commonalities. The diagrams in geometry construction programs present objects that can be viewed as abstractions of the shapes of tangible objects in the world around us, but are also, with the computer, themselves tangible and manipulatable (see Pimm, 1995, especially pp. 48–56). Geometry construction programs are designed to help students see the general in the particular. Finally, the changes to both types of problems call for the inclusion of justification as an integral part of exploration.

Tangibility of Geometrical Abstractions

Earlier, we distinguished between two views of geometrical objects: built from undefined terms, and abstracted from the shape of objects around us. We also indicated that geometry construction programs are usually used with the latter view of geometrical objects.

This is no accident. Geometry construction programs seem well suited to this view, because, like the objects in the world from which they are abstracted, the diagrams of geometry construction programs can be acted on (by dragging or "direct manipulation"). This shared characteristic, tangibility, may help students approach the abstractions of geometry. Students may come to see that the abstractions of geometry are "abstractions" by virtue of their relationship to the objects from which they were abstracted. Thus, geometry construction programs may help combat a view of abstraction as "removal and detachment" and replace it with a view of abstraction that maintains the connections between an abstraction and that which it was abstracted from (see Arnheim, 1969, especially chapters 9 and 10, for an elaborate exploration of the meaning of the process of abstraction in the context of art history). Students may then be able to appreciate the links between mathematics and experience, rather than interpreting geometry as completely disconnected from their experience.

On the other hand, the diagrams of geometry construction programs are not like pencil-and-paper diagrams or abstractions of the shapes of rigid objects; they are "more general"; they can change into other configurations that preserve the specified invariants. Ironically, it is this very difference from traditional diagrams and the shapes of rigid objects that makes it possible to stretch and manipulate these objects (other than operating on the figure as a whole). Thus the tangibility of the screen objects of geometry construction programs is related to their peculiar mix of particularity and generality.

Particularity and Generality

With geometry construction programs, either there are more diagrams than in traditional instruction (the Geometric Supposer), or, with dragging, "stretchy" diagrams are dynamic and thus themselves become general. Students write general procedures (or computer programs)[2] by working on a particular diagram.[3] Because of the generality of the procedures with which they are created, the diagrams created with geometry construction programs seem poised between the particular and the general. They appear in front of us in all of their particularity, but, at the same time, they can be manipulated in ways that indicate the generalities lurking behind the particular.

It is hoped that, as a result of interacting with these programs, students will come to understand that statements and procedures in Euclidean geometry are meant to be more general than the static diagrams that may accompany them in a text. In an effort to explicate this paradox, French researchers distinguish between a figure and a drawing: "*Drawing* refers to the material entity [drawn on paper] while *figure* refers to a theoretical object" (C. Laborde, 1993, p. 49; for more detail on this distinction, see Goldenberg, Cuoco, & Mark, chap. 1, this volume). Parzysz (1988) de-

[2]Schwartz (1993, p. 13) argued that the generality of these procedures differentiates geometry construction programs from computer-aided design (CAD). In CAD the designer creates a specific object having particular measurements, whereas with geometry construction programs the user creates a general procedure to be studied. A particular instantiation of the procedure has particular measures, but another instantiation differs in other respects.

[3]In all of the programs we have discussed, programming is done through screen objects. By way of contrast, other systems, like the Plane Geometry System (Comsed Company, 1992), allow for linguistic programming.

scribed the figure as "the geometrical object which is described by the text defining it" (p. 79). In these terms, it is hoped that students will look at a drawing and be able to "see" the figure (for some evidence that this occurs, see Yerushalmy & Chazan, 1993). Furthermore, it is hoped that when students appreciate the figure, they may then become more autonomous as learners and be able to develop their own general ideas or conjectures (for some evidence that this occurs, see Yerushalmy, 1993).

Justification as a Part of Exploration

With each of these two problem types, students are asked to justify or explain why their conjecture or construction is true. In both cases, there are different possible kinds of justification. On an almost empirical level, students may justify their construction or conjecture by reference to measurements or visual invariants (e.g., my construction always produces a square because all of the angles are always 90° and the measures of the sides always stay the same). Such justifications reveal knowledge of the defining characteristics of a square.

But, as mathematics educators, we also seek other kinds of justification. Students can also justify their conjectures or constructions by making reference to relationships between that which is already known and a new phenomenon, a different sort of answer to the question why. As Lakatos (1976) suggested, this is what a proof does: It shows us the connection between something we deem known and something that is not known.

This second kind of justification is an important one in terms of students' feelings of control, autonomy, and mathematical power. If we know that our conjectures or constructions work, but have no explanation for why they work, exploration leaves us with the sense of regularity, but without an insight into the reason for the regularity. Within Euclidean geometry (whichever proof methods are chosen), justification of conjectures and constructions is possible. There is a relatively small number of concepts (congruence, similarity, sums of angles, and so on) that can be used to explain a wide range of results. Although evidence about students' abilities to write proofs in traditional geometry instruction is discouraging (e.g., Senk, 1985), perhaps if students have a better sense of the generality of geometric claims and if proofs are approached as justifications for conjectures, students will be more successful at writing proofs (see in this volume, de Villiers, chap. 15, and Koedinger, chap. 13).

DIFFICULTIES OF IMPLEMENTATION IN TEACHING[4]

Although the approaches described earlier only change the traditional course's view of the nature of the objects of study and the representation of how mathematics is done, although the reported research seems

[4]For specific examples of the issues raised in this section, see Chazan & Houde, 1989; Lampert, 1995; Wiske & Houde, 1993; Yerushalmy, Chazan, & Gordon, 1987.

promising, and although there seems to be genuine excitement about such approaches among some teachers, the difficulties in actually carrying out such changes in teaching must be emphasized and underlined. In attempting to capture this difficulty, Lampert (1995) used a metaphor to describe the experience of teachers using the Geometric Supposers to revamp their teaching of geometry:

> The situation the Supposer teachers experienced was something like that of a tour operator whose bus (the textbook) has been replaced with a collection of glitzy motor scooters (computers booted with the Supposer) in the middle of the Place de la Concorde with roads leading out in all directions to the attractions of Paris (geometry). Many of the tourists (students) could see places they wanted to explore, but did not know how to get to them. Others did not even know where to go or how to begin to make choices. Everyone was complaining because their expectations for the tour were not being met, while at the same time they were anxious to jump on a scooter and take off down the road towards some interesting looking destination. To make matters worse, the cameras (tests), that were to serve to record where everyone had been, were left on the bus. (pp. 228–229)

What were the tensions and difficulties that led Lampert to use this extended metaphor? When students were asked to conjecture, the nature of their role in the class changed. Some students enjoyed the involvement in inquiry; others did not. Some students thought that by making conjectures they were building knowledge, whereas others thought that they were "wasting time" and that the class would have made more progress if the teacher had simply told them what they needed to know.

Moreover, for students, sharing conjectures in public was risky business. In school, students are not often asked to share uncertain ideas publicly. For teachers, creating a safe environment for sharing conjectures was one of the major challenges in creating class discussions. A second challenge was posed by the teachers' lack of familiarity with leading discussions. As mathematics teachers, most were unused to such a role. Furthermore, it was often quite a challenge to connect students' findings—some of them not anticipated by the teachers—to the traditional curriculum. On the one hand, if the teacher emphasized the traditional curriculum more than students' ideas, students wondered why they were bothering to explore the problem. However, if teachers followed students' ideas, they might be taken away from the order of the traditional curriculum. It became a challenge to orchestrate the student exploration with the curricular order in the textbook.

At the same time, engaging students in inquiry and conjecturing made the teachers dependent on students to come up with ideas. If students did not develop conjectures, then there would not be anything to work with. Some teachers, especially those teaching in lower track classes, were concerned about this on a regular basis.

These tensions caused the teachers studied by the Educational Technology Center, for example, to revise their grading procedures and to invent pedagogical strategies for helping students make conjectures and for orchestrating class discussions. (For an elaborate view of some of these strategies, see Chazan & Houde, 1989.) The teachers suggested using "educated guesses" as an initial, informal definition of a conjecture, modeling the processes of generating hypotheses, and testing with measurement data in a whole-class format before asking students to work on their own. To help their students learn to be proficient explorers and to have a language for talking about exploratory work, they distinguished between "hypotheses," which are ideas that have not yet been tested empirically; "observations," which are conclusive remarks about specific cases; "conjectures," for those general ideas that have withstood empirical testing, but that have not been proven deductively; and finally "theorems," for those that have been proven deductively. By analyzing their own exploration of geometrical topics, the teachers also developed a taxonomy of "inquiry skills" broken down into the categories of verifying, conjecturing, generalizing, proving, and conjecturing and then designed activities to help students learn these skills (Chazan, 1990a).

A QUESTION OF VALUES

Having examined the ways in which geometry construction programs have been used to alter the teaching of traditional high school geometry, we return to the questions of curriculum with which we started this chapter. What constitutes a defensible and effective high-school geometry curriculum? Such a question inevitably raises issues both of value and of strategy. What criteria are most important in deciding that a curriculum is defensible? What can be done in schools?

We have described teaching, with geometry construction programs, that focuses on Euclidean geometry. We might argue that the teaching we have described is based on an inauthentic view of the objects of study in geometry (as abstractions of the shapes around us); non-Euclidean geometries make it impossible to conceive of Euclidean geometry as the study of the space around us. Further, we could argue that in its proof methods and axioms, the teaching we have described does not represent 20th century geometry faithfully and does not illustrate the connections between geometry and other areas of mathematical endeavor; perhaps, in high school, students should learn about fractals, spherical geometry, and reasoning by continuity.

Although we agree that these areas are interesting and potentially fruitful areas for exploration with students, in thinking about the concerns raised about Euclidean geometry, it is useful to recall John Dewey's essay "The Child and the Curriculum" (1902/1990). Dewey remarked that every subject matter is viewed differently by scientists and teachers:

> The problem of the teacher [with subject matter] is a different one [than that of the scientist]. As a teacher he is not concerned with adding new facts to the science he teaches; in propounding new hy-

> potheses or verifying them. He is concerned with the subject-matter of the science as *representing a given stage and phase of the development of experience.* (pp. 200–201, italics in original)

Of course, we agree that it is important for students to learn modern mathematics; beyond even their mathematical importance, fractals and non-Euclidean geometries have both had sizable effects on our cultural landscape. But in choosing a "rudimentary version," we cannot represent all aspects of geometry: Our representation of geometry must be based on what we value most in representing the discipline. Strong arguments can be made for many curricular offerings, but ultimately choices must be made.

When we are forced to choose, we value a curriculum's representation of how mathematics is done over its other three aspects (although they all interact[5]). We are more concerned about the nature of the students' role in classroom interactions, about the authentic representation of mathematics as a human endeavor,[6] than with authentic representation of the objects of study in modern mathematical research. This is our view of what makes for a viable and defensible "rudimentary version." For this reason, we are enthusiastic about the potential of geometry construction programs to change the nature of the traditional Euclidean geometry course and are not dismayed by the prospect that Euclidean geometry might remain the focus of high-school courses (pace Malkevitch).

As Schoenfeld's (1988) research in high-school mathematics illustrates, inauthentic representations of how mathematics is done are rampant in school. The students' views of mathematics that he catalogued are the result of mathematics instruction that leaves the student a passive recipient of teacher instruction. Does this mean that we would advise against the study of modern geometrical topics in school? Certainly not. In thinking about the teaching of modern topics in school, however, our commitment to student exploration suggests a few concerns, or perhaps more accurately, issues for consideration.

Some proposals to change the objects of study in geometry seem to suggest that new or different objects in and of themselves will increase student motivation, will bring geometry "to life." Although this reasoning seems plausible, we think the issue is more complicated. When we experience the ways in which geometry construction programs help make students' learning more active, we are concerned that proposals to change the objects of study may inadvertently cause instruction to revert to reliance on more passive learning and thus defeat the desire to develop student interest in geometry. Those interested in teaching students about fractals or about spherical geometry might object that there are ways for

[5]For example, as Schwab (1978) pointed out, "*How* we teach will determine *what* our students learn. If a structure of teaching and learning is alien to the structure of what we propose to teach, the outcome will inevitably be a corruption of that content" (p. 242, italics in original).

[6]Although in this sense our views may seem very much like Moise's (1975) view, our views of how mathematics is done differ.

students to be active in learning about these objects (see Lenart, 1993), and perhaps they are right.

However, we would argue that curricular representations of the objects of study in geometry and how mathematics is done are linked, that the choice we make about the mathematical objects of study in a curriculum affect how we portray mathematics as an endeavor. In our experience, representing the objects of study in geometry on computer as almost "tangible" (the Geometry Inventor is also titled "Tangible Math") abstractions of the shapes of objects in the world around us is what makes conjecturing possible; it enables students to understand what is being discussed, develop their intuitions through experimentation, and make conjectures.[7] Perhaps an approach to geometry that accurately represents 20th century mathematics by choosing objects of study that are more distant from students' experience and less tangible might lose these benefits, which, for us, are paramount.[8]

Similarly, our experience with geometry construction programs has convinced us of the important role that explanatory, mathematical justifications play in increasing students' mathematical power. Students who have a sense of regularity without insight into the reasons for the regularity do not have feelings of autonomy or control over their inquiry. We are concerned that some proposals for changing the objects of study in geometry do not pay sufficient attention to the important role that explanatory justification plays in student inquiry. Such explanatory justification depends on developing a system of interconnected results in which some results can be used to "explain" others.

Finally, our experience also has helped us appreciate the dramatic shift that exploratory teaching requires of many mathematics teachers. Although it need not necessarily be the case, we are concerned that changing the objects of study might make it more difficult for many teachers to orchestrate classrooms where students conjecture and justify results, because teachers are less familiar with the mathematics they are teaching (see Ball, 1991; Leinhardt & Smith, 1985; National Center for Research on Teacher Learning [NCRTL], 1993; NCTM, 1991, for the importance of teachers' familiarity with the mathematical content they are teaching).

From our perspective, there are two important reasons to value the kind of active geometry learning with construction programs just described: It asks students to be people who create ideas—who make conjectures and then justify them within the context of a system of results—and it changes the nature of student–teacher interactions in classrooms for the better, by changing the balance of intellectual authority. In thinking about the role of schools in democratic societies, such changes are of paramount importance to us, even though, as illustrated earlier, they are not easy to enact. We believe that it is important for students to think of them-

---

[7]René Thom (1986) made a similar point when he argued for exploration in the context of Euclidean geometry on the grounds that it makes "constant reference to underlying intuitively understood fundamentals" (p. 70).

[8]Mandelbrot (1994, p. 80) raised a similar concern.

selves as creators of knowledge and to think critically about knowledge that is presented to them (see Schwartz, 1991, for similar arguments).

To reiterate, we close with a restatement of our values: We are content to concentrate on Euclidean geometry, even though it is "old" and even though there are other geometries that students could study, *as long as* the teaching emphasizes student conjecturing and justification of their conjectures with reference to a growing system of results.

REFERENCES

Abelson, H., & diSessa, A. (1981). *Turtle geometry: The computer as a medium for exploring mathematics.* Cambridge, MA: MIT Press.

Aleksandrov, A. D. (1963). A general view of mathematics. In A. Aleksandrov, A. Kolmogorov, & M. Lavrent'ev (Eds.), *Mathematics: Its content, methods, and meaning* (Vol. 1, pp. 1–64). Cambridge, MA: MIT Press. (Original work published 1956)

Arnheim, R. (1969). *Visual thinking.* Berkeley: University of California Press.

Balacheff, N. (1993). Artificial intelligence and real teaching. In C. Keitel & K. Ruthven (Eds.), *Learning through computers: Mathematics and educational technology* (NATO, Vol. 121, pp. 131–158, ASI Series F: Computer and Systems Sciences). Berlin: Springer-Verlag.

Ball, D. L. (1991). Teaching mathematics for understanding: What do teachers need to know about the subject matter? In M. Kennedy (Ed.), *Teaching academic subjects to diverse learners* (pp. 63–83). New York: Teachers College Press.

Ball, D. L. (1993). With an eye on the mathematical horizon: Dilemmas of teaching elementary school mathematics. *Elementary School Journal, 93*(4), 373–397.

Capponi B., & Strasser, R. (1991). Drawing—computer model—figure: Case studies in students' use of geometry software. In P. Boero (Ed.), *Proceedings of the Fifteenth PME Conference* (pp. 302–309). Genova: Universiti de Ghenes.

Chazan, D. (1990a). Implementing the standards: Microcomputer-aided student exploration in geometry. *Mathematics Teacher, 83,* 628–635.

Chazan, D. (1990b). Quasi-empirical views of mathematics and mathematics teaching. *Interchange, 21*(1), 14–23.

Chazan, D. (1993). High school geometry students' justifications for their views of empirical evidence and mathematical proof. *Educational Studies in Mathematics, 24*(4), 359–387.

Chazan, D., & Houde, R. (1989). *How to use conjecturing and microcomputers to teach high school geometry.* Reston, VA: National Council of Teachers of Mathematics.

Comsed Company and Educational Computer System Laboratory. (1992). *Geomland Plane Geometry System.* Sofia, Bulgaria: Sophia University.

Coxford, A., & Usiskin, Z. (1971). *Geometry: A transformation approach.* River Forrest, IL: Laidlaw Brothers.

Cuoco, A., Goldenberg, E. P., & Mark, J. (1995). Connecting geometry with the rest of mathematics. In P. House & A. F. Coxford (Ed.), *Connecting mathematics across the curriculum* (pp. 183–197). Reston, VA: National Council of Teachers of Mathematics.

Dewey, J. (1990). *The school and society; The child and the curriculum.* Chicago: University of Chicago Press. (Original work published 1902)

Education Development Center. (1995). *Connected geometry.* Dedham, MA: Jansen.

Gleick, J. (1987). *Chaos.* New York: Penguin.

Healy, C. C. (1993). Discovery courses are great in theory, but.... In J. L. Schwartz, M. Yerushalmy, & B. Wilson (Eds.), *The Geometric Supposer: What is it a case of?* (pp. 85–106). Hillsdale, NJ: Lawrence Erlbaum Associates.

Houde, R. (1993). How the Supposer changed my life: An autobiography. In J. L. Schwartz, M. Yerushalmy, & B. Wilson (Eds.), *The Geometric Supposer: What is it a case of?* (pp. 179–192). Hillsdale, NJ: Lawrence Erlbaum Associates.

Jackiw, N. (1995). *The Geometer's Sketchpad 3.0* [Computer software]. San Francisco: Key Curriculum Press.

Laborde, C. (1992). Solving problems in computer-based geometry environments: The influence of the features of the software. *Zentralblatt fur Didaktik der Mathematik, 4,* 128–134.

Laborde, C. (1993). The computer as part of the learning environment: The case of geometry. In C. Keitel & K. Ruthven (Eds.), *Learning through computers: Mathematics and educational technology* (NATO ASI Series F: Computer and Systems Sciences, Vol. 121, pp. 48–67). Berlin: Springer-Verlag.

Laborde, C., & Capponi, B. (1994). Cabri-geomètre constituant d'un milieu pour l'apprentissage de la notion de figure géometrique. *DidaTech V Seminar 150,* 175–218.

Laborde, J.-M., & Bellemain, F. (1994). *Cabri II.* [Computer software]. Temple, TX: Texas Instruments.

Laborde, J.-M., & Strasser, R. (1990). Cabri-Geomètre: A microworld of geometry for guided discovery learning. *Zentralblatt fur Didaktik der Mathematik—International Reviews on Mathematical Education, 90*(5), 171–190.

Lakatos, I. (1976). *Proofs and refutations: The logic of mathematical discovery.* Cambridge: Cambridge University Press.

Lampert, M. (1993). Teachers' thinking about students' thinking about geometry: The effects of new teaching tools. In J. L. Schwartz, M. Yerushalmy, & B. Wilson (Eds.), *The Geometric Supposer: What is it a case of?* (pp. 143–178). Hillsdale, NJ: Lawrence Erlbaum Associates.

Lampert, M. (1995). Managing the tensions in connecting students' inquiry with learning mathematics in school. In D. Perkins, J. Schwartz, M. West, & M. Wiske (Eds.), *Software goes to school.* (pp. 213–232). New York: Oxford University Press.

Leinhardt, G., & Smith, D. (1985). Expertise in mathematics instruction: Subject matter knowledge. *Journal of Educational Psychology, 77*(3), 247.

Lenart, I. (1993). Alternative models on the drawing ball. *Educational Studies in Mathematics, 24,* 277–312.

Logal Software. (1994). *Tangible Math: Geometry Inventor* [Computer software]. Pleasantville, NY: Sunburst Communications.

Malkevitch, J. (n.d.). Geometry in utopia. In *Geometry's future: Conference proceedings* (pp. 93–105). Lexington, MA: Consortium for Mathematics and Its Applications.

Mandelbrot, B. (1994). Fractals, the computer, and mathematics education. In C. Gaulin, B. Hodgson, D. Wheeler, & J. Egsgard (Eds.), *Proceedings of the 7th International Congress on Mathematical Education* (pp. 77–100). Sainte-Foy, Quebec, Canada: University of Laval.

Moise, E. (1975). The meaning of Euclidean geometry in school mathematics. *Mathematics Teacher, 68,* 472–477.

National Center for Research on Teacher Learning. (1993). *Findings on learning to teach.* East Lansing, MI: Author.

National Council of Teachers of Mathematics. (1989). *Curriculum and evaluation standards for school mathematics.* Reston, VA: Author.

National Council of Teachers of Mathematics. (1991). *Professional standards for teaching mathematics.* Reston, VA: Author.

Parzysz, B. (1988). "Knowing" vs. "seeing": Problems for the plane representation of space geometry figures. *Educational Studies in Mathematics, 19*(1), 79–92.

Peitgen, H.-O., & Richter, P. H. (1986). *The beauty of fractals: Images of complex dynamical systems.* Berlin: Springer-Verlag.

Pimm, D. (1995). *Symbols and meanings in school mathematics.* London: Routlege.

Scheffler, I. (1965). *Conditions of knowledge: An introduction to epistemology and education.* Chicago: Scott Foresman.

Schoenfeld, A. (1983). *Problem solving in the mathematics curriculum: A report, recommendations, and an annotated bibliography.* Washington, DC: Mathematics Association of America.

Schoenfeld, A. (1988). When good teaching leads to bad results: The disasters of "well taught" mathematics courses. In T. P. Carpenter, & P. L. Peterson (Eds.), *Learning from instruction: The study of students' thinking during instruction in mathematics* [Special issue]. *Educational Psychologist, 23*(2), 145–166.

Schwab, J. J. (1978). Education and the structure of the disciplines. In I. Westbury & N. Wilkof (Eds.), *Science, curriculum, and liberal education: Selected essays* (pp. 229–272). Chicago: University of Chicago Press. (Original work published 1961)

Schwartz, J. L. (1991). Can we solve the problem solving problem without posing the problem posing problem? In J. P. Ponte, J. F. Matos, J. M. Matos, & D. Fernandes (Eds.), *Mathematical problem solving and new information technologies: Research in contexts of practice* (NATO ASI Series F: Computer and Systems Sciences, Vol. 89, pp. 167–176). Berlin: Springer-Verlag.

Schwartz, J. L. (1993). A personal view of the Supposer: Reflections on particularities and generalities in educational reform. In J. L. Schwartz, M. Yerushalmy, & B. Wilson (Eds.), *The Geometric Supposer: What is it a case of?* (pp. 3–16). Hillsdale, NJ: Lawrence Erlbaum Associates.

Schwartz, J. L., & Yerushalmy, M. (1993). *Geometric SuperSupposer* [Computer software]. Pleasantville, NY: Sunburst Communications.

Senk, S. (1985). How well do students write proofs? *Mathematics Teacher, 78,* 448–456.

Thom, R. (1986). "Modern" mathematics: An educational and philosophic error? In T. Tymoczko (Ed.), *New directions in the philosophy of mathematics* (pp. 67–78). Boston: Birkhauser.

Tymoczko, T. (1986). The four-color problem and its philosophical significance. In T. Tymoczko (Ed.), *New directions in the philosophy of mathematics* (pp. 243–266). Boston: Birkhauser.

Usiskin, Z. (1987). Resolving the continuing dilemmas in school geometry. *Learning and teaching geometry, K–12* (NCTM 1987 Yearbook, pp. 17–31). Reston, VA: National Council of Teachers of Mathematics.

van den Brink, J. (1995). Geometry education in the midst of theories. *For the Learning of Mathematics, 15*(1), 21–28.

Wiske, M. S., & Houde, R. (1993). From recitation to construction: Teachers change with new technology. In J. L. Schwartz, M. Yerushalmy, & B. Wilson (Eds.), *The Geometric Supposer: What is it a case of?* (pp. 193–217). Hillsdale, NJ: Lawrence Erlbaum Associates.

Yerushalmy, M. (1993). Generalization in geometry. In J. L. Schwartz, M. Yerushalmy, & B. Wilson (Eds.), *The Geometric Supposer: What is it a case of?* (pp. 57–84). Hillsdale, NJ: Lawrence Erlbaum Associates.

Yerushalmy, M., & Chazan, D. (1993). Overcoming visual obstacles with the aid of the Supposer. In J. L. Schwartz, M. Yerushalmy, & B. Wilson (Eds.), *The Geometric Supposer: What is it a case of?* (pp. 25–56). Hillsdale, NJ: Lawrence Erlbaum Associates.

Yerushalmy, M., Chazan, D., & Gordon, M. (1987). *Guided inquiry and technology: A year-long study of children and teachers using the Geometric Supposer* (Tech. Rep. No. 88-6). Cambridge, MA: Harvard Graduate School of Education, Educational Technology Center.

4

Chaos in the Classroom

Robert L. Devaney[1]
Boston University

One of the most interesting applications of technology in the mathematics classroom is the fact that it allows teachers to bring new and exciting topics into the curriculum. In particular, technology lets teachers bring topics of contemporary interest in research mathematics into both middle-school and high-school classrooms.

The mathematical topics of chaos and fractals are particularly appropriate in this regard. They are timely—many ideas in these fields were first conceived during the students' lifetimes. They are applicable—fields as diverse as medicine, business, geology, art, and music have adopted ideas from these areas. And they are beautiful—there is something in the gorgeous computer-generated images of objects such as the Mandelbrot set, Julia sets, the Koch snowflake, and others that captures students' interest and enthusiasm.

Therein, however, lies the problem. Many research mathematicians cringe at the sight of "still another fractal." Most often, discussions of chaos and fractals degenerate to simple "pretty picture shows," devoid of any mathematical content. As a consequence, students get the idea that modern mathematics is akin to a video game—lots of computer-generated action, but mindless activity at best.

This attitude is both unfortunate and unnecessary. There is mathematics behind the pretty pictures, and, moreover, much of it is quite accessible to secondary-school students. Furthermore, the mathematics behind the images is often even prettier than the pictures themselves. In this sense it is a tragedy that students come so close to seeing some exciting, contemporary topics in mathematics, yet miss out in the end. Our goal in this note is to help remedy this situation by describing some easy-to-teach topics involving ideas from fractal geometry.

---

[1]Partially supported by National Science Foundation grant ESI-9255724. Note: An interactive version of this paper is available at http://math.bu.eud/DYSYS/chaos-game/chaos-game.html.

THE CHAOS GAME

One of the most interesting fractals arises from what Barnsley (1989) has dubbed "The Chaos Game." The chaos game is played as follows: First pick three points—the vertices of a triangle (any triangle works—right, equilateral, isosceles, whatever). Name one of the vertices 1,2, the second 3,4, and the third 5,6. The reason for these strange names is that we will use the roll of a die to determine the moves in the game.

To begin the game, choose any point in the triangle. This point is the *seed* for the game. (Actually, the seed can be anywhere in the plane, even miles away from the triangle.) Then roll a die. Move the seed halfway toward the named vertex: If 1 or 2 comes up, move the point half the distance to the vertex named 1,2. Now erase the original point and repeat this procedure, using the result of the previous roll as the seed for the next: Roll the die again and move the new point half the distance to the named vertex, and then erase the previous point (see Fig. 4.1).

Now continue in this fashion for a small number of rolls of the die. Five rolls are sufficient if you are playing the game by hand or on a graphing calculator, and eight are sufficient if you are playing on a high-resolution computer screen. (If you start with the seed point outside the triangle, you will need more of these initial rolls.)

After a few initial rolls of the die, begin to record the track of these traveling points. The goal of the chaos game is to roll the die many hundreds of times and predict the resulting pattern of points. Most students who are unfamiliar with the game guess that the resulting image will be a random smear of points. Others predict that the points will eventually fill the entire triangle. Both guesses are quite natural, given the random nature of the chaos game. But both guesses are completely wrong. The resulting image is anything but a random smear: With probability 1, the points form what mathematicians call the Sierpinski triangle. We denote this set by S (see Fig. 4.2).

There is some terminology associated with the chaos game that is important. The sequence of points generated by the chaos game is called

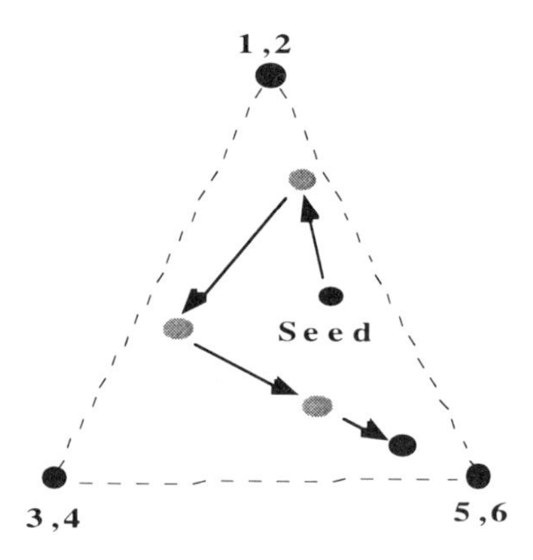

FIG. 4.1 Playing the chaos game.

FIG 4.2 The Sierpinski triangle *S*.

the *orbit* of the seed. The process of repeating the rolls of the die and tracing the resulting orbit is called *iteration*. Iteration is important in many areas of mathematics. In fact, the branch of mathematics known as discrete dynamical systems theory is the study of such iterative processes.

There are two remarkable facets of the chaos game. The first is the geometric intricacy of the resulting figure. The Sierpinski triangle is one of the most basic types of geometric images known as *fractals*. The second is the fact that this figure results no matter what seed is used to begin the game: With probability 1, the orbit of any seed eventually fills out *S*. The words "with probability 1" are important here. Obviously, if we always roll a 1 or 2, the orbit will simply tend directly to vertex 1,2. Of course, we do not expect a fair die to yield the same two numbers at each roll, so this movement does not occur in practice.

THE SIERPINSKI TRIANGLE

The Sierpinski triangle *S* may also be constructed using a deterministic rather than a random algorithm. To see this, we begin with any triangle. We then use the midpoints of each side as the vertices of a new triangle, which we then remove from the original. This procedure leaves us with three triangles, each of which has dimensions exactly one-half the dimensions of the original triangle, and area exactly one-fourth the original area. Also, each remaining triangle is similar to the original.

Now we continue (or iterate) this process. From each remaining triangle we remove the "middle," leaving behind three smaller triangles, each of which has dimensions one-half those of the parent triangle (and one-fourth of the original triangle). At this state, 9 triangles remain. At the next iteration, 27 small triangles remain; at the following iteration, 81; and at the *n*th stage, 3^n small triangles. It is easy to check that the dimensions of the triangles that remain after the *n*th iteration are exactly $1/2^n$ of the original dimensions (see Fig. 4.3).

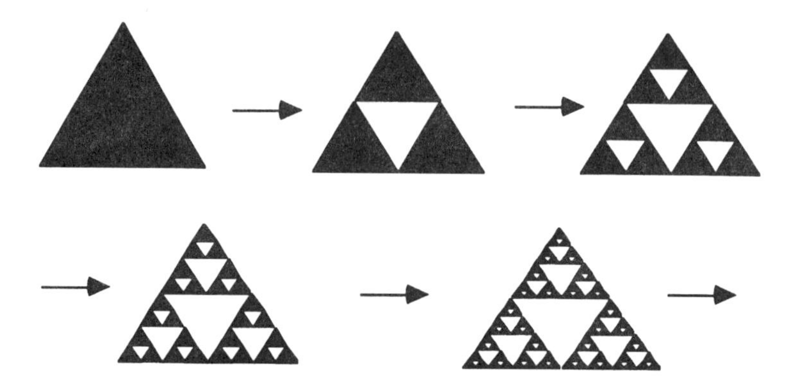

FIG. 4.3 The chaos game: deterministic construction of Sierpinski triangle *S*.

Why does the Sierpinski triangle arise from the chaos game? Students are always intrigued when they first see the Sierpinski triangle emerge from the random chaos game, but there is a simple explanation why this pattern occurs. Suppose we start with a point somewhere in the middle of the largest white (removed) triangle in the Sierpinski triangle.

Where does this point move after one roll of the die? As in Fig. 4.4, we see that this point hops into one of the three next-smaller triangles: These triangles represent all points that are half the distance to the three vertices from points in the largest removed triangle.

Now continue. At the next iteration, the point hops into one of the nine next-smaller triangles, then into the next-smaller triangles, and so on. Eventually (after very few iterations), the point enters a small triangle that is essentially invisible on the computer screen.

In actuality, the orbit of a point that starts in any of the removed triangles will never "reach" the Sierpinski triangle. Rather, it will continue to lie in successively smaller removed triangles. Of course, these removed triangles very quickly become microscopic in size, so for all practical purposes the orbit looks like it lies on *S*. Mathematicians says that the orbit of the seed is *attracted* to *S*. Sometimes *S* is called a *strange attractor*—strange, because it is not a simple object such as a point or circle.

PLAYING THE CHAOS GAME IN CLASS

The number of individuals in a class (Peitgen, Jurgens, & Saupe, 1991) provides a wonderful and easy way to play the chaos game without using technology. Give each student (or group of students) an overhead transparency onto which you have copied the vertices of the original triangle (so they all are playing the same chaos game). Have the students choose the numbers corresponding to the vertices (two numbers for each vertex) and the original seed. Instead of using a marking pen to indicate the seed, have the students use the small circles that form the output of a three-hole punch. This allows them to perform the first few iterations by simply moving the "dot" rather than erasing.

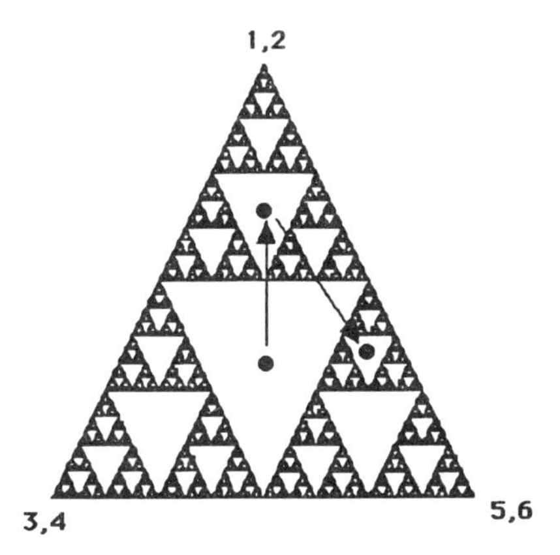

FIG. 4.4 The chaos game: the results of rolling 1 and then 6 with the seed in the middle of the largest triangle.

Select one student to roll the die and another to take charge of moving the "point" half the distance to the indicated vertex. A wonderful device for moving the point exactly half the distance is the "half-way ruler" invented by Henri Lion (see Fig. 4.5).

After seven or eight iterations with the movable point, have the students begin to track the orbit by recording the points on the orbit using a marking pen. Depending on the class size, 10 iterations of the chaos game by each group should be sufficient. Gather all of the transparencies and align them correctly on the overhead. Provided nobody has goofed, you will see the beginning of a good representation of the Sierpinski triangle. (It is important to use transparencies that are "see-through"; some transparencies become opaque when several are piled on top of each other.)

Another interesting and valuable classroom activity (originally conceived by Kevin Lee) involves target shooting. Draw one of the images in the deterministic construction of S on the board or the overhead projector. Actually, it is best to begin at a relatively low level, say the second or third stage (when either 9 or 27 triangles remain). Select a seed, perhaps one of the three vertices. Then color one of the remaining triangles at the lowest stage of the construction (before the smallest triangles have been thrown out or subdivided). This triangle is the target. The game is to move the seed into the interior of the target in the smallest number of moves. This is not a random game any longer: The students have to determine which rolls of the die put the seed in the target in the smallest number of rolls (see Fig. 4.6).

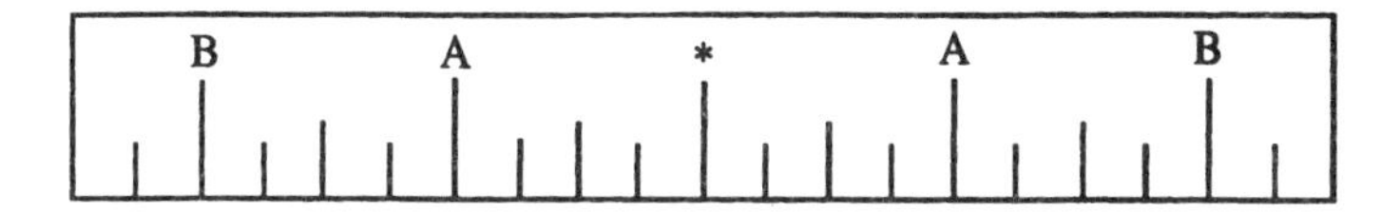

FIG. 4.5 The half-way ruler (Lion).

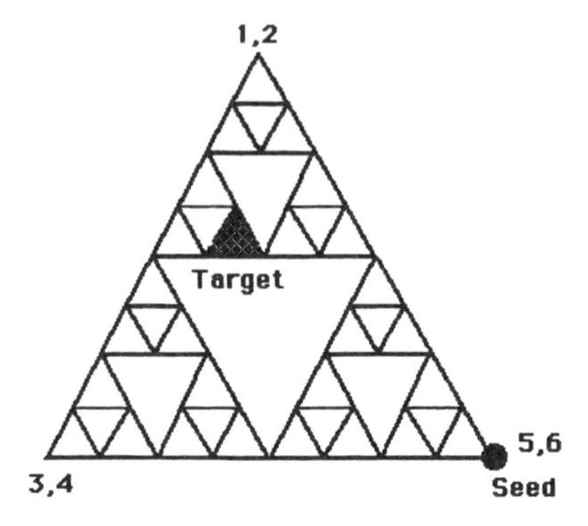

FIG. 4.6 Playing the chaos game: target shooting.

There is no unique, most efficient method to hit the target, but rather several alternatives. However, there is a least possible number of moves that cannot be beaten. At first, students seem to have a great deal of fun challenging one another with this game. Later they turn their attention to finding the algorithm that yields the most efficient solution. It is a wonderful exercise for students to try to find this winning strategy, and an even more beneficial exercise for them to explain their strategy once they have found it.

SELF-SIMILARITY

One of the basic properties of fractal images is the notion of self-similarity. This idea is easy to explain using the Sierpinski triangle. Note that S may

FIG. 4.7 Decomposing the Sierpinski triangle.

be decomposed into three congruent figures, each of which is exactly half the size of S (see Fig. 4.7). If we magnify any of the three pieces of S (shown in Fig. 4.7) by a factor of 2, we obtain an exact replica of S. That is, S consists of three *self-similar* copies of itself, each with *magnification factor* 2.

We can look deeper into S and see further copies of S. The Sierpinski triangle also consists of 9 self-similar copies of itself, each with magnification factor 4. Or we can chop S into 27 self-similar pieces, each with magnification factor 8. In general, we can decompose S into 3^n self-similar pieces, each of which is congruent, and each of which may be magnified by a factor of 2^n to yield the original figure. This type of self-similarity at all scales is a hallmark of the images known as fractals.

FRACTAL DIMENSION

Students (and teachers) are often fascinated by the fact that certain geometric images have fractal dimension. The Sierpinski triangle provides an easy way to explain why they possess this dimension.

To comprehend the concept of fractal dimension, it is necessary to understand what we mean by dimension in the first place. Obviously, a line has dimension 1; a plane, dimension 2; and a cube, dimension 3. But why is this? It is interesting to watch students struggle to articulate why these facts are true. They often say that a line has dimension 1 because there is only one way to move on a line. Similarly, the plane has dimension 2 because there are two directions in which to move. Of course, there really are two directions in a line—each the reverse of the other—and infinitely many directions in the plane. What the students really generally mean is that there are two linearly independent directions in the plane. Of course they are right. But the notion of linear independence is quite sophisticated and difficult to articulate. Students often say that the plane is two dimensional because it has "two dimensions," meaning length and width. Similarly, a cube is three dimensional because it has "three dimensions"—length, width, and height. Again, this is a valid notion, although not expressed in particularly rigorous mathematical language.

Another pitfall occurs when trying to determine the dimension of a curve in a plane or in three-dimensional space. An interesting debate occurs when a teacher suggests that these curves are actually one-dimensional. But they have two or three dimensions, the students object.

So why is a line one-dimensional and the plane two-dimensional? Note that both of these objects are self-similar. We could break a line segment into four self-similar intervals, each with the same length, and each of which, magnified by a factor of 4, would yield the original segment. We can also break a line segment into 7 self-similar pieces, each with magnification factor 7, or 20 self-similar pieces with magnification factor 20. In general, we can break a line segment into N self-similar pieces, each with magnification factor N.

A square is different. We can decompose a square into four self-similar subsquares, and the magnification factor here is 2. Alternatively, we can break the square into 9 self-similar pieces with magnification factor 3, or 25 self-similar pieces with magnification factor 5. Clearly, the square may be broken into N^2 self-similar copies of itself, each of which must be magnified by a factor of N to yield the original figure (see Fig. 4.8). Finally, we can decompose a cube into N^3 self-similar pieces, each of which has magnification factor N.

From this we see an alternative way to specify the dimension of a self-similar object: The dimension is simply the exponent of the number of self-similar pieces with magnification factor N into which the figure may be broken.

So what is the dimension of the Sierpinski triangle? How do we find the exponent in this case? For this, we need logarithms. Note that, for the square, we have N^2 self-similar pieces, each with magnification factor N. So we can write

$$\text{Dimension} = \frac{\log(\text{number of self-similar pieces})}{\log(\text{magnification factor})}$$

$$= \frac{\log N^2}{\log N}$$

$$= \frac{2 \log N}{\log N} = 2$$

Similarly, the dimension of a cube is

$$\text{Dimension} = \frac{\log(\text{number of self-similar pieces})}{\log(\text{magnification factor})}$$

$$= \frac{\log N^3}{\log N}$$

$$= \frac{3 \log N}{\log N} = 3$$

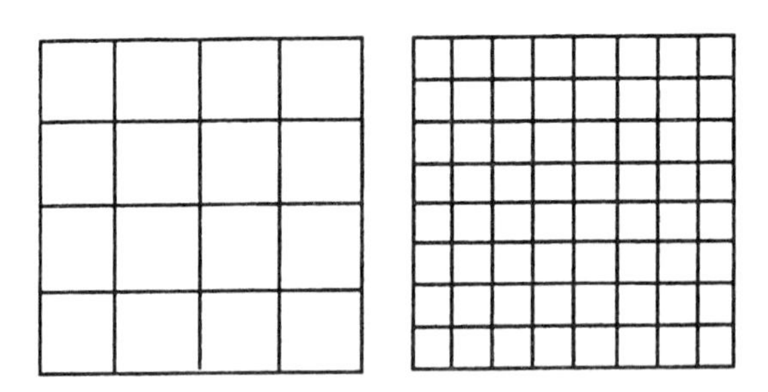

FIG. 4.8 Decomposing a square: N^2 self-similar pieces, each with magnification factor N.

Thus, we take as the *definition* of the fractal dimension of a self-similar object:

$$\text{Fractal dimension} = \frac{\log(\text{number of self-similar pieces})}{\log(\text{magnification factor})}$$

Now we can compute the dimension of S. The Sierpinski triangle consists of three self-similar pieces, each with magnification factor 2. So the fractal dimension is

$$\frac{\log(\text{number of self-similar pieces})}{\log(\text{magnification factor})} = \frac{\log 3}{\log 2} \approx 1.58$$

The dimension of S is somewhere between 1 and 2, just as our "eyes" tell us.

But wait a moment, S also consists of nine self-similar pieces with magnification factor 4. No problem—we have

$$\begin{aligned}\text{Fractal dimension} &= \frac{\log 9}{\log 4} \\ &= \frac{\log 3^2}{\log 2^2} \\ &= \frac{2 \log 3}{2 \log 2} \\ &= \frac{\log 3}{\log 2} \approx 1.58\end{aligned}$$

as before. Similarly, S breaks into $3N$ self-similar pieces with magnification factors $2N$, so we again have

$$\begin{aligned}\text{Fractal dimension} &= \frac{\log 3^N}{\log 2^N} \\ &= \frac{N \log 3}{N \log 2} \\ &= \frac{\log 3}{\log 2} \approx 1.58\end{aligned}$$

Fractal dimension is a measure of how "complicated" a self-similar figure is. In a rough sense, it measures "how many points" lie in a given set. A plane is "larger" than a line, where S sits somewhere in between these two sets.

On the other hand, all three of these sets have the same number of points in the sense that each set is uncountable. In this sense, somehow, though, fractal dimension captures the notion of "how large a set is" quite nicely, as we see in the next section.

CHANGING THE RULES IN THE CHAOS GAME

The chaos game provides a wonderful opportunity for students to develop their geometric insight as well as their understanding of the geometry of linear transformations. As we have seen, if we choose three points (the vertices of an equilateral triangle) and play the chaos game, moving half the distance toward the appropriate vertex at each stage, then the Sierpinski triangle S results. Note that these numbers are reflected in the geometry of S: This figure consists of three self-similar copies of S, each half the size of the S (or, as we said earlier, with magnification factor 2).

We really should have called S the Sierpinski equilateral triangle, because there are other Sierpinski triangles. If we begin with vertices on a right triangle (see Fig. 4.9a) or on a triangle with an obtuse angle (see Fig. 4.9b), different fractals result. However, each of these images consists of three self-similar copies, each with magnification factor 2.

Now let's vary the rules of the chaos game. Suppose we again start with three points (the vertices of an equilateral triangle). We will again move toward the vertices depending on the roll of a die. This time, if either of two vertices are chosen, we again move half the distance toward the appropriate vertex. But if the top vertex is chosen, we move two-thirds of the distance toward that vertex. A better way to say this (for reasons that will be clear in a moment) is that we compress the distance from the point to the top vertex by one-third. Equivalently, we move our point on a straight line to the top vertex so that the new distance is one-third of the old. If we play the chaos game with these rules, a very different image re-

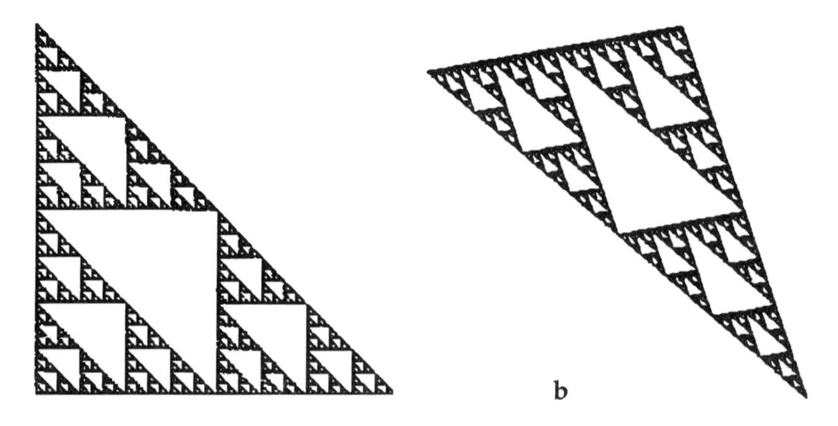

FIG. 4.9 Other Sierpinski triangles: the chaos game with (a) a right triangle and (b) a triangle with an obtuse angle.

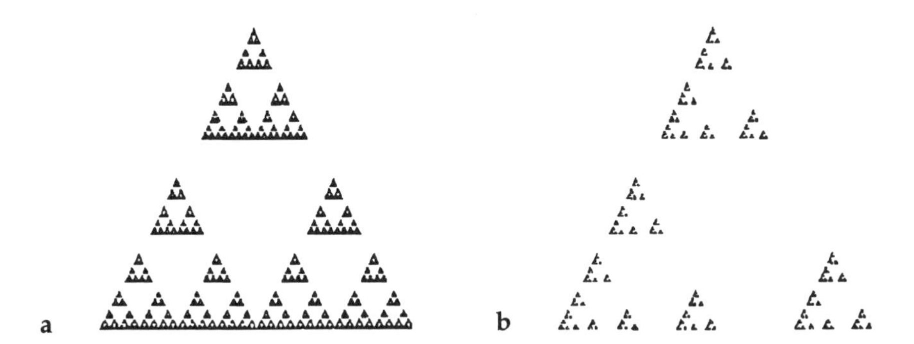

FIG. 4.10 Playing the chaos game: three vertices with (a) magnification factors 2, 2, 3 and (b) magnification factors 2, 3, 4.

sults (see Fig. 4.10a). Note, however, that this image consists of three self-similar pieces, two of which are exactly half the size of the entire image, whereas the other is one third the size. Again, the number of basic copies is equal to the number of vertices we started with, and the magnification factors at each vertex correspond as well.

If we further refine the rules of the chaos game, we see that we can still read off the original rules we used playing the game from the geometry of the resulting figure. For example, in Fig. 4.10b, we used three vertices and magnification factors 2, 3, and 4 to create the image (i.e., we moved half the distance toward one vertex, one-third the distance toward another, and one-fourth the distance to the final vertex).

We may also vary the number of vertices used. For example, to produce Fig. 4.11a, we chose six points on the vertices of a regular hexagon and moved one-third the distance toward the appropriate vertex at each roll of the die. Note that this is a "natural" chaos game in the sense that we can number the vertices 1 to 6 and use the number that comes up at each roll to determine the vertex to move toward. We call this image the *Sierpinski hexagon*; note that it consists of six self-similar pieces, each with magnification factor 3. So the fractal dimension of this image is

$$\frac{\log 6}{\log 3} \approx 2.58$$

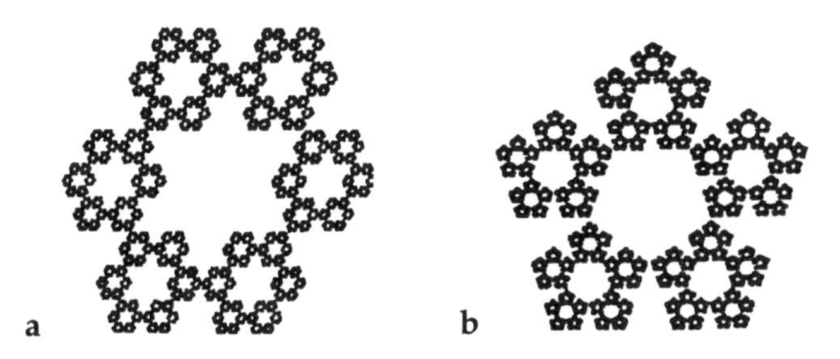

FIG. 4.11 Playing the chaos game: the Sierpinski hexagon (a) and pentagon (b).

There is another famous fractal buried inside the Sierpinski hexagon. Note that the inner boundary of this figure is the well-known Koch snowflake curve (see Devaney, 1989).

If we choose five vertices on a regular pentagon and compress toward the vertices by a factor of three-eighths at each roll of a five-sided die, the image in Fig. 4.11b results—a Sierpinski pentagon. Again we find five self-similar copies, each one-third the size of the original.

At this point it is interesting to guess what figure results when we play the chaos game with points on the vertices of a square and a magnification factor of 2. Most people expect a "Sierpinski square." Wrong! The points fill out the entire square when the chaos game is played with these rules. But that is no surprise: After all, a square is a self-similar image that may be broken into four self-similar pieces, each with magnification factor 2. So we can again follow the rules of the chaos game, reading from the resulting figure.

Rotations

Another, more complicated type of chaos game results when we allow rotations as well as contractions toward specific vertices. For example, suppose we start with the vertices of an equilateral triangle. Again, when either of the lower two vertices are selected, we simply move half the distance toward them. But when the upper vertex is chosen, we first move half way toward that vertex, and then we rotate the point about this vertex by 90° in a counterclockwise direction. Note that we could equally well rotate first, then move half the distance toward that vertex. If we now play the chaos game with these rules, the image in Fig. 4.12a results. Note that this figure consists of three self-similar pieces, each half the size of the original, but that the topmost piece is rotated by 90° in a counterclockwise direction. Again, it is not so easy to predict what figure will result from playing the chaos game, but once we see it, there is no question what game we played to generate it.

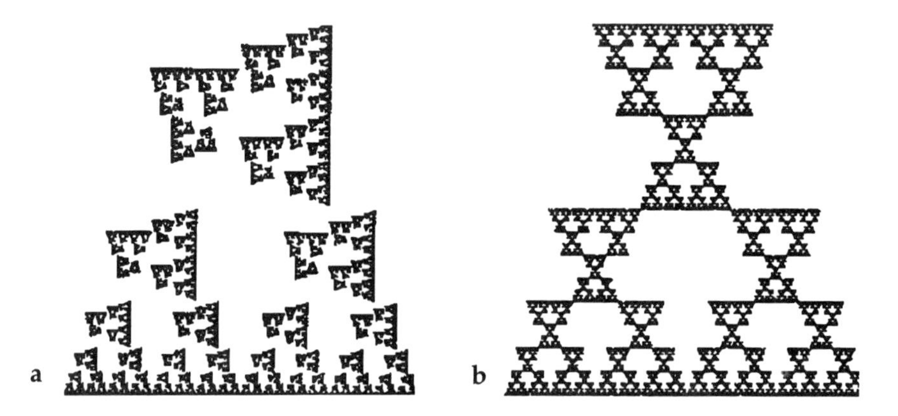

FIG. 4.12 Playing the chaos game: three vertices with magnification factor 2, featuring (a) 90° point rotation and (b) 180° point rotation.

In Fig. 4.12b, we show the result of playing the chaos game in which movement to the top vertex features a point rotation of 180°.

Rotations complicate the task of determining the chaos game that produced a given fractal image. It is a good test of geometric intuition to figure out some of these games. For example, how did we generate the fractals in Fig. 4.13?

One rather fun exercise for students who have access to good computing equipment is to make a fractal movie. This can be done by computing the images of a number of fractals generated by chaos games where each "frame" of the film is generated by changing only slightly the parameters (rotations or magnification factors) in the previous frame. In Fig. 4.14, we show a "clip" of such a film. How did we generate this movie?

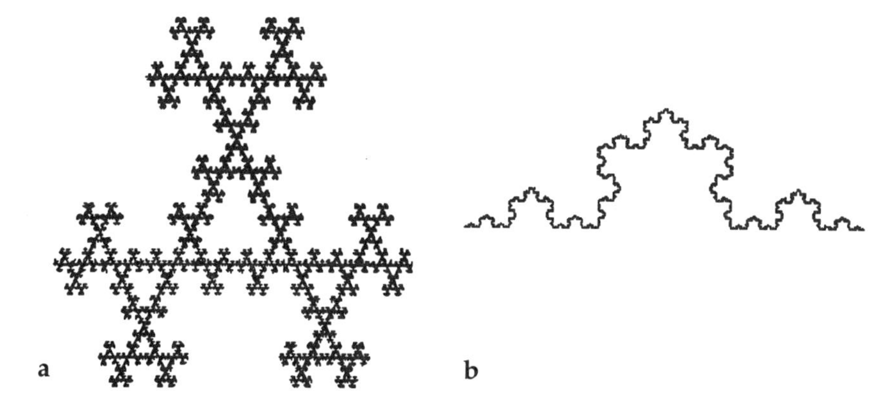

FIG. 4.13 A test: How were these images generated?

FIG. 4.14 A fractal film.

SUMMARY

The mathematics involved in constructing and understanding chaos games runs the gamut from elementary to linear algebra, from Euclidean to fractal geometry. Algorithmic thinking is a prerequisite for writing the simple graphing calculator program necessary to play the game with technology. Geometric transformations are at the root of the game. And probability and randomness lurk in the background. In short, chaos games provide the student with a wealth of different mathematical ideas and, at the same time, a glimpse of contemporary mathematics.

REFERENCES

Barnsley, M. (1989). *Fractals everywhere*. Boston. MA: Academic Press.
Devaney, R. (1989). *Chaos, fractals, and dynamics: Computer experiments in mathematics*. Menlo Park, CA: Addison-Wesley.
Peitgen, H.-O., Jurgens, H., & Saupe, D. (1991). *Fractals for the classroom*. New York: Springer-Verlag.

Part II

Studies of Conceptual Development

From the questioning of basic assumptions of how we teach traditional geometric concepts to the increased interaction of reasearchers and teachers in creating research-based, teacher-designed activities, researchers with their colleagues, classroom teachers, have begun to rethink what happens in mathematics (in particular, geometry) classrooms.

In this section, we sample that range of work, looking at models of student thinking; studies of interactive, hands-on and computer-enhanced methods of exploring geometric concepts; and the emergence and growth of students' reasoning from the primary grades through high school. In chapter 5, Pegg and Davey graft the SOLO taxonomy of modes and levels of thinking (Biggs & Collis, 1991) onto the main trunk of the van Hiele cognitive model. They examine what is perhaps the predominant theory in K–12 geometry education—the model of "levels" of teaching and learning proposed by the van Hieles—and take the position that although the van Hiele theory offers a broad framework in which to view cognitive growth in geometry, the theory is not sufficiently fine grained to account for individual differences. The synthesis they propose suggests a way to fine-tune how we look at cognitive reasoning and conceptual development.

Lehrer, Jenkins, and Osana, in chapter 6, report on a 3-year longitudinal study of elementary-grade students that questions (as do Pegg and Davey in chap. 5) the adequacy of the van Hiele model as a description of the progression of children's thinking. Their broad-scale portrait of children's emergent spatial reasoning skills suggests that current curricular practices in elementary school promote little conceptual development and that, for many children, the opportunities to develop and integrate spatial reasoning skills may substantially diminish before they leave elementary school.

In the next chapter, Lehrer, Jacobson, et al. suggest both that geometry education begins in students' informal knowledge and grows in a classroom culture that depends on skilled teachers, teacher-developed models of student cognition, and a comfortable interface between research (and researchers) and teachable moments (and teachers). This particular collaboration between researchers and elementary-grades teachers worked to design activities that stimulated children's development of spatial reasoning and their ability to conjecture. The authors trace the growth and development of children's conceptions of area and its measure as teachers structure instruction, based on children's informal knowledge of space and reallotment, to allow children to experience the conflict between appearance and reality. In closing, Lehrer, Jacobson, et al. note the key roles that invented notations, classroom conversation, and teachers' practices and understanding of children's thinking play in the development of conceptual understanding.

Many of the contributors to this section explore ways to use students' informal knowledge as scaffolding for more abstract concepts like structure, stability, measurement. Clements, Battista, and Sarama report on a curriculum development project that focused on spatial reasoning, in particular children's conceptual development of measurement. They propose that children's conceptions of space are based in physical motions like walking, measuring by hand, and so on. They trace the abstraction of length and turn and their measure from situations involving paths (especially the paths made with Logo), noting that this action-based perspective affords ready integration of space and number.

Battista and Clements, in chapter 9, investigate students' strategies for spatial structuring and their exploration of volume measure. They find that children's reasoning about volume evolves through cognitive restructuring of the shape of space. In this instance, children learn to coordinate and integrate two- and three-dimensional cube representations, working with cube arrays and developing models of cubes as composite space-filling "layers."

Middleton and Corbett encourage children to explore not only traditional, basic geometric properties like linearity and congruence, but also the more abstract notions of stability (resistance to deformation) and force. Building on children's informal notions of structure and stability, they develop engineering contexts that facilitate children's understanding and abstraction of relationships between geometry and structure.

Working from the perspective that area and volume measure are essential for the investigation of nature, Raghavan, Sartoris, and Glaser explore the ways that students "mathematize" their explorations of contexts involving immersible objects that students float and sink. Reporting on the MARS curriculum project, which uses computer modeling to assist integration of mathematical and scientific concepts, they note that students need extensive hands-on activities and call for more attention to the interplay between mathematics and science.

Dennis and Confrey also investigate the interactive roles of tools and representation. They report a case study of the "methods, voice, and epistemology" of a high-school student as he investigates analytical geometry with curve-drawing devices that reincarnate forms of thought that have been abandoned for several centuries. The student's investigations constitute an extended argument for physical exploration of geometric concepts, supported by attempts to representationally and symbolically redescribe what his hands produce with the curve-drawing tools.

One issue explored in several chapters in this volume is the development of argumentation. Koedinger proposes a cognitive model, reflecting that conjecture and argument serve not only discovery, but also problem solving and recall. In his study of secondary-school students, he finds that student arguments are often limited by excessive reliance on overly specific cases, perhaps because student conceptions of form are often rooted in visual prototypes, as noted in the Pegg and Davey and Lehrer et al. investigations. Koedinger also explores the consequences of different forms of teaching assistance and the role of tools for the arguments students develop and strongly suggests, as does Gravemeijer in Part I (chap. 2), that students be guided through their reinvention of concepts.

Collectively, the research reported in this section of the volume takes a wide-angle view of instructional environments and classroom cultures that attempt to promote the development of students' understanding. Again we underline the emphasis in these studies on the importance of building (and valuing) students' informal knowledge and on encouraging students to physically explore geometry in contexts that support reflection, conflict, and generalization.

REFERENCE

Biggs, J., & Collis, K. (1991). Multimodal learning and the quality of intelligent behavior. In H. Rowe (Ed.), *Intelligence, reconceptualization and measurement* (pp. 57–76). Hillsdale, NJ: Lawrence Erlbaum Associates.

5

Interpreting Student Understanding in Geometry: A Synthesis of Two Models

John Pegg
University of New England

Geoff Davey
Christian Heritage College

There is increasing evidence that many students in the middle years of schooling have severe misconceptions concerning a number of important geometric ideas (see, e.g., Burger & Shaughnessy, 1986; Dickson, Brown, & Gibson, 1984). There are many possible reasons for this. A clear divergence of opinion exists in the mathematics community about the methods and outcomes of geometry, and, as a result, textbook writers and makers of syllabuses have failed to agree on a clear set of objectives. Anecdotal evidence suggests many teachers do not consider geometry and spatial relations to be important topics, which gives rise to feelings that geometry lacks firm direction and purpose.

To some extent these problems may be due to the relatively small quantity of research (as compared with, say, research in number) that has been undertaken into students' thinking in geometry at the school level, which, in turn, may stem from a perceived absence of a theoretical framework. Even though Piaget and his coworkers published two significant works relating to this area, *The Child's Conception of Space* (Piaget & Inhelder, 1956) and *The Child's Conception of Geometry* (Piaget, Inhelder, & Szeminska, 1960) and these have been followed by various studies in the field of spatial cognition, little impact on classroom practice has resulted. Part of the problem lies with Piaget's "topological primacy theory," on which it has proven difficult to build a school syllabus and about which there have been some fundamental doubts (see Darke, 1982).

It was a combination of these problems and their own classroom experiences in the Netherlands in the 1950s that caused husband-and-wife team van Hiele and van Hiele-Geldof to put forward a theoretical perspective for the teaching and learning of geometry. This theory is referred to universally as the van Hiele theory, and the reader is referred to *Structure and Insight: A Theory of Mathematics Education* (van Heile, 1986) for a detailed account of this work.

In summary, the van Hiele theory is directed at improving teaching by organizing instruction to take into account students' thinking, which is described by a hierarchical series of levels. According to the theory, if students' levels of thinking are addressed in the teaching process, students have ownership of the encountered material and the development of *insight* (the ability to act *adequately* with *intention* in a *new situation*) is enhanced. For the van Hieles, the main purpose of instruction was the development of such insight.

Reaction to observed inadequacies in Piaget's formulations also inspired Biggs and Collis to explore ways of describing students' understanding more deeply than was offered by current quantitative and qualitative methods. Their work focused attention on students' responses rather than on their level of thinking or stage of development. This focus arose, in part, because of the substantial décalage problem associated with Piaget's work when applied to the school learning context, and the need to describe the consistency observed in the structure of responses from large numbers of students across a variety of learning environments in a number of subject and topic areas. Their research resulted in the development of a categorization system referred to as the Structure of the Observed Learning Outcome (SOLO; Biggs & Collis, 1982). Although the SOLO taxonomy has its roots in Piaget's epistemological tradition, it is based strongly on information-processing theories and the importance of working memory capacity. In addition, students' familiarity with content and context plays an influential role in determining the response category.

Although at first glance there may appear to be irreconcilable differences between the two theoretical stances (namely, van Hiele was concerned with underlying thinking skills and SOLO with observable behaviors), a closer examination reveals that the two stances have much in common and that the models are complementary. A synthesis provides a fresh perspective in considering student growth in understanding. The last part of this chapter takes up this theme. First, we give a brief but inclusive review of both models. Second, we consider related research to help develop these ideas further by enabling a comparison of common features. Finally, implications of the findings are explicated and discussed.

THE VAN HIELE THEORY

Since the early 1980s, the van Hiele theory has been under extensive investigation in the western world (for a detailed summary see, e.g., Clements & Battista, 1992; Hoffer, 1981). Although there are two fundamental aspects to the theory, namely, levels of thinking (a hierarchical series of categories that describe growth in student thinking) and five teaching phases (which help guide activities that lead students from one level to the next), the main focus of the research has been on the nature or existence of the levels and the assumptions that underpin them. The initial conception envisaged five levels (see van Hiele, 1986), described here as they relate to two-dimensional figures:

Level 1. Students recognize a figure by its appearance (i.e., its form or shape). Properties of a figure play no explicit role in its identification.
Level 2. Students identify a figure by its properties, which are seen as independent of one another.
Level 3. Students no longer see properties of figures as independent. They recognize that a property precedes or follows from other properties. Students also understand relationships between different figures.
Level 4. Students understand the place of deduction. They use the concept of necessary and sufficient conditions and can develop proofs rather than learn them by rote. They can devise definitions.
Level 5. Students can make comparisons of various deductive systems and explore different geometries based on various systems of postulates.

Although these descriptions are content specific, van Hiele's levels are actually stages of cognitive development: "the levels are situated not in the subject matter but in the thinking of man" (van Hiele, 1986, p. 41). Progression from one level to the next is not the result of maturation or natural development. It is the nature and quality of the experience in the teaching/learning program that influences a genuine advancement from a lower to a higher level, as opposed to the learning of routines as a substitute for understanding.

It is this focus on teaching that pervades the ideas inherent in the van Hieles' writings—so much so that the "theory" is perhaps better described as pedagogical rather than psychological, as many (or most) of the problems identified in students' learning have their basis in teaching practices rather than in the cognitive processes that may underlie performance.

Growth Through the Levels

Central to these learning experiences is that the nature of higher levels is found in the analysis of the elements at the lower levels. To explore growth through the levels, van Hiele used the terms *symbol character* and *signal character* to describe the thinking processes that occur in the "periods" between the levels.

In Period 1 (between Levels 1 and 2), the object of learning is that two-dimensional figures become bearers of properties. Symbol characters first emerge when students at Level 1 are able to identify shapes by name.

As the properties are identified, they act as a unit in representing the figure. Initially, the symbol has a strong visual content (i.e., the figure is needed to prompt the properties), but over time this is gradually replaced by a verbal content, in which properties can be stated (not by rote learning) in the absence of a figure. When this occurs, the symbols (figures) are said to take on a *signal character*, and figures can be identified by combinations of properties (Level 2 thinking).

During Period 2 (between Levels 2 and 3), the ordering of properties is the object of learning, and relationships between figures need to take on first a symbol character and later a signal character. Symbol characters are developed by discussing relationships between pairs of figures (e.g., whether squares are rectangles). The signal character of such a relationship might be that a square is a rectangle with a pair of adjacent sides equal. As symbol and signal characters emerge for numbers of relations, students begin to be able to see an *order* in the properties, and certain properties (or combinations of properties) can be implied from other properties. Other relationships, such as congruence, similarity, and parallelism, also take on symbol, then signal, characters. At Level 3 thinking, it is the ability to discuss reasons for various class inclusions that is significant, not the verbalization of statements such as "Squares are rectangles," which can be learned too easily by memorization. The futility of true/false questions in testing at this level is highlighted if we consider the question "Are parallelograms trapezoids?" Clearly a "yes" or "no" response tells nothing of the thinking processes used.

Van Hiele wrote little of Periods 3 and 4 or of Levels 4 and 5 because he believed very few students ever passed (or needed to pass) Level 3. Nevertheless, he did provide some suggestions into what that growth in thinking would be like. The object of study of Period 3 is the ordering of theorems. Initially, students notice that, in many cases, the notion of implication that arises from certain aspects "following from" other aspects extends only in one direction. As students explore this, they begin to develop multiple definitions of figures, without reliance on rote learning, and begin to use deduction to investigate and develop their understandings of geometric figures and concepts. Students begin to apply their knowledge of congruence and similarity in nonprompted situations. As van Hiele (1986) noted, however, "It takes nearly two years of continual education to have the pupils experience the intrinsic *value* of deduction, and still more time is necessary to understand the intrinsic meaning of this concept" (p. 64). Level 4 thinking represents an upper ceiling on what legitimately can be expected from better students in secondary school.

Although the levels may be associated with certain age groups (i.e., many students pass through similar cultural/learning experiences at similar ages), van Hiele offered no developmental timetable for growth through the levels. In particular, he questioned the notions of growth being linked with biological maturation. Instead, in ways that have much in common with Vygotsky (1978), he saw development in terms of students' confrontation with the cultural environment, their own exploration, and their reactions to a guided learning process.

Discontinuity Between Levels

Discontinuity was specifically identified by van Hiele (1986) as the "most distinctive property of the levels of thinking" (p. 49), implying that there is no gradual change, but a sudden leap from one level to the next, an assumption common to cognitive models using levels. Burger and Shaughnessy (1986), however, stated that they failed to detect this and found instead that "The levels appear dynamic rather than static and of a more continuous nature than their discrete descriptions would lead one to believe. Students may move back and forth between levels quite a few times while they are in transition from one level to the next" (p. 45). They found that the levels were complex structures involving the development of both concepts and reasoning processes.

Two other studies also called into question the assertion that the levels are discrete. Fuys, Geddes, and Tischer (1988) reported finding a significantly sized group of students in both Years 6 and 9 who made some progress toward Level 2 with familiar shapes, such as squares and rectangles, but who encountered difficulties with unfamiliar figures, for example, parallelograms and trapezoids: "Progress was marked by frequent instability and oscillation between levels" (p. 120). Gutiérrez, Jaime, and Fortuny (1991) also concluded that the levels were not discrete and that "people do not behave in a single, linear manner, which the assignment of one single level would lead us to expect" (p. 250). They identified students who could be coded 100%, 85%, < 40%, and < 15% for Levels 1, 2, 3, and 4, respectively, the implication being that students develop more than one level at the same time, although the thinking at the lower level is more complete than that at the upper level.

This conflict can in part be explained in terms of the closeness of the "microscope" to the analysis. When pulled back and taken in broad context, van Hiele's (1986) view of the discreteness of levels makes "logical" sense in that it represents an "ideal" theory. Students who are at, say, Level 3, can speak about relationships between figures and how properties are ordered, are able to describe these features, and can act with them in ways that are not understood by a student at, say, Level 2 (where properties are seen to be independent and relationships have not been explicitly formed). If the "microscope" is moved in more closely, a blurring occurs, and the starkness of the discrete nature is no longer apparent. There are a least four reasons to support this:

1. The growth in the periods between the levels is continuous.
2. The objects of an earlier level are not subsumed completely by a higher level.
3. Students do not always have a fully developed set of objects at a given level before they move to the next.
4. Level reduction, when provided by the teacher, can compound the issue (i.e., objects of study at a higher level are presented in a form appropriate to a lower level, and students learn to operate with these objects

in very specific, or taught, contexts but are unable to apply their knowledge broadly or understand the true nature of the objects).

When the preceding points are taken into account, van Hiele's (1986) broad statements are not as black and white as they are often portrayed. Discussions about the discontinuity of levels cannot, for example, be based solely in a single dimension. Both other aspects (e.g., periods, phases, level reduction) and the context of van Hiele's research (i.e., empirical observations made in classrooms in which the teaching plan was developed in accordance with the phases) must be taken into account. When seen from this perspective, the richness and diversity in van Hiele's observations become more apparent.

We also need to note that van Hiele (1986), in similar vein to Piaget, was interested in general features, and his model reflects this. Van Hiele offered a "universal," "ideal" path of learning/teaching painted in broad strokes. What researchers such as Burger and Shaughnessy (1986), Fuys et al. (1988), and Gutiérrez et al. (1991) attempted to do was to approach the theory at a more local level: incorporating the notion of individual differences into the van Hiele model. (We would argue that this is only possible once a broad framework has been established.) There is clearly great variability in the ways students learn, the structures on which students build their understanding, the roles teachers play, and the techniques teachers use. What is needed is a closer examination of student understanding with an eye to the seeking and documenting of diversity.

Numbering and Numbers of Levels

Another criticism of van Hiele (1986) theory concerns the numbering and number of levels. Van Hiele made two major important modifications, the second having very serious consequences.

The first modification concerns the numbering of the levels. Initially, van Hiele referred to Level 1 as the Basic Level. Level 2 was then Level 1, and so on. This change is not an exercise in semantics but came about because of "our not having seen the importance of the visual level" (p. 41). This change also has a historical base: The van Hieles worked with secondary students in the Netherlands, and geometry did not commence until the secondary school (i.e., with 12-year-olds). If van Hiele's ideas were to be extended to younger students, however, there was a need for a possible Level 0 and Period 1 (and possibly Period 0). Support for this can be found in studies in the United States (e.g., Mayberry, 1981; Usiskin, 1982) where students 12 years of age and older can be described as not having attained Level 1. This change in numbering of levels from 1 to 5 (adopted in this chapter) is consistent with the general research direction being undertaken by a number of investigators.

The second modification created a problem for the model. Van Hiele (1986) flagged an alternative set of levels that he believed was "suitable to a structure of mathematics and [that] perhaps mathematicians will be able to work with" (p. 53). In this formulation, three (alternative) levels were relevant to students: (a) *visual level*, at which students' decisions are guided by a visual network; (b) *descriptive level*, at which students can describe elements and relations; and (c) *theoretical level*, at which deductive coherence is prominent and at which geometry generated according to Euclid is considered. Although van Hiele went on to state that he chose not to use these alternative levels but to "characterize levels in a different manner" (p. 53), terms such as visual, descriptive, and theoretical levels are found throughout his writings and have been picked up by other workers in the field.

Of most interest is the question, how do the alternative levels articulate with those we have described earlier? Unfortunately, the question has generated not one unique answer, but a difference of opinion. Some writers (e.g., Olive, 1991; Fuys, 1988, cited in Clements & Battista, 1992) have chosen to link the visual level with Level 1, the descriptive level with Level 2, and the theoretical level with Levels 3 and 4. However, Clements and Battista (1992) stated that the new visual level has aspects of both the old Levels 1 and 2 and that "the mapping of one model to the other is not unambiguous" (p. 431). Pegg (1992b), working from the descriptors provided in van Hiele (1986), linked the visual level with Level 1, the descriptive level with Levels 2 and 3, and the theoretical level with Level 4.

Such controversy is disconcerting and does muddy the waters. Van Hiele (1986) justified his changes as part of an evolutionary move in which he saw his views as a theoretical model that grew as more information and newer ideas became available. In addition, he argued that the number of levels was unimportant: He "was never much interested in the questions of how many levels can be identified in a certain topic because it is possible to improve teaching without answering this question" (p. vii). Also he saw the fifth level and perhaps higher levels as only satisfying "one's theoretical lusts" (p. 47) and offering little to improve teaching and learning at secondary school.

Nevertheless, the uncertain nature of the three alternative levels and their articulation with van Hiele's previous levels create an unwanted and unnecessary distraction—and a problem with no real resolution within van Hiele's writings. It appears that the levels themselves had little value to him outside the cues they gave to appropriate teaching/learning situations, and he feared that teaching could be too easily directed to specific goals by rote learning or level reduction, with the focus of instruction directed too closely to the product at the expense of the process.

Summary

In overview, it is important to stress that the van Hiele model represents "a psychology of learning" (van Hiele, 1986), the underlying purpose of which is to see the role of instruction as the development of insight in students. The model represents a broad unidimensional approach to learning and does not take into account how individuals may proceed, other than by their rate of progress. Van Hiele did not endeavor, as have later investigators, to describe the sophisticated and varying intellectual competencies that students exhibit in their development. As a result, the van Hiele model is more descriptive than prescriptive. The structural invariants identified are an important (even necessary) first step in understanding cognitive growth, but new tools or a different framework is needed to move the model to its next phase. SOLO taxonomy appears to offer that potential.

THE SOLO TAXONOMY

The General Model

SOLO taxonomy was first described as a general model of intellectual development in Biggs and Collis (1982), and later modifications can be found in numerous papers (e.g., Biggs & Collis, 1991; Pegg, 1992a). It has much in common with the broad raft of neo-Piagetian formulations of writers such as Case (1992) and Fischer and Knight (1990). The theory resulted from an analysis of students' responses to questions posed in a variety of subject/topic areas, including number and operations, history, geography, poetry, and so on. However, it is interesting to note that Biggs and Collis never attempted to apply the taxonomy to geometry or spatial thinking.

SOLO taxonomy postulates that all learning occurs in one of five *modes of functioning*. These correspond reasonably closely to Piaget's stages of development. A major modification to the original is in placing Piaget's *early formal* stage into the earlier group of stages covered by *concrete operations*. This is consistent with Collis's (1975) earlier work, in which he claimed that most children between 13 and 15 years of age are "concrete generalizers," not "formal thinkers." By this term he implied that students in this age range are still tied to their own concrete experience: A few specific instances satisfy them of the reliability of a rule.

However, this apparent isomorphism between SOLO modes and Piaget's stages does not extend to Piaget's structural tradition. The SOLO classification does not imply that the way students perform in different situations is typical of their stage of cognitive development, nor that this is necessarily related to their age. In particular, student growth in understanding is not seen in terms of stages related to some overall logical structures that exist within the mind. Instead, Biggs and Collis viewed understanding as a more individual characteristic that is both content and context specific: The amount of information that can be retained, and features specific to the task, are important determining variables. Coding a response is dependent on its nature or abstractness (referred to as the mode of thinking) and on the individual's ability to handle, with increased sophistication, relevant cues (referred to as the level of response).

The five modes of thinking in the SOLO model (with an indication of the age at which they generally begin to appear) are the following:

1. *Sensorimotor* (soon after birth). The individual reacts to the physical environment. For the very young child, it is the mode in which motor skills are acquired.
2. *Ikonic* (from 2 years).The individual internalizes actions in the form of images. It is in this mode that the young child develops words and

images that can stand for objects and events. This mode of functioning later assists in the appreciation of art and music and leads to a form of knowledge referred to as intuitive.

3. *Concrete symbolic* (from 6 to 7 years). The individual is capable of using or learning to use a symbol system such as written language and number systems, which have an empirical referent. This is the most common mode addressed in learning in the upper primary and secondary level.

4. *Formal* (from 15 or 16 years). The individual can consider more abstract concepts and work in terms of "principles" and "theories." The individual is no longer restricted to a concrete referent.

5. *Postformal* (possibly at around 22 years). The individual is able to question or challenge the fundamental structure of theories or disciplines.

It is important to note there is no implication here that a student who is able to respond in the concrete symbolic mode in one context is able or would wish to respond in the same mode in other contexts. Nevertheless, the list implies that most students in elementary and secondary school are capable of operating within the concrete symbolic mode (Collis, 1975), usually the target mode for instruction in primary and secondary school, and teaching techniques are generally adapted to suit learners at this level. Some students may, however, still respond to stimuli in the ikonic mode, whereas others may respond with formal reasoning in some topics. Paramount is that each mode has its own identity, its own specific idiosyncratic character, and the potential for continuing development over the life of an individual.

The second characteristic of the model is a series of levels that measures an increasing sophistication in handling certain tasks within a particular mode. The five levels of response applicable to each mode of functioning are the following:

1. Prestructural (below the target mode). The learner is frequently distracted or misled by irrelevant aspects of the situation and does not engage the task in the mode involved.

2. Unistructural (U). The student focuses on the domain/problem, but uses only one piece of relevant data.

3. Multistructural (M). The student uses two or more pieces of data without perceiving any relationships between them. No integration occurs.

4. Relational (R). The students can now use all data available, with each piece woven into an overall mosaic of relationships. The whole has become a coherent structure with no inconsistency within the known system.

5. Extended abstract. The student goes beyond the data into a new mode of reasoning and can generalize from new and abstract features.

Instructional practice is also influenced by the types of responses that are indicative of moving from one level to the next within a mode.

This can be explored by considering the unistructural–multistructural–relational–extended abstract levels associated with the concrete symbolic mode.

The period between the unistructural to multistructural levels of response is one that shows a distinct increase in complexity and during which students begin to identify a number of relevant data rather than only one. Students are able to perform step-by-step algorithms and to follow routines. (Students' ability to provide a multistructural response also implies some form of organizational ability that identifies and orders relevant data.) The next period, between multistructural and relational levels, is more demanding than the previous and requires not only the finding of facts but the development of an overview of those facts: Students must be able to integrate the elements at the multistructural level into a coherent system. The third period, between relational and extended abstract, is the most demanding and most difficult to achieve. The individual must move outside of the known content by challenging or supplementing the generalizations established at the relational level.

The SOLO Model

Although the levels of increasing sophistication are useful if a single target mode is the subject of consideration, it is often more valuable to consider the model in which all modes and levels are in evidence. Relationships between the levels of response and the modes of functioning are shown in Fig. 5.1. It should be noted that the extended abstract response of one mode becomes a level (possibly the unistructural level) of the next mode, and the prestructural level can sometimes be interpreted as a level (possibly the relational level) of an earlier mode. For simplicity both the prestructural level and extended abstract level have been omitted. Figure 5.1 also highlights several important features of the model. Perhaps the most significant aspect is the cyclical nature of learning identified (as indicated by the levels), which recurs in each mode.

Three features stand out from Fig. 5.1, and these are indicated by the different paths illustrated: optimal development, multimodal functioning, and unimodal functioning.

Optimal Development. This path represents optimal cognitive development from soon after birth, through schooling and tertiary study. It has a unidimensional character of development in which earlier modes are subsumed. Most of the research into learning hierarchies is focused on this area. The work of Piaget and several neo-Piagetians (including van Hiele) could be described by this single pathway. The organization of most education systems also mirrors this course of development. Collis and Biggs (1983) have also used the path of optimal development to explain the role of various institutions concerned with educating society. For example, the primary or elementary school's role is to focus on unistructural and multistructural aspects within the concrete symbolic mode. Secondary education (up to say 17- to 18-year-olds) has the responsibility of encouraging

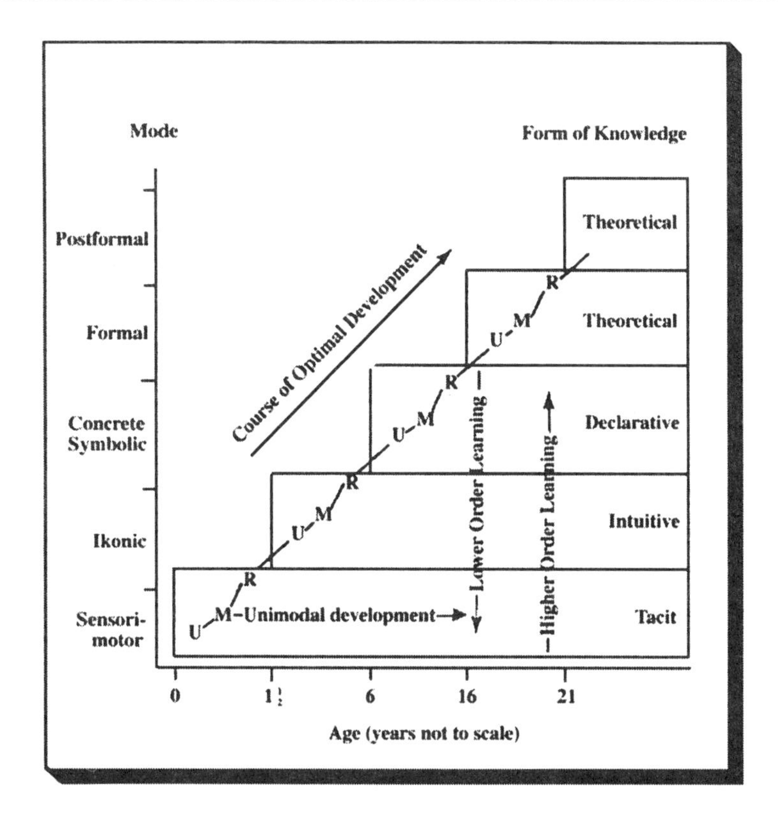

FIG. 5.1 The SOLO model: modes, learning cycles, and forms of knowledge. From Biggs and Collis (1991). Adapted by permission.

performance at relational level (concrete symbolic mode) and unistructural level (formal mode). University education is concerned with the unistructural, multistructural, and relational levels within the formal mode, with the postformal mode being appropriate for research degrees.

Multimodal Functioning. This feature is in direct contrast to that proposed by Piaget and those who subscribe solely to single-path development. In the multimodal perspective, a mode of functioning does not subsume or replace an earlier mode. Instead, development in earlier modes continues to support the development of later modes, and growth in later modes is often linked with actions or thinking associated with earlier ones. There have been some initial attempts to describe students' responses in terms of multimodal functioning (see, e.g., Collis & Romberg, 1991; Watson & Collis, 1994). These attempts have been based on a problem-solving path identified in Collis and Romberg (1991) within the ikonic and concrete symbolic modes. The mapping allows consideration of students' formalized representations and processes as well as the use of images, intuitions, beliefs, and "workplace" mathematics.

Multimodal functioning has important implications for instructional practice. Two examples illustrate these features. An athlete wishing to improve performance (the sensorimotor level of response) could practice the necessary skills and participate in the activity (sensorimotor mode) but derive benefit from (a) watching elite performances in action, which assists in building mental images (ikonic mode); (b) reading about ways to improve related aspects of the skill in order to build a better understanding of the nature of what is required (concrete symbolic mode); and (c) analyzing performance problems or developing general principles about performance or competition (formal mode). The target mode continues to be the sensorimotor mode because it is the actual performance that is the focus of attention, but the other modes make important contributions. They may also, however, create problems in the short term. If the response requires quick actions, then "images," "understandings," or "theory" will need to become "automatic." This process takes time and can account for identified slumps in performance as athletes build skills or incorporate new ones.

The second example is the reverse of the one just given: The target mode is supported by learning in earlier acquired modes. Numerous examples exist in mathematics. The use of MAB or Dienes blocks to support the development of rules involving the four operations is one familiar case involving the ikonic mode to support concrete symbolic understanding. Having students act out curve sketching on axes drawn in the playground (Lovitt & Clarke, 1988) makes use of both the sensorimotor and ikonic modes. Using concrete materials to assist in the early development of algebraic ideas is another. More generally, Bruner's (1964) three modes of representation—enactive, ikonic, and symbolic (which can be seen as a learning cycle)—can be described by multimodal functioning.

However, as with the first example, this approach, when translated into an instructional or teaching process, is not without problems. The main focus of instruction must be the target mode. Too great a reliance on supportive aspects of other modes can result in students carrying too much "baggage." For example, when students are asked to combine like terms in algebra, it is a disadvantage, in the long haul, for them to need to recast the thought process into concrete terms. Similarly, if there is too much emphasis on the supporting modes, two quite independent structures can be created, one based on the supporting mode and one on the target mode. In such a case the benefits of understanding in earlier modes are lost.

Unimodal Functioning. Most of the investigations undertaken in SOLO taxonomy have concerned unimodal functioning, and the majority of the effort in this work has been with regard to the concrete symbolic mode, with two main focuses of attention. The first focus was concerned with accumulating empirical evidence concerning the specific nature of the levels in a variety of subject and topic areas. Biggs and Collis (1982) provided particulars in topics in history, mathematics, English, geography, and modern languages. Other investigations have broadened the approach.

The second focus has been to explore the nature of student responses within a mode over an extensive range of questions within a topic. The finding of several investigations (see, e.g., Campbell, Watson, & Collis, 1992; Levins & Pegg, 1993; Pegg, 1992a) have identified at least two unistructural (U)–multistructural (M)–relational (R) cycles within the concrete symbolic mode.

Figure 5.2 represents an intramodal development pattern in which at least two cycles of growth are identified. The most noticeable characteristic is that the relational response (R_1) in the first cycle becomes the unistructural element (U_2) in the second cycle. This evolution of the SOLO model allows the developmental sequence for a concept to be interpreted across a broad range of activities. Although current research has identified only two cycles, there are probably a number of cycles within a given mode. The determining factor might be the narrowness of the question's focus and the experience of the student. It seems highly likely that further cycles will be identified. If this happens, then responses beyond R_2 can either be coded in the next mode (formal), if they involve a qualitatively different way of thinking, or at a new unistructural level (U_3) within the existing mode. This new unistructural level (U_3) in the concrete symbolic mode would incorporate the elements of R_2 into a unit, but the response would still contain declarative knowledge.

Of value is the question, how do the two-cycles-per-mode categorizations articulate with the original one-cycle-per-mode formulation? This can only be answered in the context of the research that identified the two cycles: In these cases, the first cycle (U_1, M_1, R_1) is a building block for U_2 or, alternatively, it offers a way of explaining some of the variability within U_2 responses. Hence, within this context, a unistructural response for a one-cycle model (Biggs & Collis, 1982, 1991) incorporates the result of a U_1, M_1, R_1 development in U_2 of the two-cycle model. The original multistructural level for a one-cycle model would be seen as M_2, whereas the original relational level would incorporate R_2.

Overall, SOLO taxonomy offers enormous potential as an evaluation tool and as a model to explore and explain student growth. Although there are differences of focus between the SOLO model and the van Hiele theory, there is much common ground. This similarity has already been noted in the literature and is discussed in the next section.

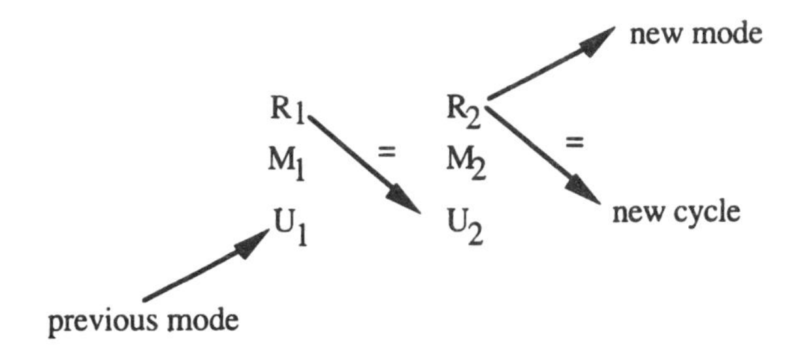

FIG. 5.2 Intramodel development pattern: two cycles within concrete symbolic pattern.

Attempts at Comparison: van Hiele Theory and SOLO Taxonomy

There have been at least three attempts to consider the van Hiele theory in light of SOLO taxonomy, namely, (a) Olive (1991), who considered student experiences in a Logo environment; (b) Jurdak (1989), who reinterpreted the findings of a study by Burger and Shaughnessy (1986); and (c) Pegg and Davey (1989), who initially considered student descriptions of two-dimensional figures, but later further developed the notion into other aspects of geometry.

Olive (1991) did not attempt to synthesize the two frameworks in terms of levels. Part of the reason for this may have been his adoption of van Hiele's visual, descriptive, and theoretical framework and his linking of these alternative levels with the earlier Level 1, Level 2, and Levels 3 and 4, respectively. Of significance, however, is his use of SOLO taxonomy to help explain and explore student responses in the van Hiele teaching/learning phases. His overview of the possibilities of level response provided a deep insight into the dynamic nature of the phases, which was not particularly clear in van Hiele's writings. Olive's work also takes on more significance when the concept of a number of unistructural-multistructural-relational cycles of growth within a mode is considered.

Jurdak (1989) classified examples of student responses collected by Burger and Shaughnessy (1986) in terms of the SOLO taxonomy and then compared them with the van Hiele classifications. He found a one-to-one correspondence between both models (i.e., Level 1 → unistructural, Level 2 → multistructural, Level 3 → relational, and Level 4 → extended abstract). There are, however, problems with this work. In particular, the responses coded unistructural and multistructural should share common elements (i.e., the unistructural response should be able to be seen within a multistructural response). In the Jurdak comparison, there is no common element between the two levels: The unistructural response has a visual component, whereas the multistructural response is based in using a number of properties.

The third study, Pegg and Davey (1989), investigated the levels of geometric understanding of 274 children in two Australian states. The results were in some ways similar to Jurdak (1989) except for two important aspects. First, Level 1 (van Hiele) was seen to share common notions with thinking in the ikonic mode (SOLO). Responses were in terms of a visual prototype, (it's like a . . .) or referred to features, such as flat or pointy. Second, there did not appear to be a response category similar to the unistructural level (concrete symbolic mode) in the work of van Hiele. However, Pegg and Davey found that the majority of children between Years 3 and 7 (i.e., 7- to 12-year-olds) appear to be somewhere midway between Levels 1 and 2 (van Hiele). Typical examples of responses of students in answer to the question, "How would you describe this shape to a friend so that (s)he could easily identify it?" focused on a single property, usually sides, but did not provide references to parallel lines, angles, symmetry, or diagonals, even though the older students were familiar with these concepts when the ideas were tested individually. Such responses are described naturally by the unistructural level in the concrete symbolic mode.

TABLE 5.1 COMPARISON BETWEEN VAN HIELE THEORY AND THE SOLO TAXONOMY

van Hiele Theory	*SOLO Taxonomy*	
	Mode	*Level (Single Cycle per Mode)*
Level 1	Ikonic	Relational
Level 2	Concrete symbolic	Unistructural/multistructural
Level 3	Concrete symbolic	Relational
Level 4	Formal	Unistructural/multistructural
Level 5	Postformal	Unistructural/multistructural/relational

As a result of this work Pegg and Davey were able to hypothesize the comparison of the two models (see Table 5.1), which is used in this chapter as the basis of a synthesis of the van Hiele theory with SOLO taxonomy.

A SYNTHESIS

The first half of this chapter spelled out the major ideas associated with the two models. Superficially, the models are different; in substance, however, they share many common traits and are mutually supportive. Thinking skills can only be traced by the individual's response to stimuli, but responses might not often represent accurately the degree of thinking involved. Such aspects as rote learning, motivation, and familiarity with tasks need to be accounted for before a one-to-one correspondence can be drawn. Alternatively, the van Hiele model may be more appropriately viewed as a theoretical construct providing some overall global view of the thinking process in geometry, whereas the SOLO model may better describe the variances in individual behavior.

This section addresses these issues by considering the two models in light of empirical evidence gathered from several studies into students' understanding in geometry undertaken both jointly and independently by the authors. For convenience, the modes and levels of SOLO are used as the organizational structure.

Ikonic Mode (SOLO): van Hiele Level 1

There is general agreement in the literature that students pass through a stage of development in which they form mental images of objects and events with which they have had contact. The SOLO term *ikonic*, borrowed from Bruner, refers to that imaging process that enables the individual to picture familiar objects even in their absence.

In the ikonic mode, SOLO taxonomy (one-cycle model) predicts the existence of three broad, distinct hierarchical levels: unistructural, multistructural, and relational. At the unistructural level, the imaging process focuses on one isolated aspect of a situation, such as the sharpness of a point. At the multistructural level, two or more unrelated aspects are recalled. Only in the last level of this mode, the relational, does the individual have full control over the imaging process and is able to classify shapes correctly and consistently.

At the relational level, accurate images of objects (geometric and other) formed in the mind enable the objects to be recognized, constructed from simpler objects, and drawn. The mental image, however, is often idealized with respect to orientation and proportion. Hence in Fig. 5.3, A may be identified as a square, but B in most cases is not. Similarly, C may be identified as an oblong, D overlooked, and F possibly rejected as a triangle.

Van Hiele, on the other hand, makes no allowance for growth before Level 1 (at which the individual can recognize, e.g., two-dimensional figures by their appearance), although other researchers have suggested a Level 0. We believe that earlier levels do exist and that they can be identified within the postulated unistructural and multistructural levels of the ikonic mode in SOLO taxonomy.

Two general characteristics of the ikonic mode of functioning are the individual's ability to communicate orally and the ability to make qualitative judgments on a perceptual basis. The individual is able to use and remember names (e.g., "It's a square") and is able to recognize differences in the sizes of objects, but is unable to explain why two figures do or do not look the same (Olson & Bialystok, 1983, pp. 203–204). Although decisions made in the ikonic mode are based on valid reasoning using intuition, the basis of that reasoning may change from one case to the next, and the individual's reasoning may appear to be inconsistent. In this mode, students associate "abstract" shapes with familiar objects (visual prototypes) and communicate their associations in expressions such as "It's like a box" (square); "like a roof" (triangle); "like a blackboard" (rectangle); "like a jewel" (rhombus); and so on.

Some students compare the shape of less familiar figures with others they have met earlier or which seem more significant to them in some way: A rectangle becomes a "stretched-out square"; a rhombus, "a squeezed square"; and a parallelogram, "a slanty rectangle" or "like a rhombus only longer." These examples also illustrate the use of informal and imprecise terms common at this level, but used regularly even at higher levels. Other similar words used include fatter, skinnier, lopsided, and corner.

If students are asked to compare or contrast two different figures, typical ikonic responses may be tautological (e.g., "They are a different shape") or may be inconsistent (e.g., "That one is a square; the other is not"), or may refer to orientation or size (e.g., "This one's fatter [bigger,

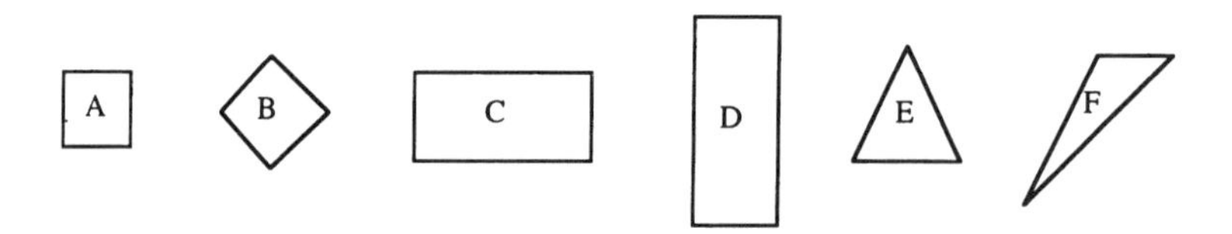

FIG. 5.3 The relational level (ikonic mode): recognizing two-dimensional shapes.

horizontal, turned over, made of plastic ...]"). Although language in the ikonic mode lacks the accuracy and precision associated with later modes, construction and drawing skills are usually reasonably well developed at the relational level. Students responding within the relational level of the ikonic mode can draw reasonably accurate figures, particularly for the square and the rectangle. Especially interesting is the students' ability to include equal sides and angles of 90° implicitly in their diagrams but their inability to express ideas in either unprompted verbal or written descriptions. In the case of parallelograms and rhombuses, drawings are less accurate. It is clear that the better drawings of the rectangle and square are due not only to the relative familiarity of these two shapes, but also to the natural human bias to the vertical axis and the right angle (Olson & Bialystok, 1983, p. 185).

Provided that students are familiar with a certain figure, they can certainly construct it on a geoboard, often by trial and error, moving the rubber band until it "looks right." Even younger children, say, 6- and 7-year-olds, are capable of solving certain pentomino and SOMA cube puzzles by simply rearranging the pieces until they fit. Indeed, adolescents and adults often use the same approach, indicating that the ikonic techniques are not abandoned in early childhood. Because skills are not due to any overt analysis of the properties of the figure, however, students in the ikonic mode fit within the definition of van Hiele Level 1.

Concrete Symbolic Mode (SOLO): van Hiele Levels 2 and 3

SOLO taxonomy predicts that students who have attained the relational level of the ikonic mode will, with appropriate experience, move to the next mode of thinking, namely, the concrete symbolic. Within this new mode, students are capable of linking concepts and operations to written symbols, providing that the entire context falls within their own personal, real-world experience. This move involves a major shift in abstraction because the symbolic systems have a logic and an order both internally and in relation to the particular context.

We have identified two cycles in the concrete symbolic mode from students' responses to geometric questions. This first cycle appears to be an important transition between the ikonic mode and formalized geometry and involves the identification of a single property, whereas the second cycle involves the acquisition of several properties and subsequent use of the relationships between them.

First Cycle—U_1, M_1, R_1 (SOLO)

The qualitative difference noticed in responses can best be categorized into three levels: (a) U_1, focusing on one prominent visual aspect; (b) M_1, focusing on one aspect and trying, unsuccessfully, to qualify it; and (c) R_1, focusing on one aspect and qualifying it successfully. This last category leads to the concrete concept of the particular property involved. These three levels can be interpreted as an early cycle in the concrete symbolic mode (see Fig. 5.1).

Students' responses at the U_1 level, when asked to describe a square or a rectangle, often use the same description: "four sides," "four straight sides," or "a flat shape with four sides."

An M_1 level response attempts, albeit unsuccessfully in mathematical terms, to qualify these figures. A rectangle may be said to have four unequal, uneven, or different sides. Other variations include "different from a square," "the bottom and top are small, but the sides are long," and so on. Many students at this level will say that a rectangle has two long sides and two short sides.

At the R_1 level, students correctly describe the relative lengths of the sides: A square or rhombus has "four equal [or even] sides" whereas a rectangle or parallelogram has "two sets of equal sides," "two of the sides are the same and the other two are the same," or "two sides one length and the other two a different length."

In addition to the responses just described, a large number of students use forms of ikonic support. They will supply visual references; use informal, everyday adjectives such as long, squashed, or slanted; or refer to the orientation of a figure (e.g., some students will note that a rhombus is merely a "tilted square," no doubt consistent with the view that a rectangle is a "stretched-out square").

Many students at this level mention "four corners" in addition to sides in relation to quadrilaterals. Although it may appear that this indicates thinking about an additional property such as angles (and hence should be classified higher), in most cases with young students this does not appear to be true. A closer examination of the concept "corner" reveals that it is not necessarily a right angle or, indeed, any angle at all. Many, if not all, young students have great difficulty in explaining what they mean by the term *corner*. Some seem to be referring to the vertex; others, to the abrupt change in direction of the eye or the finger when the figure is explored visually or by touch. Laurendeau and Pinard (1970) stated that it is the corner that enables children to differentiate a curved from a rectilinear boundary and that this is due to certain topological properties of the corner. *Corner*, therefore, is a device that offers ikonic support rather than a reference to a property (angles). Lesh and Mierkiewicz (1978) supported this finding: "the perceptual literature indicates that 'angles' are among the features that perception selects to focus on in many objects" (p. 25). They also suggested, quoting Piaget, that the "primitive notion" of corner is the precursor of the concept of line, not vice versa.

The continued use of ikonic references, even though the student has moved on to a new level of functioning, appears to be contrary to the van Hiele notion of plateau-like levels, but is consistent with Collis's suggestion that human beings are multimodal in their approach to many situations (Collis, 1988). Students themselves, although able to operate in a more abstract way, may use an earlier mode in appropriate situations where the task is set in such a mode.

Summary. In our studies, several features of this first cycle were evident. First, the style of geometrical thinking has changed from viewing a figure as a simple, complete entity: Responses appeared to shift away from an almost complete reliance on physical examples of shapes toward an ability to work with, and reason about, diagrams and sketches, which in one sense are the "symbols" of geometry.

Second, the shift did not appear to be dramatic (e.g., students may see a rhombus in ikonic terms whereas they may describe squares or rectangles in terms of a property).

Third, the majority of students' responses in this cycle in our studies focused on one idea, sides. There was little use of the concepts of angle, parallelism, diagonals, or symmetry when considering a geometric shape. Later interviews revealed, however, that (with prompting) these concepts were recognized by many students.

Fourth, the language used by the students operating in this cycle usually consisted of simple sentences involving only one relationship or idea. Such sentence structures of the type "the opposite sides are equal" connect two relationships, (a) "opposite to" and (b) "equal to," and appeared to be beyond their level of thinking. Some students used negatives in their replies in a manner that sometimes hinted at, but did not actually describe, the real properties of a figure (e.g., rectangles were said to have sides that are not equal or sides that are not the same length).

Finally, in the concrete symbolic mode, the student interprets a figure as a type of class representative and comments on the figure itself as the object of discussion. In addition to this, there is a growing awareness that the actual shape of a geometric figure depends on certain simpler notions or preconcepts and, in some cases, on the relationship of equality.

Second Cycle — U_2 (SOLO)

Once an R_1 response (in the example given earlier, a response describing the relative length of sides, i.e., focused in a general way on one property) is attained, this response becomes the new element of analysis to the student. This change can be coded as U_2, a unistructural response in the second cycle, and is often subtle and difficult to detect except by interview.

Evidence of the robust nature of U_2 can be found in the difficulty many students have in responding at the next level (M_2). When probed, students often give a transitional response, making a reference to an additional feature, but not qualifying that feature. A typical transitional response might be that a rhombus "has all sides equal . . . has two obtuse angles and two acute angles," or that a parallelogram "has four sides—two of them are equal to each other and the other two are equal. If you run the lines forever, they will never meet." Clearly, such responses fail to qualify the second property fully.

It is not clear whether van Hiele was unaware of the existence of this level or the previous cycle of levels of geometric thinking, or whether he believed them to be merely a transition between Levels 1 and 2 or a subset of Level 2. Other writers (e.g., Burger & Shaughnessy, 1986; Fuys et al., 1988) have used the term *transitional* to describe these responses. Their widespread existence and relevant nature lead us to believe that they are indicative of a significant level on the usual developmental path through which all students pass. Furthermore, we believe that this level (U_2) lasts for a considerable period of time for many students in elementary and early secondary school.

Second Cycle — M_2 (SOLO): Level 2 (van Hiele)

Significantly fewer students have been identified at M_2 level at the elementary-/secondary-school interface than at the U_2 level. Students' responses at this level are characterized by the use of at least two properties of a given figure. There does not appear to be any sudden abrupt discovery of the need to include all of the characteristics of the figure, but rather a gradual realization of a need to refer to additional properties, possibly with the intention of excluding what the student perceives to be non-examples of that particular class.

Typically, students at this level are coming to terms with the concepts of angle and parallelism. We have not been able to determine any universally preferred order in this process, and it is probably a function of classroom experience. Hence, with the rhombus, for example, many students seem to add something about the angles first, whereas with the parallelogram, the second property added often concerns parallel lines.

The language used by students (aged between 10 to 13 years) at this level often lacks the precision and accuracy that is commonly linked in the literature to van Hiele Level 2. In working with a parallelogram, for example, as with U_2 responses, few students use the structure, "The opposite sides are equal [or parallel]." Their natural response is more likely to be "sides are parallel," or "two pairs of parallel sides," and it is not uncommon for them to include redundant information (e.g., "two pairs of opposite sides"). The sentence "Opposite angles are equal" is also rare, and students are unlikely to talk with precision about angles: "Angles are different to a rectangle"; "no right angles"; "not all the angles are the same"; "two are the same and two are different."

Again, as with first cycle responses, ikonic support is often used by students to qualify or add to their descriptions of plane figures: A parallelogram "is just like a rectangle tilted over ... similar to a rhombus," or a rhombus "is just like a square turned on its side."

Three additional comments can be made about responses at this level. First, although the language of the students is improving, many still use imprecise words such as "even" and "corner." Second, most responses specifically exclude the possibility of class inclusion, for example, "The angles [of a rhombus] are different from a square." Third, older children at the multistructural level often give very long, repetitious, descriptions that add little new information.

Second Cycle — R_2 (SOLO): Level 3 (van Hiele)

Progression to the next level is marked by the understanding that geometrical properties and shapes are not separate, discrete entities, all independent of one another. Some properties are related to one another, and some shapes are related. There is a clear correspondence between the "integration of the separate features identified at the previous level" for

SOLO taxonomy and the notion of class inclusion and the ordering of properties described by van Hiele.

These concepts represent an important step in geometrical understanding and a quantum step forward for students. The value of such knowledge lies in its ability to strengthen and deepen understandings of figures and their properties. This is not, however, an easy task for students to accomplish. For students to come to terms with the idea, for example, that all squares are rhombuses, they have to overcome powerful visual cues and rely on a relatively high level of logical reasoning that runs counter to everyday-world beliefs that squares have a fundamental status and that squares, rectangles, and parallelograms are distinct classes of figures.

When students (11–13 years of age) who responded within the ikonic mode (Level 1) or within the first cycle in the concrete symbolic mode (lower Level 2) were asked to describe a square using the word *rectangle* in their description, the answers appeared to rely totally on imagery: "A square is half a rectangle"; "If you cut a square in two, you get two rectangles"; "A rectangle is wider than a square"; and so on. They noted that a diamond (rhombus) was a square that had been rotated, or that a parallelogram was a rectangle "standing on one of its points," or "a rectangle but sides are slanted."

Some students, whose responses for descriptions of figures were more detailed (i.e., they provided a M2 response), did appear to be close to understanding class inclusion, but none in our study succeeded completely (e.g., "A rectangle is a form of parallelogram because of the parallel and equal sides, but it has 90° degree angles"; "A square fits the rhombus description, but the rhombus does not have right angles"; "A square is a figure that resembles a rhombus, only the square is upright," etc).

One very capable 12-year-old student, who had not quite sorted out inclusion relationships between various quadrilaterals, made the following attempt:

> A square is a kind of rhombus ... it is not really a rhombus because a rhombus does not have 90° angles.... A rectangle is the same as a parallelogram, but it's not tilted ... you just have to push the parallelogram back ... it fits the description ... it's a special case.

Such a student could be considered to be at the interface between Level 2 and Level 3 (van Hiele) and the response between M_2 and R_2 (SOLO).

Formal Mode (SOLO): Level 4 (van Hiele)

Biggs and Collis (1982, 1991) were not as explicit about this mode as they were about the concrete symbolic, although the broad unistructural–multistructural–relational cycle was hypothesized. Van Hiele was also less explicit about this level than about the previous three. Pegg, in exploring this mode with senior secondary students (Pegg & Faithfull, 1995; Pegg & Woolley, 1994) and final-year university undergraduates (White & Pegg, 1994), identified two cycles of growth within the formal mode. The following discussion is limited to the first cycle because it is most relevant to secondary students.

First Cycle—U_1, M_1, R_1, and U_2 (SOLO)

In order to explore more closely the responses that could be considered to belong to the formal mode or Level 4 thinking, a series of deductive geometrical exercises was given to over 70 senior secondary students. The students undertook the activities in a one-to-one situation in which the interviewer adopted a form of "dynamic assessment" (Lidz, 1987) to explore their understanding. A typical question is provided in Figure 5.4. To solve the questions, students could rely on various strategies, but the two most commonly used were based on proving sets of congruent triangles and the application of necessary and sufficient conditions associated with the figures involved.

One pattern of development that emerged from the study can be summarized in terms of SOLO levels within the first cycle of the formal mode:

Unistructural 1 (U_1). Students demonstrated an ability to use the concept of congruence, but the application of congruence in a nonprompted situation, which seemed to place the students under pressure, resulted in their losing sight of the purpose of the question.

Multistructural 1 (M_1). Students did not have a clear overview of what was required to complete the question. They often lost track of the solution process, or, alternatively, they gave a sequential series of steps, but were not clear about the endpoint (of those steps) until it was reached. They seemed to have little trouble with the concept of congruency, but had difficulty keeping in mind all relevant elements of the question-and-solution process.

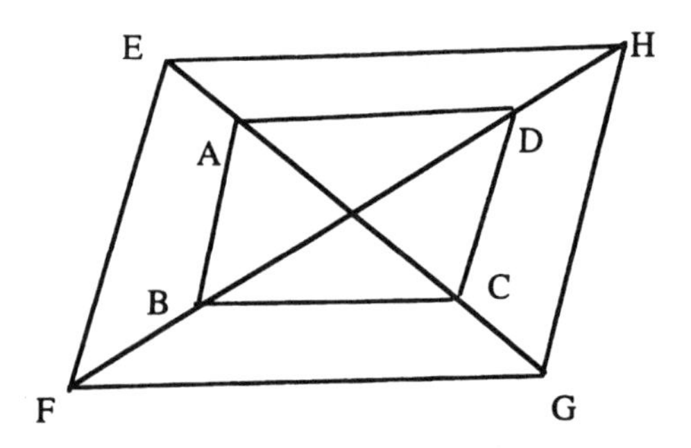

ABCD is a parallelogram,
AE = CG BF =DH.
Prove EH = FG.

FIG. 5.4 An example of a geometric exercise requiring a deductive strategy.

Relational 1 (R_1). Students in this category chose to use congruent triangles and gave no indication of alternative strategies. They were confident with congruency and applied the idea efficiently and accurately. The responses were very much within the context of the question. There was, however, no indication of an underlying general principle.

Unistructural 2 (U_2). Students had a clear overview of the question. Responses were concise, precise, and indicative that students were "comfortable" with the sufficiency of various properties (e.g., lines that bisect each other can be considered the diagonals of a parallelogram). Also impressive was the way they saw the solution in terms of a general principle outside the question posed.

These categorizations highlight in a clear way important characteristics of early formal responses in geometry. In particular, the deductive process is evident, albeit in a limited way at the U_1 level. The ability to consider what amounts to a second-order relationship (i.e., the application of congruency as a tool in a nonprompted situation) is also an important indicator of formal thinking. A U_2 response shows a degree of abstraction beyond that of simply applying congruence conditions.

A further finding of the study was that qualitative differences existed between these students' responses and the responses of those whose thinking was associated with the concrete symbolic mode. Examples of concrete symbolic responses identified follow.

Relational 2 (R_2). The students in this group were convinced that the result "has to be true." Overall, these responses were characterized by a form of natural logic although no attempt at deductive reasoning was made. Students spent much of the time making assumptions about the figures they were meant to be establishing.

Multistructural 2 and Unistructural 2 (M_2, U_2). This group of responses focused on several properties or a single property of the given figure. There was no attempt to address the question, and the answers were usually left as a series of statements about equal angles, equal sides, and parallel lines.

Further support for this classification emerges from a description provided by van Hiele (1986) of Level 4 thinking. Here he addressed the theorem "that the three bisectors of the angles of a triangle have a point in common" (p. 46), suggesting two approaches. The first involved the use of necessary and sufficient conditions and centered on the idea that a point is equidistant from two sides of a triangle if it lies on the bisector of the included angle. The second method involved the use of congruent triangles to (a) establish the equality of distance to the three sides by using the bisectors of two angles and (b) using this information to establish, again with congruent triangles, that the third angle is also bisected. He noted that

> Mathematically the second proof does not differ from the first. But the language of the second proof is much easier for a pupil than that of the

> first. In the first proof, you will have to exert yourself more in order to get to the bottom of it. In the second proof, every step can be easily traced. Perhaps the pupil will make a mistake when proving the last congruence because he is distracted by the first congruence apparently being the same. But even if everything proceeded without difficulty, he, at the end, will be astonished that the proof is now finished: He has carried out the steps without understanding their meaning, the whole remained mysterious to him, the proof is a mousetrap.
>
> A teacher who has to explain a proof of the first kind must take care not to make of it a proof of the second kind. His explanation must be focused on the points "necessary and sufficient," on the fact that in the first part of the proof the existence of a point equidistant to all three sides has been proved. (p. 46)

Van Hiele's discussion can be interpreted within the structure identified earlier. He is referring to the M_1 responses (formal mode) when the student "is distracted by the first congruence" and R_1 (formal mode) when the student has worked successfully to achieve the answer using congruence but has not seen some overall principle. On the other hand, to be able to identify and apply the appropriate necessary and sufficient condition implies the student has achieved an overview (i.e., the response can be described as at the U_2 level [formal mode]).

CONCLUSION

This chapter was written from the perspective that the van Hiele theory offers a broad framework in which to view cognitive growth in geometry. When past research into aspects of the model was directed at global or "average" features, the findings indicated substantial support for the theory. Trouble occurred when the "microscope" of analysis moved in closer to the student and took on a more "local" perspective. Here differences associated with individuals began to play an important role, and the décalage phenomenon became apparent.

SOLO taxonomy, although ostensibly very different in form, nature, and purpose from the van Hiele model, shares a great many common features. Indeed, given the necessary qualifications that must be made between the two notions of responses and thinking, the van Hiele theory can be interpreted within the structure identified in the SOLO model. In effect, although van Hiele continually focused on thinking, his judgments were made based on the responses students gave.

SOLO, however, has some clear advantages over the van Hiele model in describing students' understanding. Because of the overt focus on student responses, the SOLO taxonomy is much more context sensitive than the van Hiele theory. Also, implied within this focus is the notion that although the overarching framework is a "naturally" occurring one, the fine detail provided in specific examples of level responses may not be unchanging. The large commonality in response detail reported in studies may be more a consequence of homogeneous learning/teaching experiences than some underlying invariant features. As such, the detail associated with each level may change. The key point for future research is that, as an evaluative tool, SOLO is sufficiently robust and fine-grained to assist such explorations.

Another major strength of the SOLO model lies in its advocacy of continued development of earlier acquired modes. Instead of being replaced by later developing modes, these earlier modes continue to evolve, complement, and support development in other modes. Research is only beginning to address this aspect of the model. Of particular interest is (a) the nature of growth in the ikonic mode and (b) how the SOLO model can be used to document integrative development involving more than one mode. Information, for this last point, would offer a valuable alternative to a strict hierarchical view of learning.

SOLO also provides the means for the van Hiele model to move to a new phase of development, namely, the exploration and explanation of individuality in geometry education. The challenge for researchers and teachers is to investigate and identify individual paths of development: to seek out variability. It is only in this way, with a move away from a single-dimension learning path, that a true understanding of the nature of individual cognitive growth in geometry can be achieved. With this more proactive and relevant stance taken, students will potentially achieve a greater understanding of geometric concepts, and many more students may be challenged in their own way, as Einstein was when, at the age of 12, he was presented with a textbook on Euclidean geometry. As he recollected years later:

> Here were assertions, as for example the interaction of three altitudes of a triangle in one point, which—though by no means evident—could nevertheless be proved with such certainty that any doubt appeared to be out of the question. This lucidity and certainty made an indescribable impression on me. (Little, 1980, pp. 626–628)

ACKNOWLEDGMENT

The authors wish to acknowledge the support that they received from the Australian Research Council toward carrying out research reported in this chapter (ARC Ref. No. A79231258).

REFERENCES

Biggs, J., & Collis, K. (1982). *Evaluating the quality of learning: The SOLO taxonomy.* New York: Academic Press.

Biggs, J., & Collis, K. (1991). Multimodal learning and the quality of intelligent behavior. In H. Rowe (Ed.), *Intelligence, reconceptualization and measurement* (pp. 57–76). Hillsdale, NJ: Lawrence Erlbaum Associates.

Bruner, J. S. (1964). The course of cognitive growth. *American Psychologist, 19,* 1–15.
Burger, W. F., & Shaughnessy, J. M. (1986). Characterizing the van Hiele levels of development in geometry. *Journal for Research in Mathematics Education, 17,* 31–48.
Campbell, K., Watson, J., & Collis, K. (1992). Volume measurement and intellectual development. *Journal of Structural Learning and Intelligent Systems, 11,* 279–298.
Case, R. (1992). *The mind's staircase: Exploring the conceptual underpinnings of children's thought and knowledge.* Hillsdale, NJ: Lawrence Erlbaum Associates.
Clements, D., & Battista, M. (1992). Geometry and spatial reasoning. In D. Grouws (Ed.), *Handbook of research on mathematics teaching and learning* (pp. 420–464). New York: Macmillan.
Collis, K. (1975). *A study of concrete and formal operations in school mathematics: A Piagetian viewpoint.* Melbourne: Australian Council for Educational Research.
Collis, K. (1988). The add up or take syndrome. In J. Pegg (Ed.), *Mathematical interfaces* (pp. 66–75). Adelaide: Australian Association of Mathematics Teachers.
Collis, K., & Biggs, J. (1983). Matriculation, degree structures, and levels of student thinking. *Australian Journal of Education, 27 (2),* 151–163.
Collis, K., & Romberg, T. (1991). Assessment of mathematical performance: An analysis of open-ended test items. In M. C. Wittrock & E. L. Baker (Eds.), *Testing and cognition* (pp. 82–130). Englewood Cliffs, NJ: Prentice Hall.
Darke, I. (1982). A review of research related to the topological primacy thesis. *Educational Studies in Mathematics, 13(2),* 119–142.
Dickson, L., Brown, M., & Gibson, O. (1984). *Children learning mathematics: A teacher's guide to research.* East Sussex, UK: Holt, Rinehart and Winston.
Fischer, K. W., & Knight, C. C. (1990). Cognitive development in real children: Levels and variations. In B. Presseisen (Ed.), *Learning and thinking styles: Classroom interaction* (pp. 43–67). Washington, DC: National Education Association.
Fuys, D., Geddes, D., & Tischler, R. (1988). The van Hiele model of thinking in geometry among adolescents. *Journal for Research in Mathematics Education Monograph, 3.*
Gutiérrez, A., Jaime, A., & Fortuny, J. M. (1991). An alternative paradigm to evaluate the acquisition of the van Hiele levels. *Journal for Research in Mathematics Education, 22,* 237–251.
Hoffer, A. (1981). Geometry is more than proof. *Mathematics Teacher, 74,* 11–18.
Jurdak, M. (1989). van Hiele levels and the SOLO taxonomy. *PME Proceedings, 13,* 155–162.
Laurendeau, M., & Pinard, A. (1970). *The development of the concept of space in the child.* New York: International Universities Press.
Lesh, R., & Mierkiewicz, D. (1978). Perception, imaging and conception in geometry. In R. Lesh & D. Mierkiewicz (Eds.), *Recent research concerning the development of spatial and geometric concepts* (pp. 7–28). Columbus, OH: ERIC.
Levins, L., & Pegg, J. (1993). Students' understanding of concepts related to plant growth. *Research in Science Education, 23,* 165–173.
Lidz, C. S. (1987). *Dynamic assessment: An interactional approach to evaluating learning potential.* New York: Guildford.
Little, J. (1980). The uncertain craft of mathematics. *New Scientist,* 626–628.
Lovitt, C., & Clarke, D. (1988). Codifying and sharing the wisdom of practice. In J. Pegg (Ed.), *Mathematical interfaces* (pp. 1–21). Adelaide: Australian Association of Mathematics Teachers.
Mayberry, J. (1981). *An investigation of the van Hiele levels of geometric thought in undergraduate preservice teachers.* Unpublished doctoral dissertation, University of Georgia. (University Microfilms No. DA 8123078)
Olive, J. (1991). Logo programming and geometric understanding: An in-depth study. *Journal for Research in Mathematics Education, 22,(2),* 90–111.
Olson, D. R., & Bialystok, E. (1983). *Spatial Cognition.* Hillsdale, NJ: Lawrence Erlbaum Associates.

Pegg, J. (1992a). Assessing students' understanding at the primary and secondary level in the mathematical sciences. In J. Izard & M. Stephens (Eds.), *Reshaping assessment practice: Assessment in the mathematical sciences under challenge* (pp. 368–385). Melbourne: Australian Council of Educational Research.

Pegg, J. (1992b). Students' understanding of geometry: Theoretical perspectives. In B. Southwell, B. Perry, & K. Owens (Eds.), *Space—The first and final frontier* (pp. 18–36). Sydney: Mathematics Education Research Group of Australasia.

Pegg, J., & Davey, G. (1989). Clarifying level descriptions for children's understanding of some basic 2D geometric shapes. *Mathematics Education Research Journal, 1 (1)*,16–27.

Pegg, J., & Faithfull, M. (1995). Analyzing higher order skills in deductive geometry. In A. Baturo (Ed.), *New directions in geometry education* (pp. 100–105). Brisbane: Queensland University of Technology Press.

Pegg, J., & Woolley, S. (1994). An investigation of strategies used to solve a simple deductive exercise in geometry. In G. Bell, B. Wright, N. Leeson. & J. Geake (eds.) *Challenges in mathematics education: Constraints on construction.* (pp. 471–479). Lismore, New South Wales: Mathematics Education Research Group of Australasia, Southern Cross University.

Piaget, J., & Inhelder, B. (1956). *The child's conception of space.* London: Routledge & Kegan Paul.

Piaget, J., Inhelder, B., & Szeminska, A. (1960). *The child's conception of geometry.* New York: Basic Books.

Usiskin, Z. (1982). *Van Hiele levels and achievement in secondary school geometry (Final report of the Cognitive Development and Achievement in Secondary School Geometry. Project)* Chicago: University of Chicago, Department of Education.

van Hiele, P. M. (1986). *Structure and insight: A theory of mathematics education.* New York: Academic Press.

Vygotsky, L. S. (1978). *Mind in society.* Cambridge, MA: Harvard University Press.

Watson, J., & Collis, K. (1994). Multimodal functioning in understanding chance and data concepts. *PME Proceedings, 18*, 369–376.

White, P., & Pegg, J. (1994). A description of student responses to restrictions on assumed knowledge. In G. Bell, B. Wright, N. Leeson, & J. Geake (eds.), *Challenges in mathematics education: Constraints on construction* (pp. 691–698). Lismore, New South Wales: Mathematics Education Research Group of Australasia, Southern Cross University.

6

Longitudinal Study of Children's Reasoning About Space and Geometry

Richard Lehrer, Michael Jenkins, Helen Osana
University of Wisconsin–Madison

The development of children's ideas about qualities of space is of longstanding interest to researchers and educators alike. For the former, reasoning about space provides a window to issues of mind, such as how children represent images or process configural information (Anderson, 1983; Eilan, McCarthy, & Brewer, 1993; Kosslyn, 1980). For the latter, because effective instructional design ideally begins with children's prior knowledge (Bruer, 1994; Freudenthal, 1973), children's reasoning provides the foundation for instruction about the mathematics of space.

Although the teaching and learning of geometry in the United States is traditionally reserved for high-school curricula, recent recommendations of national educational teaching organizations suggest that geometry instruction should begin in the primary grades (e.g., Goldenberg, Cuoco, & Mark, chap. 1, this volume; National Council of Teachers of Mathematics, 1991). The pioneering efforts of Piaget (Piaget & Inhelder, 1948/1956; Piaget, Inhelder, & Szeminska, 1960) and van Hiele (1959, 1986) remain the most extensive sources of information about school-age children's initial conceptions about space and corresponding trajectories of change. Much of this work, however, was conducted several decades ago, and more contemporary efforts have not embraced the wide range of mathematical concepts and mental skills characteristic of the earlier work. Consequently, our task was to develop a contemporary and widespread portrait of children's emerging skills in reasoning about space. Our purpose was to describe the development of prototypical forms of spatial reasoning in ways that teachers would find accessible and useful for guiding instruction about geometry in the primary grades so that teachers could use students' knowledge of mathematics as the building blocks of instruction (Carpenter, Fennema, Peterson, Chiang, & Loef, 1989). In chapter 7, we report on the altered course of the development of spatial reasoning in three classrooms where teachers designed instruction in light of our findings.

We designed a 3-year longitudinal investigation to characterize trajectories of intra-individual change, a strength of the longitudinal design (Baltes, Reese, & Nesselroade, 1977). Because the development of geometry and spatial reasoning encompasses a wide range of mathematical concepts and mental skills, we examined the development of children's conceptions of (a) two- and three-dimensional Euclidean forms, including angle; (b) the measurement of length and area; and (c) related skills such as mental manipulation of images, drawing, and graphing. The design also facilitated investigation of the ways in which growth in one strand (e.g., drawing) might support or inhibit growth in another strand (e.g., form). The study involved multiple measures and occasions of measurement. Hence, we report only the highlights of findings of most interest to readers.

CONCEPTIONS ABOUT TWO- AND THREE-DIMENSIONAL SHAPE

The major research agenda was a test of the adequacy of the van Hiele model as a description of the progression of children's thinking (for excellent reviews of the van Hiele model, see Burger & Shaughnessy, 1986; Clements & Battista, 1992; Fuys, Geddes, & Tischler, 1988; Hoffer, 1983; Pegg & Davey, chap. 5, this volume). According to the van Hiele model, children's conceptions of shape evolve in an ordered, stageline sequence. Children first find salient the overall appearance of shapes (e.g., rectangles look like the fronts of cereal boxes) and then gradually come to recognize shapes as carriers of properties such as the number of sides of a figure or the measure of its angles. The model predicts that, over time, properties of shapes are coordinated and ordered, so that conventional classifications of shape emerge (van Hiele, 1959, 1986). Unlike Piaget's model of development, the van Hiele model ties development not to the growth of general mental structures, but to particular forms of instruction. Hence, the van Hiele framework suggests that in the absence of designed instruction, development may be arrested. It also predicts consistency in reasoning, so that, for instance, a child who notes that a triangle "looks like a boomerang" might also be inclined to say that a trapezoid "looks like a ramp." To provide an adequate test of the generality and differentiation of children's thinking about form, we employed a wider range of forms than in previous studies, and children were asked to compare and contrast triads of forms in order to provide sufficient context for eliciting spatial reasoning.

Method

Children were interviewed individually in a small room in their school. Interviews were videotaped and transcribed for later analysis. Each child was interviewed six times during the course of the school year (once each about form, drawing, graphing, angle, length, and area measure), and each interview was repeated during each of 3 years of "waves." Hence, each child participating was interviewed a total of 18 times over a span of 3 years.

Participants

Thirty-seven children were selected randomly from an elementary school in the Midwest: 13 in first grade (7 boys and 6 girls), 12 in second grade (7 boys and 5 girls), and 12 in third grade (6 boys and 6 girls). During the second year of the study, 4 children left the study (3 moved). Teachers selected children most like those who left and solicited their participation. During the third year of the study, 3 more children left the study (2 moved). In the end, the 3-year longitudinal sample consisted of 30 students: 10 children in the youngest grades (first to third grade), 9 in the middle grades (second to fourth grade), and 11 in the oldest grades (third to fifth grade). This design allowed comparisons between grades and enabled researchers to track patterns of change in children as they moved from grade to grade.

Description of Interview

Eliciting Conceptions About Two-Dimensional Shapes. We developed nine triads of two-dimensional shapes[1] to assess children's conceptions of planar figures; reduced-scale versions are displayed in Fig. 6.1. Each triad was printed on 8.5 × 11 in paper and displayed on a stand about 2 ft in front of the child. The triads were designed to invited multiple forms of reasoning, according to the van Hiele hierarchy. For example, a child might reason that the second and third shapes in triad 2 (chevron and triangle) are most similar because both looked like triangles or that the first and second shapes (rectangle and chevron) are most similar because both have four sides.

For each triad of figures, children determined which two were most alike and provided a verbal justification for their choice. Occasionally the interviewer probed children for further information in order to clarify a response or when it seemed that children might have more to say if given the opportunity to do so. Children's justifications for their choices were recorded. We developed a categorization system based on analysis of 10 cases and informed by previous research (Lehrer, Knight, Sancilio, & Love, 1989). This resulted in 21 categories[2] of reasoning, ranging from resemblance-based reasoning (e.g., "It looks like a ramp") to reasoning by appeal to class membership (e.g., "They are both rectangles"). Responses were categorized by two independent raters with an interrater agreement of 86%. In the second year of the study, the fourth triad was modified so that both

---

[1]Some of the shape triads were developed in collaboration with Michael Battista and Douglas Clements.

[2] Category descriptions and examples available on request or via the web version of this document.

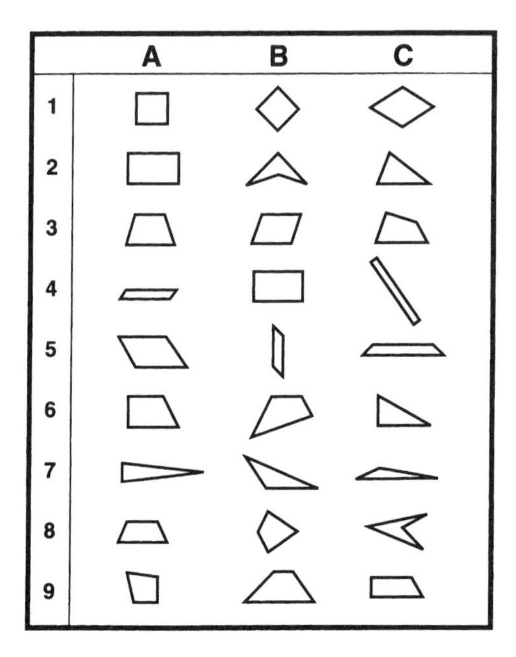

FIG. 6.1 Stimulus triads used in Year 1 for similarity judgments.

rectangles were identical but the area of the contrasting parallelogram was increased (see Fig. 6.2). The purpose was to test the effects of minor variations in context on the form of children's reasoning; we expected children's reasoning about the rectangles to change as a result of the comparatively minor modifications made to the contrasting parallelogram.

Eliciting Conceptions About Three-Dimensional Solids. Two triads consisting of wooden models of three-dimensional objects were included in

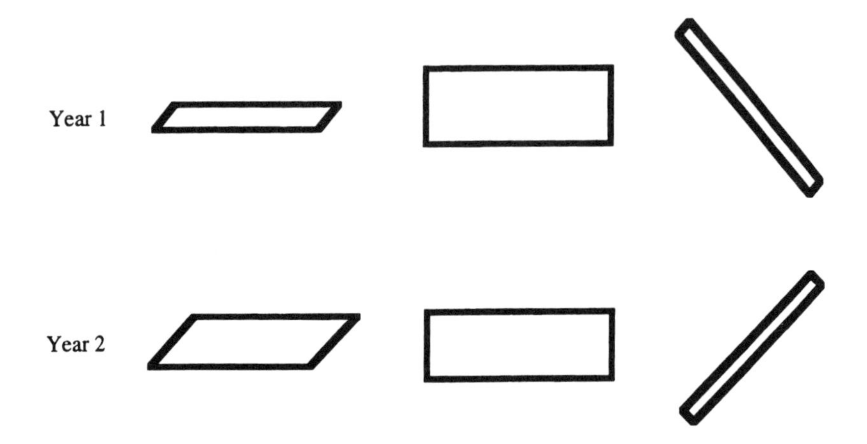

FIG. 6.2 Stimulus triad used in Year 1 and its modification in Year 2 for assessing context effects.

the second and subsequent years. One triad consisted of a cube, a cone, and a pyramid; the other consisted of a cube, a triangular prism, and a rectangular prism. Children's responses were categorized in much the same manner as before. New categories for classifying children's responses included number of vertices, number of edges, and number of faces.

Results

Reasoning About Two-Dimensional Form

Children's reasoning varied from triad to triad and also varied within triads. Across all nine triads, children's justifications fit, on average, eight different categories of reasoning. Children's justifications varied within each triad as well. Table 6.1 displays the percentage of children in the first wave of the sample (Grades 1–3) who provided justifications of each type for each of the nine planar figure triads. Inspection of Table 6.1 makes it clear that children's reasoning most often involved the visual appearance of figures, as suggested by van Hiele (1986). This appearance-based reasoning, however, encompassed many distinctions. Sometimes children compared figures to prototypes of other figures ("It's squarish") or to prototypes of real-world objects ("It looks like a ramp"). At other times, children focused on the size of the figure ("It's skinny") or to attributes that resulted from properties of angles ("It's pointy" or "It's slanty").

TABLE 6.1 MODAL CATEGORIES OF JUSTIFICATION USED BY STUDENTS IN GRADES 1, 2, AND 3 (FIRST WAVE): BY PLANAR TRIAD

	Triad								
Justification	*1*	*2*	*3*	*4*	*5*	*6*	*7*	*8*	*9*
Visual									
A		38	68	35	24	51	30	27	41
B		38			30		43	32	
C								27	24
D	22	24	35		32	24	35	27	33
E				68	62	22	54		
F				35		22	41		46
G	32								
H	81	38	32		22	24		24	24
I								38	
Property									
J		22						27	
Class									
K		32							
L	41								

Note: All values are percentages; percentages less than 20 not shown. Key to response categories: A = slanty/diagonal, B = pointy, C = looks like object, D = morphing, E = skinny/fat, F = big/small, G = orientation, H = looks like prototype, I = miscellaneous/visual, J = number of vertices, K = is a rectangle, L = is a square.

Some children regarded figures as malleable objects that could be pushed or pulled to transform them into other figures, without any concern for the effects of these transformations on the properties of the figures (i.e., nonrigid transformations to produce *morphing*—see Hoffer, 1983). Some children, for example, decided that a chevron and a triangle were alike (see triad 2 in Fig. 6.1) because "if you pull the bottom [of the chevron] down, you can make it into this [the triangle]." Children mentally animated an action on a side of one figure, often accompanied by appropriate gestures analogous to a "drag" operation with a computer mouse, to morph it into another figure. Significant percentages of children (more than 20%) provided morphing explanations of similarity among forms for all but one of the nine triadic comparisons. Consequently, although we might describe children's reasoning as primarily "visual" (van Hiele, 1986), children's justifications involved many distinctions about form that appeared to involve several different types of mental operations, ranging from detection of features like fat or thin, to comparison to prototypical forms, to the action-based embodiment of pushing or pulling on one form to transform it into another. These distinctions appear to defy description by a single, "visual" level of development.

Moreover, many children also referred to properties of figures, such as the number of sides (24%) or the number of vertices or "corners" (43%), as well as to classes of figures, (e.g., squares and rectangles). Of the 37 children in this first wave of the study, only 4 (3 in the third grade) never referred to either conventional properties or classifications of shape in their justifications of similarity. Level mixture was therefore the most typical pattern of response, and children's justifications often "jumped" across nonadjacent levels of the van Hiele hierarchy.

The Effects of Changing Context on Reasoning. Because the range of responses suggested that variations in context (the contrast set) affected how children reasoned about forms, we tested this conjecture further by modifying one of the triads in the second year (see Fig. 6.2). In the first wave of the study, 76% of the children judged the first (parallelogram) and last (rectangle) figures as most similar. They reasoned that the figures were "skinny" and/or "slanty," meaning that they perceived the shorter sides of the parallelogram and the shorter sides of the last rectangle to be the same or similar. In the second year, however, 60% of the children judged the first (parallelogram) and second (rectangle) figures as most similar, many again reasoning along the fat–skinny dimension. Over 40% of these children also talked about pulling on the "corners" to morph the first figure into the second or reasoned about the parallelogram as a "stretched" rectangle. Hence, minor shifts in context changed the saliency and relative weight assigned by children to different features of these shapes.

Profiles of Justification. To examine profiles of reasoning for individuals, we derived eight summary categories (summaries of the 21 categories coded): resemblance (looks like or resembles a figure or object), size (bigsmall or skinny/fat), angle (slanty, pointy, diagonal), orientation, morphing, counting (number of vertices or number of sides), property (angle measure, parallelism, or other property descriptor), and class (triangle, trapezoid, parallelogram, rectangle, square). We calculated the proportion of use for each of the eight categories. A score of 0 indicated that the child never used the category of response in any of the nine triadic comparisons (e.g., never suggested anything about resemblance) and a 1 indicated that the child used every possible opportunity to use that category of response (e.g., mentioned something about resemblance in every triadic judgment).

A cluster analysis of these profiles based on a measure of Euclidean distance is displayed in Fig. 6.3. The cluster solution suggests that the morphing justification (pushing and pulling on corners and sides to transform a shape) was not consistently accompanied by other forms of reasoning. On further analysis, we found that in every year of the study, children who tended to think about morphing one figure into another as the basis for similarity (i.e., they made these judgments more than one third of the time) proved the exception to the general trend of "change the context, change the form of reasoning"; these children were more consistent in their justifications from triad to triad.

Reasoning About Three-Dimensional Forms

Inspection of Table 6.2, obtained 1 year later during the second wave of the study (Grades 2–4), suggests that children's reasoning about solids

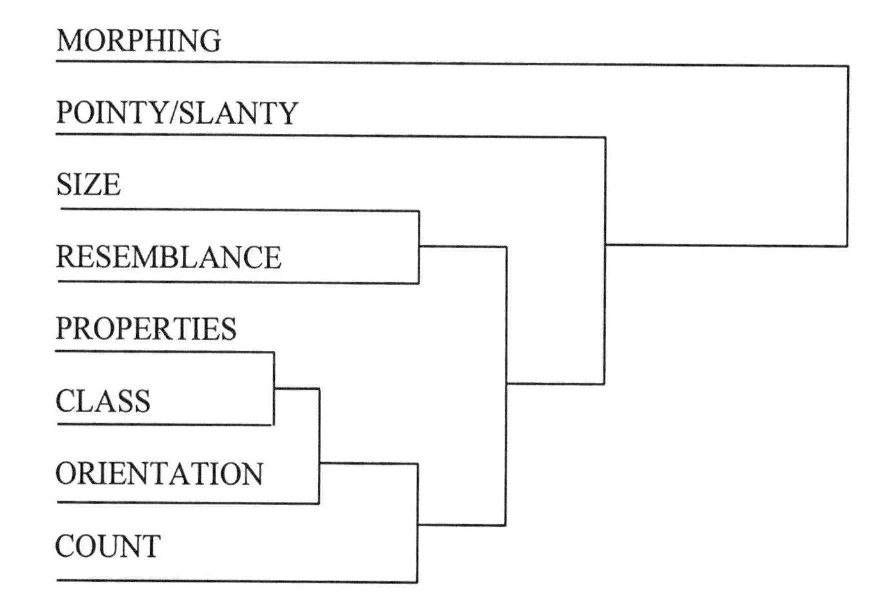

FIG. 6.3 Cluster analysis of justifications for the shape triads.

TABLE 6.2 MODAL CATEGORIES OF JUSTIFICATION USED BY STUDENTS IN GRADES 2, 3, AND 4 (SECOND WAVE): BY SOLID TRIAD

Category	*Triad 10*	*Triad 11*
Pointy	70	
Morphing	35	32
Skinny/fat		32
Big/small		51
Resembles prototype	51	54
Number of vertices	22	
2-D shape	46	22

Note: Values are percentages; percentages less than 20 not displayed.

was much like that about plane figures. The 10th triad (cube, pyramid, cone) invited reasoning about "pointiness," just as did the second triad of plane figures. In a like manner, the 11th triad (cube, rectangular and triangular prisms) invited children to reason about comparative size and slenderness of the objects, a form of reasoning that they had often demonstrated for the planar figures. Surprisingly, students also treated the solid wooden figures as malleable (morphing category), suggesting, for instance, that the pyramid could be "shaved" into a cone or that the rectangular prism could be transformed into a cube by "sitting on it."

Longitudinal Trends

The van Hiele model predicts longitudinal decreases in the categories of reasoning involving visual attributes and increases in those based on conventional properties (e.g., number of sides) and classes (e.g., rectangles, parallelograms). To analyze individual patterns of growth in reasoning, we focused on each of the eight components of reasoning about shape, conducting a multivariate profile analysis for each component with orthogonal polynomial contrasts to test linear and quadratic change during the three years of the study. The general pattern predicted by the van Hiele model was not confirmed. Significant trends were detected for justifications based on (a) resemblance, (b) morphing, (c) orientation, (d) size, and (e) class. Growth in orientation and size justifications was related to the manipulations of stimulus materials (e.g., triad 4) noted previously. The other longitudinal trends resulted from increasing propensity to (a) relate figures to known figures (e.g., "looks like a pushed-in rectangle") and (b) view figures dynamically (morphing via nonrigid transformations by "pushing" or "pulling" at vertices). The growth in children's use of class membership appeared to be due to explicit instruction about new classes of figures, like parallelograms, in Grades 4 and 5. Children's propensity to view figures as "slanty" or "pointy," to count sides or vertices, or to describe figures in terms of other properties such as parallelism remained stable.

Computer Simulation of Shape Similarity Judgments

As noted previously, the van Hiele model suggests qualitative shifts in children's thinking, implying fundamental differences in mental representation among levels. To assess this claim, we developed a computer simulation of children's judgments of similarity among the nine two-dimensional shape triads employed in this investigation (Towell & Lehrer, 1995). We represented children's encoding of each figure with a set of primitives, like "pointy," "tilted," and "has four sides," arranged as an associative (parallel, distributed) network in memory. We then presented each stimulus triad to the model and obtained its estimate of the most similar pair. Like the children in this study, the model's judgments changed with minor variations in context and reflected accurately the judgments children made about most similar pairs. Somewhat to our surprise, when the model processed examples of form found in many popular mathematical texts used in elementary schools, little learning occured. With exposure to more varied examples of forms, the model tended to learn about the diagnostic significance of conventional mathematical descriptions of figures, like the number of sides, without any qualitative shift in the fundamental nature of the mental representation. We concluded that although the van Hiele levels may provide rough benchmarks about phenomenological transitions in children's reasoning about form, there is good reason to call into question the notion of a level or combination of levels as an adequate description of mental representation.

Brief Discussion

The children's thinking about shape can be characterized as appearance based, as suggested by van Hiele (1986). However, children distinguished among many different features of form, and the nature of the distinctions children made varied greatly with the contrast set involved in the similarity judgment. There was little consistency in the use of any particular form of reasoning, with the exception of the children who viewed relationships between figures in terms of the nonrigid transformations that could morph one figure into another.

Children's thinking was much more sensitive to context than to grade; age-related trends suggested that, if anything, experiences in school and in the world did little to change children's conceptions of shape throughout the course of the elementary grades. One exception to this trend was an increasing propensity to reason about similarity between and among figures by mentally animating the action of pulling or pushing on a vertex or side (face) of a two- (or three-) dimensional form. Interestingly, this mental animation falls within the range of the dynamic geometries made possible by computer tools (see in this volume, Goldenberg, Cuoco, & Mark, chap. 1, and Olive, chap. 16) and was perhaps engendered by experience with video games and related media.

CONCEPTIONS OF ANGLE

Two major views of children's conceptions of angle have dominated previous research. The first, Piaget's (Piaget et al., 1960), ties children's ability to recognize and measure planar angles to operations on the plane, so that angle measure is enabled by the mental equivalent of a Cartesian coordinate system. No matter how simply we define an angle, the "measurement of an angle still implies a special kind of coordination between the length of the arms and the distance which separates them." The second view, van Hiele's (1986), also regards children's development as constrained by their level of thought. According to van Hiele, thinking of figures at the first, visual level precludes separating ideas about angle from perception of the figure as a whole.

In contrast to these researchers, who regarded angle as a unitary concept, Freudenthal (1973) suggested that multiple mathematical conceptions of angles were appropriate targets for instruction. Henderson (1996) notes that an angle can be defined from at least three different perspectives: (a) angle as movement, as in rotation; (b) angle as a geometric shape, as a delineation of space by two intersecting lines; and (c) angle as measure, a perspective that potentially encompasses the other two. Because children's conceptions of angles potentially include multiple models, one purpose in this study was to investigate children's models of angles in both dynamic-rotational (e.g., a hinge or a straw opening and closing) and static-shape (pipe cleaners bent at different angles) contexts.

A second purpose was to develop an information-processing analysis of angle. We relied on Siegler's (1981) rule-assessment approach to characterize children's ideas about angles in a problem-solving, construction task. In contrast to other approaches that rely on children's verbal reports, the rule-assessment analysis infers children's conceptions from their performances. In addition, the task we posed helped us determine whether or not children could "distance" (Sigel, 1993) their conceptions of angle from the overall appearance of a figure.

The rule-assessment approach is based on two related assumptions. First, children's problem-solution processes can be modeled as the application of one or more rules. Development is a product of the acquisition of increasingly complex rules. Second, we can infer which rules children use by posing problems, the solutions for which show distinct patterns of correct answers or errors depending on the rule applied (Siegler, 1986). Accordingly, we designed a shape construction task that required children to encode angular relations. This task and approach helped us understand the ways in which children of different ages encoded angles.

Method

Participants

Participating students were those described previously. However, because the rule-assessment task was refined each year, making longitudinal comparisons difficult, we supplemented the longitudinal sample with an additional sample (n = 60) of 14 first graders, 19 second graders, 16 third graders, and 11 fifth graders. These children were all students at the same school as the children in the longitudinal study.

Description of Interview

Interview items were designed to assess children's models of angles in both static and dynamic contexts. We examined a number of potential conceptions about angles, ranging from simple departures from horizontal or vertical lines to notions of angles as sweeps. These conceptions, and our measures, are described here.

Angles as Departures from Straight Lines. We first asked children to draw a straight line and to explain why it was straight. We then presented children with a line oriented 50° from vertical and asked if this line was also straight, and why.

Comparing Angles. To see what features were most relevant in comparing and contrasting angles, each child judged the most similar pair of pipe cleaners for each of the four triads displayed in Fig. 6.4. The first and third

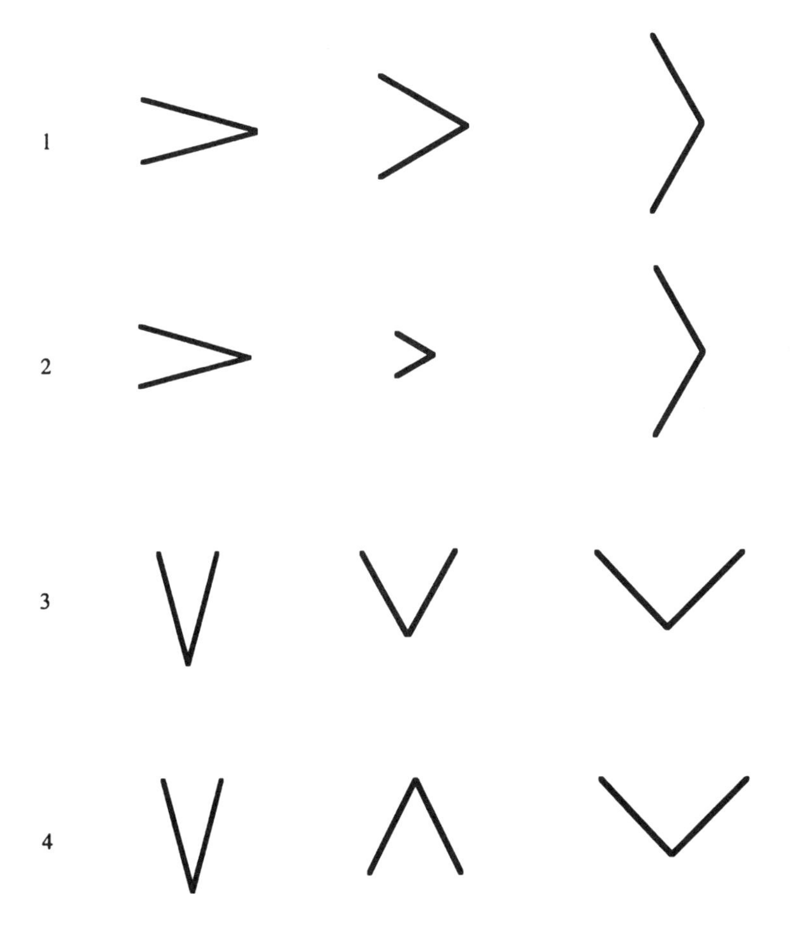

FIG. 6.4 Stimulus triads for angle comparisons.

triads were designed to see if children could distinguish (a) acute (30°, 60°) angles from obtuse angles and (b) acute (30°, 60°) angles from 90° angles ("straight corners"). The second and fourth triads were designed to assess the effects on children's judgments of (a) varying the length of the line segments but not the measure of the angle and (b) varying the orientation of the angles but not the angle measure or the length of the line segments. For each judgment, children explained their reasoning, and the bases of the justification were coded. Dimensions of coding included (a) orientation (similarity based on similar orientation), (b) length of segments (similarity based on length), (c) width (similarity based on distance between legs of the angle), (d) sweep or arc (similarity based on notion of similar sweeps or turns), (e) overall appearance ("these two look the same"), and (f) conventional angle measure.

Measuring Angles. The interviewer asked each child to find a way to measure the "opening," alternately phrased as the "bending," in a set of wooden jaws (two hinged pieces of wood) when open or closed, between an open door and the actual doorway, in a bent straw, and in a bent pipe cleaner. In each case, we coded the part of the angle that children measured (ray length, distance between endpoints, rotation, conventional angle measure, or other).

Drawing Angles. Children drew three different angles and explained how each angle was different from the other two. The number of actual (conventional) angles drawn was scored, and the differences given by the children were categorized in the same way as the responses for the angle triads.

Shape Construction Task: "Pipes and Connectors." We designed a construction task to assess children's understanding of angle in nonverbal context. Children pretended that they were in a store to order the parts needed to build each of a series of similar shapes: large square, small square, large equilateral triangle, small equilateral triangle, rectangle, parallelogram, large rhombus, small rhombus, and trapezoid. The parts consisted of "pipes" (hollow straws, 77 mm and 154 mm long) and "connectors" (solid pieces bent at angles of 30°, 60°, 90°, and 120°). Children's selections of parts were recorded, and rules describing what they were encoding about angles were developed to describe the pattern of response. For example, if a student did not seem to be able to separate angles from perception of the whole figure, then we expected that random guesses would dominate his or her choice of connectors. In principle, such a student would not be able to select the appropriate pieces for any of the nine figures. By contrast, a student able to distinguish 90° angles from others would be expected to construct the squares and the rectangle but no others. In all, we developed three additional encoding rules: (a) distinguishing 90° or 120° angles from angles of 30° or 60°, meaning that students differentiated "big" angles from "small" angles but did not distinguish types of big or small angles; (b) distinguishing 90° and 120° angles from acute angles (30° or 60°); and (c) distinguishing all angles.

Results

In the first year of the study (Grades 1, 2, 3), children's intuitions about straight typically involved ideas like "no bumps," "no wiggles or bends," "no corners," and "no zigzags." When asked to describe things they knew about that were not straight, children often talked about bends or gestured with their hands in ways to indicate departure from a straight line. However, half of the first- and second-grade children believed that a line oriented 50° from vertical was not straight. They characterized the line as "slanted" or "bent." Apparently, children's initial ideas about angles as departures from straight lines included nonhorizontal lines as exemplars. Children in the third grade and above, however, rarely (less than 5 %) claimed that the straight line 50° from vertical was not straight.

Conceptions of Angles in Static Contexts

In the first year of the study (Grades 1, 2, 3), 67% of the children judged the two acute angles in the 30°, 60°, 120° triad as most similar (triad 1 in Fig. 6.4), but when the length of the line segments of the 60° angle was decreased, only 43% of the children then judged the acute angles as most similar. Thus, the length of the line segments had a substantial influence on children's judgments of similarity, indicating that many children's conceptions of angle-as-form included delineation of relative size and area, and not simply arc subtended by the two intersecting lines. The effects of length on children's judgments about angles did not diminish during the 3 years of the study.

Children also judged the most similar pair among a triad of pipe cleaners representing 30°, 60°, and 90° angles. About half of the children (49%) chose the acute angles as most alike. This percentage declined to 38% when the angles of 30° and 90° were shown in the same orientation (see triads 3 and 4, Fig. 6.4), suggesting that for the youngest children in the sample, orientation of angles influenced decisions about measure. However, over time, the effects of orientation decreased. For example, 62% of children in the third wave of the study (Grades 3, 4, 5) explicitly mentioned that changing the orientation did not change anything important about the angles. This "no change" justification was increasingly mentioned as children grew older; nearly half of the children shifted from not using this reasoning in the first wave of the study (Year 1) to using it as an explanation for their choice in the last wave of the study (Year 3).

In the first wave of the study, children's justifications for their choices of similar angles commonly included appearance (e.g., "They just look that way") or the linear distance between the endpoints of the pipe cleaners. Although the incidence of appearance-related explanations decreased longitudinally, explanations involving the use of the distance between the endpoints of the angle's rays increased during the course of the study. Perhaps this trend explains why manipulations involving orientation had less and less effect on children's judgments, but children's judgments continued to be affected by manipulations involving the length of the line segments. Although orientation does not affect distance between endpoints of line segments, increasing or decreasing the length of these line segments does.

Conceptions of Angles in Dynamic Contexts

We anticipated that dynamic contexts might provoke models of angles involving turns or sweeps. Children rarely described (less than 10% in any of the 3 years of this study) or otherwise represented angles in these dynamic contexts as turns, sweeps, or arcs of rotation. When asked to measure the amount of opening in hinged wooden jaws, between an open door and its threshold, or in a bent straw, children overwhelmingly measured length between end points of the objects (over 95% of the time), regardless of grade, task, or year of study, just as they did for the static contexts.

Drawings of Angles

Although most children (73% the first year of the study and 92% in the third year) reported knowing about angles, their drawings of different angles often included solitary lines oriented off the vertical or horizontal (generally first and second graders), angles of identical measure but with different lengths of line segments, and identical angles oriented differently on the page. Children's drawings of angles generally reflected their judgments in the other tasks and thus served as an additional source of corroborative evidence. Changes in children's judgments about angle triads were usually reflected in their drawings. For example, children who came to reject orientation as salient for judgments about which two angles were most alike also tended to discontinue representing drawings of identical measure in different orientations as different.

Angle: Rule Assessment

Children's production of angles and their judgments about relevant criteria for comparing angles or measuring angles, in both the dynamic and static contexts, indicated that children's models of angles derived from their conceptions of straightness and were influenced by properties of form (e.g., orientation and length). Children's performance in the shape-construction task provided a different window to their conceptions, one less subject to verbal or production task demands. Performance on this task demanded that children mentally decompose figures into component angles ("connectors") and line-segment lengths ("pipes").

Fit of the Scoring System. The rules developed to account for children's performance could be applied to 84% of the students; the remainder of the children seemed either to be characterized by rule mixture, or by the strategy of choosing one angle correctly and then using the symmetry of the figures to infer that the remaining angles were the same. A small number of children (2%) performed in ways that could not be described syntactically.

Rules Used by Children for Angles. Table 6.3 displays the percentage of children using each rule by grade in the cross-sectional sample. The majority of first- and second-grade children did not consistently distinguish among the four angles, so their strategy was largely one of trial and error. Significant minorities of the youngest children, however, successfully encoded one or more of the distinctions among these angles. In the third grade, most children (69%, rules 1–4) mentally decomposed figures to allow encoding of one or more angles; the modal performance (use of rule 3) indicated ability to reliably differentiate and coordinate relationships among all but the two most similar angles (the 30°, 60° pair). By the fifth grade, children reliably encoded multiple angles, but nevertheless only 9% of these children could reliably distinguish and coordinate relationships among all angles.

Brief Discussion

Children's tacit knowledge about angles seemed to grow significantly between the second and third grades, as might be expected from the Piagetian tradition (see Olson, 1970, about the child's construction of diagonality and other manifestations of the construction of mental "coordinates"). Results obtained from the rule-assessment task suggested that even some of the youngest children encoded and coordinated relationships among some of the angles in the figures, and all the children did so by the fifth grade. This finding belies the presumed fusion of figure-component relations postulated by the van Hiele model. Nevertheless, there were persistent effects of angle-length fusion on children's models of angles, manifested in two ways. First, many children found the lengths of segments constituting angles more salient than their respective angle measures. Second, across a variety of contexts, most children believed that length was an adequate measure of angle. Even in dynamic contexts involving movement, children at all ages typically failed to generate models of angles as sweeps or turns. In-

TABLE 6.3 ENCODING OF ANGLES BY RULE AND GRADE FOR THE CROSS-SECTIONAL SAMPLE

Grade	*Rule 0*	*Rule 1*	*Rule 2*	*Rule 3*	*Rule 4*
1st	64%	0%	22%	7%	7%
2nd	55%	5%	22%	11%	11%
3rd	31%	6%	19%	38%	6%
5th	0%	36%	18%	36%	9%
All students	40%	10%	20%	22%	8%

Note: Rule 0: Does not differentiate among angles. Rule 1: Differentiates 90° angles; uses trial and error for all other angles (30°, 60°, 120°). Rule 2: Differentiates the two smallest (30° and 60°) from the two largest (90° and 120°); uses trial and error in determining which angle to use in a group (e.g., a square has either 90° or 120° angles, but not 30° or 60° angles). Rule 3: Distinguishes acute, right, and obtuse angles. Rule 4: All angles properly distinguished; errorless performance.

stead, refinements of early but static models of angles as deviations from "straightness" dominated children's views about angles.

LENGTH AND AREA MEASURE

We were most interested in children's implicit theories of length and area measure because these implicit theories constitute the building blocks for developing a mathematics of measure. Consequently, we designed interviews to elicit children's conceptions of (a) the relationship between the attribute being measured and the unit of measure (e.g., area measure demands some unit of cover, not simply a unit of length), (b) the relationship between characteristics of the unit of measure and quantity (e.g., smaller units will yield higher quantities), (c) the need for iteration of units (i.e., measure units are concatenated to produce a quantity, a process that Piaget et al. [1960] suggested involved mental operations of additive subdivision [subdividing a line] and order of position [e.g., moving a standard unit along a line]), and (d) the need for identical units of measure (for meaningful measure, like units must be combined). For length measure, we also examined children's conceptions of the zero point of a scale. The zero point refers to the origin of a measurement scale and an associated assumption of equal intervals, such that, for example, the difference between 10 and 20 is the same as that between 45 and 55.

Method

The participants were the students described previously for the longitudinal sample. Each year, students were interviewed once about length and its measure and once about area and its measure. We report here abbreviated versions of the interviews.

Length Interview. We assessed children's understanding of iteration of units and the need for identical units of measure. Children were provided two 7-in. rulers: one marked at equal intervals, so that every unit was identical, and one marked at unequal intervals. Children had a choice of rulers for measuring the length of a 7-in. stapler and a 9-in. book. We recorded children's choices, the way they measured each object, and their justification for their choices and methods. In another task, assessing the identity-of-unit construct, the interviewer placed one 2-in. paper clip and four 1.25-in. paper clips in a continuous line along a piece of wood, covering the length of the wood. The interviewer probed children's conceptions of unit of measure, asking, for example, whether or not 5 was a good measure for the length, and why.

Children's understanding of the zero point of measure was probed by showing a 7-in. piece of wood placed against a 12-in. ruler with edges of the wood aligned against the 2-in. mark and the 9-in. mark of the ruler. Children measured the length of the wood.

Finally, children's ideas about measuring nonrectilinear forms were assessed. Each child was shown a curved track drawn on an 8.5 × 11 in. sheet of paper and was provided a 12-in. ruler, many paper clips (all the same length), and a piece of string. Each child was asked to measure the length of the curved track and was offered the use of the string, paper clips, and ruler. Then the child explained why he or she used (or did not use) each of the three items offered.

Area Interview. We began by exploring children's conceptions of the unit-attribute relationship by eliciting their spontaneous ideas about how to find "the amount covered by" a square 6 cm on each side and a right isosceles triangle with 6-cm side. Variations of the wording included "how much this square covers," "the amount covered by this square," "area," and "the amount inside." These variations were introduced according to the interviewer's perception of the meaningfulness of the questions to the child. We then reexamined the children's conceptions of the unit-attribute relationship when provided with measuring tools in the form of plastic squares, triangles, rectangles, and circles (DMP pieces; see Romberg, Tornrose-Dyer, & Harvey, 1994).

To assess children's understanding of identical units, the interviewer covered the cardboard square with three plastic rectangles, two plastic squares, and two plastic triangles. The interviewer asked the child if an answer of 7 was a good measure, using this method. Children justified their answers. Probes were structured as a dialogue: "Bert says that 7 is a good measure of the amount of space covered by this square. But Ernie says, no, he's not so sure. What do you think? Why?"

The interviewer also filled the cardboard square with nine plastic circles and asked if the circles were a good measure of the area. In each case, the child's answer was scored based on whether he or she recognized any potential difficulty with this method of finding area measure. We used this item as a measure of children's understanding of space-filling units of area measure.

In the next portion of the interview, children's conceptions of additivity and composition of area were assessed (Wagman, 1975). Children were presented with the plastic manipulatives described previously (squares, triangles, rectangles, and circles) as well as with the cardboard cutouts of the square and right triangle. They then were asked to predict the area of a cardboard cutout of a rectangle 6 cm × 12 cm (twice the area of the square). We recorded children's strategies with a focus on whether or not they used their previous knowledge about the square and triangle to make predictions about the rectangle. Children were shown a parallelogram with the same area as the rectangle and again made predictions about its area. We recorded their strategies.

Last, children were shown a plastic square and a plastic parallelogram, each composed of two identical right triangles. The diagonal line made by the joint was visible. We asked children to compare the areas of the square and the parallelogram and to justify their answers.

Results

Length

Iteration. Use of a ruler to measure length assumes an understanding of iteration and of the identity of units. We presented children with two rulers of equal length, both with the same numbers as indicators, but one ruler had equal intervals and the other did not. Children measured the length of a book longer than either ruler. In the first year of the study (Grades 1, 2, 3), most children (89%) chose the equal-interval ruler, but their use of the ruler varied greatly with age: First-grade children typically joined the two rulers together and simply read whatever number coincided with the endpoint of the second (typically unequal-interval) ruler. About 70% of the second graders also used this strategy or one like it. In contrast, most third graders (67%) used the equal-interval ruler iteratively, marking the end point of the ruler and then using this end point as the starting point to continue the count of units. Analysis of longitudinal patterns confirmed clear age-related differences in the ability to iterate a unit of length measure; only third-grade and older children were likely to employ iterative strategies.

Identical Units. Children's notions of identical units were assessed by their explanation of whether or not it was acceptable to mix units of length measure (different-length paper clips). In the first year of the study, nearly all children (over 80%) suggested that the use of two different units was acceptable. However, over time, identical units of length measure were judged necessary; for example, 80% of fourth and fifth graders noted that units must be the same.

Zero Point. Children's strategies for measuring a 7-in. piece of wood with a ruler beginning at the second interval (2-in. mark) and ending at the ninth interval (9-in. mark) indicated their conception of the zero point of the scale. Twenty-one percent of the children focused solely on the endpoint of the measure (9) or added the two marks together (11), and 41% used 1 as the "zero point" of the scale, responding "8." The use of 1 as the zero point was equally prevalent in all three grades (first, second, and third). Longitudinal analysis indicated significant growth in the ability to use the zero point of a ruler; by the third year, 80% of the sample treated 2 as the zero point, obtaining 7. The remaining children, however, persisted with treating the origin of the scale as 1, obtaining 8.

Area

We present results for the first wave of the study (Grades 1, 2, 3), again with an eye toward describing children's implicit theories of measure. Longitudinal comparisons are then presented.

Spontaneous Ideas About Unit-Attribute Relationships. The children's ideas about area measure were elicited first by providing cardboard cutouts of a square and a right triangle. We asked: "How could you find out how much is inside this?" (or alternatively, "How much space does the cutout cover?"). As noted previously, we asked additional questions to probe children's understanding of the meaning of a numerical quantity (i.e., 9) in this measurement context.

Responses to these questions were quite varied. For the square, the modal response across grades (41%) was to use a ruler or some other instrument to measure length and that the quantity (9) could refer to inches. The second most popular response was "Don't know" (22%). Only four children (11%) indicated that they would find some type of unit of cover. For example, one child suggested using the eraser-end of a pencil to serve as a unit of measure and then attempted to iterate this unit to obtain a measure of area. For the triangle, the modal response (41%) again was to use a ruler. However the second most popular response was to use a unit of cover (19%). In summary, for both figures, the modal response was to use a unit of length to measure area. A minority (11%) of children conceived of the idea of using a unit of cover. Approximately one third of the children did not invent or suggest any way of measuring area.

Unit-Attribute Relationships with Tools. Children were provided manipulatives: squares, right triangles, circles, and rectangles. Two triangles had the same area as one square, and two squares had the same area as one rectangle. Nine manipulative squares completely covered the cardboard cutout of a large square. They were asked the same questions about measuring "how much inside" or "the amount covered by" the cardboard cutout of the square. The modal response across grades (78%) was to use some combination of the manipulatives to cover the square. Hence, with manipulatives, the majority of children resorted to the idea of measuring area with units of cover. A minority (22%) persisted with the idea of area as a length, using the length of a side of a manipulative as a measure of cover. For these children, area was obtained by matching the manipulative to the length of one or more of the sides of the square. Six children (16%) held a notion of area as an *iterative length.* They suggested measuring the length of a side of the square, recording the value, then moving the ruler over slightly, but on a line parallel to the side, and then adding the value of this length to the previous value. They continued this iteration until they reached the other side of the figure, with the number of iterations controlled (sometimes) by the thickness of the ruler (or some other arbitrary criterion).

When asked to measure the area of a square, children who used manipulatives as a unit of cover were evenly split between those who adopted squares for measure (45%) and those who supplemented squares with some other unit, like triangles (55%). The tendency to match the measuring unit to the target shape was more pronounced for triangles: Seventy percent of the children preferred triangles as the sole unit of measure, and the other 30% supplemented the triangles with squares. For rectangles, the matching of the unit to the shape was also more pronounced:

Sixty-eight percent of the children used rectangles alone, and only 16% supplemented the rectangles with squares. The remaining children used another unit of measure—squares (13%), triangles (3%), or circles (3%). Overall, although all children did not exclusively use units of measure that resembled the figure being measured, there was a pronounced tendency to match units to target figures. In short, children tended to use *resemblance* as a criterion for selection of the unit of measure.

Invariance. Children saw two different-looking figures—a square and a parallelogram—composed of identical congruent right triangles with a visible joint. Forty-three percent of the children judged the areas as equal, even though the shapes did not look alike. Of these, 10 (62%) justified their choice of noting the identical components of the shapes, and 6 (38%) appealed to a compensation principle, viewing the area as fluid, so what was "lost" from the region of the parallelogram was gained in another.

Additivity. After using manipulatives to measure the areas of a square and a right triangle, we asked children then to *estimate* the area of a rectangle (composed of two squares) and of a parallelogram (composed of the rectangle and two right triangles). The cardboard cutouts of the previous figures were arranged immediately above the focal figure (i.e., a square and right triangle were placed immediately above the rectangle). The rectangle's area was twice that of the square; the parallelogram's area could be decomposed as the sum of the areas of the right triangle (applied twice) and the rectangle.

Children's modal response was to estimate the area by mentally imposing a model of one of the physical manipulatives, or one of their own invention, on the figure, and then counting the number in the interior region of the figure. For example, one student used the eraser at the head of his pencil as a unit whereas another used a "pretend" square (a square outlined by the student's fingers and placed repeatedly over the figure).

Only three students (8%) decided to use the square as a unit of measure for the rectangle, concluding that the rectangle was equivalent to two squares. However, 27% of the students decided to represent the area of the parallelogram as a composition of the rectangle and the right triangle. This shift in strategy indicated the ability of some children within a single session to adopt the idea of area as a unit of cover and to reason about its decomposition.

Identical Units. To assess children's ideas about identical units, we presented a cardboard cutout of the square and covered it completely using three rectangles, two squares, and two triangles. Children were told, "Someone I know decided to measure how much was inside [or how much was covered by] the square this way. What do you think about what they did?" Most children (84%) saw no difficulty with the method and simply said that the area was 7 (the number of manipulatives used). Other children noted that there was a problem; three suggested that the area was "three rectangles, two squares, and two triangles." Children's responses here were consistent with their own use of different combinations of units in the tasks mentioned previously (e.g., finding the area of a square with triangles and squares and counting the total number of objects as the measure of area). Hence, both when provided tools and again in this more structured context, most children failed to employ identical units to measure area.

Space-Filling (Tiling). To assess children's ideas about the importance of completely tiling the area, we placed nine circles within the area of the cardboard cutout of the square. Children were informed that "Someone decided to measure how much was inside [or "how much was covered by"] the square this way." We asked: "What do you think about what they did?" Most children (73%) did not find this problematic. They justified the count by "It doesn't spill over" and other statements that led us to conclude that children were employing a rule about boundedness rather than about space-filling. Even when probed about the "space in between the circles," children generally did not find this problematic. Hence, we concluded that children generally use a boundedness rather than space-filling criterion for area measure, although in many of the contexts in which children regularly use manipulatives to measure area, the difference can be difficult to detect.

Strategies for Measure of Irregular Figures. Up to this point in the interview, we had employed only polygons. To assess children's understanding of how to measure the area of an irregular shape like a leaf, we provided a figure of a simple closed curve. We provided the same manipulatives again and added string and a ruler.

The modal strategy (43%) used by the children in response to this task was to fit as many manipulatives as they could into the interior of the figure without violating its boundaries. They most often chose a circle as the manipulative of choice. This modal response suggests again the propensity to use units that resemble the shape being measured and to use boundedness rather than complete cover as a salient attribute of area measure. A number of children (19%), however, used a unit like a square and adopted a compensation strategy: They estimated the part of the square enclosed in the interior and then iterated this process over all the parts and wholes to estimate the area of the figure.

Grade-Related Differences in the First Wave. Children's ideas about area were generally not ordered by grade: Their use of manipulatives to measure area and their ideas about cover, identity, space-filling, and other concepts did not vary by grade. The only trend that appeared reliably related to grade was that second- and third-grade children were more likely to conserve area mentally, recognizing that the square and parallelogram, each composed of the same congruent right triangles (although arranged differently), had the same area.

Longitudinal Change. Longitudinal growth for children's conceptions of area is displayed in Fig. 6.5 for the second graders in the sample, tracing the endpoint of their development for area concepts 2 years later (in

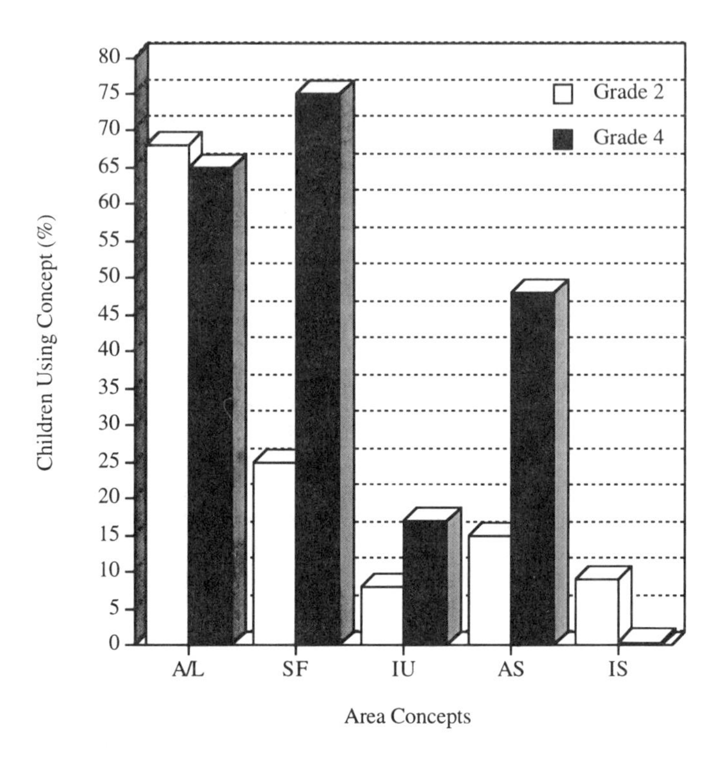

Key
A/L: Avoids area/ length confusion
SF: Recognizes space-filling
IU: Uses identical units
AS: Uses additive strategies for measure
IS: Finds measure of irregular shape

FIG: 6.5 Longitudinal change in children's conceptions of area measure, from Grade 2 to Grade 4.

Grade 4). (We display the second-grade portion of the longitudinal sample to facilitate comparisons with second graders participating in a "design experiment" that aimed to foster the development of spatial thinking that is reported in the next chapter.) Inspection of Fig. 6.5 suggests that although over time more children came to appreciate the importance of space-filling for units of area measure and as a group increased in their propensity to view area in terms of additive composition, there was little change in other major components of measure theory. For instance, a significant proportion of the children persisted in thinking about length units (e.g., iterative length strategy) as appropriate units of area measure.

Brief Discussion

Examining patterns of growth with respect to components of children's implicit theories of length and area measure suggests that some aspects of measure are readily susceptible to change during the course of everyday experience whereas others change little. With respect to length measure, younger children were not likely to understand the need to iterate a unit of measure, despite their apparent proficiency with rulers. In short, although rulers are designed to make iteration transparent to users, such "transparency" does not mean that young children understand this quality of measure. Similarly, young children often did not attend to the need for identical units of length measure, another quality made "transparent" to the users of rulers. However, over time, children's experiences seemed to promote growth about these components (iteration, identical units) of measure. The concept of the zero point proved more resistant to change; even by the end of the third year of this study, about 20% of the children still had difficulty understanding that numbers other than zero could fulfill this function (see also Ellis, 1995).

Growing recognition of the importance of constructs like identical units in the context of measuring length was often not reflected in children's ideas about measuring area. For example, even children who insisted that length be measured with identical units often did not use identical units to find the measure of an area. For a small but significant minority of children in every year of the study, area measure was best accomplished with length—length was perceived as having space-filling properties. Younger children were also prone to use *resemblance* as the prime criterion for selecting a unit of area measure (e.g., squares should be measured with square units, rectangles with rectangles, etc.), and this tendency was also pronounced for older children as well. Children's conceptions of other components of area measure, however, changed substantially during the 3 years. For example, although only a small number of younger children reasoned about additive composition of areas (e.g., how regions that look different can have the same area measure), this skill developed substantially during the course of the 3 years (although it was still not recognized by nearly half the children). Similarly, although younger children did not think of space-filling as very important, most children came to recognize its importance for area measure. Nevertheless, when we consider that children often use tools proficiently (e.g., rulers to measure length, grid paper to measure area), the gulf was wide between proficient use of these tools and the understanding of the essential qualities of length and area measure.

CHILDREN'S SKILLS IN RELATED AREAS: DRAWING AND SPATIAL VISUALIZATION

Drawing and spatial visualization skills are often cited as fundamental resources for developing an appreciation of space. For some, drawings are indicators of the nature of children's organization of space and form (Goodnow, 1977; Olson, 1970). For others, drawings are simply tools for representing space but aren't privileged windows to the mind's apprehension of space (Kosslyn, Heldmeyer, & Locklear, 1977). The relationship between spatial visualization and reasoning about space also provokes variable views: Some view spatial visualization as a fundamental constraint on reasoning about space, whereas others conjecture that spatial visualization may be the product rather than the precursor of reasoning about space. To investigate these viewpoints, we examined relationships among children's drawings of objects, their spatial visualization skills, and the other realms of spatial reasoning that we examined in this study.

Drawing Tasks. We presented children with six sets of objects, one at a time (a cube, a cylinder, a pyramid, a pyramid on top of a cube, a pyramid next to a cube, and a pyramid in front of and partially occluding a cube). Each child sat at a table approximately 3 ft from the objects and on a blank sheet of 8.5 × 11 in. paper drew what he or she saw so that "other people would know what it is, too." Scoring criteria for each drawing followed from previous research and from our preliminary analysis of the drawings of 10 children. To illustrate the scoring system, the scoring exemplars for a cube are displayed in Fig. 6.6. A drawing composite score was created by combining the scores across the six drawings. Because the more difficult drawings had a wider scoring range, these drawings contributed more to the composite score.

Spatial Visualization Tasks. Three spatial visualization tasks were used as potential correlates to geometric thinking: two measures of spatial working memory (Case's 1985 Mr. Cucumber and spatial matrix tasks) and one measure of mental imagery (Kosslyn, Margolis, Barrett, Goldknopf, & Daly, 1990). The Mr. Cucumber and spatial matrix tasks established individual differences in the number of spatial items (e.g., body parts for Mr. Cucumber and shaded cells for the spatial matrix task) that children could simultaneously hold in memory. Computerized versions of both tasks were administered in the first year of the study (see Lehrer & Littlefield, 1993, for more detailed descriptions of the Mr. Cucumber and spatial matrix tasks).

In the second and third years of the study, computerized versions of mental imagery tasks (Kosslyn et al., 1990) designed to assess component processes involved in imagery were also administered to all participants. The tasks measured component processes of (a) image generation (capacity to form an image), (b) image maintenance (memory of an image), (c) image scanning (search of the image for a part or property), and (d) image rotation (mental simulation of the movement of an image). Error rates and reaction times were recorded for tasks assessing each component process (for descriptions of tasks, see Kosslyn et al., 1990).

Results

Drawing. Drawing skill, especially knowledge of conventions for depicting dimension and occlusion (one form in front of another), increased

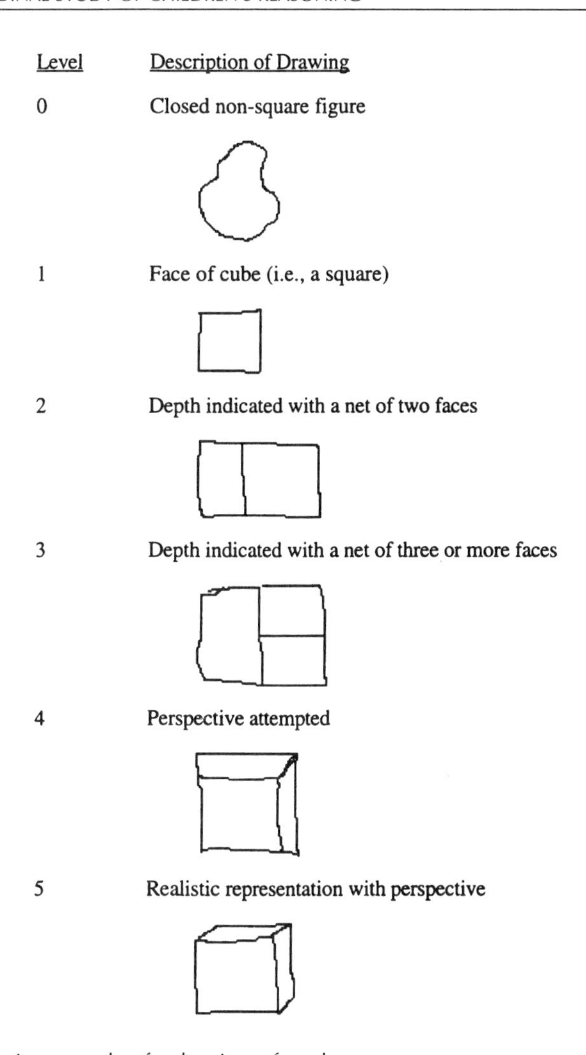

FIG. 6.6 Scoring exemplars for drawings of a cube.

with age, as expected from previous research (Braine, Schauble, Kugelmass, & Winter, 1993; Willats, 1977, 1984). Figure 6.7, which displays mean composite score by grade, clearly indicates that the largest amount of growth occurred between the first and second grades.

Although we expected systematic relationships between drawing skill and reasoning about shape, we found little evidence for this. Drawing skill was unrelated to children's reasoning about shape and angle. One exception was that in the longitudinal sample, more skillful drawers in the third year of the study were also more likely to encode all four angles in the shape-construction task (i.e., angle rule-assessment).

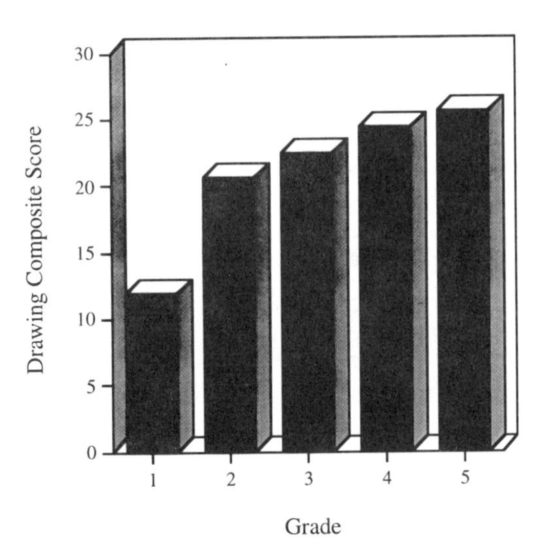

FIG. 6.7 Drawing composite score for the longitudinal sample, by grade.

Spatial Visualization. As expected, we found age-related differences in working memory span and in performance on most components of image processing. These results paralleled previous research (Case, 1985; Kosslyn et al., 1990). Children who tended to reason about two- and three-dimensional shape by mentally morphing one form into another also tended to have higher spatial working memory ($r = .40$). In addition, children who performed well on the scanning component of visual imagery (children who could search an image relatively quickly) also tended to use properties to justify their perceptions of similarities and differences among and between forms. In general, however, we found few reliable relationships between general spatial visualization skills and spatial reasoning.

GENERAL DISCUSSION

The longitudinal design employed in this study allowed us to track continuity and change in children's reasoning about two- and three-dimensional forms, implicit theories of length, area, and angle measure, and related skills like drawing and spatial visualization. Patterns of change in children's reasoning (or lack thereof) suggest the advisability of amending several long-standing accounts of children's reasoning in each of these realms of spatial thinking.

Shape and Form

The van Hiele (1986) model suggests a systematic progression from reasoning about the appearance of forms as wholes to reasoning that embraces figures as bearers of properties. In this study, children's concern with "how it looks" (the lowest level of the van Hiele hierarchy) often included distinctions that suggested differentiated processing of attributes, not global calculations of similarity based on overall appearance, as suggested by the van Hiele model. Thus, although the logic of children's responses often was not mathematically conventional, neither was it a sign of thought anchored to or fused with appearance as a whole, as predicted by the van Hiele model. Instead, children's embrace of unconventional attributes like "pointiness" seems very relevant when we consider children's activities. Pointy things are uncomfortable to sit on, objects slide off slanty forms, and fat things take up a lot of room.

Children's judgments about forms call into question the descriptive adequacy of the van Hiele levels. As noted previously, children's justifications involved many distinctions about forms that were accomplished through a range of mental operations on these forms, including comparing forms to visual prototypes, mentally animating actions to morph one form into another, and detecting a variety of attributes of shape that varied with the contrast set involved. Moreover, these variations in judgments need not involve qualitative differences in mental representation, a proposition supported by our computer simulation and by related work in visual perception (see, e.g., Smith, 1989).

Perhaps the chief difficulty with the van Hiele theory is its reliance on static descriptions of knowledge (levels) and its attendant emphasis on discontinuities among these levels. More contemporary accounts portray conceptual change as "overlapping waves" of development, where thinking is characterized by variation in forms of reasoning, with attendant selection of any particular form adapted to task, learning history, and related elements of context. Development is characterized then by which "waves" or forms of reasoning are most dominant at any single period of time, without necessarily implying the extinction of other forms (Siegler, 1996). This view better describes the pervasive variation in reasoning about form that we found in this study. Moreover, the van Hiele theory does not account in any principled way for the ecologies constituted by instructional activity and the history of learning in any classroom. Classroom ecologies are not given, but are continuously generated and achieved by participants, so that conceptual change in reasoning about space (or anything else) is apt to depend on the dynamic qualities of interactions among students, teachers, tasks, tools, conversations, and so on. In Lehrer, Jacobson, et al. (chap. 7, this volume), we portray how classroom ecologies influence the development of reasoning about space in the primary grades: To foreshadow, children's reasoning about space is more variegated and elaborate than the van Hiele framework suggests when they participate in classrooms designed to promote understanding of space and geometry.

Although Piaget's topological primacy hypothesis (i.e., children understand topological relations before any other) has been refuted consistently (e.g., Miller & Baillargeon, 1990), the results of this study suggest that attention to topological thinking should enjoy a revival. Children tended increasingly to think of two- and three-dimensional forms topologically: They increasingly (with age) considered morphing one form into another by pushing or pulling vertices or sides. Because most of these proposed transformations did not involve cutting or joining previously unconnected points, such transformations could be considered topological. The dynamic geometries made feasible by new software tools offer tools for exploring these forms of relationship (see Goldenberg & Cuoco, chap. 14, Goldenberg, Cuoco, & Mark, chap. 1, and Olive, chap. 16, this volume). Moreoever, the potential role of topology in the elementary curriculum should be reconsidered.

Angle

This study also suggests the need to revise conventional views of children's reasoning about angle. Although there was a clear age-related transition in children's conceptions of angle, consistent with Piaget's theory, the results of the rule-assessment analysis suggested that significant proportions of even the youngest children in the sample (Grades 1 and 2) were consistently encoding and separating angles from the larger context of the figures within which they were embedded. This capacity suggests that many young children are doing more than simply noting that figures "have corners." Instead, they appear to reliably distinguish among (at least some) angles of different measure, and they can mentally decompose a two-dimensional figure into separate attributes of length and angle. Nevertheless, because most children have intuitions about angle measure that are anchored in length, significant hurdles remain for the development of a mathematical description of angle. Moreover, most of the children in this sample did not entertain multiple models of angle: Models of angle based on movements like turns or sweeps were comparatively rare.

Length and Area Measure

Our analysis of children's understanding of the foundations of length and area measure suggests that conventional instruction is not fostering the development of critical ideas in measure, like the relationship between the unit of measure and the attribute being measured. Perhaps because so much of children's experiences with measure focuses on length, units of length assumed a role as a universal, all-purpose measure, even for area and angle. For example, some children persisted in ascribing space-filling qualities to length, measuring the area of a square by finding the length of a side, moving the ruler and adding the length of the side to the previous measure, and so on until terminating this iterative process at some arbitrary point. In a similiar vein, children's use of tools like rulers, especially in the primary grades, often masked (and overestimated) their understanding of foundational ideas in measure, like those involving iteration, identical units, and space filling.

Conceptual Development and Relationships Among Realms of Spatial Reasoning

Conventional views regarding the intimate relationship between drawing, spatial visualization, and many forms of geometric reasoning found little support in the results of this investigation. For example, over the course of 3 years, there was little evidence to suggest that poor drawing was any indicator of ability to reason about spatial form (see Kosslyn et al., 1977). Conversely, many good drawers could not articulate their hands' visions: Their justifications about relationships among and between two- and three-dimensional forms were no more detailed or mathematically elaborated (e.g., with respect to the van Hiele model) than those of their peers. Moreover, consistent relationships among children's conceptions of two- and three-dimensional form, and their conceptions about measurement of length or area, were not evident in this sample.

In summary, current curricular practices in the elementary school tend to promote little conceptual change. Neglect here is not benign: In the first wave of the sample, first-grade children were more likely than older children to think of contrasting forms by counting the number of sides or vertices. Over time, children were *less* likely to notice these attributes, given conventional instruction of geometry in the elementary grades. Consider also the relative paucity of conceptual change about area measure (see again Fig. 6.5). Although most of the children in this sample came to understand the need for identical units of measure for length, their understanding was grounded in length only and did not encompass a more general theory of measure. As a result, understanding the need for identical units of area measure was an independent and unlikely accomplishment. As suggested by the van Hieles, in the absence of systematic instruction, children's opportunites for developing a mathematics of space languish (and for many, are extinguished), making potential revival in later years of schooling problematic. The results of this study also suggest that different facets of spatial reasoning develop independently; their integration awaits instruction. In chapter 7, we describe forms of instruction that build on the promise of children's early understandings and everyday experiences in developing a mathematics of space.

REFERENCES

Anderson, J. R. (1983). *The architecture of cognition.* Cambridge, MA: Harvard University Press.

Baltes, P., Reese, H. W., & Nesselroade, J. R. (1977). *Life-span developmental psychology: Introduction to research methods.* Monterey, CA: Brooks/Cole.

Braine, L. G., Schauble, L., Kugelmass, S., & Winter, A. (1993). Representation of depth by children: Spatial strategies and lateral biases. *Developmental Psychology, 29,* 466–479.

Bruer, J. T. (1994). *Schools for thought: A science of learning in the classroom.* Cambridge, MA: MIT Press.

Burger, W., & Shaughnessy, M. (1986). Characterizing the van Hiele levels of development in geometry. *Journal for Research in Mathematics Education, 17,* 31–48.

Carpenter, T. P., Fennema, E., Peterson, P. L., Chiang, C., & Loef, M. (1989). Using knowledge of children's mathematics thinking in classroom teaching: An experimental study. *American Educational Research Journal, 26,* 499–531.

Case, R. (1985). *Intellectual development.* New York: Academic Press.

Clements, D. H., & Battista, M. T. (1992). Geometry and spatial reasoning. In D. A. Grouws (Ed.), *Handbook of research on mathematics teaching* (pp. 420–464). Reston, VA: National Council of Teachers of Mathematics.

Eilan, N., McCarthy, R., & Brewer, B. (1993). *Spatial representation.* Cambridge, MA: Blackwell.

Ellis, S. A. (1995, April). *Developmental changes in children's understanding of principles and procedures of measurement.* Paper presented at the biennial meeting of the Society for Research in Child Development, Indianapolis, IN.

Freudenthal, H. (1973). *Mathematics as an educational task.* Dordrecht, the Netherlands: Reidel.

Fuys, D., Geddes, D., & Tischler, R. (1988). The van Hiele model of thinking in geometry among adolescents. *Journal for Research in Mathematics Education Monograph,* No. 3.

Goodnow, J. (1977). *Children drawing.* Cambridge, MA: Harvard University Press.

Henderson, D. W. (1996). *Experiencing geometry on plane and sphere.* Upper Saddle River, NJ: Prentice Hall.

Hoffer, A. (1983). Van Hiele–based research. In R. Lesh & M. Landau (Eds.), *Acquisition of mathematics concepts and processes* (pp. 205–227). New York: Academic Press.

Kosslyn, S. M. (1980). *Image and mind.* Cambridge, MA: Harvard University Press.

Kosslyn, S. M., Heldmeyer, K. H., & Locklear, E. P. (1977). Children's drawings as data about internal representations. *Journal of Experimental Child Psychology, 23,* 191–211.

Kosslyn, S. M., Margolis, J. A., Barrett, A. M., Goldknopf, E. J., & Daly, P. F. (1990). Age differences in imagery abilities. *Child Development, 61,* 995–1010.

Lehrer, R., Knight, W., Sancilio, L., & Love, M. (1989, March). *Software to link actions and descriptions in pre-proof geometry.* Paper presented at the annual meeting of the American Educational Research Association, San Francisco, CA.

Lehrer, R., & Littlefield, J. (1993). Relationships among cognitive components in Logo learning and transfer. *Journal of Educational Psychology, 85,* 317–330.

Miller, K. F., & Baillargeon, R. (1990). Length and distance: Do preschoolers think that occlusion brings things together? *Developmental Psychology, 26,* 103–114.

National Council of Teachers of Mathematics. (1991). *Professional standards for teaching mathematics.* Reston, VA: Author.

Olson, D. R. (1970). *Cognitive development: The child's acquisition of diagonality.* New York: Academic Press.

Piaget, J., & Inhelder, B. (1948/1956). *The child's conception of space.* London: Routledge and Kegan Paul. (Original work published 1948)

Piaget, J., Inhelder, B., & Szeminska, A. (1960). *The child's conception of geometry.* New York: Basic Books.

Romberg, T., Tornrose-Dyer, M., & Harvey, J. G. (1994). *Developing mathematical power* (Vol. 2). Chicago: Encyclopaedia Britannica Educational Corporation.

Siegler, R. S. (1981). Developmental sequences within and between concepts. *Monographs of the Society for Research in Child Development, 46* (2, Serial No. 189).

Siegler, R. S. (1986). *Children's thinking.* Englewood Cliffs, NJ: Prentice Hall.

Siegler, R. S. (1996). *Emerging minds: The process of change in children's thinking.* New York: Oxford University Press.

Sigel, I. E. (1993). The centrality of a distancing model for the development of representational competence. In R. R. Cocking & K. A. Renninger (Eds.), *The development and meaning of psychological distance* (pp. 141–158). Hillsdale, NJ: Lawrence Erlbaum Associates.

Smith, L. B. (1989). A model of perceptual classification in children and adults. *Psychological Review, 96,* 125–144.

Towell, G., & Lehrer, R. (1995). A knowledge-based model of geometry learning. In T. Petsche, S. J. Hanson, & J. Shavlik (Eds.), *Computational learning theory and natural learning systems* (pp. 55–74). Cambridge, MA: MIT Press.

van Hiele, P. M. (1959). Development and the learning process. In *Acta Paedogogica Ultrajectina* (pp. 1–31). Groningen, the Netherlands: Wolters.

van Hiele, P. M. (1986). *Structure and insight: A theory of mathematics education.* Orlando, FL: Academic Press.

Wagman, H. G. (1975). The child's conception of area measure. In M. F. Rosskopf (Ed.), *Children's mathematical concepts; Six Piagetian studies* (pp. 71–110). New York: Teachers College Press.

Willats, J. (1977). How children learn to represent three-dimensional space in drawings. In G. Butterworth (Ed.), *The child's representation of the world* (pp. 189–202). New York: Plenum.

Willats, J. (1984). Getting the drawing to look right as well as to be right: The interaction between production and perception as a mechanism of development. In W. R. Crozier & A. J. Chapman (Eds.), *Cognitive processes in the perception of art* (pp. 111–126). Amsterdam: Elsevier.

7

Developing Understanding of Geometry and Space in the Primary Grades

Richard Lehrer, Cathy Jacobson, Greg Thoyre,
Vera Kemeny, Dolores Strom, Jeffrey Horvath,
Stephen Gance, and Matthew Koehler
University of Wisconsin–Madison

Our approach to geometry with young children begins with students' informal knowledge about situations, followed by progressive mathematical reinterpretation of these experiences, an approach consistent with the Dutch approach to "realistic mathematics education" (see Gravemeijer, chap. 2, this volume). Young children's everyday activities—looking, walking, drawing, building, and manipulating objects—are a rich source of intuitions about spatial structure (Freudenthal, 1983; Piaget & Inhelder, 1948/1956; Streefland, 1991; van Hiele, 1986). By looking at pattern and form in the world, children develop informal knowledge about geometric constructs like perspective, symmetry, and similarity. For example, preschoolers pretend that miniatures are small-scale versions of familiar things, and even infants distinguish contour and symmetry (Fantz, 1958; Gravemeijer, chap. 2, this volume; Haith, 1980). By walking in their neighborhoods, children learn to reason about landmarks, routes, and other elements of large-scale space (Piaget, Inhelder, & Szeminska, 1960; Siegel & White, 1975). By drawing what they see, children represent form (Goodnow, 1977). By building structures with blocks, toothpicks, or Tinkertoys, children experience first-hand how shape and form play roles in function (e.g., objects that roll vs. those that do not) and structure (e.g., sturdiness; see Middleton & Corbett, chap. 10, this volume).

Everyday experiences like these, and the informal knowledge children develop over time by participating in them, constitute a springboard into geometry. For example, the ideas that children develop about position and direction while walking in their neighborhood can be elaborated mathematically in a variety of ways—as coordinate systems, as compass directions, as maps, and as dynamic Logo models. Each of these mathematical forms of thought has antecedents in children's experiences (e.g., maps in children's drawings, coordinate systems in city blocks), and collectively these experiences constitute a good grounding for making mathematical sense of the spatial world.

Although rooting mathematics in children's experiences is consistent with theories of learning that emphasize the importance of situation (Lave & Wenger, 1991), it is essential that teachers also establish a classroom culture that grounds student activity in mathematical reflection and generalization (Cobb, Yackel, & Wood, 1995; Vygotsky, 1978; Watt, chap. 17, this volume; Wertsch, 1991). Consequently, developing student understanding of geometry relies on classoom culture as much as it does on mathematically fruitful situations. Skilled teachers develop models of student cognition and its typical trajectories of change (Clark & Peterson, 1986; Fennema & Franke, 1992; Schifter & Fosnot, 1993). Such models help teachers recognize "teachable moments" and other worthy landmarks in the ebb and flow of classroom activity.

Teaching and learning, then, are best viewed in tandem. It is important to identify mathematically important ideas and to build on children's experiences in ways that help children see mathematics as a way of making (more) sense of their experiences. Yet it is equally important that teachers understand landmarks in the progression of children's learning because, without a model of student learning, teachers must rely exclusively on curriculum and its associated scripts. Curriculum, however, cannot be designed to meet the manifold of possibilities inherent in a classroom. Hence, no matter how soundly designed, and no matter how sensitive to children's informal knowledge, curriculum alone cannot result in significant conceptual change.

Because many of the developmental trajectories we observed in the longitudinal investigation described in the previous chapter were comparatively "flat" or incremental, we decided to design classroom environments that would promote development of student reasoning about space and geometry. To establish a robust coordination between teaching and learning about space for young children, we collaborated with a small group of primary-grade teachers to develop a primary-grade geometry based on children's everyday activity related to (a) perception and use of form (e.g., noticing patterns or building with blocks), leading to the mathematics of dimension, classification, transformation; (b) wayfinding (e.g., navigating in the neighborhood), leading to the mathematics of position and direction; (c) drawing (e.g., representing aspects of the world), leading to the mathematics of maps and other systems for visualizing space; and (d) measure (e.g., questions concerning how far? how big?), leading to the mathematics of length, area, and volume measure. Our selection of these strands of experience was guided partly by the developmental trends in children's reasoning evident in the results of the longitudinal investigation described in Lehrer, Jenkins, and Osana (chapter 6, this volume) and partly by our intuitions about fruitful continuities between children's experiences and early geometry. The longitudinal investigation provided starting points for development and some potential signposts of progress, but for the most part we designed instructional environments incrementally, conjecturing about appropriate instructional activities and then testing them in the crucible of the classroom. Collectively, teachers and researchers conducted "design experiments" (Brown, 1992) that revealed typical patterns and progressions of student thinking when students were immersed for prolonged periods of time in classroom activity that supported the development of spatial reasoning. Each year, we revised our instructional design by updating our selection of curriculum tasks and tools, our models of student thinking, and our assessment practices in light of what we learned. In this chapter, we summarize some of the design principles and outcomes of a 3-year study of teaching and learning geometry in several second-grade classrooms.

INSTRUCTIONAL DESIGN

Our instructional design was multicomponential. The key constituents of the design (see Fig. 7.1) included (a) researcher descriptions of student thinking derived from the longitudinal study of development (Lehrer, Jenkins, & Osana, chapter 6, this volume), (b) teacher–researcher collaborative investigation of student thinking in the context of classroom instruction, (c) professional development workshops, and (d) a parent program that enlarged the learning community beyond the walls of the classroom. We briefly describe the first three components; a fuller description of the parent component is available in Lehrer and Shumow (in press).

Portraits of Student Thinking

We developed text and video descriptions of the growth and change in student thinking that we had observed during the longitudinal study and supplemented these descriptions with other research findings, as needed

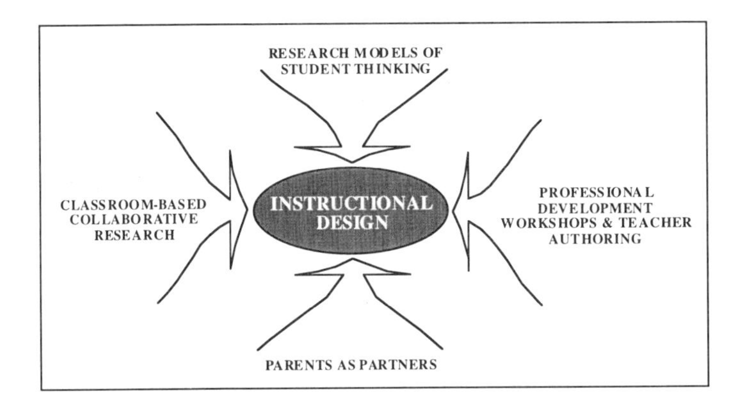

FIG. 7.1. Instructional design for the teaching and learning of geometry.

(see Clements & Battista, 1992, for a summary of related research). The resulting portraits of student thinking traced landmarks in reasoning about (a) structure and form, (b) large-scale space (wayfinding), (c) representations of space including drawings and maps, and (d) measure of length and area (Lehrer, Fennema, Carpenter, & Ansell, 1992). We also collaborated with a small group of teachers to develop descriptions of transitions in student thinking observed in their classrooms. This collaboration resulted in substantial modification and revision of the documents developed by reseachers. What started as a researcher-centered effort was transformed during the course of the 3 years into a collaborative research effort in which teachers and researchers jointly developed portraits of student thinking.

Professional Development

A major portion of the instructional design was devoted to professional development, including in-service workshops, teacher design of curriculum tasks, and teacher-authored descriptions of student thinking related to those tasks.

In-Service Workshops. During the first year, in-service workshops were anchored in Cognitively Guided Instruction (CGI; Carpenter & Fennema, 1992). The CGI program provides teachers with examples of developmental benchmarks in children's acquisition of knowledge about number, especially about situations involving addition and subtraction or multiplication and division. CGI focuses explicitly on the interactive roles of problem structure (e.g., the semantics of types of arithmetic word problems) and solution strategies (e.g., how children solve different types of problems) and, more tacitly, on the role of the teacher as a facilitator of student discourse about mathematical problem solving. In CGI classrooms, teachers commonly base instructional decisions on knowledge of student thinking (Fennema, Franke, Carpenter, & Carey, 1993). Part of the teacher's knowledge is knowledge of the individual child—his or her learning history—and part is knowledge of typical developmental trajectories. Based on developmental benchmarks, teachers decide what is potentially fruitful for "rousing to life" (Tharp & Gallimore, 1988) and what is likely to be pointless and even frustrating to a student. We redesigned CGI to include descriptions of children's spatial reasoning and geometric thinking (Lehrer & Jacobson, 1994).

Ten primary grade teachers participated in approximately 60 hours of in-service workshops devoted to developing their knowledge about key transitions in children's reasoning about space and number. The workshops were led by a CGI mentor teacher, Annie Keith, and a CGI project researcher, Ellen Ansell. Workshop participants read text about the development of children's understanding, saw videos of children solving problems in clinical interview settings, attempted to track the development of thinking in three "target" children in their classrooms, solved problems from the perspective of their children through role playing and related techniques, and discussed issues related to classroom implementation. Because teachers taught in different buildings, an electronic network facilitated communication among teachers, workshop providers, and researchers.

Tracking the development of selected children during the course of the year proved especially powerful, perhaps because teachers had the opportunity to reason about specific cases rather than reasoning about more general, and perhaps more ambiguous, groups like "children" or "the classroom." Case-based reasoning (Williams, 1992) helped teachers see the relationship between instances of thinking embedded in clinical interviews and children's ongoing thinking in their classrooms. Classroom cases helped teachers instantiate the more general benchmarks about children's thinking presented and discussed during the workshops.

The geometry workshops acquainted teachers with new elements of curriculum (e.g., wayfinding and three-dimensional form), new tools(e.g., magnetic compasses, Polydrons, and Logo), new forms of mathematical notation (e.g., two-dimensional "nets" to represent three-dimensional structures), and what research suggested about developmental benchmarks of student reasoning. Problems that teachers solved were designed to illustrate the potential interplay among instructional tasks, tools, and the notations that we anticipated that students might invent or appropriate. For example, to help teachers recast their understanding of spatial pattern (conceived by most teachers as the result of arranging wooden "pattern blocks" representing familiar forms like squares and trapezoids to form sequences like square, triangle, square, triangle, etc.), during one teacher workshop we presented a task that included the creation of a pattern as a repetition of identical elements (Senechal, 1990):

> Write directions to tell someone how to draw one of the shapes from the pattern-block bucket. What pattern do you notice about the shape, and how is the pattern represented in your description of the drawing? Now write a procedure in Logo to draw the shape. How does the change in representation from paper-and-pencil to Logo change the directions? What stays the same?

While solving this problem, teachers explored simple shapes as a repetition of identical elements and considered the roles that different forms of representation (i.e., drawings, directions, Logo procedures) might play in children's thinking about form and pattern. Logo procedures, for instance, make repetition of identical elements explicit (Lehrer, Randle, & Sancilio, 1989). For example, a square in Logo is the fourfold repetition of a movement of the turtle forward or backward followed by a 90° turn by the turtle. During another workshop, teachers created a map of the school library, a model of a large-scale space. While constructing the map of the library, teachers considered the interplay among spatial configuration, scale, and measurement. Teacher discussion connected this experience to related ideas, such as perspective for depicting objects in the map (bird's-eye view vs. object view) and the utility of symbols for maps (e.g., icons, photos, pictures). Teachers also talked about standard units of measure, what they might be "good for," and what ideas children might have about standard measure.

The in-service workshops helped teachers develop greater understanding of student thinking about number and space, and they introduced teachers to pedagogical issues, especially the importance of multiple forms of representation and the need to help children develop a language about space. The workshops were forums for constructing knowledge, for enlisting the support of peers and of researchers, and for developing a community for the reform of mathematics education in the participating schools. In subsequent years, teachers organized their own forums for professional development, where they gathered to share ideas, reflect on their experiences, and initiate others in their practices. During the third year, this resulted in a teacher-led series of afternoon and weekend workshops for colleagues, with researchers in an advisory role.

Teacher Authoring. After the first year of workshops, five teachers elected to participate in a week-long summer institute, during which they designed tasks to help students develop understanding of the mathematics of space. For each task they designed, teachers conjectured about conceptual landmarks or "ways of thinking" that students would encounter as they worked with the task. Ideally, tasks allowed multiple points of entry for different forms of background knowledge and skills. Tasks also included appropriation of curriculum units developed elsewhere, with revisions guided by the goal of making student thinking more visible. Besides helping teachers track student understanding, making student thinking visible is a necessary first step in creating shared reference for classroom discussions.

Tasks were refined during group conversation. For example, one teacher, Jennie, proposed incorporating length measure into a thematic unit on "How big am I?" The teacher's original idea was to use a story, "How big is a foot?" to illustrate how using a different unit of measure can change the measured length of an object. She proposed that students use their own feet to measure the length and width of their classroom and then talk about why their measured lengths were different. During the group discussion, several colleagues noted that the task as designed might, in accord with the research portraits of student thinking, raise other issues in measure theory, such as the need for identical units, iteration of units (some teachers thought that students might leave spaces between footprints), and the relationship between the object being measured and the unit of measure. This discussion resulted in redesign of the task to include other personal units of measure and more opportunities for developing the need for standard units (to reconcile different but valid measures obtained with personal units). Jennie then used the task during the course of instruction the following year, documenting benchmarks in student reasoning. The same teachers also participated in a second summer institute the following year, editing and revising their work in light of student performance during the course of the second year. During this second year, teachers' descriptions of student "ways of thinking" became richer and more complete, resulting in more articulated portraits of student thinking.

RESEARCH DESIGN

Our research design reflected the complexity of the instructional design. Accordingly, we developed multiple sources of data about student learning, teacher beliefs and teacher practices. We focused on three Grade 2 classrooms. At the beginning of the study, one teacher had taught for 3 years, another for 7, and the third for 5 years. We tracked change for 3 years in two of the classrooms, and for 2 years in the other classroom (the teacher moved at the beginning of the third year). Two of the classrooms were in the same elementary school; the third was in an elementary school located nearby. The students came from a mixture of middle- and working-class families.

Teacher Practices

We observed each teacher's classroom practices three times each week during the course of each school year. Each observation lasted approximately 2 hours and included audiotape and/or videotape of classroom activity, interviews with individuals or groups of children, (often) interviews with teachers about their goals and their views about each lesson, and artifacts of classroom work. Consequently, we were able to amply document teacher practices and transition during the 3-year span. The classroom observations also provided a window to student understanding and to conceptual change.

Student Learning

Each year, we assessed student learning about (a) two- and three-dimensional shapes, including transformations of the plane; (b) position, direction, and perspective; (c) notations and representations of space, including drawings, plan views, maps, and nets; and (d) measure of length, area, and volume. We also assessed student learning about number, especially student ability to solve a variety of word problems. We employed two forms of individual assessment, administered at the beginning and end of each school year. The first consisted of paper-and-pencil measures of problem solving. The second consisted of clinical interviews that allowed us to examine children's reasoning and strategies in greater depth. Sample items and interview questions are displayed in the Appendix. In addition to this individual level of analysis, classroom observations provided a window to collective student learning as revealed by children's whole-group conversations and by ongoing researcher–student interviews of selected children in each class.

CLASSROOM IMPLEMENTATION

Although the three teachers we observed all participated in the same series of workshops and engaged in the summer institutes devoted to curriculum design, each developed a unique stance in relation to teaching children about space, so that the view from each classroom was characterized more by variation than commonality in activity. For one teacher, Ms. C., the predominant theme for coming to know about space was to measure it, so in her classroom, many of the tasks posed to students involved measurement. For example, students designed their own tape measures for length, investigated and invented units for area, and, given a single sheet of 8.5 × 11 in. paper, designed popcorn containers that would hold the most and the least popcorn. In contrast, for Ms. S., the predominant theme for coming to know about space was to experiment with form, and many of the tasks she posed to students involved contrasting and comparing different two- and three-dimensional forms, finding and constructing the Platonic solids, and designing quilts and other patterns. For Ms. J., children's learning about space was anchored in wayfinding and representation. Accordingly, she most often posed tasks to children that involved wayfinding, mapping, and graphing.

Transitions in Task Structure

Despite the difference in the strand of spatial mathematics that teachers chose as their "leading edge" or emphasis, all teachers shared some core practices. Over time, all three teachers went from posing tasks in isolation to developing sequences of tasks that provided children with opportunities for progressive elaboration of core concepts. For example, in Ms. J.'s first year, children wrote directions to find a spot marked "x" in their classroom, with revisions emerging as children negotiated the measuring of length and turns in their directions. By the third year, writing directions was merely the starting point for a series of tasks involving wayfinding, mapmaking, and Logo. Consequently, children in her class were afforded progressive opportunity to reconstitute their initial understanding of position and direction from intrinsic notions of position and direction, to understandings that encompassed perspective, scale, and extrinsic frames of reference.

Transitions in Representational Fluency

During the course of the study, teachers also increasingly emphasized representational fluency; children invented or appropriated multiple forms of representation. For example, in Ms. S.'s classroom, students reconstituted their natural language for two- and three-dimensional forms as properties; they transformed their understanding of form to include (consensually agreed on) lists of these properties, and amplified their knowledge of form through drawings (e.g., top, front, and side views of solids), Polydron constructions (e.g., trying to find "perfect" [Platonic] solids), and Logo procedures. A related development was a proliferation of tools (and associated practices), which children used to sustain investigation of space (ranging from magnetic compasses and Polydrons to student-invented means for measuring length, area, and volume).

Transitions in Classroom Discourse

A noticeable shift in conversational patterns emerged during the 3 years of the study. Initially, teachers were more likely to encourage classroom talk when it concerned numeric solution strategies: They asked children to compare and contrast the solution strategies of their peers and emphasized that a number of different strategies could be used to find an answer. In contrast, their initial scaffolding of conversations about space generally consisted of simple elicitation of multiple ideas from children, a "stirring" of the pot of ideas,with little attempt to compare and contrast ideas or to guide children toward selecting some of their ideas for further exploration.

During the course of the study, the nature of classroom conversations about space changed dramatically. First, there were many more of them as the proportion of time children spent exploring the mathematics of space increased, especially during the second year of the study. Second, teachers became increasingly adept at discerning patterns in children's thinking about space, so that they often helped children talk about similarities and differences among their ideas, in contrast to their previous pattern of simple elicitation without comparison. Third, teachers became much more adept at helping children develop a coherent language about space to supplement their initial, near-exclusive reliance on gesture and shared visual regard to communicate about space. Children's talk about figures came to include descriptions of properties and ways to generate instances, in contrast to their earlier reliance on single-word descriptions (e.g., "skinny") and gesture. Fourth, talk became intimately connected to justification and argument. Classroom conversations often revolved around the need of students to convince others of the validity of their own viewpoints.

In summary, during the years of the study, teachers either designed or appropriated a number of different tasks as initiators of student learning. Tasks became progressively more interconnected, and teachers used them to revisit or "spiral" important ideas. Representational fluency was increasingly emphasized, so that students rarely ever talked without drawing or building, or measured without designing a tool. Perhaps the most noticeable change was in the nature of classroom talk about space. Children's initial talk about space was nearly always gestural and rarely intersubjective; in later conversations, gesture supplemented linguistic description, and language was often taken as shared.

STUDENT LEARNING

Each year of the study, we measured students' problem solving in both space and number, and we noted significant transitions in student thinking each year in all classrooms. We found significant growth in children's number sense and in children's spatial sense, as indicated by the number of problems correctly answered at the beginning and end of each year.

For number sense, individual interviews suggested that this change in performance could be attributed to two main factors. First, children developed more sophisticated strategies for solving number problems during the year. To measure this increase in sophistication, strategies were assigned levels: 0 for No Solution, 1 for Direct Modeling (e.g., children uses counters to represent quantities following the action sequence in the word problem), 2 for Counting (e.g., children solve an addition problem by counting on from the larger addend), 3 for Recalled Fact, Derived Fact, and Algorithm (e.g., children invent or use algorithms, like $27 + 27 \rightarrow 20 + 20 = 40$ and $7 + 7 = 14 \rightarrow 40 + 14 = 54$; see Fennema et al., 1996). The growth in the highest level strategy available to children at the beginning and end of each year is displayed in Fig. 7.2a. Second, over time, children were able to apply a wider range of strategies to problems. To track progress in the range of strategies available to children, strategies were assigned to one of four classes: Direct Modeling, Counting, Recalled and Derived Facts, and Algorithms. The significant increase in number of classes of strategies that children demonstrated to solve arithmetic word problems is illustrated in Fig. 7.2b.

We also noted significant pre–post conceptual change in each of the four strands of spatial sense, as indicated by student scores on problems

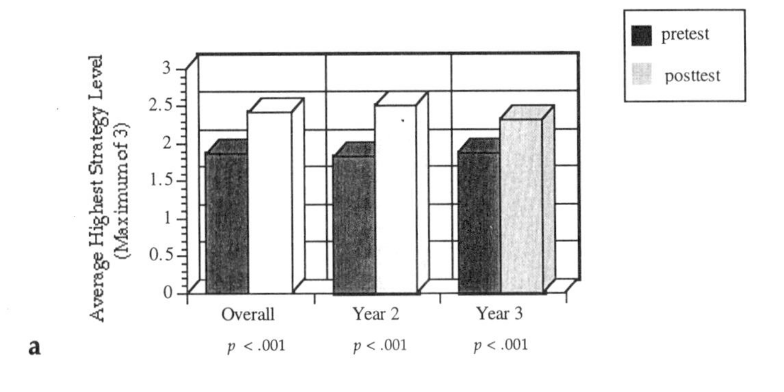

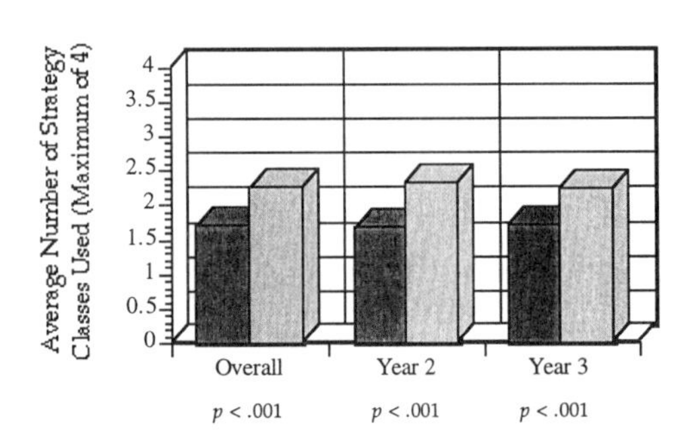

FIG. 7.2. Change in students' use of arithmetic problem-solving strategies over the 3-year study, by year.

administered at the beginning and end of each school year. For example, Fig. 7.3 shows the improvement in students' performance on problem-solving assessments during the third year of the study. The proportion correct was calculated for each of the four strands, and these four proportions were then combined with equal weight to arrive at a total score. Individual interviews suggested major transformations in how children thought about the structure of two- and three-dimensional space, measurement, representations of space, and position and direction in a large-scale space. For example, children's initial conceptions of shape and form were dominated by resemblance to familiar objects and other aspects of appearance (see Lehrer et al., chap. 6, and Pegg & Davey, chap. 5, this volume), but they changed during the course of the year to include reasoning about properties across a variety of contexts. Similarly, children's initial ideas about length or area measure often confused the two, but every year, their understanding of key ideas in length and area measurement, such as the need for identical units of measure, far exceeded that of fifth-grade children in their school. Although these transitions were expected, they were unusual considering the relatively static patterns of growth and development that we observed in the longitudinal study (Lehrer et al., chap. 6, this volume).

TEACHING AND LEARNING IN THE CLASSROOM

To illustrate the interactions between teaching and learning in these classrooms that could account for the striking pattern of conceptual change noted previously, we focus on two strands of learning in Ms. C.'s classroom. In the first, children learned about transformational geometry and symmetry as they designed a quilt. Not all of children's learning was domain specific; children also explored issues in epistemology, especially the the limits of case-based induction. In the second, children learned about area and its measure. The lessons on area illustrate how teachers wove

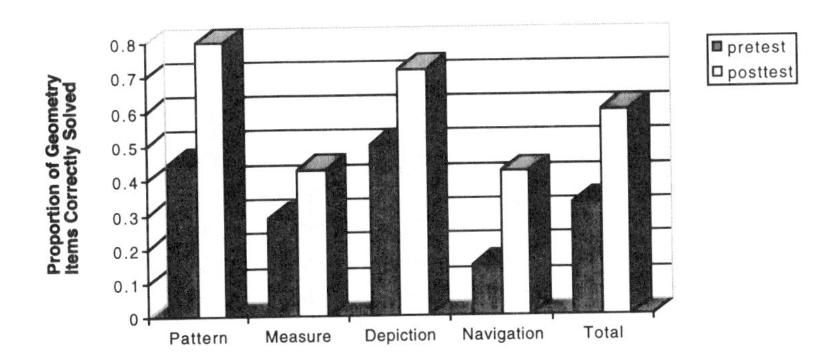

FIG. 7.3. Improvement in students' performance on problem-solving assessments during the 3-year study, by strand of geometric reasoning.

tasks, notations, and conversations to guide children's exploration of the foundations of measure theory.

Designing Quilts[1]

Quilts and quilting were part of the cultural heritage of most of the children in this classroom; children were interested in the quilt designs they saw in books, video, still photos, and pieces brought to the classroom by quilters in the community. Quilt design provided children the opportunity to explore important mathematical ideas like symmetry and transformation, and to develop conjectures about how these mathematical ideas might contribute to their aesthetic experiences of the artistry of quilt designs. Here we trace the progression of student thinking in Ms. C.'s class over a 5-week period during the third year of the study, by focusing on "snapshots" of classroom activity and conversation. More extensive discussion of connections between children's thinking about quilt design and algebraic concepts are discussed in Strom and Lehrer (in preparation).

Partitioning Space. Students first designed a paper-and-crayon (replaced later by paper cutouts and then computer-screen objects) "core square," the basic object subjected to geometric transformations to produce a quilt (see Fig. 7.4). A core square was composed of an array of squares, each partitioned into two right isosceles triangles. For some students, the apparently transparent idea of partitioning the square into triangles was somewhat problematic, partly because the task presupposes acquisition of diagonality (Olson, 1970).

After creating the core square, children used isometries of the plane (flips, turns, and slides) to arrange four identical copies of their core square into 2 × 2 designs. These designs were then composed to create the final pattern for the quilt. Other quilts were designed by transforming "strips" of core squares: A strip was a row of core squares arranged by application of the three transformations.

The complexity of the core square could be varied by using a mosaic of different forms (e.g., smaller triangles that tiled the region) and colors. Efforts to redesign quilts by changing the core square led children to consider a variety of ways of partitioning the same region of space. Moreover, these design efforts helped children explore the consequences of different transformations and combinations of transformations for properties of form like color adjacency and symmetry. In the sections that follow, we describe these and related forms of thinking about the plane.

Distinguishing Physical from Mathematical Motion. Because children constructed two-sided paper core squares or core squares composed of Polydrons, they could physically enact flips, slides, and turns. The curriculum confined these initial experiences to translation, vertical and hor-

[1] The quilt design curriculum was developed by Education Development Center. The classroom teacher described here also contributed to the design of the curriculum during field testing.

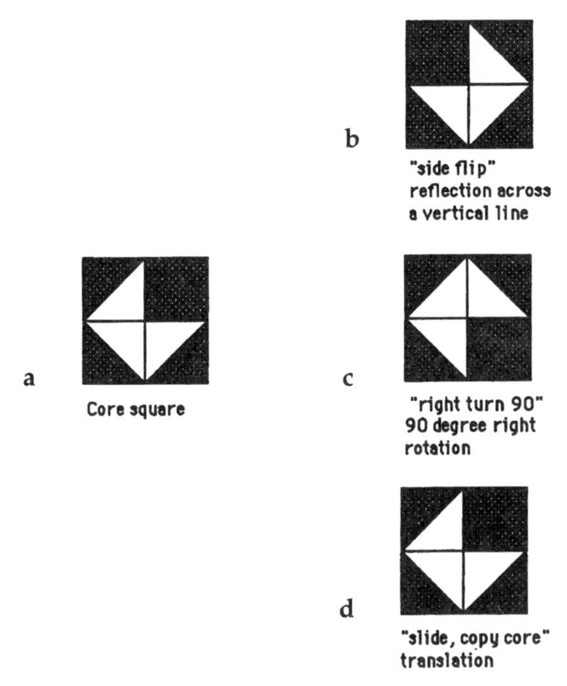

FIG. 7.4 Sample core square and three possible transformations. (a) Core square. (b) "Side flip": reflection across a vertical line. (c) "Right turn 90": 90° right rotation. (d) "Slide, copy core": translation.

izontal reflection (flips), and rotation in increments of 90°. To help children mathematize these physical motions, children developed a notation to describe each motion so that they could easily write directions for other children to follow when replicating a particular quilt design. Through discussion and consensus building, children developed the following notations for flips: UF for *up-flip* (reflecting about the horizontal axis by grabbing the bottom edge and flipping the square over), DF *down-flip* (reflecting about the horizontal axis by grabbing the top edge and flipping the square over), LF *left-flip* (reflecting about the vertical axis by grabbing the right edge and flipping the square over), and RF *right-flip* (reflecting about the vertical axis by grabbing the left edge and flipping the square over). Note that children's notations for reflections included some that could be distinguished in physical motion but that had no mathematical counterparts.

When conjecturing about what steps a fellow student might have taken to create her 2 × 2 design from the original core square, some students thought that one of the actions taken was an up-flip, and other students thought that the same action was a down-flip. Upon testing these conjectures, the class discovered that both flips produced the same result—it was impossible to distinguish one from the other.

Because no consensus was reached after the first example, Ms. C. went on to test other examples, using Polydron models of children's core squares:

> Maybe that was just their core square. Maybe they just had a weird core square. Let's try it with this core square. OK, do they look the same, the way I have them now?

Note that Ms. C.'s comment about "maybe that was just their core square" invited children to consider multiple cases, and, by implication, a search for a negative case, not just single confirming instances in their justification. At this point, Ms. C. had established a routine for testing the conjecture on a given core square. She placed two copies of the same core square side by side, then flipped one up and the other down. When she finished these flips, the core squares were no longer side by side; the one that was flipped up was above the original position; the one that was flipped down, below the original position. It was visually apparent that the actions performed on the two squares were physically different; it was also obvious that both actions resulted in exactly the same pattern on both squares. After several more examples, children became convinced that they needed to change their notation to UDF for *up–down* flip and SF for a *sideways* flip because different physical actions (up or down, right or left) led to identical results. This process helped children distinguish the plane of action from the plane of mathematics.

Viewing the World Through Notation. The use of notation for communicating design also helped children reexamine physical motion from a mathematical point of view. When children wrote directions for 2 × 2 designs that involved compositions of transformations (successive transformations, like a turn followed by a flip), they often only represented one of the motions in notation. For instance, a child wrote SF for a composed motion, like SF TR 1/4. After prolonged discussion and extended exploration, some members of the class proposed that the source of the difficulty was that their wrists could "flip and turn" at the same time—the differentiation implicit in the mathematical notation was not well differentiated in the motion of the wrist. Children used the notation to recast continuous physical motion as discrete steps. More generally, this discovery is a microcosm of the relationship between a model (here, the motions on the plane) and the world (here, the physical movements of hands): Models fit (to a degree) the world, but the world also changes as students view it through the model. Hence, model-fitting is a dual relation.

Generalization. Ms. C. often invited students to make conjectures about transformations by considering the generality of a case generated by one or more students. The signature phrase "Do you think this is true all the time?" was usually followed by a search for other confirming cases and for counterexamples. The grounds of evidence usually consisted of a larger number of cases that clearly fit what the class called a "rule" (a generalization), accompanied by lack of (a failure to find) counterexamples. This type of conjecture–evidence cycle can be illustrated by a conjecture about the number of flips of a core square that would return it to its original position (the order of the up–down flip). The original position was distinguished by a small "x" in the upper left corner, a convention introduced by the teacher to facilitate discussion. When one child suggested that it would take two up–down flips, Ms. C. replied, "Two? Let's try it. Watch. Memorize Katie's core square. This is what it looks like. One [flipping the core square]. Two [flipping it again]."

At this point, children established that the order of the up–down flip was two. Students suggested that 0 and 4 flips would also work, and they tested this conjecture with the core square. Br then suggested that any *even* number would have the same result. Children went on to explore this conjecture, testing a number of cases before one student, Ke, suggested that counting by 2s "as high as you wanted" would have the same result.

The Limits of Case-Based Generalization. Although the class norm for evidence about a conjecture consisted primarily of inductive generalization from positive and negative cases, the limitations of cases were discovered by this class. Ms. C. noted that all children had designed at least one asymmetric core square, and all of these asymmetric core squares were used to create at least one symmetric 2 × 2 design. Despite Ms. C.'s appeal to students' experiences and their positive cases, some students in the class remained unconvinced about the generality of the conjecture. Ms. C. decided to probe children's thinking about the number of cases that might serve:

Ms. C.: So if we shared 20 or more ways together today that you could start with an asymmetrical core square, and still every time make a symmetrical 2 × 2 design, how many more do you think we'd have to test before we could say you could always do it?
Ni: Hundreds of hundreds of thousands.
Na: We would have to test all the core squares in the world.
Ms. C.: Could we do that?
Class: No, no.
Na: We'd have to test all the core squares in the world that are asymmetric.

Children decided that they would have to test every case (an exhaustive procedure), and ultimately decided that this would be neither practically feasible nor even, in principle, possible because, as one child noted, "People are probably making some right now." We believe that reasoning about the limits of induction, a theme that resurfaced later in this classroom, sets the stage for (informal) proof as a form of argument. We are currently working with teachers on forms of instruction that build on this foundation.

Art and Geometry. During the course of 5 weeks, children's appreciation of the aesthetics of quilt design changed markedly. One measure of this transformation was their talk about what they found interesting when they saw a video accompanying the unit, which displayed a variety of quilt designs. Initially, children's talk about design was weighted heavily toward "cool" colors. Over time, their comments began to shift, so that by the end of the unit their talk about "cool" quilts included consideration of the roles played by lines of symmetry, complexity of form (e.g., number of partitions of the core square), and transformations that produced different types of color adjacency. Students also began to notice the constraints inherent in certain design considerations. One child (Danny), for example, cautioned a peer about the constraints certain symmetrical core square designs put on the variation of 2×2 designs that can be constructed from them:

> Don't make one color diamond in the middle and all the corners one other color because no matter if you flip it or do anything with it [meaning transformations to the core square to make the 2×2 design] it won't work [to produce multiple 2×2 designs]... because it is symmetrical all the way.

Fig. 7.5 displays one instance of Danny's general principle. Note that any transformation will produce, in Danny's words, "the only 2×2 you can make" from the core.

Summary

Quilt design was a fertile ground for developing and exploring a mathematical model of the plane. Informal knowledge of drawing and aesthetics served as a springboard to mathematical notation and argument, which were mutually constituted in the ongoing activity of the classroom. Children's explorations of transformations and symmetry resulted in reorganization of their informal knowledge of aesthetics and design, thus

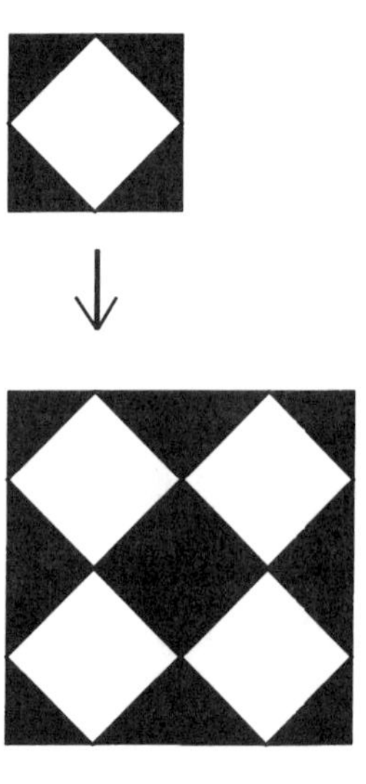

FIG. 7.5. An example of constraint caused by symmetrical design of core square (Danny's general principle). Note that no transformation in design occurs no matter which way the core square is "flipped."

completing a feedback loop in which the mathematization of informal knowledge was incorporated into the body of that knowledge. In the next section, we describe a similar process of progressive mathematization of everyday experience: how children's informal knowledge of appearance and the amount of space covered by a shape was successively transformed into the mathematics of measure.

DEVELOPING AN UNDERSTANDING OF AREA MEASURE

Ms. C. designed a spiral of tasks, all related to thematic units, to help children develop their ideas about area and its measure. The design of each task was guided by her knowledge of how children think about principles of area measure like space-filling, additivity of areas, and identity of units (Lehrer et al., chap. 6, this volume). The progression of tasks illustrates how teachers used their knowledge of children's thinking as a guide for designing and adapting instruction. Children's responses to each task indicate again the interactive roles of tasks, notations, and classroom conversation.

Three Rectangles

Ms. C. first asked children to judge which of three "quilt pieces" (rectangular strips of construction paper tacked on the blackboard) "covers the most space." The dimensions of each quilt piece were unknown to children, but the pieces were designed to correspond to different arrangements of the core squares of the quilting unit (1×12, 2×6, and 4×3 core squares, respectively). The core squares were not demarked in any way. She labeled each quilt with a letter—A, B, and C (see Fig. 7.6).

The design of the task reflected Ms. C.'s understanding of student thinking. She wanted children to construct a unit of area measure by building on their informal knowledge of cutting and rearranging pieces (see Lehrer et al., chap. 6, this volume). She chose these shapes expecting that the conflict between perception (the rectangles appear to cover different amounts of space) and conception would lead to eventual construction of a unit of measure:

> Once they make predictions [about which covers the most space], I expect they might say, "Well, I think Shape C covers more space." "No, no, no, it's A—Look how long it is." But when I ask them, "How can we find out?," what are they going to say? Will they suggest covering it? Will they suggest measuring around the outside? Will they suggest folding it in half? [And I will tell them,] "You are looking at these three shapes. You have different ideas about which might cover more space, but how are you going to prove to someone what you think might be true?" By the end of the lesson, some of them will say, "Well, it looks like it takes up more space, but really you could just push that space around and make it fit."

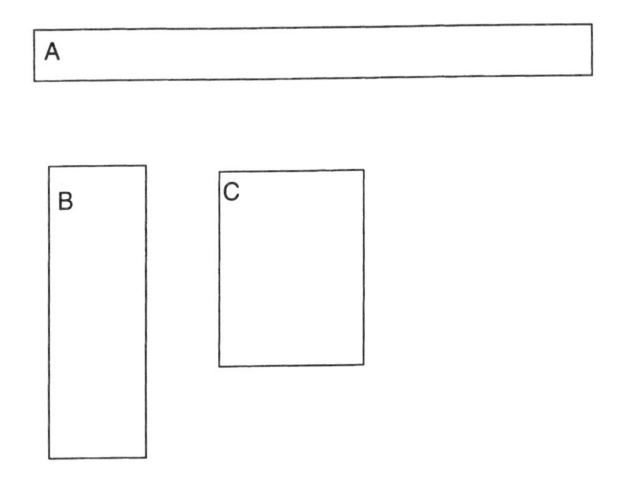

FIG. 7.6. Constructing units of area measure: "Which quilt piece covers the most space?"

Additive Congruence. Most students began by claiming that shape B or shape C would cover the most space because they were "fatter" or "looked bigger." However, some of the students in the class disagreed and thought that perhaps shapes B and C were really the same size. Mi's work is shown in Fig. 7. 7. She decided to fold B in half because she saw that if she folded it in half and rotated it, it would cover exactly half of C. She also noted that if half-B was flipped horizontally, it would cover C. Mi's knowledge of transformations facilitated this exploration. Children went on to explore other partitions of B and C that could lead to this result.

Constructing Units of Measure. Eventually satisfied that B and C did indeed cover the same amount of space, the class turned its attention to A.

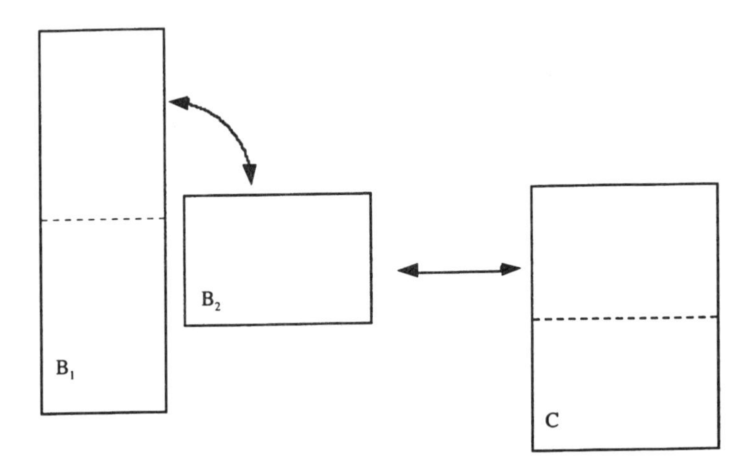

FIG. 7.7. Mi's solution: folding B to measure C.

Mic claimed, "You make C into A," and proceeded to demonstrate by folding C into four equal parts, each the width of A (see Fig. 7.8a). He placed this "strip" at the top edge of A and used his finger to demark the bottom edge. Then he moved the entire strip down to his finger, and proceeded in this way to iteratively mark off four equal segments of A.

Another student, Ti, folded C into three long strips instead of four short strips (see Fig. 7.8b). When Ti finished with his demonstration of additive congruence, and C was unfolded, the fold lines clearly divided the rectangle into a 4 × 3 array of squares (see Fig. 7.8d). This was not noticed by the class until Ms. C. asked: "How come it took Mic four strips and only took Ti three?"

While the class pondered this problem, Ti was counting "1, 2, 3, . . ., 12." When Ms. C. asked him to clarify what he meant, Ti replied, "Twelve squares! That makes a quilt!"

A second student pointed excitedly to A and said, "Then it takes 12 squares to make *that*." Ms. C.'s question instigated a transition in strategy from additive congruence to units of measure. Before, the children were talking in terms of cutting up shapes and rearranging figures, but at this point they were talking about the number of core squares in a quilt. The children went on to verify that each of the three forms could be composed of exactly 12 square units or core squares. Ms. C. then posed an additional problem of designing as "many shapes as you can" composed of 12 core squares. She used computer software as a tool for letting children freely explore the idea that appearances can be deceiving (e.g., different looking forms can cover the same amount of space) and recognized units of measure as good conceptual tools for addressing this problem.

Ms. C. also used this task as a forum for considering other forms of argument. For example, in another class, one student, Sa, proposed a form of transitive inference—by folding and covering, the class had established that rectangle A and rectangle B each covered the same amount of space (A = B), and also that rectangles B and C each covered the same amount of

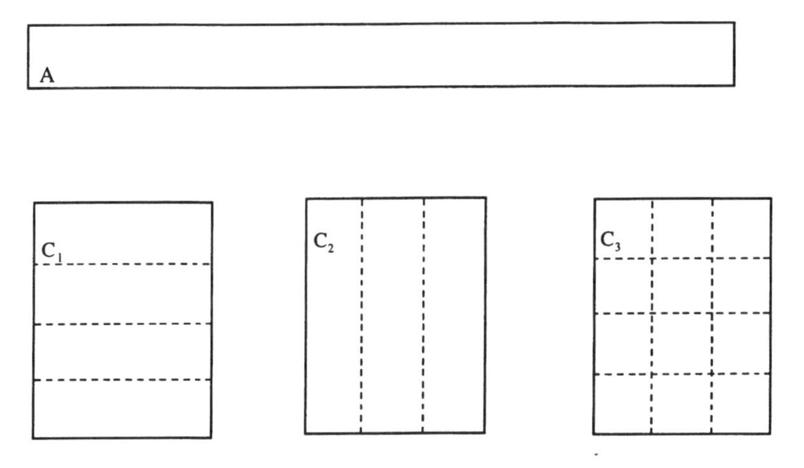

FIG. 7.8. Folding C to measure A.

space (B = C). Consequently, Sa suggested that this must mean that A and C also covered the same amount of space. Ms. C. then asked the class to consider this "without testing," meaning that she wanted them to establish how to verify a conjecture based on a chain of propositions, rather than simply to test its truth empirically.

Ms. C. noted that, at the end of the lesson, children "were using a square in the conventional way (of measuring area) but they were using it in a meaningful way and then they were able to make sense of other area problems ... it's a kind of bridge between a standard unit of measure and a personal unit of measure." Here Ms. C. was drawing a contrast between textbook problems that assume that conventional square units of area measure are conceptually transparent versus her recall of the constructed nature of the square unit in her classroom.

Area of Hands

Following the construction of a unit of measure for the area of a familiar form like a rectangle, Ms. C. posed the problem of rank-ordering the "amount of space covered" by individual students' hands to help children "cement" the utility of a unit of area and to explore the qualities of different potential units of measure. Ms. C. suggested that "the hand doesn't lend itself to thinking about squares of space, as quilts do." (She knew from our research that children prefer units of area measure that perceptually resemble the figure being measured.) She also indicated that the task provided opportunities for children to develop strategies for finding area when the form does not lend itself to easy partitioning into subregions, especially the problem of "what to do with the leftovers" (units that are fractional pieces). The task was ill-structured, in that size, material (real hands or representations), and method were all unspecified.

As Ms. C. anticipated, the task provoked considerable discussion about appropriate units of measure. Children first tried to solve the problem by adapting the strategy that had worked to determine the congruence of the three rectangles: They superimposed pairs of classmates' hands. They discarded this approach as both too time-consuming and perhaps fatally flawed. As one child said, "What do we do with fat thumbs and thin fingers?," indicating that handprints are not uniform, and direct comparisons are, therefore, difficult. This prompted children to consider working from a representation of a hand, rather than working directly with hands. A prolonged discussion eventually led to adoption of a classroom convention about tracing hands on construction paper. Thereafter, children worked with paper representations in lieu of hands.

Children then attempted to develop a unit of measure. As Ms. C. anticipated through her knowledge of children's thinking, most of the children's inventions resembled perceptual features of the hand in some way: beans, fingernails, spaghetti, and rope were all considered and subsequently rejected. Children's reasons for rejection helped them better understand fundamental properties of area measure. For example, children found that using beans wasn't satisfactory: Two different attempts to use beans as units to measure Gl's hand led to two widely different quantities; beans "leave cracks" (a reference to the space-filling principle of area measure); "beans are not all the same" (a reference to the identical-units principle of area measurement); and so on. As children deliberated about their choices, it became clear that they found none of their inventions satisfactory—all violated one or more of the canons of measurement that they had decided were important.

At this point, Ms. C. held up a piece of grid paper marked off in squares and asked children: "Could we use this as a tool?" Most replied very emphatically no, suggesting that squares did not look anything like hands. One child disagreed with the others and said yes, because the squares were all the same (identical units), and they didn't have any "cracks" (space-filling). At this point, another child agreed but noted, "I see a problem—there will be leftovers." After explaining to her peers the nature of the problem, another child proposed a solution: Use different colors to estimate parts that would constitute a whole. For example, $^7/_8$ of one "leftover" might be combined with approximately $^1/_8$ of another to constitute one whole. Both pieces were marked with the same color (e.g., purple), and then other pieces would be identified and marked with different colors (e.g., $^1/_2$ and $^1/_2$ marked red). This system of notation did not quantify the part–whole relation (children never wrote $^1/_8$ or $^1/_2$), but it did help children keep track of their estimates.

By the end of this lesson, children had confronted some fundamental issues involved in constructing a unit of area measure, most especially the need for combining identical units, the importance of space-filling, and the irrelevance of resemblance for judging the merits of a unit of measure. Children invented a system of notation to record their estimates about combining the "leftovers," a process that helped many see that area measure need not be confined to integer values. Most especially, children were able to see measurement do some real work; their deliberations led to a rank-ordering of all of the handprints in the class.

Ms. C. noted that the task also led children to reflect again about area measure, "distinguishing between that kind of measurement [some children first proposed using length measure, e.g., the span of the hand] and what area really is, and they quickly saw that they had to have a way to quantify how much space that [the handprint] covers." As she anticipated, the lesson helped children reflect about the need and functions of units of area measure: "I find that kids, when you do this with them, want something that is going to fill in the tips of the fingers, like beans or fingernails. They wrestle with beans and fingernails, then they figure out that they need things the same size and with complete cover." Ms. C. pointed out that the latter properties of units of area measure (that they be identical and space-filling) were tacit in the core squares of the three rectangles task, but were made explicit in this task.

Area of Islands

Several weeks after rank-ordering the area of hands, children each drew their own "islands," during a unit on geography. They then attempted to rank-order the area of each island. Ms. C. chose this task as a follow-up to the area of the hand, partly because she believed that children needed to explore further the fundamental properties of area measure that they had discovered in the previous lesson ("another context for making sense of why area measure might matter"), and partly because she wanted to introduce children to a notational system for keeping track of the pieces invented in another classroom ("They needed a chance to work with those parts of squares again"). She also believed that the strategy of simply superimposing objects would be more obviously unwieldy, and children would, therefore, be more likely to consider units-of-measure strategies rather than congruence strategies.

Although this lesson was also rich in mathematical talk and provided children further opportunities for exploring principles of area measure, the lesson was perhaps most noteworthy for children's use of their previously invented notations to measure the area of each island. Children's first ideas about measuring area mobilized (Latour, 1990) their previous "color matching" strategy to deal again with the problem of the leftovers (the fractional pieces of area measure), and they reached consensus quickly about the virtues of again employing square-grid paper as a measurement tool. At this point, Ms. C. introduced a new notational system, invented in another class, where students symbolized all part–whole relations (e.g., 1/4, 1/2, 1/3), filled all equivalent fractions with the same color, and then combined the pieces to make whole units. This alternative notational system put another cast on composed congruence: A whole unit could be constituted in a variety of ways, but each of these ways could be shared symbolically, not just indexically. In the first system, there were also multiple ways of making a whole unit, but each composition had to be considered by shared visual regard and represented the judgments of individuals. In the second notational system, each composition was more easily communicated because it could be shared symbolically, and it appealed to conventional representations (i.e., fractional pieces) rather than idiosyncratic representations of part–whole relations.

Ms. C. continued to emphasize understanding and reflection, not simply doing: "[The task provided opportunities for students]" to deepen their understanding of why it mattered for them to think about those leftover parts. And what it was that they were actually doing, which was making those parts into whole units so that you could account for all the space."

Area of Zoo Cages

Children designed a zoo and investigated ways to redesign the city zoo (a project being undertaken by the city). Within this context, Ms. C. posed a problem of comparing the areas of different zoo enclosures on a large sheet of paper displayed on the blackboard (see Fig. 7.9). She suggested that the task provided opportunities for students to revisit the conception–perception mismatch of the three-rectangles task and to re-represent the idea of area measure symbolically, as a multiplication of lengths. In this instance, students were not provided any tools except a ruler.

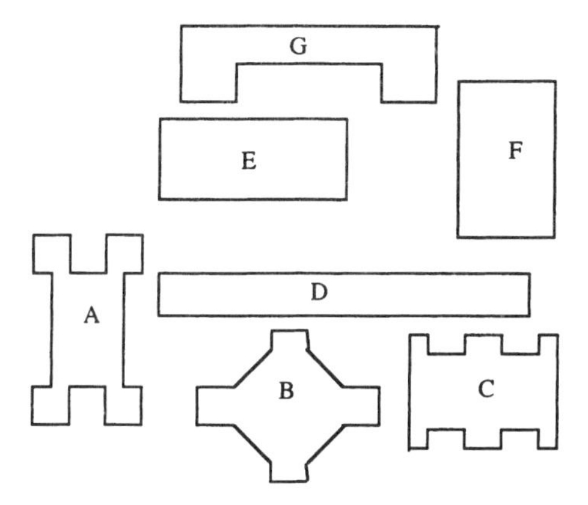

FIG. 7.9. Redesigning a zoo: units of area measure.

The conversation began with one student's assertion that two of the rectangles were "exactly the same." Other students took this as an assertion that the two rectangles were congruent. To test this idea, students proposed superimposing one rectangle on the other, but Ms. C. did not permit students to cut out the rectangles. Another student proposed that if two rectangles were congruent, then their corresponding sides would have the same measure. This student used the ruler and found that the measure of one rectangle was 5 × 8 in. and the other, 4 × 10 in.:

Ms. C.: So the whole Shape E and the whole Shape F—you're saying that Shape E and Shape F are the same?

Ca: Not shapewise, but they take up... they both take up the same amount of space.

Ms. C.: OK. Ca is saying that he thinks Shape E and Shape F cover the same amount of space. He's saying they're not the same shape exactly—he says they look different, and we just measured two sides and showed that they're different.

Ca: Four times 10 is 40 so that means it covers up 40 inches, and then 8 times 5 is 40, so it covers up 40 inches.

Classmates asked Ca what he meant by "covers 40 inches" because what they saw was a pair of rectangles whose longest side was 10 inches. Another child got up and worked with Ca, and together they partitioned each rectangle into 40 square inches, demonstrating two forms of array multiplication: 4 groups of 10 and 10 groups of 4. The conversation then turned to how multiplication of length resulted in units of area, and the class discussed whether or not this principle was true for all rectangles. By the end of the lesson, children had again recast their knowledge of units of area measure—what was formerly known primarily by finding appropriate material means for covering a space had been reconstituted symbolically as multiplicative length.

Student Learning

At the end of each year, we administered items like those posed in the Lehrer et al. (chap. 6, this volume) clinical interviews to children in the three classrooms. The first wave of the Lehrer et al. longitudinal sample provided a baseline for describing second-grade children's conceptions of area measure, and the last wave provided a look at typical patterns of development for these second graders (then in the fourth grade). Against this baseline, we contrasted children's development within the target classrooms as indicated by their performance at the beginning and end of the second grade with respect to their strategies for finding the area of irregular forms, as well as important ideas in area measure, like space-filling (area units should tile the plane). Measurement of these components of area measure at the beginning of the year indicated marked similarity in profiles for the longitudinal and target classroom samples. However, inspection of Fig. 7.10 suggests significant differences by the end of the year; the average performance of children in the target classrooms exceeded that of both waves of the longitudinal sample.

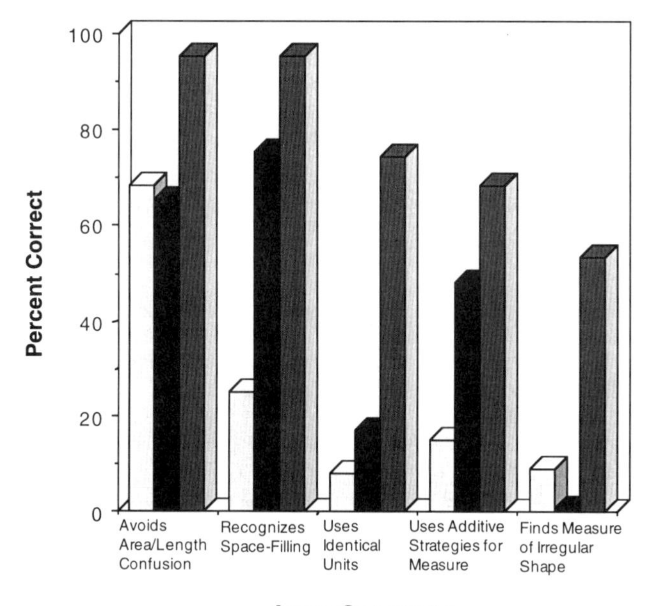

FIG. 7.10. A comparison of children's understandings of area concepts.

Reprise

The sequence of tasks invented by Ms. C. helped children progressively elaborate and mathematize their informal knowledge about area and its measure. The first task, involving comparisons among three rectangles, invited a conflict between perception ("it looks like") and conception (the mathematics of congruence). Ms. C. exploited this tension to motivate practical inquiry built around children's ideas about additive congruence (reallotment, i. e., different arrangements of the same spatial regions do not change area). This practical inquiry took the form of folding and rearranging subregions of the three rectangles, with the eventual emergence of the idea of a unit of measure as children folded one of the rectangles into three congruent pieces vertically and four congruent pieces horizontally. Interestingly, Ms. C. made the equivalence of these two different forms of folding problematic for children; measure emerged from children's resolution of the impasse.

The next two tasks designed by Ms. C. point to the important roles played by tools and notations in the development of understanding of space. When children attempted to rank-order the area of their hands, the very availability of a resource like beans seduced children into investigating its measure properties. Children's discovery of difficulties with beans led to greater understanding of two fundamental principles of area measure: identical units and space-filling (tiling the plane). During their investigation, children invented a notational system that helped them both estimate and record fractional pieces. This notational system was applied again during their investigations of the area of islands, but this time an alternate notation was introduced (symbolic representation of fractional pieces) by Ms. C. Children adopted the alternative notation because it proved easier to conventionalize (the rules about what counted as one unit were less idiosyncratic and more communal) and, therefore, easier to communicate with others. Consequently, children had the opportunity to reflect on the uses and purposes of mathematical notation even as they elaborated their ideas about area and its measure. The last task in the sequence (zoo cages) provided children further opportunity to confront once again the conflict between perception and conception, but this time the resolution was developed at a symbolic, and hence generalizable, plane. Rather than reallot units of measure, children could simply multiply lengths to obtain a single quantity that revealed unambiguously whether or not two shapes "covered the same amount of space."

Collectively, these tasks illustrate a spiral of design that started with children's informal understanding and built successively on the history of the understandings they developed as they solved the problems posed by these tasks. The tasks provided frequent opportunities for emergent goals (e.g., in the three-rectangles task [the first task], issues about squares and rectangles emerged; in the last task, issues about the associativity property of multiplication emerged) even as they provided sufficient structure and constraint for the development of productive mathematical thinking. The tasks also afforded forums (e.g., appearance vs. reality) for the invention of systems of notation in the service of progressive mathematization of everyday experiences. All of the tasks provided frequent opportunties for classroom conversations centered around sharing strategies, making and justifying conjectures, and exploring different forms of mathematical argument.

CONCLUDING COMMENTS

We embarked on a program to redesign geometry education in the primary grades in such a way that young children had the opportunity to develop a mathematics of space even as they were developing a mathematics of number. The cornerstones of our work were commitments to mathematizing children's informal knowledge about space and supporting teachers' professional development. Each teacher devoted substantial resources to develop a professional identity congruent with instruction rooted in understanding children's thinking, and every teacher participated in a small, but generative, community that provided opportunities to elaborate those professional identities. The development of a community of practice was fostered by a series of in-service workshops that emphasized cases of student reasoning, by the conduct of collaborative research, and by teacher authoring.

Teachers' practices reflected a consistent set of design principles. First, teachers invented or appropriated problems and tasks that were rooted in children's informal understandings of space, a practice consistent with the Dutch realistic mathematics education (see Gravemeijer, chap. 2, this volume). For example, children all had ideas about what makes designs like quilts interesting, and their interest in form and pattern provided a rich springboard for the mathematics of transformation and symmetry. Similarly, all children at one time or another experienced conflict between appearance and reality, and teachers skillfully helped children develop the mathematics of area as an explanation for why some forms looked different, yet covered the same amount of space.

Second, teachers continually promoted children's inventions of ways to depict and represent space. These depictions were not merely displays; rather, they were tools for developing mathematical arguments. For instance, children's invented notations for horizontal flips ("up–down" flips) helped them distinguish and describe which aspects of a physical motion were worth preserving mathematically. Other notations connected spatial and number sense: Recall, for instance, children's discussions of odd and even numbers of flips, and their explorations of array multiplication as they thought about measuring areas of rectangles through multiplication of length and width.

Third, teachers promoted forms of classroom conversation that helped children develop understanding about space. In these classrooms, the roles of talk were many and diverse. Talk served to help children develop a mathematical language that fixed and anchored mathematically important elements of space, (e.g., properties of figures). This function of talk played an important role in mathematical generalization: What was first known primarily through perception, common visual regard, came to be known and shared through talk. Classroom talk was also a vehicle for argument and justification. Much of the classroom talk supported what logicians refer to as suppositional argument: reasoning about "true" propositions purely for the sake of argument. This form of reasoning is sustained by adopting a supposition "for the sake of the argument" and then considering what its consequences would be (Levi, 1996). For example, during the conversation about three rectangles in Ms. C.'s class, one student proposed that form C was a square. She responded, "If C is a square, what must be true?" Suppositional argument requires maintenance of relationships among what is judged true, what is judged false, and what "hangs in suspense." In this instance, children were uncertain about the status of C, although it did look like a square. So they decided to treat C as if it were a square, and then decided that the student's conjecture was false because they found that C's properties were not consistent with those of a square. (One child drew a line congruent with one side and rotated the figure to test for congruence of sides.) Reasoning about the tripartite balance among true, false, and conditional beliefs can be difficult to establish and sustain, even for adults and older children. Yet this form of argument is indispensable to mathematical modeling: What-ifs serve as axioms, and mathematicians explore their consequences.

Classroom conversations also played a key role in helping children articulate a sense of the history of their thinking: Teachers often asked students to narrate how they came to know something. Such narratives helped children develop identities as mathematical thinkers whose mental efforts led to a progressive elaboration of understanding about space. Classroom conversations were the primary means by which mathematical instruction became dialogic, in the sense intended by Bakhtin (1981) and Wertsch (1991). Children came to know as they participated in dialogue, either directly with a peer or with the teacher, or indirectly in relation to some shared supposition (usually established by past practice in the classroom). Classroom talk helped children develop voice (a speaking consciousness) about what they understood first informally and intuitively.

Last, teachers' orchestration of curriculum tasks, tools, notations, and classroom talk was guided by their continually evolving understanding of student thinking. Ms. C.'s design of tasks to promote children's understanding of area and its measure and her continued attempts to help children reflect on their thinking about area and its measure suggest a form of teaching that hinged on her "reading" of student understanding. In each of the classrooms we observed, the evolution of student thinking was matched by a corresponding evolution in teachers' understanding of the pedagogical implications of student thinking.

Our observations of these classroom-based cases of the progressive elaboration of student understanding about space suggest the need for a reexamination of pedagogical policies and practices that ignore the mathematics of space in the primary grades. The opening chapters of this volume indicate that spatial reasoning and visualization are essential to mathematics. In addition to its central role in mathematics, for many children, spatial reasoning provides a more accessible entrée to powerful mathematical ideas like conjecture, proof, and refutation. The lesson seems clear: Space and geometry are best introduced early in schooling and thereafter maintained as a central part of learning and understanding mathematics. But whatever the point of departure, space and number can be mutually constituted only by children who are afforded opportunity to reason about them jointly.

REFERENCES

Bakhtin, M. M. (1981). *The dialogic imagination.* Austin: University of Texas Press.

Brown, A. L. (1992). Design experiments: Theoretical and methodological challenges in creating complex interventions. *Journal of the Learning Sciences, 2,* 137–178.

Carpenter, T. P., & Fennema, E. (1992). Cognitively guided instruction: Building on the knowledge of students and teachers. *International Journal of Educational Research, 17*(5), 457–470.

Clark, C., & Peterson, P. (1986). Teachers' thought processes. In M. Wittrock (Ed.), *Handbook of research on teaching* (3rd ed., pp. 255–296). New York: Macmillan.

Clements, D. H., & Battista, M. T. (1992). Geometry and spatial reasoning. In D. A. Grouws (Ed.), *Handbook of research on mathematics teaching and learning* (pp. 420–464). Reston, VA: National Council of Teachers of Mathematics.

Cobb, P., Yackel, E., & Wood, T. (1995). The teaching experiment classroom. In H. Bauersfeld & P. Cobb (Eds.), *The emergence of mathematical meaning.* Hillsdale, NJ: Lawrence Erlbaum Associates.

Fantz, R. L. (1958). Pattern vision in young infants. *Psychological Record, 8,* 43–47.

Fennema, E., Carpenter, T. P., Franke, M. L., Levi, L., Jacobs, V. R., & Empson, S. B. (1996). A longitudinal study of learning to use children's thinking in mathematics instruction. *Journal for Research in Mathematics Education, 27,* 403–434.

Fennema, E., & Franke, M. (1992). Teachers' knowledge and its impact. In D. A. Grouws (Ed.), *Handbook of research on mathematics teaching and learning* (pp. 147–164). New York: Macmillan.

Fennema, E., Franke, M. L., Carpenter, T. P., & Carey, D. A. (1993). Using children's mathematical knowledge in instruction. *American Educational Research Journal, 30,* 555–583.

Freudenthal, H. (1983). *Didactical phenomenolgy of mathematical structures.* Dordrecht, the Netherlands: Reidel.

Goodnow, J. (1977). *Children drawing.* Cambridge, MA: Harvard University Press.

Haith, M. M. (1980). *Rules that infants look by.* Hillsdale, NJ: Lawrence Erlbaum Associates.

Latour, B. (1990). Drawing things together. In M. Lunch & L. Woolgar (Eds.), *Representation in scientific practice* (pp. 19–68). Cambridge, MA: MIT Press.

Lave, J., & Wenger E. (1991). *Situated learning: Legitimate peripheral participation.* Cambridge: Cambridge University Press.

Lehrer, R., Fennema, E., Carpenter, T., & Ansell, E. (1992). *Cognitively guided instruction in geometry.* Madison, WI: University of Wisconsin, Wisconsin Center for Education Research.

Lehrer, R., & Jacobson, C. (1994, Winter). Geometry in the primary grades. *NCRMSE Research Review: The Teaching and Learning of Mathematics,* 3(1), 4–14.

Lehrer, R., Randle, L., & Sancilio, L. (1989). Learning pre-proof geometry with Logo. *Cognition and Instruction, 6,* 159–184.

Lehrer, R. & Shumow, L. (in press). Aligning the contruction zones of parents and teachers for mathematics reform. *Cognition and Instruction.*

Levi, I. (1996). *For the sake of the argument.* Cambridge: Cambridge University Press.

Olson, D. R. (1970). *Cognitive development: The child's acquisition of diagonality.* New York: Academic Press.

Piaget, J., & Inhelder, B. (1956). *The child's conception of space.* London: Routledge and Kegan Paul. (Original work published 1948)

Piaget, J., Inhelder, B., & Szeminska, A. (1960). *The child's conception of geometry.* New York: Basic Books.

Schifter, D., & Fosnot, C. T. (1993). *Reconstructing mathematics education: Stories of teachers meeting the challenge of reform.* New York: Teachers College Press.

Senechal, M. (1990). Shape. In L. A. Steen (Ed.), *On the shoulders of giants* (pp. 139–181). Washington, DC: National Academy Press.

Siegel, A. W., & White, S. H. (1975). The development of spatial representations of large-scale environments. In H. W. Reese (Ed.), *Advances in child development* (Vol. 10, pp. 9–55). New York: Academic Press.

Streefland, L. (1991). *Realistic mathematics education in primary school.* Utrecht, the Netherlands: Center for Science and Mathematics Education.

Strom, D., & Lehrer, R. Springboards to algebra. In J. Kaput (Ed.), *The development of algebraic reasoning in the context of everyday mathematics.* In preparation.

Tharp, R., & Gallimore, R. (1988). *Rousing minds to life.* New York: Cambridge University Press.

van Hiele, P. M. (1986). *Structure and insight: A theory of mathematics education.* Orlando, FL: Academic Press.

Vygotsky, L. S. (1978). *Mind in society* (M. Cole, Ed.). Cambridge, MA: Harvard University Press.

Williams, S.M. (1992). Putting case-based instruction into context: Examples from legal and medical education. *Journal of the Learning Sciences, 1,* 367–427.

Wertsch, J. V. (1991). *Voices of the mind.* Cambridge, MA: Harvard University Press.

APPENDIX: EXAMPLES OF ITEMS FOR INDIVIDUAL ASSESSMENT

Area of a Rectangle

Figure 7.A1 displays a paper-and-pencil item designed to assess children's understanding of finding the area of a rectangular polygon.

4 cm.

2 cm.

1 cm.

How many squares will cover the rectangle? ______

FIG. 7.A1. Paper-and-pencil item for finding area of regular polygon.

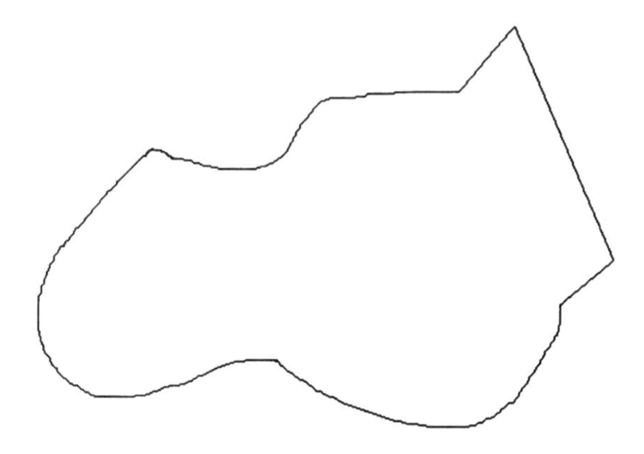

FIG. 7.A2. Interview item on finding area of irregular figures.

Area of Irregular Figure

Students found the area of the irregular figure shown in Fig. 7.A2. Students were offered tools such as a graph paper transparency (divided into squares 2 cm on a side), an overhead marker, a ruler, and a length of string.

Included in the interview protocols were scaffolds to assist students in developing a strategy for solving the problem:

Int: Can you find out how much area this shape has? If the student does not recognize the word "area," ask how much it would take to "cover" exactly this shape.

Scaffold: Offer the student the graph transparency and pen and ask, "Would this help you find the answer?"

Figure 7.A3 shows a student's solution (at the end of the school year) to the area problem shown in Fig. 7.A2. Note her strategy of aggregating

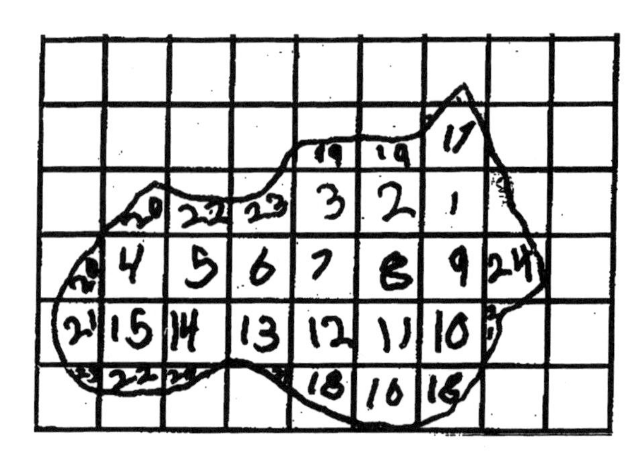

FIG. 7.A3. Example of student solution to finding area of irregular figure.

pieces of the figure that individually did not cover an entire square (e.g., two 19s at the top of the figure, each numbered inside a half-covered square.) By this student's estimate, the area of this figure is 24 squares.

Transformations

Figure 7.A4 displays a paper-and-pencil item designed to assess production of one-step transformations on a simple core square. To probe students' understandings of transformations, we presented strips and core squares (see Fig. 7.A5) and then asked students to compose the transformations that would make the strip.

This item also included several levels of scaffolds—a verbal prompt, an identical core square as a manipulative, and modeling of the movement:

Here is a picture of one quilt square from a quilt we are making.

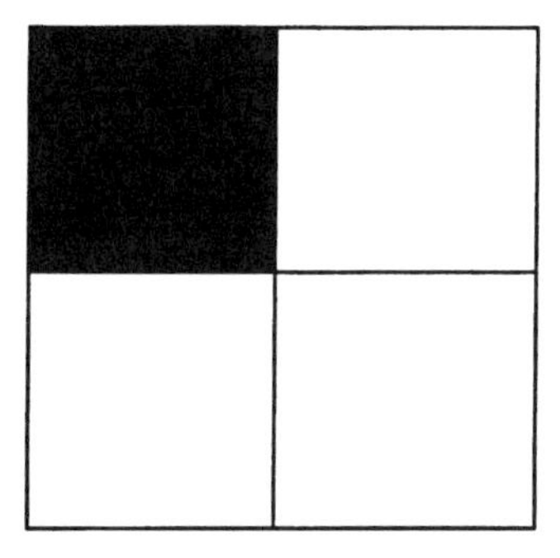

Color in the figure below what the quilt square looks like after an "up-down flip."

Color in this figure below what the quilt square looks like after a "right turn of 90."

FIG. 7.A4. Pencil-and-paper item on one-step transformations.

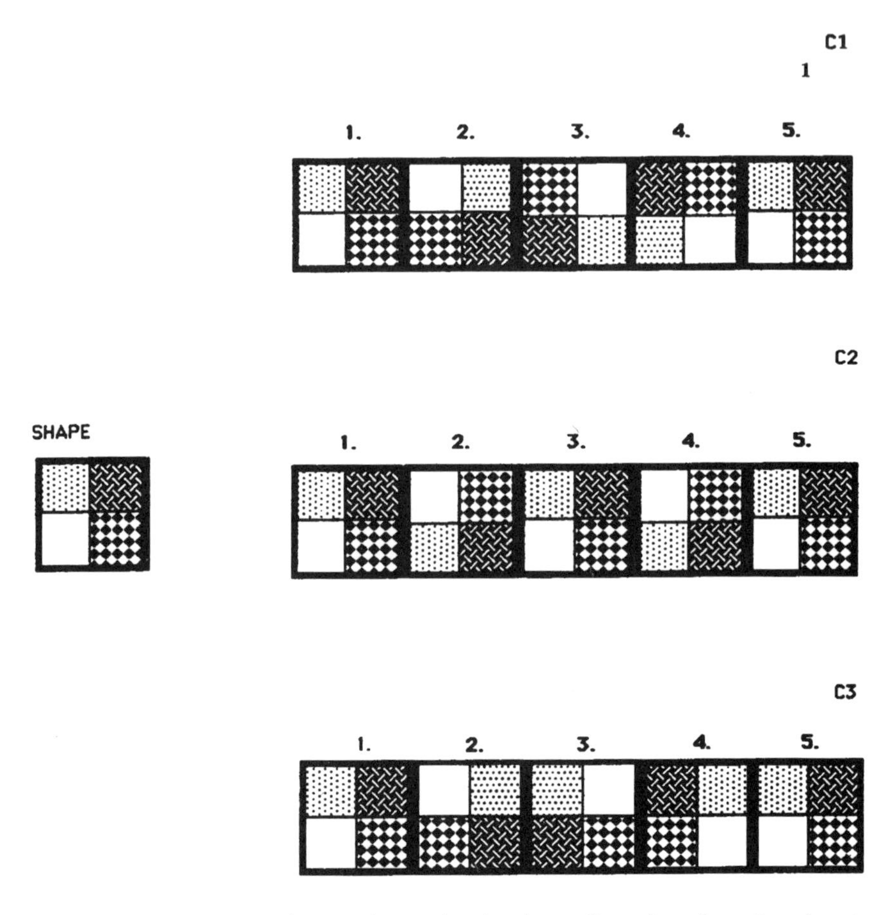

FIG. 7.A5. Interview item that examines students' understandings of transformations, based on recognition rather than on performance.

Int: For each of the motion strips, ask the student, "What did we do to make this design with this core square?" or "Can you find the pattern we used to make this long quilt?"

Scaffolds

Level I "What kinds of motions do you know about?"
"Could it be turns (C1)? flips (C2)? or both (C3)?"

Level II Offer the core square to the student.
"Would this help you to figure it out?"

Level III Model the first movement (only) for the student.
"Do you think you could show me how the rest are made?"

8

Development of Geometric and Measurement Ideas

Douglas H. Clements
State University of New York at Buffalo

Michael T. Battista
Kent State University

Julie Sarama
Wayne State University

The separation of curriculum development, classroom teaching, and mathematics educational research from each other has vitiated each of these efforts. We are working on several related projects, the aim of which is to combine these efforts synergistically. The first[1] is a large-scale curriculum development project that emphasizes meaningful mathematical problems and depth rather than exposure. Our responsibility (and goal) in this project is to develop the geometry and spatial-sense units in this curriculum based on existing research on children's learning of mathematics as well as our own classroom-based research on children's learning within the context of formative evaluations of the curriculum. The second project[2] has the related goal of conducting research on children's learning of geometric and spatial concepts in computer and noncomputer environments.

The curriculum unit, discussed in this chapter, *Turtle Paths* (Clements, Battista, Akers, Woolley, Meredith, & McMillen, 1995), engages third-grade students in a series of combined geometric and arithmetic investigations

---

[1]*Investigations in Number, Data, and Space: An Elementary Mathematics Curriculum*, a cooperative project among the University of Buffalo, Kent State University, Technical Education Research Center, and Southeastern Massachusetts University (National Science Foundation grant no. ESI-9050210).

[2]*An Investigation of the Development of Elementary Children's Geometric Thinking in Computer and Noncomputer Environments* (National Science Foundation Research grant no. ESI 8954664).

exploring paths and the lengths of paths. To achieve our two related goals, we investigated students' learning as they worked through this unit, focusing on their development of geometric and measurement ideas.

The theoretical basis of both the study and the curriculum unit is that students' initial representations of space are based on action rather than on passive "copying" of sensory data (Piaget & Inhelder, 1967). This notion implies that noncomputer and Logo turtle activities designed to help students abstract the notion of path (a record or tracing of the movement of a point) provide a fecund environment for developing students' ideas about two-dimensional figures and their measures. Students' actions in these environments may be (a) perceptual—scanning paths in the classroom and watching the turtle draw paths, (b) physical—walking and drawing paths, or (c) symbolic—interpreting the turtle's movement as physical motions as their own and creating computer programming code to control those motions (Clements & Battista, 1992).

For example, encouraging students to scan visually the side of a building, run their hands along the edge of a rectangular table, or walk a straight path will help them develop an intuitive idea of straightness. Their ideas become more sophisticated and explicit as they work with Logo path activities; they easily use the turtle to discover that a straight path is one that has no turning. As this type of activity is extended to more complicated figures such as quadrilaterals, students must analyze the visual aspects of these figures and the ways their component parts are put together, an activity that encourages analytical thinking and awareness of the properties of geometric figures (Clements & Battista, 1990, 1992).

Performing the activities and then consciously connecting path and movement experiences may also help students build units of length by segmenting (i.e., dividing into parts) continuous motion (Steffe, 1991). The computer experience can also play a pivotal role in helping students forge those connections (Clements & Meredith, 1993). Further, connections between mathematical symbols and graphics may catalyze in students a belief in the necessity of equal-interval units (i.e., "partitioning," dividing or intending to divide a length into equal-interval parts), another significant conceptual advance (cf. Petitto, 1990). In addition, these connections may help students build mental connections between numerical and geometric ideas. There is some evidence that supports these claims, although the types of items on which researchers have found effects are limited (e.g., relative lengths of horizontal and oblique lines on a grid or use of arithmetic in a real-world geometric setting; Clements & Battista, 1989, 1992; Noss, 1987).

In a similar vein, because turns (and angles) are critical to the path perspective on shapes and because the intrinsic geometry of paths is closely related to real-world experiences such as walking, path activities may be efficacious in developing students' conceptualizations of turn and turn measurement. There is supporting evidence for this view. Logo activities are beneficial to students' development of turn ideas and turn measurement (Clements & Battista, 1989; Kieran, 1986a; Olive, Lankenau, & Scally, 1986). Students nevertheless conceive many different schemata regarding angle and angle measure. Third graders, for example, frequently relate the size of an angle to the lengths of the line segments that form its sides, the tilt of the top line segment, the area enclosed by the triangular region defined by the drawn sides, the length between the sides (from points sometimes, but not always, equidistant from the vertex), the proximity of the two sides, or the turn at the vertex (Clements & Battista, 1989). Middle-school students often possess one of two schemata: (a) a "45–90 schema" wherein students associate oblique lines with 45° turns and horizontal and vertical lines with 90° turns and (b) a "protractor schema" wherein students base inputs to turns on an image of a protractor in "standard" position (e.g., no matter what the turtle's initial orientation, if the end goal is a heading corresponding to 45° on a protractor in canonical position, a "rt 45" command is issued; Kieran, Hillel, & Erlwanger, 1986). Children who have experience with Logo, however, have notions about larger angles that are more likely to reflect mathematically correct, coherent, and abstract ideas, possibly because children using Logo develop an intrinsic, personally meaningful mental model instead of, for example, a protractor schema (Clements & Battista, 1990; Kieran, 1986b; Noss, 1987). These benefits are educationally significant because students generally have considerable difficulty with angle, angle measure, and rotation concepts, ideas central to the development of geometric knowledge (Clements & Battista, 1992; Krainer, 1991; Lindquist & Kouba, 1989).

Such experiences, however, do not replace previous ideas of angle measure (Davis, 1984). For example, students' misconceptions about angle measure and difficulties coordinating the relationship between the turtle's rotation and the constructed angle have persisted for years, especially if students are not properly guided by their teachers (Hershkowitz et al., 1990; Hoyles & Sutherland, 1986; Kieran, 1986a, 1986b; Kieran et al., 1986).

Most studies, however, have not focused on the mechanisms of learning and have often been narrow in scope (e.g., having students work in computer microworlds designed to teach only angle and turn ideas). In our investigation we studied the development of these ideas in the context of instructional units that reflect recommendations of recent reform documents.

Our goal in these projects was to investigate the development of linear-measure ideas through an instructional unit on geometric paths, observing the roles of both noncomputer and computer interactions. We adopted a multifaceted methodology, emphasizing an openness to emergent conjectures and theories (for an emic perspective, see Strauss & Corbin, 1990). Four themes emerged: (a) ideas of length and unit of length, with an emphasis on segmenting and partitioning; (b) combining and decomposing processes in a length context; (c) connections between students' number and spatial schemes; and (d) the role of the computer environment in students' development of processes and ideas regarding length.

RESEARCH METHODS

Procedure

Third graders worked on the unit of instruction *Turtle Paths* in two different situations (Clements, Battista, Sarama, & Swaminathan, 1996; Clements,

Battista, Sarama, Swaminathan, & McMillen, 1997). In the first situation, a pilot test, a graduate assistant taught the unit to four students during spring 1992. The researchers conducted pre- and postinterviews and wrote case studies of each student. The second situation, a field test, involved two of the authors teaching the unit to two Grade 3 classes the following autumn. Data collection for this field test included pre- and postinterviews, paper-and-pencil pre- and posttests, case studies, and whole-class observations.

Students

Participants in the pilot test were four 9-year-old volunteers from a rural town: two girls, Anne and Barb, and two boys, Charles and David. Participants for the first field test were 38 students in two heterogeneously grouped inner-city third-grade classes whose teachers had volunteered their classes; students were 85% African American with most of the remainder Caucasian. Each class was a heterogeneous group representing the school population; 80% of the students qualified for Chapter 1 assistance in mathematics. Researchers intensely studied one student in each class, students identified by their teacher as talkative students with good attendance (Luke and Monica). When students worked in pairs, the researcher also observed Monica's partner, Nina. Luke's partner had such low attendance that meaningful data collection was not possible. Participants for the second field test were all students from two classrooms in the same school the following year.

Turtle Paths

Turtle Paths engages third-grade students in a series of combined geometric and arithmetic investigations. The unit teaches geometric figures such as paths (including properties such as closed paths), rectangles, squares, and triangles; geometric processes such as measuring, turning, and visualizing; and arithmetic computation and estimation. Throughout the unit, students explore paths and the lengths of paths. An outline of activities from Investigations 1, 2, and 3 (paths and lengths of paths, turns in paths, and paths with the same length, respectively) follows:

Investigation 1

1. Students walk, describe, discuss, and create paths. They give Logo commands to student-robots, specifying movements that create paths. This activity starts the development of a formal symbolization, which is built on during the remainder of the unit.

2. In an off-computer game, Maze Steps, students count steps in a maze to find certain paths (a path that is 14 steps in length with 2 corners) and play a similar Maze Paths game. The activity emphasizes problems that have more than one possible solution.

3–4. Students give movement commands to the Logo turtle in order to create paths in a Get the Toys game (see Fig. 8.1). This activity uses only 90° turns. The computer activity promotes thoughtful use of the com-

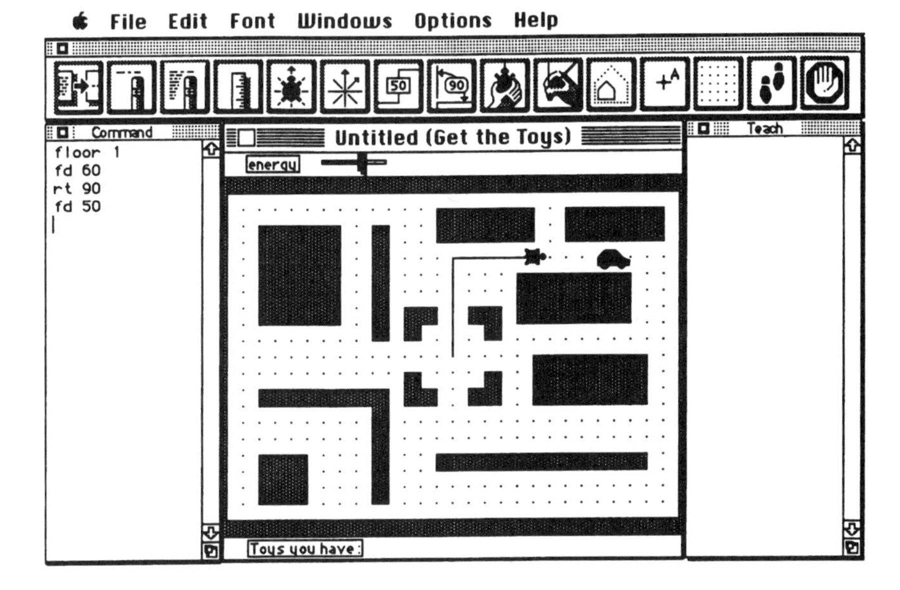

FIG. 8.1 The "Teach" window: In the Geo-Logo game Get the Toys, students instruct the turtle to get three toys (one on each "floor"), before its battery runs out of energy. Each command uses one unit of energy, regardless of the size of the input to that command, and any command entered is immediately reflected in the geometric figure on screen.

mands (as opposed, e.g., to nonreflective trial and error). An on-screen battery's limited energy decreases with each command used.

Investigation 2

1. Students turn their bodies, discuss ways to measure turns, and learn about degrees. The teacher introduces another computer game, Feed the Turtle, in which students must use their knowledge of turns to direct the turtle through channels of water (turns in this game are multiples of 30°).

2. Students play Feed the Turtle and thus continue to estimate and measure turns. They discuss the nature of triangles and build their own descriptions of those geometric figures.

3. Students write Logo procedures to draw equilateral triangles.

4–5. Students find the missing measures (lengths and turns) to complete partially drawn paths. The students must figure out that the turtle must move "fd 10" (forward 10 turtle steps) to finish this side and so on. In this and in more complex problems, students must analyze geometric situations and use mental computation in meaningful settings.

Investigation 3

1–2. Students write as many procedures as possible to draw different rectangles of a certain overall length or perimeter—200 steps. As with the missing-lengths-and-turns problems, students must analyze geometric situations and use mental computation.

3–4. Students design and use the computer to draw a picture of a face, for which each part (e.g., ear, mouth) has a predetermined length or perimeter; the shape (rectangle, square, or equilateral triangle) is the students' choice.

Geo-Logo Environment

A modified Logo environment, Geo-Logo[3] (Clements & Sarama, 1996), is an intrinsic component of the instructional unit. Geo-Logo's authors based the program on curricular considerations and on a number of implications for the learning and teaching of geometric ideas with turtle graphics (Clements & Sarama, 1995). For example, although research indicates that Logo experiences can help students learn geometry, research also indicates that students often continue to use visually based, nonanalytical approaches (here we explicitly do not mean approaches that include dynamic imagery; rather, we mean approaches that are limited to visually based guessing, as implied by the van Hiele levels; van Hiele, 1986). The authors designed Geo-Logo on five principles abstracted from research (Clements & Sarama, 1995): (a) encourage construction of the abstract from the visual, (b) maintain close ties between representations, (c) facilitate examination and modification of code (and thus explorations of mathematical ideas), (d) encourage procedural thinking, and (e) provide freedom within constraints (e.g., various activities restricted turns to multiples of 90° or 30°).

A critical feature of Geo-Logo, and one that illustrates how these principles are put into practice, is that students enter commands in "immediate mode" in a command window (though they can also enter procedures in the "Teach" window; see Fig. 8.1). Geo-Logo automatically reflects any change to these commands. For example, if a student changes "fd 20" to "fd 30," Geo-Logo immediately makes a visible corresponding change in the geometric figure on the screen. The dynamic link between the commands and the geometry of the figure is critical; the geometry of the figure and its representation on screen always precisely reflect the commands in the command window. Other features include a variety of icon-based tools. Tools for writing and editing code include tools for easy definition of procedures (the icon on the left of the tools bar shown in Fig. 8.1), for editing (two tools, one to erase one command and a second to erase all commands in the command window, both with dynamic links to the geometric figure), and for inspecting and changing commands (the "walking feet," a Step icon that allows students to "walk through" any sequence of commands simultaneously highlighted, then executed). Tools to enhance measurement include measuring tools (a ruler to measure length; two other tools to measure turns and angles) and labeling tools (to label lengths and turns).

---

[3]Geo-Logo is disseminated as a stand-alone as follows: Clements, D. H., & Meredith, J. S. (1994). *Turtle math* [Computer software]. Montreal, Quebec: Logo Computer Systems, Inc. (LCSI).

Data Collection

Case Studies. One researcher sat next to and observed each case-study student throughout each session. If the researcher could not determine the student's thinking through passive observation, he or she would ask the student to think aloud or would pose specific questions. One of the field-test teachers identified two additional students, Oscar and Peter, as among the lowest performers in mathematics and placed them in the front of the class so that she could "keep an eye on them." The researcher who taught the activities and the researcher who observed the class as a whole observed these two students frequently.

Interviews. Researchers interviewed the 4 pilot students informally following the completion of the unit; they interviewed 13 students from the field test (originally teachers had selected 7 students: 3 average, 2 above average, and 2 below average; 1 student left the school in the middle of the study) using a structured protocol on either the first or second day of the unit and then again after completion of the unit. The case-study students, Oscar and Peter, were interviewed. We describe individual interview items in the following section.

RESULTS AND DISCUSSION

Creating Units of Length

The first major area we investigated was students' development of ideas and their processing regarding length measurement (Clements, Battista, Sarama, Swaminathan, & McMillen, 1997). We consistently observed many students drawing dots or hash marks to create line segments of a certain length, to assign a measure to an already drawn line segment, or to label a given line segment with a given measure. These segments were usually not units. For example, two students drew a segment of 62 units (see Fig. 8.2). These students did not establish and maintain connections between the numbers for the measures, the dots they drew to indicate a measure of length, or the shape and size of the geometric figures.

We also observed this apparent need for physical marks in estimating distances in computer contexts. In Feed the Turtle, half of the pilot stu-

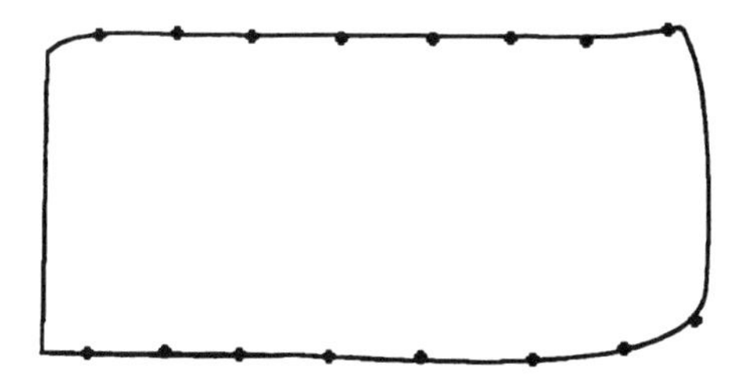

FIG. 8.2 Student work: closed line segment 62 units long.

dents drew dots on a paper replica of the computer screen to help them determine an input to the forward command.

Although students continued to use hash marks on the missing-measures activities, they also began to use superordinate units (units greater than 1), initially informally and then with increasing sophistication. For example, David figured out the length (40) for the one segment by matching the width of his figure to a different segment labeled 40, explaining, "It's about 40 steps because I measured it with my finger. It's the same as this one." On the final interview, however, David found the missing distances for each figure by using 10 as a unit and iterating this unit over the segment to ascertain its length.

Most students operated at a more abstract level by the end of the curriculum unit. The ability to impose a number meaningfully on a line segment without physically partitioning the segment represents a significant advance. A student must step back and do anticipatory quantification, applying his or her number competencies to measurements of a geometric figure. The student who has not constructed an abstract conception of length and length measure may need to partition the segment physically.

We can infer additional support for this theoretical position from the size of the steps these students (during student-robot activities) actually walked and drew. The students who played the role of student-robot, following others students' directions, usually changed the size of their actual steps, the step length decreasing with each successive line segment constituting the figure. When students drew dots on figures, they generally attempted to keep the intervals between the dots equal in length. But when the path changed direction, students often changed the scale, with the interval between the dots decreasing toward the end of the last line segment. This suggests that the students had an intuitive sense that the steps should be equal in length. If the motion was extended or (especially) interrupted, however, the students did continue to make (or walk) equal intervals. In these instances, they appeared to focus on the discrete number of the steps.

Composing and Decomposing Lengths

Many activities encouraged students to compose and decompose measures. In Get the Toys, the first computer activity, the failing energy of the battery caught the students' attention and motivated them to combine old commands to create new, more efficient ones. Anne told her partner, "You should go back 50. Because 30 + 10 + 10 is 50. It's less commands to type one "bk 50" than three backwards, 30, 10, and 10."

Growth was evident in most students' work. During their second round of the game, these girls began to combine commands before the second command was entered (e.g., "We need 20 more—80 plus 20—change 'fd 80' to 'fd 100' "). Some students, however, could not think of a change in a new command before they actually saw the new command. For example, Oscar and Peter used successive approximation to place the turtle:

"fd 50," "fd 10," "fd 10." Then they edited the commands to produce a more efficient solution, "fd 70." They were never able to anticipate the combination. In contrast, Anne and others gave evidence that they had generalized their number and arithmetic schemes to include situations of length and connected line segments (Steffe, 1991). Some students, such as Monica and Nina, also generalized their part–whole schemes. They could maintain a 70-step length as a whole while simultaneously maintaining 60- and 10-step lengths as parts of this whole. Oscar and Peter did not appear able to disembed part of the length from the whole length in this way. Rather, they used trial and error and combined steps after the fact. They appeared to combine *numbers* (inputs to Geo-Logo commands), rather than *lengths*. These lengths—measured line segments that the person can place in a part–whole schema—were linked to both spatial and numerical mathematical objects—"connected lengths."

Using connected lengths in different spatial situations was not easy for these students. They did not immediately use analytical strategies to solve missing-measures problems, as shown in the following classroom dialogue:

Teacher: Try this one [drawing the first three sides of a 20 by 70 rectangle, then only 20 units of the final side, leaving a space of 50].
Anne: But there's no dots! How do we know how far to make the turtle go?
Dave: How do you count 20 if you don't know where the 20 is at?
Aaron: Why is the part on the bottom 20?
Nina: Because it's small.
Charles: Because the space looks about the same as that side that is 20. [We call this the *imagistic move, segment, numerically adjust* strategy.]
Nina: Because it says 20 right there, and it looks like 20.
Anne: This side on the left would be 20 too, because it's the same length as that one on the right.
Dave: The top would be 70, because it says it right there.
Anne: Seventy for the bottom too, even though it is unfinished.
Benjamin: Twenty plus 50 equals 70.
Anne: You need another 50 on that last side. Because the sides of a rectangle need to be the same length.
Dave: Just do "fd 70." That will do the 20 and the 50.
Teacher: All the sides?
Anne: No, just the sides here and here [pointing].
Teacher: The sides opposite one another?
Anne: Yes.

The teacher saw that most students were making visual estimates, although some were analyzing the figure. Overall, students seemed to understand. The teacher was a bit surprised, then, at the results when the students worked on the other figures independently:

Aaron: You can't get this one! [Draws the bottom segment of Fig. 8.3a] [The teacher drew hash marks to show how the missing part along the bottom could be broken down into three segments.]
Aaron: Oh, I see! You just add the three numbers. So, 15 + 20 + 20 is 55.

The teacher was taken aback that Aaron just added the first three numbers, from left to right, including one vertical segment (just as Barb had done earlier). Nina ignored the existing number labels, and other students used different techniques, such as measuring with their fingers. The teacher decided to move them to the computers immediately so that the feedback would lead them to reflect on their strategies. This helped. Aaron, for example, found his error after entering "fd 55." By the end of the activity, students were assigning numbers to figures as measures of length, abandoning early guessing strategies.

Some took considerable time to develop sophisticated strategies. On Luke's first missing-measure problem, he used visual estimation: "Try 50. I don't know how much for sure." Even after observing others using different, successful strategies, he returned to guessing. Luke placed his finger down on a 20-unit line segment (see Fig. 8.3a) and then marked off three such units to determine the missing length of the figure.

Similarly, Luke had not often combined commands in the Get the Toys and Feed the Turtle, preferring instead to erase commands—"That didn't work"—and try new numbers. In both situations, Luke did not abstract the mathematical problem to a strictly numerical form, but preferred to take the "safer" and often more perceptually based course. There was one notable exception: In missing-measures problems involving rectangles, Luke operated on numbers. For example, given a rectangle with one side of length 100 and an opposite segment of 78, he said, " 'fd 22' because 78 and 22 is 100." This revealed the competence Luke possessed but

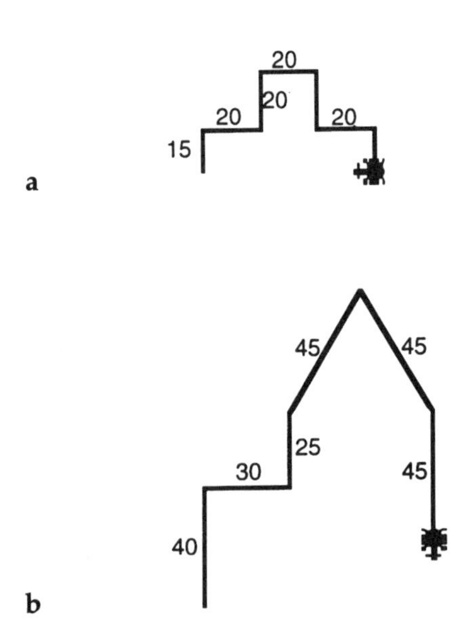

FIG. 8.3 Missing-measures problem: (a) "Sliding" one line segment onto corresponding segment (perpendicular sides as possible guides for spatial transformation). (b) Aligning two segments separated by a third segment.

did not use in more complex figures. To solve rectangle problems, Luke mentally moved one line segment to coincide with the others. He motioned with his fingers and said, "I dropped that down." He then placed those two lengths into a part–part–whole schema and used arithmetic procedures to produce a valid answer (the same *move, segment, part–whole* strategy that David and Charles used). For more difficult missing-lengths tasks, Luke also used mentally conceived motions; however, he usually moved any convenient labeled segment over the missing length (without precision) and used the former as a unit with which to estimate the measure of the missing line segment—the *move, segment, numerically adjust* strategy. This was Luke's fallback strategy for complex figures. Luke may have been able to use a more sophisticated strategy for rectangles because the perpendicular sides may have served as guides for the spatial transformation in his mind of sliding one line segment over to correspond to the other (the opposite side). In contrast, in situations in which there was an interceding segment (between, and perpendicular to, the two segments that had to be combined, as in Fig. 8.3b), students had to bring the two segments into alignment and may have had problems assigning a role to the interceding length (30, in this case) that lay between the segments (40 and 25) but added nothing to their combined length.

Connections Between Number and Spatial Schemes

After gaining experience playing Get the Toys, some students forged a connection between the spatial-geometric form of the turtle's path and the symbolic-numeric representation of the Logo commands that created it.

Did most children make such connections? The answer to this question differs by task and by student. Charles, for example, gave signs of simultaneously applying numeric and geometric schemes. On a missing-length problem, he had entered "fd 40" for the missing bottom side, but had come up short. He mentioned that he forgot the "20 part" at the top:

Teacher: Do you have to go back and do the fd 20 before the fd 40?
Charles: No! I can just add the 20 now. It doesn't matter. I should have put 60. My command was wrong. 'Cause the 40 only takes care of this part on top. [For the 200-step rectangle task, Charles talked about his plan.]
Charles: First, I wanted 10 and 10, then I couldn't get it, so I made it 20, 20, 80, 80.
Teacher: What if you had wanted it 10 and 10?
Charles: Well, then that would have to be... 90 and 90.
Teacher: How did you do that?
Charles: I just added the 10 onto the two long sides... because you can't leave it out.

Charles was saying that because the total perimeter had to be maintained and the difference of 10 between the original side length (20) and the one proposed by the teacher (10) had to be maintained, the only way to do this was to add the "lost" length to the other sides.

Not all students gave evidence of seeing such connections in all situations. When Barb and Anne initially tried to "get the numbers to add up" to 200, they drew rectangles without establishing a noticeable connection between the numbers and the figures they drew. For example, they drew a rectangle and labeled the sides 70, 30, 60, 40. In an activity in which they interpreted numbers as inputs to movement commands (e.g., Get the Toys), researchers observed them maintaining the numeric/spatial connection. When they believed a path to be static (especially when the path was almost incidental to their interpretation of their task), such as in the 200-steps problem, they gave no signs of establishing, maintaining, or using such a connection. Their intentions in this case were not to make sense of the numbers as measures of length in a setting of constructing certain geometric figures. Rather, they were using algorithms in an attempt to find numbers that "added up."

Further, many students who assimilated the problem as an algorithmic scheme did so with a vengeance. It was common, for example, for students to check the total length of a rectangle of sides 70 and 30 by using a vertical, written algorithm. They did not "see" or use the "70 + 30 = 100" shortcut in either mental or paper-and-pencil computations. Given that there were signs that the students invented numbers to sum to 100 (to meet the 200-steps criterion), this is a particularly painful example of subverting number sense to algorithmic procedures.

In the Face activity, in which students planned a drawing consisting of several figures of given perimeters, connections were more complex. Anne, when planning, drew the shapes she wanted and only afterward tried to make the numbers fit. She did not ignore scale completely, but neither did she adjust the drawing. Even though she did not draw to scale, she was conscious of the need to fit figures inside one another and, thus, that the numbers had to be relatively ordered.

Concept of Turn

The students had a more difficult time conceptualizing turns (Clements, Battista, Sarama, & Swaminathan, 1996). As an example, in the first lesson the teacher asked the difference between turning along U- and V-shaped paths on which the children had worked. Luke talked with classmates only about the "space in the middle," not mentioning the turning actions. During the same lesson, the following conversation (in which the students struggle with the idea of turn) occurred:

Teacher: Are there any paths you would consider similar in some way?
Maria: This one [an "S"] and this one [a similar figure with straight line segments] are the same 'cause they're almost like Ss.
Teacher: What's different about them?
Aaron: The difference is that you had to turn on the first one.
Teacher: Don't you turn on the curvy S?
Aaron: On the one with square corners, you have to make bigger turns.
Samir: What?
Aaron: You turn a lot in one place.

Tamara: On the curvy one, you go right and left, real quick.
Teacher: Explain that more.
Tamara: You turn right and right, then immediately left and left.
Teacher: Are you just turning right or left?
Ray: No, you're going forward and turning—at the same time.
Teacher: What is a turn?
Elena: It's when you go around.
Valerie: You change the direction you're heading.
Teacher: What could you say about a path made without any turns?
Valerie: It would have to be a straight path.

With much teacher guidance, the students came to an apparent understanding. Even so, at the computer some children would enter two forward commands in a row, forgetting a necessary turn command in the middle. Turn commands may be less conspicuous than movement commands because their representation does not involve a single graphic object, but a *relationship* between two line segments. Of course, it is true that the turtle makes a motion for each command, and in Geo-Logo, turns are made slowly so that students can see them. After the action had taken place, however, the records of turn actions (bends in paths) were not as salient to students as records of forward movement (line segments).

Luke made slow progress in discriminating turns from forward or backward movements. His first Geo-Logo commands were: "fd," "stop," "rt," and "stop." He took the idea of "turning on a point" seriously, feeling the need to end the forward motion before turning. The teacher drew a rectangle on the board and asked for Logo commands. Luke's were " 'fd 7,' then right face, then go right, then left." If the turtle was not facing straight up, Luke became less precise, even allowing a single right or left command to fulfill the dual role of "turn and move," an idea not uncommon for Logo beginners (Clements & Battista, 1991; Singer, Singer, & Zuckerman, 1981). When he typed his commands in the computer, however, Luke discovered the need to discriminate between movement ("fd," "bk"), turns and their related commands. When he entered a turn command, he was surprised the turtle did not move. He then reflected on the discussion and on body movements previously made in class, to conclude that "the 'rt' command doesn't *move* to the right."

Monica and Nina appeared able to tie the notion of turn to their notion of corner. However, the latter notion—a corner of a closed shape—was initially the more salient. The act of turning was still less "real" to them than were forward or backward movements, as indicated by their omission of turn commands in their first attempts to write Logo code. They also did not take turns into account when counting the number of commands needed in Get the Toys, claiming—against visually displayed evidence—that "turns don't use up the battery." Only after a researcher asked them to keep an eye on the battery as the turtle turned did they agree that the turn used energy, and they then counted the turns. The combination of typing the Logo turn commands and seeing the turns reified by the battery appeared to lead to Monica and Nina's increased attention to turns. After this activity, they did not neglect turns and turn commands.

Turn Measure

The limited salience of turns, combined with less familiarity with turn measurement, may have accounted for children's slow and uneven development of ideas for turn measure as well. After measurement in degrees was introduced in the pilot test, the teacher asked the students to estimate a certain turn. Charles guessed "rt 60" (correct); Anne, "rt 40." Barb placed a pencil on her student sheet across the rays of an angle:

Barb: See these stripes on the pencil? I'm using those to find out how much to turn.
Teacher: How does that help you?
Barb: I just count the stripes. See?—10, 20, 30.
Teacher: Let me try it. [She purposely places the pencil farther from the vertex] 10, 20, 30, 40, 50—
Barb: No! That's too much. You did it wrong. Hold it closer.
Teacher: How close? Here? [putting it very close]
Barb: I guess it doesn't work. The stripes don't measure it well. They worked for forwards! Where's the turtle turner?
Teacher: Why won't they work as well for rights?
Dave: 'Cause those are turns. And turns are in a circle. Turtle turners... See those lines on the turtle turners, Barb? They go out from the center. They always show the right turn.

In almost all succeeding situations, Barb gave evidence that she needed something figure-like to use to measure the turn. We inferred that she could not mentally develop an image of turns and of turning through degrees.

In Feed the Turtle, the girls in the pilot test frequently tried a turn command (e.g., "lt 50") and then typed small increments, such as "lt 10," "lt 10," "lt 10," "rt 5." Observations indicated that such guessing, followed by successive approximation, was the norm for the pilot-test students, despite the students' introduction to degrees and physical practice turning their bodies various measures. For this reason, the authors altered the game for the field test so that it accepted only multiples of 30°.

Anne and Barb defined a Logo equilateral triangle before going to the computer: "fd 30," "rt 90," "fd 30," "rt 90," "fd 30," "rt 90." When the teacher asked, "Can you draw what the computer will draw?" the students drew what they expected to see. The first turn was just a bit more than 90°, but the second one was much more—in order to make a triangle:

Teacher: That's not a square corner, is it?
Valerie: No, I'll draw it again. [She draws the corner square, but then curves the line so that the figure is closed.]

Sounding unsure of herself, Valerie said: "Well, I guess that would work." Fortunately, Logo provides feedback that is more constrained—the turtle does exactly what the student tells it to do. At the computer, the girls issued the "fd 30" command. They then tried "rt 60", "rt 10," "rt 20," and finally figured out that this was 90°! New attempts with 100 and then

110 didn't make a closed figure. They debated whether to try 120 or 130 or both, and ended up agreeing that 120 was correct. These students achieved only a limited knowledge of turns in this context. Other students built images of turns earlier in the process:

> *Samir*: [puts a hand up, rotates it slowly] It's 90 [turns hand a bit more], 100!
> *Dave*: It didn't turn enough! See, we need a sharper corner, so we need more of a sharp turn.

Samir then tried 110, then 120. After trying 120, he appeared satisfied.

Luke developed a sense of turn measurement only tenuously and only after a protracted time. Early in the curriculum, when asked, "How many 'rtfs' does the turtle have to do to face back where he started?" he answered, "Two." To explain, he drew two right angles, connecting them to form a rectangle. His idea of turn included line segments, and his depictions of turns were like textbook examples of right angles. He was accurate on turns only when they began at a 0° heading.

When Luke worked on an activity that required him to guess and then measure turns, he initially determined the amount of turn by using the length of the line segments. For example, he said that because one segment was half the size of the line segment indicating 90° on the protractor, the angle had to be 45°. Luke saw this activity as a competition between himself and his partner and was frustrated that most of his answers were incorrect. Under this mild pressure, he fell back to manipulations of numbers and measures of length, with which he felt more competent.

Later, when the researcher repeated the activity just with Luke, Luke began using language that indicated an awareness of motion: "It's going way over" for a 120° turn (his guess was 150). Luke stood up and enacted the turn for the last two items, and he was more accurate in his estimations of the turn measures.

During the following session, the researcher asked Luke to enact various turn commands. For "rt 45" he said, "45 is half of 90," and he physically turned less than 90°. For a "rt 180" he said that it would be two 90s, and he described a "rt 360" as a "big turn." On paper tasks, however, Luke tended to use his finger to measure turns, or to guess, without using any identifiable reference unit. Although he did indicate a turning motion, he fell back to his competence with arithmetic to produce a measure. The numbers he employed in such arithmetic manipulations were usually not meaningfully connected to quantities, but were numbers "pulled out from" the situation.

At one position on the Feed the Turtle game, Luke turned "rt 30." He then remarked, "10 more." Although this was inaccurate (it was 30° more), his notion that the turn had to be greater showed more competence than he had shown on previous activities. Similarly, a few moves later, he said, "Right [pause]." The teacher asked, "Is it more than 90 or less than 90?" "More—90—120!"

Although Luke showed some progress in the Feed the Turtle game, he still had quite a bit of difficulty with turns. When his class reviewed body turns, Luke turned into the opposite direction for the command "rt

150." Two days before, he had stated that halfway around would be two "rt 90s," or 180. When asked why he was facing the same way for the "rt 150" as he would for a "rt 180," he acted confused. For "rt 120," Luke said, "I don't know it—wait. [performing the turn correctly] Now I do."

On a subsequent attempt to draw an equilateral triangle, Luke entered 90 for all "fd", "rt", and "lt" inputs. The researcher asked him to stop entering commands at this time "so that I can write them down in my notebook." Luke sat back in his chair and stared at the screen. He noticed that the turtle was facing the opposite direction. He stated, "I ended up the other way, too. Isn't that 180?" He then observed that the last two commands did combine to make 180. This was a significant insight for Luke. He then used his body to estimate the correct turn of 120. When asked why 120 worked, he drew a line at a 90° heading and said, "Because you go up, and that would be 90, and you need 120 to go down here." Luke was consistently successful with turns (other than 90° turns) if, and only if, he used his body to estimate the turns.

Monica and Nina developed greater knowledge of turn measure. Initially, Monica seemed to associate each right-face or left-face turn with facing specific walls in the classroom. Soon, however, both she and Nina appeared to think of 90° as a certain amount of turn not associated with any particular part of their environment. Often, when walking to the computer lab, they would ask the researcher to give them directions "like the turtle." They were consistently accurate about turning 90 to the right or left in this context. Similarly, Monica could easily say where she would be heading after one, two, or three right 90s.

Deciding turn measures on paper or computer was more difficult. Monica typically gestured to indicate a 90° turn before giving an estimate of any turn greater than 90°. Even though the teacher and curriculum stressed the motion of turning, Monica and Nina sometimes attended only to the final heading of the turn. That is, if the turtle started at 30° and ended at 180°, they saw the turn as "rt 180"—a feature shared with the "protractor" schema for turns (Kieran et al., 1986). After receiving advice from the teacher, classmates, and Geo-Logo, Monica changed her strategy: She would swing her arm 90°, then visually gauge whether a turn was equal to, greater than, or less than 90°. Later, she began to use a second benchmark, commenting, "That's 'rt 30.' It's a small turn."

Monica and Nina's class solved the first equilateral triangle problem as a group. After trying turns of 50, 90, 100, and 110, Monica was sure 120 was the correct amount. Later, however, she maintained that she needed to turn more to make a bigger triangle. Only after defining several triangles of different sizes and examining their Logo definitions did she change her mind.

Combining Turns

Luke did not combine turn commands as frequently as he combined length commands, and he gave more of an indication that what he was combining was not *rotations*, but numbers. He once examined the last three commands he had entered, and, while the researcher was writing notes, asked if he could change the two "rt 90s" to a "rt 180." The researcher asked him if he could combine all three ("rt 90," "rt 90," "lt 30"), and he replied, "rt 210." He tried it, noticed that the turtle was not at the same heading, and said, "I'll have to leave it." The researcher told him to "stand up and try it." Luke did so, while saying each command, then announced, "rt 150." Again, while working on the problem without turning his body, Luke focused on numbers (only) rather than on quantities.

In the Feed the Turtle game, Monica and Nina usually estimated turn measures accurately; therefore, they rarely had need to combine turn commands. Other children in the class would combine turn commands, especially after prompting by the teacher. However, students' hesitation to do so and their insistence on checking the results indicate that these combinations were more algorithmic and less meaningful to them than were combinations of forward commands. Although the curriculum asked students to combine turns less frequently than it asked them to combine lengths, students' low-level ideas about turn measure also may have limited their confidence and competence in combining turns.

Undoing

A final process related to movement and rotation commands that showed considerable growth in some students was that of "undoing." One boy, Dave, entered "rtf" instead of "ltf." Jimmy, his partner, exclaimed, "We'd have to turn all the way around!" They neither thought of entering "ltf" nor of merely using "bk," even though they knew these commands.

After only a day or two, most children were using inversion. For example, David said, "We have to get it back to the elevator. So do the same commands, but backwards." This revealed undoing that was at a higher level of abstraction than that they failed to use the first day. Here they were not merely undoing, but reversing the command.

Although children's development of ideas about *turn* measurement often lagged behind ideas about *length* measurement, undoing appeared at approximately the same time in both realms:

Teacher: So if you had "ltf," "rtf," "rtf," what did you change?
David: We just used one "rtf." Otherwise, the turtle's just dancing!
Teacher: How did that help you?
David: It uses up the battery just once, not three times!

Note, however, that the use of the "rtf" and "ltf" commands, without numerical inputs, may have facilitated the recognition of undoing in the domain of rotations.

CONCLUSIONS

These studies investigated the development of geometric and measurement ideas within an instructional unit on paths, with an emphasis on the role of noncomputer and computer experiences. Perhaps the most striking result regarding length problems concerned the different strategies children used. We observed three main strategies:

1. Some students did not segment lengths. Moreover, these students did not connect the number for the measure with the length of the line segment. Rather, they either used visual estimation or guessed numbers that appeared disconnected from any visual or spatial quantity. We never observed them making statements that would indicate that they were operating on *quantities*. When an adult asked them to draw a figure with certain dimensions, there was no discernible use of those dimensions in their drawings. On the positive side, these students did make sense of the problems in their own way, rather than merely trying to discern the answer the teacher wanted. They occasionally used memorized algorithms in situations in which other strategies such as mental computation seemed to us to be more appropriate. They were, nonetheless, solving a (numerical) problem that was meaningful to them.
2. Students drew hash marks, dots, or line segments to partition lengths or to segment lengths without maintaining equal-length parts. (Most students used this strategy.) A turtle step is a small unit. In contrast, for most students, objects 100 units in length were "large." Students often had difficulty abstracting the turtle step quantitatively and therefore marked off lengths in units that made sense to them, usually units of 10. They needed to have perceptible units such as these to quantify the length (Steffe, 1991). This process led to interesting use of numerical units of units. In addition, this may explain some students' resistance to making 200-step rectangles, the side lengths of which were not multiples of 10. Initially, 10 may have been a unit they could not decompose in that context.
3. Students used quantitative ideas in discussing the problems, drew proportional figures, and sighted along line segments to assign them a length measure. They used a move, segment, part–whole strategy on most missing-measures tasks. They had interiorized units of length and had developed a measurement sense that they could impose mentally onto figures. These observations support Steffe's (1991) argument that these students have created an abstract unit of length. This is not a static image, but rather an interiorization of the process of moving, visually or physically, along an object, segmenting it, and counting the segments. When students considered consecutive units as a unitary object, they built a "conceptual ruler." They could then project this conceptual ruler onto unsegmented objects (Steffe, 1991). Students, like adults, use such processes without conscious awareness.

We conjecture that once students using the first type of strategy have had sufficient physical measurement experience iterating and partitioning into units, they build mental processes that allow them to partition unsegmented lengths, but only on the physical level. They need to use physical action to create partitions. In solving problems, these partitioning schemes develop the constraint that equal intervals must be maintained. This constraint leads to the construction of an anticipatory scheme because the equal-interval constraint can be realized most efficiently when it is done in imagery. At this point, strategies of the third kind emerge.

Turns, as a relationship, may be less salient for children than lengths and the motions that create them. The turning motion itself usually does not leave a trace. Also, people do not clearly distinguish turns from other movements in most real-world contexts. For instance, when we turn a car or bike, we are also moving forward. Therefore, students have to work harder to internalize the relationships of turns in order to impose the dynamic transformation onto static figures.

In the case of Luke, we see most clearly how students work to learn about turns. Luke relied heavily on physical turns, especially turns of his body. He also gained some knowledge of assigning numbers to certain turns. He built up landmarks in the domain of degrees. Like Piaget's (Piaget & Inhelder, 1967) children, Luke originally focused only on the beginning and ending states of turns rather than on the movement in the middle. Moreover, the beginning and ending states were represented by line segments, which, for Luke, then became more salient than the relationship between the segments, or the transformation that would move the turtle from one segment's heading to that of the other.

His static graphics depictions and corresponding need for correctness interfered with his developing dynamic, rotation-based ideas about turns. Eventually, with help from adults and computer experiences, Luke started to connect the two ideas of turn-as-body-motion and turn-as-number. Under stress, however, he lost that connection, usually falling back to operating on numbers without a corresponding image of a turning motion. Further, this connection and the resultant idea of turn measurement were tenuous, even at the conclusion of the unit of instruction. Nevertheless, Luke's development of imagery and his initial connection to numerical ideas suggest that a synthesis of these two domains might be an important aspect of learning about turns.

In many aspects, Monica developed greater knowledge of turns than Luke. She still had difficulty, however, with turns in which she and the turtle were at or nearly at opposite headings. She built internal frameworks for 90° turns. Students estimated turns that measured more or less than 90° based on another unit, 30°. However, this use may have been constituted as imposing a mental protractor onto geometric situations requiring an amount of turn, rather than actually estimating, using units of turn. For other students, however, we have evidence that 30° was an iterable unit of turn.

To think about a turn and its measure, children have to maintain a record of mental images of both the initial and final headings of an object, using a frame of reference to fix these headings. They have to represent to themselves the motor activity of rotation of the object from the former to the latter and then compare that new image to one or more iterations of an internalized image of a unit of turn—for our students, units of 90° or 30°—or partition the represented turn into these units.

The complexity of these mental processes may explain why students compared, combined, and reflected on turns less than they did on lengths. Creating turns and turn measures as mathematical objects is a significant construction, one to which we probably give too little attention in the elementary and middle school years (see Lehrer, Jenkins, & Osana, chapter 6, this volume). Many students do not internalize turns as mathematical objects. Such students would not be able to visualize, or represent to themselves, a turn in the absence of physical objects. Most never interiorize turns and turn measures. They cannot abstract these mathematical entities in their most general form, nor can they isolate, coordinate, and operate on them in mathematical situations. The result is that students' performance in a wide range of activities in mathematics, other sciences, and geography is limited by their inability to apply these ideas in different settings.

All students in this series of studies had much to learn about turn measure. However, they did make advances. In addition, the common misconception of assuming a connection between rotation and length measures, as in thinking of angle measure as the distance between two rays (Clements & Battista, 1989, 1990; Krainer, 1991), was not observed. This supports the usefulness of units of instruction and software such as that employed here.

In both lengths and turns, students' predispositions and capabilities in connecting numeric and spatial schemes affected their choice of strategies. Some students were skilled with numbers and computation, including mental computation. Not all of these students, however, operated on spatial representations and linked their computations to quantity.

The Geo-Logo environment played an important role in allowing meaningful exploration and providing needed feedback. Compared to regular Logo, a new command that is the same as the immediately previous ones does not have to be re-entered; more important, a new command that is different from the previous one is not added onto the previous command, but is seen as a new command to the figure, with change in the figure reflecting change made by the new command. This helps students encode contrasts between different turn commands. Geo-Logo's feedback consists largely of the graphic results of running Logo code precisely, without human interpretation. This was important to the girls who, seeing the difference between their own interpretations of their Logo code for an equilateral triangle and the Logo turtle's implementation of that code, immediately examined and altered the code. Giving commands to a non-interpretive agent, with thorough specification and detail, has been identified as an important advantage of computer use in facilitating the learning of mathematics (Johnson-Gentile, Clements, & Battista, 1994).

The computer context was also highly motivating for these students. Teachers consistently remarked that their "lower level" students seemed genuinely interested in solving mathematics problems while working on the computer. Further, these students learned that accuracy was necessary and abandoned their usual guessing strategies.

Geo-Logo also aided the construction of mental connections between symbolic and graphic representations of geometric figures and between these representations and number and arithmetic ideas. When this occurred, the students spoke not of spatial extent or length, but about dynamic movement. For example, Nina and Monica used arithmetic to solve problems, but their language was about the turtle: "It's going on 50, then 30 more, so that's ..." This emphasis is not on the geometric figure as much as it is on the turtle's movements. Thus, the emphasis on physical action and the dynamic connections between the symbolic and graphic representations in Geo-Logo facilitated students' development of such connections for themselves. We must, however, temper this conclusion with a recognition that these connections were tenuous and situation bound in many instances.

Educational Implications

We could view those students who did not connect spatial and ideas, but solved the problem with arithmetic, as having effective solution strategies. We believe, however, that such a position is invalid and even dangerous for these students. They would benefit from activities that guide them to synthesize these two domains for four reasons:

1. Students need to pay attention to the scale provided for certain figures. This is important not only for their future learning of geometry, but also in many other situations, such as those in which they use maps and graphs.
2. This study's data indicate that those students with connected ideas had more powerful, flexible solution strategies for solving spatial problems at their disposal. In the domain of turns, students without connected ideas were not developing concepts critical to many areas of mathematics and other sciences.
3. The activity of connecting mathematical ideas is a valuable mathematical activity. Developing students' predisposition and ability to do so, therefore, is a valuable educational goal.
4. The combination of arithmetic and geometry is an efficient, effective way to teach both domains. Porter reported whole districts in which fourth- and fifth-grade teachers spent "virtually no time teaching geometry" (Porter, 1989, p. 11). Teachers can help students learn often-neglected geometry and spatial sense while still developing arithmetic ideas and skills.

Recall that students using the first and third kinds of strategy for solving length problems did not use hash marks to segment or partition line segments. Teachers should take care to observe these students' interpretations of the task, for they need to engage in quite different types of activities. Teachers should challenge students who use the third kind of strategy by assigning more difficult tasks. Students using the first kind of strategy need to engage in partitioning and iterating lengths, continually tying the results of that activity to their counting schemes. Tasks in which applying only numerical schemes is ineffective may be especially useful. For this reason, we changed several missing-lengths problems. For example, in an early version of the missing-measures task shown in Fig. 8.3a, all segments were 20 units; in a later version, the authors changed the two vertical segments to 15 units. With the latter figure, students who chose the 15-unit length segments showed themselves, through their own computer work, that their solutions were not adequate.

Even though these third-grade students did make significant progress, turns were still difficult. If curriculum writers and teachers integrated turns throughout the mathematics curriculum, difficulties might diminish and benefits increase. Our results imply that static representations of turns, no matter how cleverly designed, may not only be inadequate for children's learning, but may delay their development of dynamic ideas of turn and turn measurement because static representations limit students' construction of notions of physical motion and the integration of these notions with geometric figures and numerical measures. Instead, teachers need to emphasize students' physical motions (e.g., actually using their own bodies to enact a turn), records of the transformation (e.g., holding out one arm to maintain the initial heading, then moving the other arm through the turn), reflection on these motions (e.g., by contrast, the apparent uselessness of the students' nonreflective hand twitching), and connections between such physical activities and other dynamic activities such as those using computer or paper. Further, teachers should encourage students to return to the physical motions as an aid during such computer-and-paper activities. They should emphasize the establishment of perceptually salient units of turn (e.g., 90°, 180°, 30°, or 45°, and so on, whether invented by students or suggested by the teacher) and relationships among the units. Finally, in concert with other researchers (Hoyles & Sutherland, 1989), we have found that activities such as those featured in the curriculum (featuring well-defined goals, often with pictorial representations) are critical in developing ideas of turn and turn measurement.

Curriculum Development, Teaching, and Research

The isolation of curriculum development, classroom teaching, and mathematics educational research from each other negatively affects these three aspects of mathematics education. Integrated research and curriculum evaluation efforts have a better chance of contributing to a progressive change in mathematics curriculum than do traditional procedures. The likely alternative to such integration is the continuation of cycles of production of new curricula that are distinguished only by changes dictated by conformity to social and political whims.

From a research perspective, we believe that studies that use interviews or teaching experiments based solely on a theoretical perspective are worthwhile but limited. We believe that to fully understand children's learning, researchers and teachers must have optimal educational environments in which teachers (and researchers) can simultaneously observe and facilitate children's mathematical thinking. In addition, this approach brings values and goals of mathematics education to the forefront. In too many curricula and studies, these values and goals are only implicit.

A critic might argue that this is an egocentric approach to research—that it limits implications of the research to a specific curriculum. Certainly, there is some truth in this position. We would respond, however, that in every study, children respond within the context of the environment, even if it is so rudimentary as to be nearly invisible. This caveat, therefore, applies to all studies. Further, many of the environments (e.g., those in clinical studies) have been limited in educing that mathematical thinking of which children are known to be capable.

We argue, then, the benefits of multiple foci: researched-based curriculum development, curriculum-guided research on children's thinking, and acknowledgment of the inextricable issues of computer and noncomputer environments, and teacher beliefs and knowledge. Such perspectives are, palpably, demanding. Requirements of curriculum development (e.g., deadlines from publishers and funding agencies) and of the classroom (e.g., teachers' need to "cover" other material; often tight, inflexible schedules) must be balanced with those of research. Controlling all the variables is absurd. A flexible, multifaceted approach can achieve a balance that synergistically contributes to curriculum development, classroom teaching, and mathematics educational research.

ACKNOWLEDGMENT

The work described in this report was funded in part by the National Science Foundation (grant no. ESI-8954664 and grant no. ESI-9050210). Opinions expressed are those of the authors and not necessarily those of the foundation.

REFERENCES

Clements, D. H., & Battista, M. T. (1989). Learning of geometric concepts in a Logo environment. *Journal for Research in Mathematics Education, 20*, 450–467.

Clements, D. H., & Battista, M. T. (1990). The effects of Logo on children's conceptualizations of angle and polygons. *Journal for Research in Mathematics Education, 21*, 356–371.

Clements, D. H., & Battista, M. T. (1991). *The development of a Logo-based elementary school geometry curriculum* (Final Report, NSF Grant No. MDR–8651668). Buffalo, NY/ Kent, OH: State University of New York at Buffalo/Kent State University Presses.

Clements, D. H., & Battista, M. T. (1992). Geometry and spatial reasoning. In D. A. Grouws (Ed.), *Handbook of research on mathematics teaching and learning* (pp. 420–464). New York: Macmillan.

Clements, D. H., Battista, M. T., Akers, J., Woolley, V., Meredith, J. S., & McMillen, S. (1995). *Turtle paths.* Cambridge, MA: Dale Seymour.

Clements, D. H., Battista, M. T., Sarama, J., & Swaminathan, S. (1996). Development of turn and turn measurement concepts in a computer-based instructional unit. *Educational Studies in Mathematics, 30,* 313–337.

Clements, D. H., Battista, M. T., Sarama, J., Swaminathan, S., & McMillen, S. (1997). Students' development of length measurement concepts in a Logo-based unit on geometric paths. *Journal for Research in Mathematics Education, 28*(1), 70–95.

Clements, D. H., & Meredith, J. S. (1993). Research on Logo: Effects and efficacy. *Journal of Computing in Childhood Education, 4,* 263–290.

Clements, D. H., & Sarama, J. (1995). Design of a Logo environment for elementary geometry. *Journal of Mathematical Behavior, 14,* 381–398.

Clements, D.H., & Sarama, J. (1996). Geo-Logo [Computer software]. Logo Computer Systems, Palo Alto, CA: Dale Seymour Publications.

Davis, R. B. (1984). *Learning mathematics: The cognitive science approach to mathematics education.* Norwood, NJ: Ablex.

Hershkowitz, R., Ben-Chaim, D., Hoyles, C., Lappan, G., Mitchelmore, M., & Vinner, S. (1990). Psychological aspects of learning geometry. In P. Nesher & J. Kilpatrick (Eds.), *Mathematics and cognition: A research synthesis by the International Group for the Psychology of Mathematics Education* (pp. 70–95). Cambridge: Cambridge University Press.

Hoyles, C., & Sutherland, R. (1986). *When 45 equals 60.* London: University of London Institute of Education, Microworlds Project.

Hoyles, C., & Sutherland, R. (1989). *Logo mathematics in the classroom.* London: Routledge.

Johnson-Gentile, K., Clements, D. H., & Battista, M. T. (1994). The effects of computer and noncomputer environment on students' conceptualizations of geometric motions. *Journal of Educational Computing Research, 11,* 121–140.

Kieran, C. (1986a). Logo and the notion of angle among fourth and sixth grade children, *Proceedings of PME 10* (pp. 99–104). London: City University Press.

Kieran, C. (1986b). Turns and angles: What develops in Logo? In G. Lappan & R. Even (Eds.), Proceedings of the Eighth Annual Meeting of the North American Chapter of the International Group for the Psychology of Mathematics (pp. 169–177). East Lansing, MI: PME–NA Program Committee.

Kieran, C., Hillel, J., & Erlwanger, S. (1986). Perceptual and analytical schemas in solving structured turtle-geometry tasks. In C. Hoyles, R. Noss, & R. Sutherland (Eds.), *Proceedings of the Second Logo and Mathematics Educators Conference* (pp. 154–161). London: University of London Press.

Krainer, K. (1991). Consequences of a low level of acting and reflecting in geometry learning—Findings of interviews on the concept of angle. In F. Furinghetti (Ed.), *Proceedings of the Fifteenth Annual Conference of the International Group for the Psychology of Mathematics Education* (Vol. II, pp. 254–261). Genoa, Italy: Program Committee, 15th PME Conference.

Lindquist, M. M., & Kouba, V. L. (1989). Geometry. In M. M. Lindquist (Ed.), *Results from the Fourth Mathematics Assessment of the National Assessment of Educational Progress* (pp. 44–54). Reston, VA: National Council of Teachers of Mathematics.

Noss, R. (1987). Children's learning of geometrical concepts through Logo. *Journal for Research in Mathematics Education, 18,* 343–362.

Olive, J., Lankenau, C. A., & Scally, S. P. (1986). *Teaching and understanding geometric relationships through Logo: Phase II. Interim Report: The Atlanta–Emory Logo Project.* Atlanta, GA: Emory University Press.

Petitto, A. L. (1990). Development of numberline and measurement concepts. *Cognition and Instruction, 7,* 55–78.

Piaget, J., & Inhelder, B. (1967). *The child's conception of space.* New York: Norton.

Porter, A. (1989). A curriculum out of balance: The case of elementary school mathematics. *Educational Researcher, 18,* 9–15.

Singer, D. G., Singer, J. L., & Zuckerman, D. M. (1981). *Teaching television*. New York: Dial.

Steffe, L. P. (1991). Operations that generate quantity. *Learning and Individual Differences, 3,* 61–82.

Strauss, A., & Corbin, J. (1990). *Basics of qualitative research: Grounded theory procedures and techniques*. Newbury Park, CA: Sage.

van Hiele, P. M. (1986). *Structure and insight*. Orlando, FL: Academic Press.

9

Students' Understanding of Three-Dimensional Cube Arrays: Findings from a Research and Curriculum Development Project

Michael T. Battista
Kent State University

Douglas H. Clements
State University of New York at Buffalo

As part of our work in developing the geometry strand in the *Investigations in Number, Data, and Space* elementary mathematics curriculum project, we developed three-dimensional geometry instructional units for each of Grades 3, 4, and 5 (Battista & Berle-Carman, 1995; Battista & Clements, 1995a, 1995b). A major portion of these units focuses on exploring three-dimensional configurations of cubes, especially enumerating cubes in three-dimensional rectangular arrays. Concurrent with our curriculum development, we conducted a research program investigating students' understanding of three-dimensional configurations of cubes (see Battista & Clements, 1996). This chapter gives an overview of both our research and our curriculum development efforts.

We first describe the various conceptual structures that students construct in enumerating three-dimensional cube arrays and suggest the mental operations that underlie these constructions. We then describe instructional activities that target the development of these conceptual structures, as well as the rationale we had in using these activities. Finally, we illustrate how learning occurs in students using these instructional activities.

STUDENT MEANINGS FOR THREE-DIMENSIONAL CUBE ARRAYS

According to the constructivist view of learning, which we have adopted for both our research and curriculum development, as individuals interact with the world, they develop sets of cognitive structures that reflect the way their current knowledge and thought processes are organized. Individuals learn as they create or change these structures, as the structures become more sophisticated and powerful. The goal of instruction is to help students build cognitive structures that are more complex, abstract, and powerful than they currently possess. Effectiveness in designing instructional tasks and in assessing and guiding students' learning progress depends critically on detailed knowledge of how students construct meanings for specific mathematical topics, as well as on the conceptual advances that students can make in those topics during the course of instruction.

We conducted research aimed at understanding how students enumerate three-dimensional cube arrays by (a) individually interviewing students, (b) pilot testing instructional tasks with individual or small groups of students, and (c) having teachers test instructional units in their classrooms. In this chapter, we describe students' conceptions and enumeration strategies as they were applied in enumerating three-dimensional cube arrays presented either pictorially or with actual cubes. During our research, we observed several categories:

Category A. Students see an array in terms of layers. They determine the number of cubes in one layer and then multiply or use repeated addition to account for all the layers. The layers can be vertical or horizontal, and students often use one of the faces of the prism array as a representation of a layer.

Students also see an array in terms of dimensions. Students understand how dimensions can be used to describe and enumerate the cubes in an array. For instance, for a 4 by 3 by 3 array, a student might reason that there are three top rows each with 4 cubes, or ($3 \times 4 =$) 12 cubes, in a layer. Because the array is 3 cubes high and there are three layers, there are ($3 \times 12 =$) 36 cubes.

Category B. Students conceptualize the set of cubes as space-filling, attempting to count all cubes in the interior and exterior. Strategies range from unsystematic counting to various grouping procedures. Some students see an array as an unstructured set. Given a "building" made of cubes, they act as if they see no organization of the cubes. They usually count cubes one by one and almost always lose track of their counting. For them, the task resembles counting a large number of randomly arranged objects.

Other students locally organize subsets of the array. Although they attempt to organize the array by grouping the cubes in some way, they fail to apply a consistent, global organizational scheme. For instance (see Fig. 9.1), one student counted the cubes visible on the front face of the array (12), then counted those on the right side that had not already been counted (6). To account for the remaining cubes, the student pointed to the 6 uncounted cubes on the top and, for each, counted three times: 1, 2, 3; 4, 5, 6;. . ; 16, 17, 18. The student then added 18, 12, and 6. For this same array, another student counted all the outside cube faces, getting 66. He then said that there were 2 cubes in the middle, arriving at a total of 68.

Finally, some students see an array in terms of rows or columns. They count by ones or skip count to enumerate the cubes in rows or columns. Row or column structuring represents a consistent, global organization of the cubes.

Category C. Students see the array in terms of sides or faces. They conceptualize three-dimensional arrays of cubes by thinking only about the faces of the prisms formed by the arrays. Such students might count all or some of the cube faces that appear on the six faces of the prism. They usually double-count edge cubes and almost always omit cubes in the interior of the array. Most students with the "sides" conceptualization utilize it consistently, whether they are working with pictures, patterns for boxes, or actual cube configurations.

Category D. Students explicitly use the formula L × W × H with no indication that they understand it in terms of layers.

Category E. This category includes all other strategies, such as multiplying the area of one face by the area of another. Several of these strategies suggest that students might see the three-dimensional array strictly in terms of its faces; however, such strategies might also represent a misguided attempt to apply a previously encountered formula.

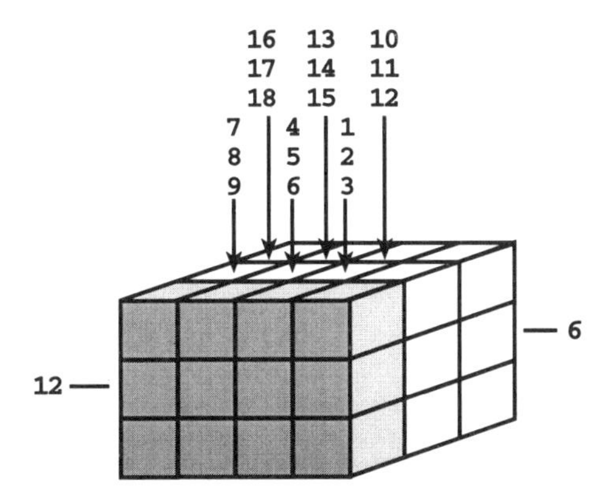

FIG. 9.1 An example of a local structuring of a cube array.

STRATEGIES USED BY THIRD AND FIFTH GRADERS

Table 9.1 shows the percentage of students in Grades 3 and 5 falling into each of the above categories, based on individual interviews on the items shown in Fig. 9.2. As can be seen from the table, the distributions of strategies were fairly consistent across problems. Utilization of strategies was not greatly affected by whether the mode of presentation was concrete or pictorial, so we cannot attribute the use of nonlayering strategies to stu-

TABLE 9.1 STRATEGIES USED BY STUDENTS TO ENUMERATE THREE-DIMENSIONAL CUBE ARRAYS, BY PERCENT[a]

	Problem 1a[b]		*Problem 1b*[b]		*Problem 2*[c]	
Category	*Grade 3*	*Grade 5*	*Grade 3*	*Grade 5*	*Grade 3*	*Grade 5*
A	16	56	20	56	18	60
B	24	17	13	8	22	13
C	58	18	62	19	53	17
D	0	3	0	3	0	4
E	2	5	4	13	7	6
Students correctly solving problem	33	71	18	55	24	63

Note. Categories are given as A, layering strategies; B, space-filling strategies; C, sides conceptualization; D, formula (L × W × H); and E, all other strategies (e.g., multiplying face areas).

[a]$n = 45$ at Grade 3; $n = 78$ at Grade 5.

[b]Pictorial representation.

[c]Concrete representation.

1. [Pointing to picture] This is a picture of a unit cube. How many unit cubes will it take to make each building below? The buildings are completely filled with cubes, with no gaps inside.

Unit Cube

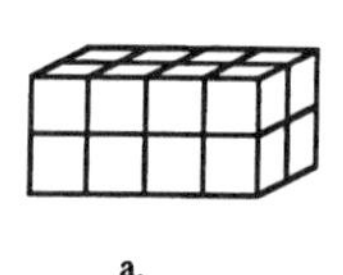

a.

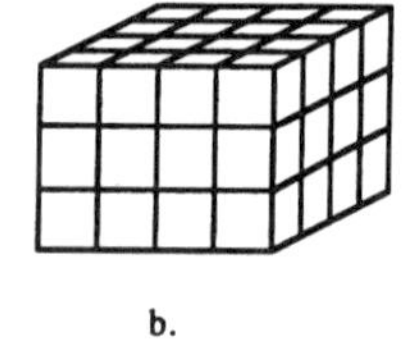

b.

2. How many cubes are in this building? It is completely filled with cubes, with no gaps inside. [Students were shown a 3 × 4 × 5 cube building made from interlocking centimeter cubes. They were permitted to handle the building, but were not to take it apart.]

FIG. 9.2 Questions from interview with students from Grade 3 and Grade 5.

dents' misinterpretation of the diagrams. There was a marked difference between the strategies used by third and fifth graders. About 60% of the fifth graders, but fewer than 20% of the third graders, used layering strategies. On the other hand, about 60% of third graders, but only about 20% of fifth graders, used a strategy that suggests that the student saw the arrays as consisting of their outer faces. Only 7% percent of third graders and 29% of fifth graders correctly used a layering strategy for all three problems. Five students in Grade 5 used the formula rotely; no student used it meaningfully (i.e., no Category A student used the formula).

Students' double-counting of cubes was also an important phenomenon. (A cube was considered double-counted if it was counted at least twice.) Although not shown in the table, 64% of the third graders and 21% of the fifth graders double-counted cubes at least once on the three interview questions; 33% of third graders and 6% of fifth graders double-counted on all three items.

The Source of Students' Difficulties with Three-Dimensional Arrays

Examples of students' difficulties in dealing with three-dimensional arrays provide us with important clues about the mental operations that underlie students' thinking about such arrays. The example that follows involves a Grade 5 student.[1]

RA. [After establishing the difference between cubes and squares and after RA has built several arrays with interconnecting cubes, the interviewer asks RA how many cubes it will take to make Building 1 (see Fig. 9.3a). RA first counts all cube faces visible in the diagram, double-counting cubes on the edges, then gives an answer of 40. When the interviewer asks RA how the number of cubes was determined, RA changes the answer.]

> *RA:* It has 4 × 3 and 4 × 4. 12 × 4 = 48, + 16 = 64. There's some you can't see in the picture. [RA counts the faces of exterior cubes, accounting for all four lateral sides and the top of the prism, but not its bottom. She then spends about 20 min trying to construct Building 1 with cubes. She builds the configuration shown in Fig. 9.3b several times, making the top (4 by 4) and the side (3 by 4), then joining them. Each time, however, RA stops and starts over after noting that the front of

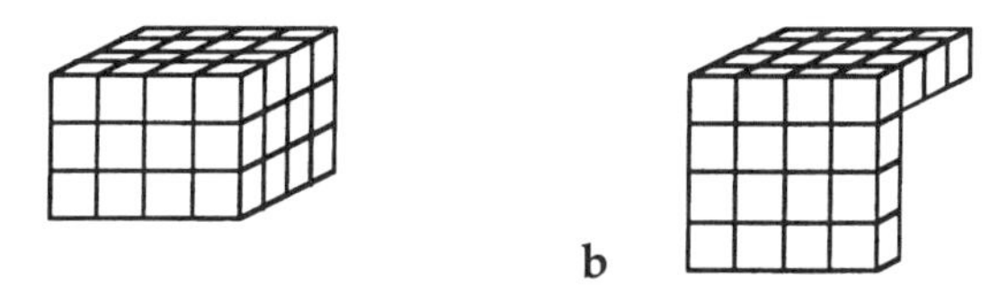

FIG. 9.3 Building 1 (a) and RA's attempt to construct it (b).

[1]The case-study examples described in this chapter happen to be females. We take them, however, to be representative of both males and females because we have observed no gender differences in performance on instructional or assessment tasks.

the cube configuration does not match the front of the picture of Building 1. The next day, the interviewer asks RA to find the number of cubes in the figure that the student had built at the end of the last session.]

RA: Four rows of 4 on this side [the top] and four on this side [the front]. So it's 16 + 16 = 32.
[So, even in this perceptually simpler situation, RA still double-counts cubes; she still fails to coordinate the counting of cubes in the top with the counting of the front faces of the cubes. The interviewer now separates the 4 by 4 top array from the 3 by 4 front array and lays them flat on a table, one above the other.]

I: How many cubes?

RA: 16 + 9 = 25 [enumerating the 3 by 4 array as 3 × 3].
[The interviewer reconnects the two flat arrays in their original orthogonal position and asks again how many cubes there are.]

RA: 16 + 16 = 32.

I: [again taking apart the two arrays] How many of these cubes [holding up a single cube] would it take to make this [pointing to the flat 7 by 4 cube array]?

RA: 7 × 4, 28. [The interviewer returns the two arrays to their original orthogonal position and asks again how many cubes.]

RA: I guess it would be the same amount [28].

RA's inability to coordinate different orthogonal views of the three-dimensional array was even more striking in the attempts she made to build a rectangular prism than in her attempts to enumerate its cubes. She first constructed the top with cubes, then the front, copying each from the picture. When putting the top and side constructions together, RA recognized that the result did not match the picture, but she could not figure out how the two constructions fit together. RA could not coordinate the two different orthogonal views represented by these adjacent prism faces. Moreover, when attempting to count the number of cubes in the configuration she built, RA clung to the original, uncoordinated conception of the two faces, even after seeing them taken apart and getting a different count when they were laid flat. Although RA finally decided that her original count was incorrect, there is no evidence that she did so by spatially coordinating the sets of cubes belonging to the top and front faces. More likely, RA simply abandoned the answer given by one enumeration strategy in favor of that given by a strategy she considered more reliable.

Coordination

Following up on the hypothesis that students have difficulty integrating orthogonal views of a three-dimensional cube array into a coherent whole, we gave a problem that required students to do exactly that (see Fig. 9.4). Neither enumerating nor constructing the prism depicted by the three orthogonal views can be accomplished by the student without some type of coordination of the views.

Not one of the 15 fifth graders whom we interviewed determined the number of cubes in the array. Nine of the students did not even try to

Suppose we completely fill the rectangular box at the right with a rectangular cube building. The box is transparent: You can see the building through the box's sides.

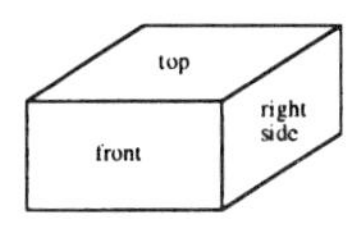

After we fill the box, we look straight at the building from its front, top, and right sides. [Indicate orthogonal viewing lines with a pencil.]

From the front, the building looks like this:

From the right side, the building looks like this:

From the top, the building looks like this:

A. How many cubes does it take to make the building?

B. Can you make the building with cubes?

FIG. 9.4 Interview problem: integrating orthogonal views of a three-dimensional cube array into a coherent whole.

coordinate the views (i.e., they did not attempt to figure out how the views fit together). The other 6 tried, but made errors in coordination. Only 8 of 15 students were able to correctly build the cube configuration by looking at its three views. The other 7 students attempted to construct the array by building, then connecting, two or more intact views. Six of these students recognized that the uncoordinated figure that they were constructing was incorrect, but only 2 subsequently built the prism correctly by coordinating views.

ST. ST was one of the students who used cubes to make different views and then simply connected them. Recognizing that her constructed building was incorrect and being unable to correct it, ST concluded that the task was impossible. When asked how many cubes were in the building, ST said there were 8 on the right vertical face and 6 on the front face, then multiplied 6 by 8 to get 48. ST tried to use layers of 8 cubes, but was unable to compose the entire cube configuration in terms of these layers. She could not properly coordinate the right-side view with any other view. Because ST could not "see" a global three-dimensional layering structure, multiplying the numbers of cubes showing in the front and side views seemed to be a reasonable method for enumerating the cubes.

Summary of Findings

The research just described provides a description of students' strategies and difficulties in enumerating three-dimensional rectangular cube arrays.[2] Our findings suggest that students' initial conception of a three-dimensional rectangular array of cubes is an uncoordinated set of faces. Such conceptions were seen in students who used category C strategies on the interview items. Eventually, as students became capable of coordinating views and as they reflected on experiences with counting or building cube configurations, perturbations occurred in this conception. Students began to see the array as space-filling and strove to restructure it as such. Those who completed a global restructuring of the array utilized A strategies. Those in transition, whose restructuring was local rather than global, utilized B strategies. They had not yet formed an integrated conception of the whole array that globally coordinated its dimensions. Indeed, the interview data are consistent with this hypothesized sequence of conceptions. From third to fifth grade, we saw that students made a definite move from seeing a three-dimensional cube array as an uncoordinated medley of faces toward seeing it in terms of its layers. We also saw a significant number of students in transition, with these students exhibiting a wide range of sophistication in their structuring of such arrays.

Our findings suggest that *spatial structuring,* defined as the mental act of constructing an organization or form for an object of set of objects (Battista & Clements, 1996), is a fundamental notion in understanding students' strategies for enumerating three-dimensional cube arrays. The process of spatial structuring includes establishing units and relationships between units and often requires coordination and integration operations. We assume that an individual does not "read" off a structure from objects, but creates, instead, a structure as a result of his or her mental actions concerning the objects (cf. Cobb, Yackel, & Wood, 1992; von Glasersfeld, 1991).

Our research suggests that the development of students' meaningful enumeration of cubes in three-dimensional arrays, a fundamental notion in understanding the measurement of volume, is far more complex than has previously been described. It suggests that many students are unable to correctly enumerate the cubes in such arrays because their spatial structuring of the arrays is incorrect. In particular, we found that for many students the root of such errant spatial structuring seemed to be their inabil-

---

[2]A more detailed description of this research, including a more elaborate discussion of the relationship between spatial and numerical reasoning, can be found in Battista and Clements (1996).

ity to coordinate and integrate the views of an array into a single coherent mental model of that array. Other researchers have found similar thinking among fifth through eighth graders (Ben-Chaim, Lappan, & Houang, 1985). Indeed, we have found such reasoning in high-school geometry students.

INSTRUCTIONAL TASKS AND RATIONALE

As part of our research, we developed several instructional tasks to help students construct viable understandings and strategies for dealing with three-dimensional configurations of cubes. At all three grade levels, our tasks required students to first make predictions, then to check those predictions using concrete materials such as cubes and paper boxes. Because they are based on students' current mental models of the cube configurations, making such predictions encourages students to reflect on and refine those mental models (Battista, 1994). Having students merely make boxes and fill them with cubes does not promote nearly as much student reflection because (a) opportunities for cognitive conflict arising from discrepancies between predicted and actual answers are greatly curtailed and (b) students' attention is focused on physical activity rather than on their own thinking. Promoting such reflection is a critical component of instruction because it is through such reflection and resulting abstractions that students develop increasingly more sophisticated cognitive structures for dealing with mathematical problems. What follows is a discussion of example tasks by grade level.

Grade 3

The goal of the Grade 3 activities is for students to become familiar with and begin to understand the structure of rectangular boxes and three-dimensional arrays. We do not expect students to develop strategies for predicting the number cubes that will fill a box; we only expect them to find viable ways for enumerating cubes, given cubes and paper boxes. We think that the experiences of reflecting on their actions with cubes and boxes will eventually enable them to construct appropriate mental models.

After being introduced to the idea of a box pattern by making and exploring box patterns that can hold only one or two cubes, students investigate patterns for boxes that can contain larger numbers of cubes. They predict, then determine, the number of cubes that boxes made by patterns can hold (Example 1, Fig. 9.5). Students next design patterns for boxes that hold given numbers of cubes (Example 2, Fig. 9.5).

In a final project, students make a city of buildings made from bottomless boxes. They decide on the shapes of the boxes based on the types of buildings they want to include in their cities (e.g., houses, malls, factories). They also determine the number of cubes that each box can hold and

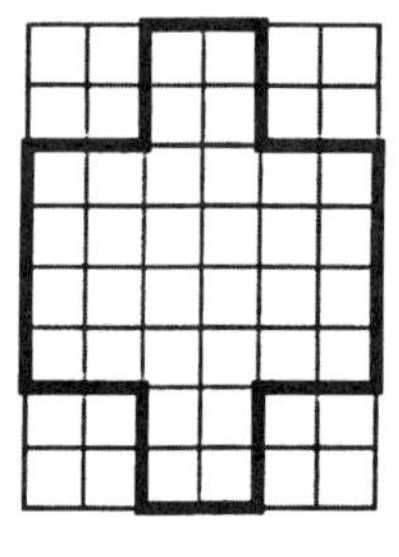

Example 1

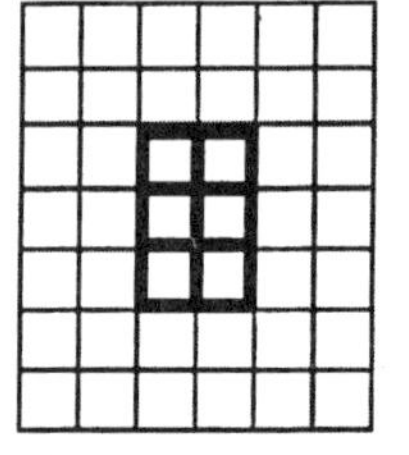

Example 2

FIG. 9.5 Grade 3 activities for discovering strategies to enumerate three-dimensional cube arrays. Example 1: Students are asked to predict how many cubes fit in the box made by the pattern. They check their answers by cutting out the pattern, building the box, and filling it with cubes. (The size of the squares in the grid matches the size of the cubes.) Example 2: Students are shown the bottom of a pattern for a rectangular box that contains exactly 12 cubes and has no top. They are to draw the sides to complete the pattern. (Some students completed the pattern only after they built the cube array and placed it on the base.)

the total space inside all the buildings in their model city. This project encourages students to use the concepts dealt with earlier in a creative, engaging project.

Grade 4

In the fourth-grade unit, one of our goals is for students to learn to communicate about three-dimensional cube configurations. Students build, visualize, describe, and draw cube configurations. They make cube configurations depicted in isometric drawings and draw front, top, and side views of their constructions. They are also asked to construct cube buildings from top, side, and front views.

When students see only three sides of a cube configuration, there is no information about how the sides are related or how they should be put together. To interpret the diagram, students must coordinate, integrate, and synthesize the diagram information into an appropriate mental model.

Grade 5

The fifth-grade unit focuses on students' development of strategies that accurately predict how many cubes or packages fill a given rectangular box. They check their predictions by making the boxes from paper and filling them with cubes. These activities encourage students to structure the boxes and cube arrays in ways that support viable enumeration strategies. Our goal is for students to learn to think about three-dimensional cube arrays in terms of layers and to use appropriate numerical strategies to enumerate the cubes based on this conceptualization (e.g., "A layer contains 12 cubes, there are five layers, so there are 60 cubes in all.")

Students start with the problem of predicting the number of cubes in boxes, given a picture of the box and/or a pattern for making the box (refer to Fig. 9.9). This problem provides students with the chance to individually construct strategies for enumerating cubes in a box. Students then solve problems like those in Fig. 9.6 in which the goal is to ensure that students are actually structuring the arrays rather than using a numerical procedure that they do not understand.

STUDENTS' CONSTRUCTION OF KNOWLEDGE WHILE WORKING ON THE INSTRUCTIONAL TASKS

The examples that follow illustrate how our instructional tasks promote the development of the type of thinking required for structuring three-dimensional arrays. These examples were taken from classrooms in which an instructional unit developed by the researchers was taught by a classroom teacher as part of her regular mathematics instruction. Research teams consisting of three to four persons observed classroom instruction throughout each unit.

Grade 3

The Grade 3 example illustrates the difficulty one student had in enumerating actual cubes. KA drew the pattern shown in Fig. 9.7 on grid paper. To determine the number of cubes in the box made by the pattern, KA counted 9 in each of the four side flaps and concluded there would be 36 cubes.

The interviewer asked KA to cut out the pattern, construct the box, and fill it with interlocking cubes. KA constructed three intact 3 by 3 layers of cubes and placed them inside the box, one on top of the other. When the interviewer asked how many cubes were inside, KA said 36. The in-

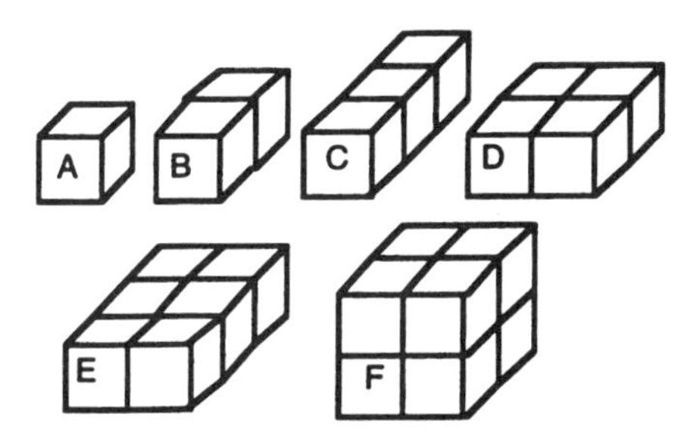

FIG. 9.6 Grade 5 instructional activity. Students are asked to predict how many packages of each size (A–F) can fit into a box 4 cubes by 6 cubes on the bottom and 3 cubes high. They check their answers by building boxes and filing them with packages.

FIG. 9.7 Grade 3: KA's box pattern.

terviewer asked KA to take the cubes out of the box and count them. She removed the intact layers from the box and placed them on top of each other. KA counted the cube faces that appeared on the four lateral sides and again obtained an answer of 36. The interviewer separated the layers and laid them flat, next to each other:

I: How many cubes?
KA: There should be the same number.

KA counted 27 cubes, one at a time, and looked puzzled. The interviewer asked her how many cubes would fit in the box. KA restacked the layers and said 36. The interviewer asked how many cubes would be in a bag if the three layers were taken apart and all the cubes put inside. KA said 36. The interviewer placed the three layers flat again and asked KA to count the cubes. She counted 27:

I: How many do you think fit in the box?
KA: 27 [sounding convinced].

During this episode, KA did not seem capable of enumerating the cubes organized by layers. Even when KA removed the cubes from the box, cubes that she herself had arranged into three layers of 9 cubes, KA maintained the belief that the strategy of counting cube faces seen in the four lateral sides of the prism was correct. Her enumeration strategy was determined by spatially structuring the array as an uncoordinated set of composite units signified by the lateral sides. It was only after twice counting the cubes one by one that KA abandoned the answer provided by the original "faces" strategy. But it is not likely that KA abandoned the original answer because she had restructured the cube array. More probably, KA believed that counting cubes was simply more reliable than her original strategy.

KA and the interviewer then discussed how many cubes would fit in a 2 by 2 by 2 open box pattern that the student had created:

KA: 4 + 4 + 4 + 4 [looking at the four side flaps attached to the base].
I: How many cubes are in the bottom layer?
KA: Four.

I: How many layers?
KA: Four [pointing to the four squares on one of the sides].

The interviewer placed a 2 by 2 cube layer on the bottom of the pattern and again asked how many layers would fit in the box (see Fig. 9.8). KA looked at one side, apparently puzzled. The interviewer asked where the bottom two squares on the side of the pattern would end up when the side was folded to create the box. KA correctly indicated the cubes where these two squares would go. The interviewer asked KA where the next two squares would go. KA thought about it for a while, then pointed, saying they would go above the 2 cubes she had previously indicated:

I: How many layers?
KA: I think two.
I: So how many cubes would there be?
KA: 8.

Initially, KA saw this three-dimensional cube array only in terms of its four lateral sides. She was unable to mentally create a horizontal layering of the cubes that filled the box. However, with sufficient guidance, the interviewer helped KA see a layering spatial structure, which seemed to enable her to restructure the array numerically as well. We suspect, however, that KA's spatial and numerical restructuring were quite fragile at this point.

Grade 4

Students' work on problems in which they were asked to construct a cube building, given its top, side, and front views, helped us distinguish between coordination and integration and illustrated the critical role that these mental operations play in dealing with cube configurations. We observed several levels of sophistication in this work. At the first level of sophistication, students made a different building for each view, even though they were familiar with the task because they had previously drawn three views of a figure. None of these students attempted to coordinate the views. At the second level, students recognized that coordination of the views was needed, but were unable to accomplish it. These students typically looked at one of the views and developed a conjecture about what the object would look like, based solely on that view. They then built an object that had that view and checked to see if the object also had the other two views. If it didn't, they made successive trial-and-error

FIG. 9.8 Open box pattern with cubes: Grade 3 (KA).

changes to the object in an attempt to make it match all three views, focusing on only one view at a time. They too seemed unable to coordinate views.

In slightly more sophisticated solution attempts, students first made a building that matched one view. But they seemed to expect that the configuration of cubes in the building they made would require changes. These students reflected on their changes before making them, trying to anticipate the effects of such changes on more than one of the views. They were often able to predict the effects of changes to the building before they made them. These students seemed to coordinate, but not integrate, the views.

Students at the highest level of sophistication truly integrated the information given in the views. They solved the problem by integrating the information given in all three views to make a mental model of the object—apparently an image something like the customary isometric pictures used to depict the objects (Cooper, 1990). These students then constructed the pictured object without error.

Grade 5

The following example from a fifth grade class gives an in-depth view of students' construction of the layering structure of three-dimensional cube arrays. The class was divided into pairs of students, each working on the *How Many Cubes?* activity sheet (see Fig. 9.9). The teacher circulated around the room, listening to students' conversations and asking questions. One of the researchers observed and recorded the work of one pair of girls, HT and AN, throughout the instructional unit:

Box A.

HT: Well, since it is like two rows up, you should at least stack two rows up [pointing to the two rows on the front face]... 4 and 4, so there would be 8 on the bottom. So, I would say there would be 8.

AN: Wait a minute! Okay, there would be 8 here [pointing to the front of the box picture]. But there are two of these [pointing to columns on the right side of box]. So wouldn't that be 16?

HT: Yeah! Okay! [The students correctly make the pattern and fill it with cubes, finding that the box holds 16 cubes. They conclude that their conjecture is correct.]

Box B.

AN: There's 6 here [front] and 8 here [right side]. And 6 here [back] and 8 there [left side].

HT: Wouldn't that 8 be with that 8 because you go up like that [indicating that the left and right sides come together]? This would be 12 and 8—it's 20!

AN: But wait, wouldn't there be another 8 over there [pointing to the left side in the box picture]?

HT: We didn't count 8 on this side [Box A] because these would just be together.

AN: Yeah!...

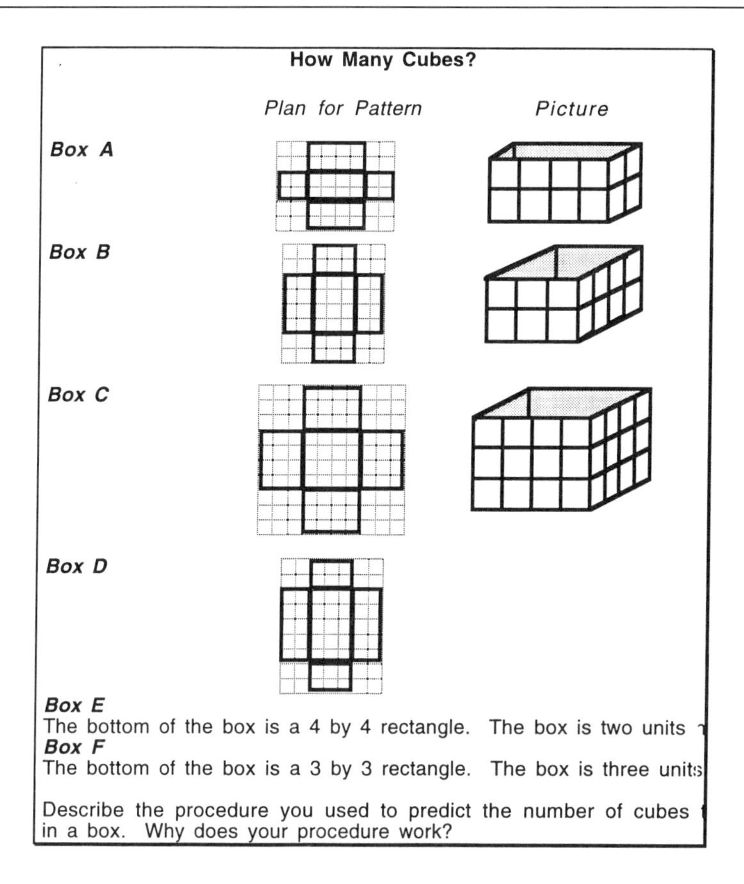

FIG. 9.9 Grade 5 student sheet. Students are asked to predict the number of cubes each box can hold. They check their answers by building the boxes and filling them with cubes.

[after making the box and finding that 24 cubes fill it]

HT: Wonder what we did wrong?

AN: [pointing to the 2 by 4 flap on the right side of the pattern picture] Right here, in this part, we said there would be 8 here, but we didn't add 8 here [on the left side].

I: So, if you had to do this problem again, how would you predict?

AN: Probably by multiplying 3 × 4. There's 3 × 4 here, and 3 × 4 here [pointing to the left column and bottom row, then the right column and top row of the pattern base].

HT: And that's 24. There's 3 [corrects herself], two sets of 12, and that would be 24. And that's how many came up. [pauses] Maybe the next time we should do that.

Note that even though AN and HT got the correct answer for Box A, their conceptualization of the cube array was not correct, as is illustrated by their prediction for Box B. For Box B, AN wanted to add the cubes showing on all four lateral sides of the box. HT thought that the right and left sides somehow folded together so that only the cubes on the right side needed to be counted. So although on Box A it appeared that AN saw two vertical layers, she may have been conceptualizing these as back and front, like she did on Box B. And even though HT said she agreed with AN on Box A, HT's comments on Box B suggest that the two girls conceptualized Box A differently.

After filling the box with cubes showed that their prediction was incorrect, AN and HT attempted to figure out what was wrong with their enumeration procedure. They conjectured that they had forgotten to add 8 for the left side, thus returning to AN's suggestion. Although the students did not see what was wrong with their spatial structuring, discrepancies between their individual predictions and between predicted and checked answers caused them to reflect on the connection between the numerical procedures they were employing and their spatial structurings of the array.

Box C.

AN: Oh, this one is 4 × 4 [pointing to the base of the pattern], 16.

HT: 16 and 16 is [pauses] 32. Okay, so we think that it is going to be 32.

I: You said it is 4 × 4 here. How did you figure out it would be 32?

AN: Um, there's 4 here [pointing to the rightmost column of the 4 by 4 pattern base] and 4 here [top row of the pattern base]. And 4 × 4 is 16. And there's another 4 here [bottom row of the pattern base] and another 4 here [leftmost column of the pattern base] and so that is 16 also.

I: Why are you multiplying 4 × 4? [Pointing the rightmost column and top row of the pattern base]

AN: Um because, what we did on the last problem is, we would have multiplied 4 × 3 there, and 4 × 3 [points to the right column, top row, left column, and bottom row of the pattern base], and then we would have come up with the correct answer. So, we're trying to do that this time.

HT and AN decided that the procedure they had used at the end of their analysis of Box B was correct, so they applied it to Box C without change. Although the students tried to connect their numerical procedure to their spatial structuring, their analysis seemed to focus on a numerical procedure that gave them correct answers. In fact, students can achieve much success in traditional school mathematics by finding and using numerical patterns that are unconnected to physical quantities. So HT and AN, not unlike many students, employed a problem-solving approach they had probably used successfully in the past. However, they were not attempting to delve more deeply into the structure of the cube array.

[The next day, HT and AN use cubes to check their prediction for Box C, finding that their predicted answer of 32 cubes fill only two of the three layers.]

AN: There's another whole layer.

HT: No, we only counted two, remember?

AN: But, there is three of them here. [Points to the side of the box.]
HT: [Looking confused] I thought we only had two layers. Oh yeah, we do need another one [running fingers along the top rows of squares on the box sides]. We forgot there were three layers.
AN: I know what we did—we didn't add the depth of it.

When the students realized that their prediction was wrong, they immediately tried to figure out why. This time, they described their mistake in terms of layers of cubes: They forgot to account for the top layer. Although the students did not mention layers when they originally made their prediction for Box C, HT now reinterpreted in terms of layers what they had done the previous day.

I: Now that you have filled the box, how would you make your prediction?
HT: I'd probably say we would look at [looks at the box picture for C, and points to the horizontal rows on one side]. There's three layers, so we would need three layers of those cubes [pointing to the filled box]. And then, plus 16 for the middle [points to the base of the pattern picture]. And that would be like 48.
I: Okay, so there's three layers?
HT: Of 4 [pointing to the front side of the box picture].
I: Three layers of 4?
HT: Yeah, 16 [looks at the pattern], plus three layers [looks at the picture of the box].
AN: There's three layers actually, not of 4, but um, of 16 [looking at the filled box].
HT: Yeah!

Although both students were using the term *layer*, they were referring to different concepts, seemingly unaware of the discrepancy in their referents. HT used *layer* to refer to rows of cubes on the side of a box, whereas AN and the interviewer used *layer* to describe horizontal layers of cubes. The communication problem was compounded because the students pointed to the same things—rows of squares on the front of the box—when referring to layers. But AN was using these rows to indicate horizontal *layers* of cubes whereas HT was using them to refer to rows of cubes.

[After HT states that there are 12 cubes in each layer (3 by 4), the interviewer asks the students how they obtained 48 from 12.]
AN: Well, then you add up all the other sides [pointing to the four sides in the box picture].
HT: That's 12, that's 12, that's 12, and that's 12, so four 12s make 48 [pointing to the picture].
I: What does a layer look like?
HT: This would be a layer [pointing to the top layer of cubes in the filled box].
AN: This whole top part that you can see right now would be one layer [pointing to the top part of the filled box].
I: How many cubes are in that?
AN: Um, 16.

HT: And then under that one would be another layer of 16. And, then another [pointing to the second and third rows on the side of the filled box].

In structuring the cube array, the students wavered, referring to the array in terms of both horizontal layers and lateral faces; they also vacillated between focusing on number patterns unconnected to spatial quantities and structuring those quantities. By the end of the episode, the students seemed to have structured the array into three horizontal layers of 16 cubes. To see if the students had made a general or specific accommodation to their enumeration scheme, the interviewer next asked them to return to Box B:

I: Okay, let's go back to this one [pointing to pattern B]. Could you look at the pattern and figure out how many cubes would be in it?
AN: Um, well it would probably be easier to look at this [pointing to the box picture for B].
HT: There's two layers [pointing to the front side of the box picture].
AN: Of 3 each, and that's 6. And there would be 6 back here, stuck in the back [pointing to the box picture].
HT: So, that's 12; they're equal.
AN: There's 8 here and 8 here [pointing to the right and left sides of the box picture].
HT: And that's 16. 16 and 12 is 28.
I: How many cubes were actually in Box B?
HT: 24 [looking puzzled].
AN: Well, um, there's only 12 in here, in this layer [pointing to the top layer in the filled box].
HT: Twelve on this one, that's 24 [pointing to the bottom row of squares on the side of the filled box].
AN: Oh, I know what we could do [pointing to the picture for Box B]. We could just add up one layer, and then we would just multiply that by two. So like for this one [Box C] we could multiply by three. So all the way around, um here [points to the Box C picture], 16, that's all the way around, we could multiply 16 by three.

At first, the students returned to their earlier strategy of enumerating cubes that appeared on the four lateral faces of the box. Both students also used the term *layer* to refer to a row of cubes. On confronting the inconsistency between their prediction and the actual number of cubes, however, they returned to a horizontal layering strategy, even though such a strategy was apparently difficult for them to implement. (Recall the difficulty RA and KA had in restructuring such an array.) But the interviewer was suspicious of AN's use of the phrase "all the way around," so he asked the students about it:

I: How about this? Are there 16 all the way around here [pointing to the Box B picture]?
AN: No, there's, um, 14 [pointing only to the visible squares on the bottom row of the box picture].
AN: So, you multiply 14 by two.
HT: Because there's two layers.

AN: And, 14 × 2 is 28. Oh, but it's 24. [HT recounts the cubes in the filled Box B, getting 24]

I: [taking one layer of cubes out of the filled Box B] How many cubes are in here?

HT: There's 12!

I: Can you show me where this layer shows up in this diagram [pattern B].

HT: Well, there would be probably one more under it [pointing to the picture of the pattern for Box B]. Like this would be this layer [the one that the interviewer has taken out] and then this one [the one in the box] would be under it.

AN: I'm trying to figure out another way to get this.

HT: There's 12 here [pointing to the top layer of cubes for Box B, which are still on the table] and 12 there [pointing to the 12 cubes that are still in Box B], that's 24. So that's how many we got.

HT: [As the students move on to Box D] I think you said there's two layers of 12, so it's 24. So let's see if that works for Box D.

At the beginning of this episode, the students incorrectly determined the number of cubes in a horizontal layer as they were examining the box picture: They enumerated only the perimeter cubes in a layer, double-counting those in the corners. They still didn't have an adequate mental model of layering that they could superimpose on this array. But during the episode, the students seemed to restructure their conception of the cube array: They properly saw horizontal layers of cubes. We conjecture that, at this time, the students abstracted a layer-structuring of the array, with this abstraction being powerful enough to enable then to apply it to future problems. In fact, the students' work on Boxes D–F supports this contention. However, their work also illustrates how their abstraction needed to be refined through further problem solving and reflection.

Box D.

HT: [pointing to the pattern picture] We know there's going to be 15 in each layer, but we don't know how many layers there are going to be.

AN: Probably two [pointing to the right side flap on the pattern picture].

HT: Yeah, because there's two right here [pointing to the two columns of the right side flap].

AN: So that's 15 in each layer.

HT: And 15 × 2 is [pause] 30.

Box E.

HT: So it's 4 by 4; that's probably 16, of two.

AN: But that would be the same as that [Box A]. No wait a minute.

HT: No. That's 4 by 2. So it will probably be 32; because it's 4 by 4 is 16, and 2 high.

HT: I think our theory is that if it's 4 by 4 on the bottom, that's 16, and if it's 2 high, that's 16 by 2 is 32.

Box F.

HT: There's going to be 9 on the bottom and 3 high. So 9 by 3 is 27.

AN: Right! That makes sense!

[After HT and AN have written their prediction strategy in their journals, the interviewer asks them to read what they have written aloud]

HT: You see how many things are in each layer, then you multiply by the number of layers there are.

Consistent with our conjecture that they restructured their thinking about cube arrays, AN and HT determined the answers to the last three problems using a layer approach. Their use of layering was by no means automatic: They still stumbled at times as they attempted to make sense of this approach for the variety of problems presented by Boxes D–F.

Summary

Throughout this account, HT and AN were trying to develop a theory of how to make correct predictions. The discrepancies between what they predicted and what they actually found caused them to reflect on their prediction strategies and their conceptions of cube arrays. At first, their enumeration strategies were based on a more primitive spatial structuring of three-dimensional arrays—seeing them in terms of the faces of the prism formed. The students seemed to focus more on numerical strategies than on a deep analysis of the spatial organization of the cubes. However, because their initial spatial structuring led to incorrect predictions, HT and AN refocused their attention on the structure of the cube arrays, which led to a restructuring of their mental models of the arrays. In fact, during their work on Box C, HT and AN seemed to develop a layer structuring of the arrays, a structuring that they verified and refined on subsequent problems.

The interviewer played a dual role in his interactions with HT and AN. Most of his questions were intended to help him understand the students' current ways of thinking. For instance, to probe whether HT and AN had changed their conception of the cubes in Box B, given their reflection and experience in dealing with Box C, the interviewer requested that the students return to their prediction for B. On Box C, because the interviewer recognized that the students' use of the phrase "all the way around" might indicate a misconception, he explicitly addressed this terminology with them, hoping that he could understand their use of the term and, if there were a misconception, that the resulting discussion would help the students discover it. Although these actions were meant to probe the students' understanding, they also served the instructional role of promoting reflection and focusing the students' attention on relevant aspects of the task. In this sense, the interviewer's questions were the same questions a constructivist teacher would ask. In fact, to support inquiry-based learning in which students construct personally meaningful concepts, after presenting appropriate tasks, a teacher's role is threefold: (a) to constantly assess students' thinking in order to properly guide their activity, (b) to promote student reflection and communication, and (c) (both in interactions with individual students and in moderating class discussions) to focus students' attention on potentially productive lines of analysis.

Consistent with constructivist accounts of the learning process, the essential components of learning for HT and AN were reflection, cognitive conflict, and abstraction. Reflection and cognitive conflict were promoted by focusing students on *predicting* the number of cubes in three-dimensional arrays. Errors in predictions, which the students themselves discovered, caused cognitive conflicts or perturbations in the students' current mental models for arrays. HT and AN attempted to resolve these conflicts by reflecting on the strategies they were using, all the while examining and restructuring their mental models of the arrays. In fact, they moved from an incorrect conception of the arrays, to a period of confusion in which they vacillated between different conceptions, to a viable conception that resolved their confusion.

The account of HT and AN's work nicely illustrates the constructivist claim that, like scientists, students are theory builders. They build conceptual structures to interpret the world around them. Cognitive restructuring is engendered when students' current knowledge fails to account for certain happenings, or results in "obstacles, contradictions, or surprises. The difference between the scientist and the student is that the student interacts with a teacher, who can guide his or her construction of knowledge as the student attempts to complete instructional activities" (Cobb, 1988, p. 92). This guidance is often covert; in the present situation, guidance came from the sequence of tasks and the interviewer's questions, not from a teacher telling students the solutions to the problems.

CONCLUSION

In this research and curriculum development project, our first task was to determine the different levels of sophistication of students' conceptualization and enumeration of three-dimensional arrays—we wanted to find out what relevant mathematical constructions were possible for students of this age. We then designed instructional tasks in which students were encouraged to assess the viability of their current theories. We put students in situations that we knew from our research were likely to create perturbations in their theories and thus promote reflection on and alteration of those theories. As can be seen from the detailed description of HT and AN's collaborative classroom work, this research-based approach to designing instruction was effective in bringing about substantial change in these students' mathematical thinking. Moreover, this detailed description also illustrates the complexity of learning as it unfolded in actual classroom practice, not only confirming that learning can take place as constructivist researchers have theorized, but bringing constructivist theory to life in a way that enables teachers to better understand and apply it.

ACKNOWLEDGMENT

Support for this work was provided by grants MDR 8954664 and ESI 9050210 from the National Science Foundation. The opinions expressed, however, are those of the authors and do not necessarily reflect the views of that foundation.

REFERENCES

Battista, M. T. (1994). On Greeno's environmental/model view of conceptual domains: A spatial/geometric perspective. *Journal for Research in Mathematics Education, 25,* 86–94.

Battista, M. T., & Berle-Carman, M. (1995). *Containers and cubes.* Palo Alto, CA: Dale Seymour.

Battista, M. T., & Clements, D. H. (1995a). *Exploring solids and boxes.* Palo Alto, CA: Dale Seymour.

Battista, M. T., & Clements, D. H. (1995b). *Seeing solids and silhouettes.* Palo Alto, CA: Dale Seymour.

Battista, M. T., & Clements, D. H. (1996). Students' understanding of three-dimensional rectangular arrays of cubes. *Journal for Research in Mathematics Education, 27*(3), 258–292.

Ben-Chaim, D., Lappan, G., & Houang, R. T. (1985). Visualizing rectangular solids made of small cubes: Analyzing and effecting students' performance. *Educational Studies in Mathematics, 16,* 389–409.

Cobb, P. (1988). The tension between theories of learning and instruction in mathematics education. *Educational Psychologist, 23,* 87–103.

Cobb, P., Yackel, E., & Wood, T. (1992). A constructivist alternative to the representational view of the mind in mathematics education. *Journal for Research in Mathematics Education, 23,* 2–33.

Cooper, L. A. (1990). Mental representation of three-dimensional objects in visual problem solving and recognition. *Journal of Experimental Psychology: Learning, Memory, and Cognition, 16*(6), 1097–1106.

von Glasersfeld, E. (1991). Abstraction, re-presentation, and reflection: An interpretation of experience and Piaget's approach. In L. P. Steffe (Ed.), *Epistemological foundations of mathematical experience* (pp. 45–67). New York: Springer-Verlag.

10

Sixth-Grade Students' Conceptions of Stability in Engineering Contexts

James A. Middleton
Arizona State University

Robert Corbett
Kahler Slater Architects

Geometry, as a curricular subject for primary- and middle-school students, provides a descriptive medium by which real-world actions and objects can be described in terms of their physical properties. Geometric models can be applied to real-world problems to simplify complex problem situations, and many algebraic and numeric ideas can be fostered by looking at them through a geometric perspective (National Council of Teachers of Mathematics [NCTM], 1989). The complex spatial patterns of the real world can be simplified into component relationships such as points, lines, angles, transformations, similarity, and dimensionality, and these physically simpler (but cognitively more abstract) ideas can be operated on mentally—changed, recombined, transformed—whereas the physical objects themselves may not be. In other words, the field of geometry can be applied to realistic situations, and conversely, realistic situations give rise to geometric thinking:

> The Euclidean space with all its objects is a rich structure, although it is poor if compared with all I perceive around, its colours, polished and rough surfaces, sounds, smells, movements. But thanks to the impoverishment it furnishes a certain context, which for some reasons suits us extremely well.... This Euclidean space has never been an aim in itself, but rather it has been the mental and mathematically conceptual substratum for what is done in it. (Freudenthal, 1983, p. 224)

As Freudenthal went on to describe, the development of students' knowledge of geometry reflects a tension between experience in an irregular world and this notational/representational system (or conceptual substratum) that explains physical reality in terms of abstract regularities. As students gain experience manipulating objects and thinking about similarities and differences of the objects, "frequently used sequences of movements establish relational links between parts of figures, between figures and their parts, and between different figures" (Battista, 1994, p. 89). These relational links become the basis for a larger network of geometric ideas. This emphasis on realistic experience and observation is prevalent in European curricula, especially that of the Dutch (see de Moor, 1991), but is only now gaining support in the United States.

Freudenthal (1983) discussed this process as the *mathematization* of physical phenomena—a process by which children create "mental objects" that they can operate on, as opposed to operating only on the physical analog (see Gravemeijer, chapter 2, this volume). Perceptually, children see certain invariants in the objects of their environment that become the properties of the mental object (Bohm & Peat, 1987, as cited in Kieran, 1993). The *primitives,* or basic geometric properties, readily abstracted from the physical world include linearity, flatness, parallelism, right angles, symmetry, circles, and congruence and similarity. To this list, we add the notion of *stability* (resistance to deformation).

Newer curricula and methods texts advocate hands-on building of geometric solids to help students visualize figures in three-dimensional space and to draw attention to the properties of these figures. However, little research has been done on young children's understanding of three-dimensional geometry, especially in a dynamic engineering context. Even less research is available on primary- or middle-school students' thinking about the relationship among shape, physical structure, and stability.

The ideas presented in this chapter offer a window into one area where realistic situations are used to help children develop beginning notions of physical structure, structural stability, and force, and connect these notions to children's burgeoning understanding of geometry. We present these ideas from a dual perspective. The first author is an educational researcher interested in understanding childrens' thinking in the area of geometry and in designing curriculum that stems from this understanding. The second author is an architect, interested in the accurate reflection of the underlying ideas of his profession in educational settings and in the creation of appropriate geometric models that assist students in construction of knowledge through engineering and architectural contexts.

DEFINITION OF TERMS

Physical structure, as it is treated in this chapter, can be thought of as the geometric features of an object, how these features relate to each other in a spatial sense, and how they interact when acted on by an outside force. The basic principles of the structural load of a building are in some ways analogous to those shown in the flow of water from the upper floors of a building under construction all the way to the ground. The water first spreads across the top floor. It then flows to the beams, from the beam to the girders, and from the girders to the columns. The columns, in essence, provide a conduit that carries the water down to the foundation. The foundation then evenly spreads the water to the ground.

This basic principle of loading is true of all types of architectural and engineering structures. The live load, such as people and furniture and the weight of the structure itself, is fairly similar for all structures whether those structures are made of concrete, wood, fabric, or pneumatic materials. The loading spreads from beams to girders to columns or cables and then is transferred to the ground.

The structural analysis of a building goes from the micro to the macro level, from studying the individual beam and column connections, to studying how an entire building reacts during an earthquake or in strong wind. The individual connections play as important a role as the beams or the complete frame design of the building. Each piece contributes to the overall design, whether that piece is a flexible frame designed to sway during an earthquake, or a stiffly braced structure designed to resist any movement.

In the real world, unique and ingenious solutions have been developed to resist structural failure, including designs like the Citicorp building in New York City. This huge skyscraper, due partly to building design and partly to economy of material, would sway beyond an acceptable level at the top were it not for a unique design solution. The architects and engineers designed the building to contain a large concrete mass at the top that will automatically roll in the opposite direction to the sway of the building. This roll results in pulling the building back in the direction opposite the sway before the sway has a chance to complete its full period.[1]

Structural stability can be thought of as the relative amount of deformation an object displays as a load is applied. This stability is fundamentally related to the physical structure of the object (as just described). The ways in which the load is transferred from member to member down to the base of the object determine whether that object is more sound than another object (given that the material makeup of the objects is equivalent in both type and amount). This transfer of load is determined by the geometric properties of the object.

In the simplest cases, cubes are minimally stable because the four equal sides in each face (a square) do not imply any set of given angles—infinitely many parallelograms can be made with four equal side lengths. If pressure is exerted on a bar model of a cube, the faces collapse (forming infinitely many parallograms as they go). Tetrahedra, on the other hand, are maximally stable because the three equal sides in each triangular face *do* imply a set of given angles—only one triangle can be made by any three given side lengths. Pushing down on the vertex of a bar model of a tetrahedron merely results in a sore finger. By testing the stability of objects,

---

[1] There are various design approaches that relate to the physical structure of a building. The structure or support of the building, whether a simple wood frame house or a stone-clad multistory skyscraper, is obscured from the viewer within the various skins, or outer layers, of the building and serves only a functional role. In other buildings, such as the Guggenheim Museum designed by Frank Lloyd Wright, the concrete structure becomes an important component in the integration of structure and space.

deformation can be seen as resulting from networks of force vectors, working either in opposition or in conjunction with each other, depending on the geometry of the structure.

TEACHING STABILITY

Stability as a physical property can be explored in this manner through children's inquisitive behavior with tools. When children are handed objects, they turn them, twist them, drop them, tip them over, and squeeze them in an attempt to discover their nature. They soon discover that objects with different physical properties behave differently when they are played with. The trick for developing appropriate mathematical activities that capitalize on this natural tendency is to group models of, or actions on, physical objects that differ in important geometric ways. In this manner, children can actively test patterns of variance and invariance in the properties of the objects.

There is some evidence that even young children can begin to learn some of the concepts mathematically related to structural stability. Lamon and Huber (1971) demonstrated that vector space can be learned better with concrete materials, and their study was cited in Clements and Battista (1992) as supporting the use of hands-on activities for the teaching of geometry. The study is informative in that it suggests that students as early as sixth grade can perform operations on matrices (which *can* be thought of as representing vectors), although whether they can develop any geometric understanding of vector (i.e., vector space) is still unclear.[2]

The approach we take in this project is quite different. Rather than teach vectors as an end product and culmination of algebraic and geometric reasoning near the end of the high-school years, we examine how children initially conceive of situations that embody notions of vectors (pre-vector notions, if you like) in a tangible manner. In this approach, the activity students engage in serves as a mediational context between the physical exploration and geometric understanding (see Gravemeijer, 1994, for a discussion of "bridging by vertical instruments," p. 451, in research and curriculum development).

Through the findings of two studies, we describe students' conceptions of stability as a mediational context between testing physical objects in an engineering sense and the abstraction of geometric properties. The first study examined the construction of knowledge of geometric stability and vector force in the social context of the classroom. The second looked in detail at two individual students' perceptions of the properties of geometric objects and their conceptual change with respect to stability.

One justification for this type of pilot research is to provide a small pattern of successes and failures to build on for others interested in stu-

[2] Lamon and Huber's study, like most studies that attempt to ascertain students' abilities dealing with vector, is decidedly not geometric. Rather, it takes a rule-based approach to operations on matrices (i.e., row substitutions, scalar operations). It is doubtful that any of their subjects thought of the operations they were performing in a spatial manner.

dents' understanding of geometry. With a larger body of research, teachers and curriculum developers can begin to apply these motivating, realistic activities to instructional situations and be alert to the possible pitfalls and misconceptions that can develop because of the highly complex nature of the tasks (Aguirre & Erickson, 1984). A second justification concerns the NCTM (1989) recommendation that understanding vectors as models for force and velocity be introduced into the core secondary curriculum for all students. If these concepts are meant to be taught for understanding, then some prevector experience in the primary and middle grades seems necessary to prepare students for more in-depth study in high school.

CLASSROOM STUDY

As part of the pilot phase of the National Science Foundation-sponsored curriculum project *Mathematics in Context: A Connected Curriculum for Grades 5–8*, 16 fifth- and sixth-grade students (8 girls and 8 boys) participated in a 2-week summer course in geometry (1.5 hr per day). The project was a collaboration between U.S. and Dutch researchers and reflected the Dutch tradition of using realistic situations, mathematization, and bridging levels of geometric thinking with models. Children were divided into four groups of four students each. The course was videotaped using two cameras that focused on two groups of children at a time with a third camera focused on the teacher and the classroom as a whole. Through these three cameras, we obtained a relatively complete picture of how the class progressed both at the group level and at the class level.

The assigned tasks began with looking at real-world examples of geometric solids, creating bar models of the objects out of toothpicks and gumdrops, and testing their stability by applying force to one or more points on the model (see Kindt & Spence, 1998). The differences between the stability of triangular shapes (e.g., tetrahedra) and square shapes (e.g., cubes) were emphasized in the activities. The results of these tests were recorded, and students were expected to use these results as heuristics in a transfer task, building a suspension bridge. We expected students to use more triangular shapes than squares in their designs. Bridges for each group were tested for stability under load until structural failure was achieved. The amount of load at failure (the number of dice the bridges could support) was used as evidence of structural integrity. The component primitives of the bridges were then decomposed and discussed in relation to the properties that enhanced or diminished stability. The bar models were then used to develop deeper understanding of the forces interacting in the system and how the geometry of the system determined possible deformation and structural failure.

In general, students' initial conceptions of stability were naive, focusing on the colloquial definition of stability—something that doesn't fall over. However, by building bar models of geometric solids and testing their stability, students began to perceive that the types of objects were different with respect to resistance to deformation, their strength under load, and the economy of materials used. Students initially tested stability by pushing on the top of their cubes and tetrahedra to see which would tip over easier. Doing this, there was no discernible difference in stability, so new strategies (dropping the objects, smashing them) had to be developed. Movement of the structural members of the figures was the most obvious piece of information keyed in on by students. Students would observe the edges of their cubes collapsing, whereas the edges of the tetrahedra stayed put. Throughout the 2 weeks as students built more complex figures, they enjoyed shaking and squashing them with the purpose of examining their stability.

Following the initial exploratory activities, students were told, "Try building a bridge between two chairs that are 30 cm [about 1 ft] apart. Use as many vertices and edges as necessary." Students were given toothpicks and gumdrops as building materials. As students became engaged in building their suspension bridges, we observed two significant problems. First, although all students verbalized that triangles were more stable than squares, half of the groups used square components in developing their bridges. Second, even the groups who used triangles in building their bridges attempted to build more strength by adding more material, thus compromising the geometric advantage of using regular triangles.

Despite this, during the testing phase, the structures built out of triangles were significantly (amazingly) stronger than the bridges made of squares, supporting roughly twice the number of dice in a cup. Moreover, overbuilt though these bridges were, they represented much more economy of materials than did the square-based bridges. In short, the differences between the stability of rectangles and triangles and the resulting solids created from these figures became more salient to the students through the building and testing of the suspension bridges.

Following the course, as we examined the videotapes, it became increasingly unclear what students' thought processes were while they were actually involved in the building and testing. It was clear that all of the students had come to the realization that triangles were more stable than squares, and that tetrahedra were more stable than cubes. It was unclear, however, that this classification had anything to do with students' thinking about the underlying geometric properties of the objects, or whether students could even talk about geometric properties in the ways we had envisioned the activities would facilitate. In other words, students knew that triangles were more stable than squares, but so what? Where could it go? What background geometry did students need to have to approach this understanding? Was this even a fruitful beginning to the introduction of geometric properties and vectors?

THE INDIVIDUAL INTERVIEWS

To gain more detailed knowledge about how students were thinking about these concepts, a second set of activities was developed, tailored toward allowing individual students to articulate, as best they could, their understanding of stability, force, and change in force as load is applied to simple frame structures. We interviewed and videotaped two students, one grade-4 girl, and one grade-6 boy, for 2 hours each as they engaged in the activities. Students were interviewed individually.

Three possible confounding variables came to mind from the results of the classroom observations:

1. The type of materials (i.e., the relative rigidity of structural members) may impinge significantly on students' understanding of stability by enhancing deformation of the structure not related to its global geometry.
2. The complexity of the figures under study may diffuse students' attention from the important properties we were trying to teach—the more complex the objects, the more complex the interactions between forces, and the more information students have to take into account simultaneously.
3. Children may not have the language to articulate their understanding of force in mathematical terms.

We attempted to overcome these hurdles by systematically varying the material used in creating frame structures, from thick, flexible, closed-cell foam to rigid bamboo skewers, and by varying the complexity of the structures from simple beams, to triangles and squares, to three-dimensional bar models of tetrahedra and cubes. The plane figures made out of bamboo used rubber bands as "vertices" to draw students' attention to the movement and change in direction of the applied forces (see Fig. 10.1).

To assist students in articulating their knowledge, we structured each task such that each student would

1. Describe the figure presented to him or her.
2. Make a verbal guess as to what would happen to the figure when force was applied at different points.
3. Draw a picture of what he or she thought would happen when the force was applied.
4. Test his or her hypothesis by applying force to the figure.
5. Amend his or her guess in the light of the results of the test.

Several notations were introduced to the students: Arrows were suggested as ways of representing force in students' drawings, and a system of parallel lines highlighting the different "cuts" or cross sections of the

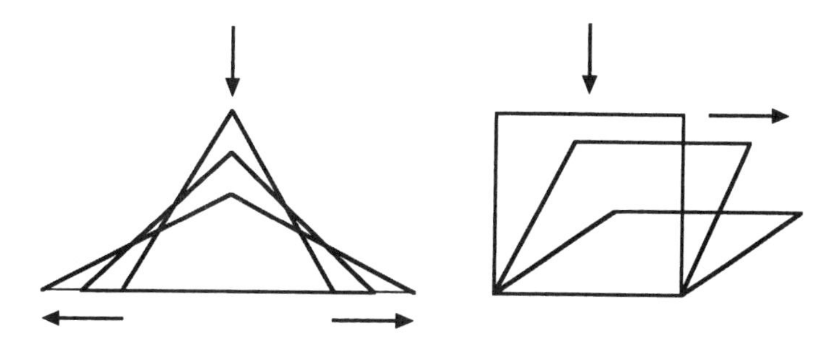

FIG. 10.1 Bar models of an equilateral triangle and a square illustrating the movement of members as force is applied.

foam beams could be used to illustrate areas of tension and compression in the drawings (see Fig. 10.2).

The parallel lines drawn on the foam beams served two purposes for this study. First, as stated earlier, they were introduced as a notation for students to illustrate tension and compression in their drawings. When the beams were deformed, the endpoints of the lines got closer together under compression and farther apart under tension. Students easily perceived this and readily used the system to denote deformation of the foam models. Second, the parallel lines also served as an embedded referrent that made the very subtle ideas of tension and compression overt. Without the system, we felt that students would be less able to perceive tension and compression. The rubber bands that functioned as the vertices of the rigid bamboo figures also served as embedded referrents that highlighted force transfer from member to member in a structure.

Research also indicates that the orientation of a figure often determines students' ability to classify the figure. Children, especially young children, often attend to the orientation of a figure in classification of geometric shapes. For example, young children often do not recognize a triangle to be a triangle if its base is not oriented parallel to the bottom of the page (Hoffer, 1983). Moreover, orientation of a figure does have a significant impact on the ways in which forces interact in a system. To account for these variables, we presented each figure in different orientations.

The final task for the students in the individual interviews paralleled the bridge-building activity of the students in the pilot class. They were to build as sturdy a suspension bridge as possible (using gumdrops and skewers) that would span 30 cm, test the structure until failure, and describe where and why the bridge failed and what could be done to make it stronger. Objects were presented to students in this order: foam beam, foam equilateral triangle, foam square, bamboo triangle, bamboo square, bamboo tetrahedron, bamboo cube.

Students' Knowledge of Force and Stability

Both students were able to describe the deformation of the objects in terms of force and direction, underlying notions of vector, and were able to ac-

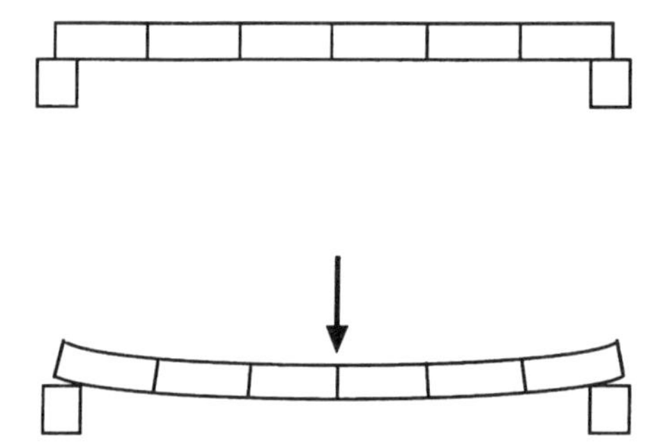

FIG. 10.2 Illustration of foam models. Notice how the system of parallel lines in the model deforms as force is applied, highlighting tension and compression.

curately reproduce the deformation in their drawings. For example, even Kari, the fourth grader, was able to describe the tension and compression resulting from deformation of the foam bars using the introduced notation of the vertical markers on the bar (see Fig. 10.3).

Both children needed to manipulate the objects and examine the change in form to come to a general conclusion about how the top of the bar compressed as the bottom stretched:

Interviewer: OK. What's happened to the lines on the bar?
Kari: What lines? These? [points to marker lines].
Interviewer: Yes, those. How have they changed? [presses]
Kari: Don't. Wait. Don't. Back up.
Interviewer: I'll let you play with it. You can press on it and test it. OK?
Kari: [presses] They get longer.
Interviewer: What do you mean longer? Can you show me?
Kari: Right now [unbended] they are this big. But when I press down, they're like this [gestures showing slant of lines from top to bottom].
Interviewer: What would happen if you bent it the other way?
Kari: The bottom wouldn't be wider, but the top one would be wider.

Both students were able to use arrows to represent the change in direction of the forces in a system. In addition, they were able to describe

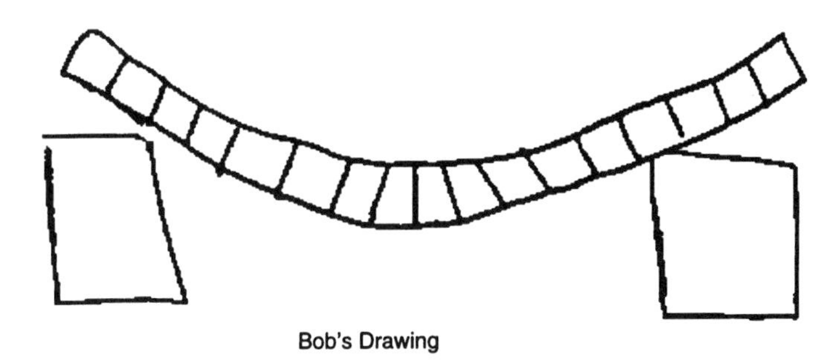

Bob's Drawing

Kari's Drawing

FIG. 10.3 Children's representations of the foam bar after applying force.

their own informal theories of how forces change direction from member to member in a structure and used the arrow notation as a natural method for describing change in direction. However, although their drawings of the forces in the triangle and tetrahedra roughly approximated the orthogonality of the actual case, in the case of the square and cube their use of the arrow notation mimicked the *movement* of the members. For the squares and cubes, arrow notation was tied to the movement of a side in the horizontal plane (sides parallel to the tabletop "slide" to the left or the right), but tied to the movement of a vertex in the vertical plane (vertices collapse in a diagonal direction):

> *Bob*: [see Fig. 10.4] Say this rubber band is pulling this rubber band this way [arrow 1; indicating on figure], then the force of this rubber band [arrow 2] will carry this rubber band towards this way like this [arrow 3]. Because [gesturing] this rubber band is pulling this way, it will make this end keep on pulling down this way so it gets more of a slant [arrow 4].

After examining the stability of triangles and squares, and tetrahedra and cubes, both children verbalized a belief that the triangular objects were more stable than the square objects, and both children had similar conjectures as to why the triangles were more stable than the squares. Ba-

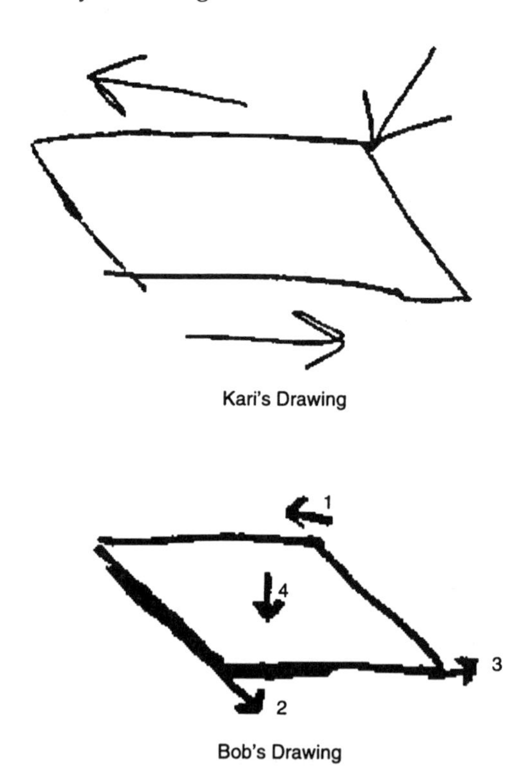

FIG. 10.4 Children's representations of the bamboo square after applying force.

sically, these conjectures focused on the parity of sides and corners existing in the faces of the square objects, and the lack of parity in the number of sides and corners in the faces of the triangular objects. The children felt that the parity existing in square objects caused the sides to fold in under load, whereas the sides of a triangle worked in opposition to hold each other up. For example, Bob accurately perceived that the two diagonal sides of the triangle would "clog" together when a force was applied, whereas with the square, the forces were transferred from one side to another in a cyclic pattern:

Interviewer: Why do you think that this [square] is less stable than this [triangle]? What's happening to the force?

Bob: [draws] OK. Well, the rubber bands are pushing out. Well, these sticks, they are pushing out this way [pulling on sides of triangle], so these sticks will be pushing out, while these ones [at top vertex, pushing down], well, one of them is uneven. Oh, I see, when you push on these two [top vertex], they clog up together.

Interviewer: Clog up together?

Bob: Yeah, they touch each other more, like when you have a regular triangle. See they aren't touching each other because this rubber band is between them. When you press down [pressing] they are touching each other.

As the figures presented to the students became more complex, it became much more difficult for them to predict how the structures would deform under load, and their use of the arrow notation became more random (see Fig. 10. 5). However, by observing actual deformation of the objects, the children were able to amend their drawings and provide reasonable descriptions of the nature of the deformation:

Interviewer: All right, good. Now, on the bottom here. What's going to happen? Can you draw me a picture of that cube?

Kari: Arrows are going to be going in all different ways [drawing]. They are all going to be facing out and there are going to be lots of arrows.

FIG. 10.5 Kari's prediction of the direction of forces on (a) a tetrahedron, (b) a cube before testing, and (c) a cube after testing.

Interviewer:	Are they all going to go in the same way or different ways?
Kari:	No, some are going to go like here, and some are going to go like here [drawing]...
Interviewer:	All right. Now I want you to press down on this and tell me if it's what you thought.
Kari:	[presses] Yeah.
Interviewer:	What's happened? Are the arrows going in all different directions?
Kari:	No. Not exactly. It's sort of what I thought because they are bending. But if I went like this [pushing to make cube collapse in opposite direction], they would go this way.

It is unclear, at this time, how students' perceptions of the deformation were related to their understanding of force.

In the bridge-building activity in particular, the sheer complexity of the structures made it difficult for the children to take their understanding of how forces interact and apply it to the task beyond the general heuristic that "triangles are stronger than squares." Like the students in the pilot classroom, both Kari and Bob produced a first approximation of their final bridge, tested this structure, and "shored up" the weak areas. Through several "guess and test" cycles, Bob was able to generate a tolerably strong bridge (see Fig. 10.6), using a minimum of materials and judicious application of the geometry of triangles.

Despite understanding the relative stability of the prototype figures, Kari did not apply this knowledge to the bridge-building task (similar to about half of the pilot students). Her bridge used right angles and an abundance of joints in an attempt to increase rigidity by adding material to the structure (also similar to about half of the pilot students; see Fig. 10.7). Throughout the testing, Kari struggled through her belief that quantity of material implied structural integrity:

Interviewer:	So why are you putting gumdrops like that?
Kari:	So I can stick these sticks into them so it will hold it more. See what I'm going to do is start off like this [skewer with six gumdrops on it] and sticks into these guys [gumdrops]. I'll stick this into here. I'm making it stronger.

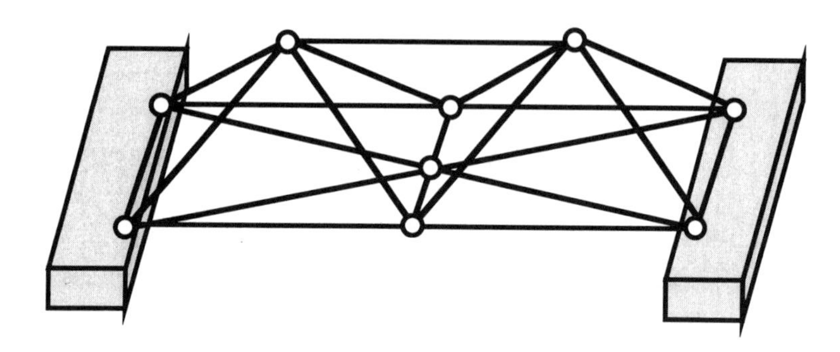

FIG. 10.6 Illustration of Bob's bridge. Notice the economy of materials and the use of triangular sections.

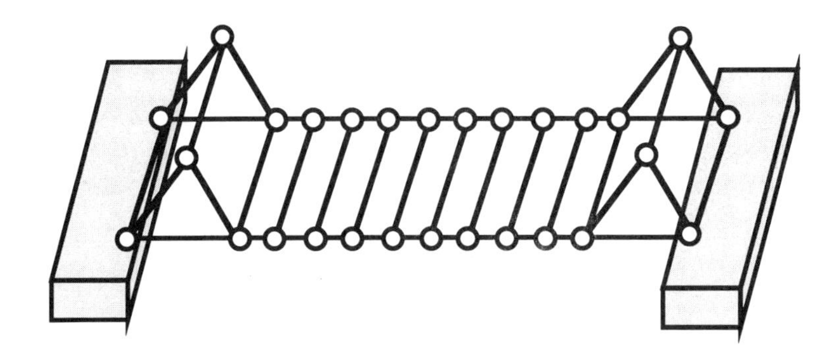

FIG. 10.7 Illustration of Kari's bridge. Notice the rectangular sections and the overuse of materials.

Interviewer:	So, why are you building it this way?
Kari:	So it will be stronger.
Interviewer:	So why is it going to be stronger?
Kari:	Because it has got more sticks on it.

In activities such as this, many of the students in the pilot classroom, like Kari, seemed to equate strength with stability, and both of these desired goals implied "more material." Even the students in the pilot classroom who did apply this understanding of geometry and stability to the bridge task tended to overbuild their structures. Bob was the one exception.

By varying the material used in building the simple figures in this study, it became clear that, indeed, the type of material used in constructing figures has a significant impact on how students initially think of stability and force. The flexible foam was particularly powerful in drawing students' attention to tension and compression, the ways in which the forces on a beam work in opposition at the center, and the distribution of force from one member of a structure to another. However, the flexibility of the structures also introduced such undesirable confounds as *torsion* (twisting), and "dead spots" (typically the corners of the foam structures were more rigid than the beams). All of these had an impact on how the children perceived the more-rigid bamboo figures.

For example, when we asked Bob to describe how the forces interact when a load is applied to the vertex of a bamboo tetrahedron, he used the analogy of the foam bars to illustrate the propagation of forces from the vertical edges to the horizontal edges:

Bob:	The [bottom] sticks might go up in the air 'cause if you press down really hard, and you press down on the ends [showing on tetrahedron; see Fig. 10.8]. Remember when we found out if you do this [presses down on ends], this will go up?
Interviewer:	Yeah.
Bob:	So the forces are pressing down on this [bottom skewer in tetrahedron] making this go up.

FIG. 10.8 Bob's prediction of the deformation of a tetrahedron before testing. Notice how he emphasizes the bottom edges being pushed up by the force exerted on the sides.

In reality, the rigidity of the bamboo allowed for no discernible bending of the base.

CONCLUSION

Students were able to abstract significant knowledge from the engineering activities, both in the pilot classroom and in the individual interviews. Students could describe in informal terms that triangles are more stable than squares, as well as the corollary that things built out of triangles are more stable than things built out of squares. Importantly, students verbalized their understanding that stability in objects has something to do with their geometry. Granted, these understandings were naive with respect to the mathematical reality of the system; however, the first approximations of theory (i.e., that the number of sides/corners has something to do with the relative stability of triangles and squares) do not seem to be a trivial abstraction.

Moreover, at least Bob and Kari developed informal theories of force propagation and transfer. They were able to describe how forces change direction as they transfer from one member in a structure to another, and they grew in their ability to perceive where weaknesses in structures existed by examining their designs. Through testing, they became successively more accurate at shoring up a structure, and better at describing how the forces interacted (even if their definitions of "forces" were limited to movement of structural members).

However, despite this knowledge (or ability to learn), it is still unclear what place such activities have in teaching geometry outside of their motivational appeal (and we must say, this appeal is significant). They are extremely time consuming, and the basic knowledge of geometric figures, stability, and force related to geometrical structure seems to be gained long before students get to the bridge activity. In addition, only about half of the students in the pilot study were able to apply this knowledge to the bridge activity because the reality of the engineering problem (the complexity of the figures and the "amount of material implies strength" confounding notion) had a significant impact on how they viewed the task.

Lastly, it is unclear where these tasks should lead in a curricular sequence. The evidence suggests that even Kari, in the fourth grade, had an informal understanding of pre-vector-like concepts, but is this a fruitful area for middle-grades geometry to pursue in more depth? The notions of vectors have historically been difficult to teach, even in a realistic physics situation in high school and college. Students very often fail to take sufficient information into account in simple motion problems and tend to retain their naive conceptions of vectors in the face of school instruction. The independence of vector components in a system seems to be particularly counterintuitive (Aguirre & Erickson, 1984; Aguirre & Rankin, 1989). At the least, given the recommendations by the NCTM (1989) that concepts of vectors be introduced into the secondary curriculum for all students, some attention should be paid to the types of precedent activities in the middle grades that could provide a "stable" base for the high-school experience. Knowledge of how students initially conceive of force and structure can assist in creating such a base.

From these investigations, we can draw some general conclusions about the nature of students' initial experiences with engineering contexts related to geometry. Each of these conclusions has a significant impact on the process of curriculum design in this area.

First, the complexity of figures confounds students' understanding of force. With simple plane figures and bar models, students were able to determine readily which ones were more stable than others through physical manipulation, and they were able to predict how this understanding of the primitives could be used to predict the stability of more complex figures. However, as the complexity of the figures increases, so does the interaction among force vectors. Even with simple cube models, the students were less able to think about how the forces interacted as they manipulated the figures. They tended to focus on the movement of vertices in their problem solving, and their descriptions of the forces mimicked this movement. With the triangular figures, students were better able to see the orthogonality of force vectors (prevector notions) and relate these to the overall integrity of the tetrahedron. When students entered into the bridge activity, the overall complexity seemed to become too great to relate to their understanding of the stability of the primitives, and students resorted to other, more naive understandings.

Second, the students in each of the investigations seemed to focus overmuch on the amount of material used in a construction as a determinant of its strength and stability. In the pilot classroom, all of the students overbuilt their bridges. Kari also overbuilt her bridge, although this cannot be separated from her use of rectangular sections. This focus on amount of material is realistic reasoning (e.g., two pencils tied together are harder to break than a single pencil). However, it seems to be so highly salient to the children's understandings of stability that they favor this aspect of engineering over the geometric aspects. This confounding notion may be a sizable stumbling block for the introduction of prevector activities into the lower grades and must be attended to carefully in any instructional design.

Third, the type of material used in building objects relates significantly to children's perceptions of force and stability. The use of flexible foam (with marked cross sections) was effective in drawing students' attention to the fact that tension and compression coexist in most simple situations. (Note: Some arches are entirely compression systems.) However, it also introduced torsion and dead spots that drew children's attention away from the direction of the forces they were applying. Moreover, because we introduced the flexible foam prior to the rigid bamboo, the children tended to apply their thinking in the foam situations to the situations involving the bamboo.

Fourth, students' use of drawings seemed to be far more advanced than were their verbal abilities to describe the stability of the figures they were manipulating.

These preliminary findings suggest that curriculum developers pay careful attention to the ways in which realistic engineering activities are designed if they intend that children abstract relationships between geometry and structure. Variation in the complexity of figures, the types and amount of materials used, and the introduced notations all influenced children's conceptions of the geometry of structure. Nevertheless, engineering and design activities that build on children's intuitions about stability appear promising for helping to expand children's connections between structure and space.

REFERENCES

Aguirre, J., & Erickson, G. (1984). Students' conceptions about the vector characteristics of three physics concepts. *Journal of Research in Science Teaching, 21*(5), 439–457.

Aguirre, J. M., & Rankin, G. (1989). College students' conceptions about vector kinematics. *Physics Education, 24,* 290–294.

Battista, M. T. (1994). On Greeno's environmental/model view of conceptual domains: A spatial/geometric perspective. *Journal for Research in Mathematics Education, 25*(1), 86–89.

Bohm, D., & Peat, F. D. (1987). *Science, order and creativity.* New York: Bantam.

Clements, D. H., & Battista, M. T. (1992). Geometry and spatial reasoning. In D. A. Grouws (Ed.), *Handbook of research on mathematics teaching and learning* (pp. 420–464), New York: Macmillan.

de Moor, E. (1991). Geometry instruction in the Netherlands (ages 4–14)—The realistic approach. In L. Streefland (Ed.), *Realistic mathematics education in primary school: On the occasion of the opening of the Feudenthal Institute* (pp. 119–138). Utrecht, the Netherlands: CD-ß Press.

Freudenthal, H. (1983). *Didactical phenomenology of mathematical structures.* Dordrecht, the Netherlands: Reidel.

Gravemeijer, K. (1994). Educational development and developmental research in mathematics education. *Journal for Research in Mathematics Education, 25*(5), 443–471.

Hoffer, A. (1983). Van Hiele-based research. In R. Lesh & M. Landau (Eds.), *Acquisition of mathematics concepts and processes* (pp. 205–227). New York: Academic Press.

Kieran, T. E. (1993). Rational and fractional numbers: From quotient fields to recursive understanding. In T. P. Carpenter, E. Fennema, & T. A. Romberg (Eds.), *Rational numbers: An integration of research* (pp. 49–84). Hillsdale, NJ: Lawrence Erlbaum Associates.
Kindt, M., & Spence, M. S. (1998). Packages and polygons. In National Center for Research in Mathematical Science Education & the Freudenthal Institute (Ed.), *Mathematics in context*. Chicago: Encyclopaedia Britannica.
Lamon, W. E., & Huber, L. E. (1971). The learning of the vector space structure by sixth grade students. *Educational Studies in Mathematics, 4*, 166–181.
National Council of Teachers of Mathematics. (1989). *Curriculum and evaluation standards*. Reston, VA: Author.

11

Interconnecting Science and Mathematics Concepts

Kalyani Raghavan, Mary L. Sartoris, Robert Glaser
University of Pittsburgh

Historically, practically, and intellectually, mathematics has played an integral role in the investigation of nature. As a study of space and quantity, mathematics directly contributes to the understanding and mastery of the real world. Physical problems, when idealized and formulated in the language of numbers and geometry, become mathematical problems. Some of the world's greatest mathematicians, including Archimedes, Newton, and Riemann, were also great scientists. In addition to being conceptually interconnected, the two disciplines share common process skills (Berlin & White, 1991; Gallagher, 1979; Kouba, 1989). It is therefore not surprising that recent educational reform movements advocate coordinating instruction to reinforce and exploit this interdisciplinary connection (National Research Council, 1994; Project 2061, 1990). Unfortunately, however, the mutually supportive nature of mathematics and science is often underemphasized or even ignored in school curricula.

Area and volume, measures of basic properties of matter, are central concepts in science, yet they are commonly presented in fifth- and sixth-grade mathematics classes. Moreover, instruction is mostly quantitative in nature, emphasizing rote application of formulas to solve a limited set of problems rather than fostering qualitative understanding that supports meaningful application of concepts within a variety of contexts. Most science textbooks incorporate a brief review of definitions and formulas into a chapter on units and measurement, assuming that students have acquired the necessary background from the mathematics curriculum. No explicit link is made to science concepts for which area and volume are components, such as surface force and mass. As a result, students must generate their own connections and devise their own strategies for using rote knowledge in nonroutine situations.

An alternative approach would be to interconnect mathematics and science instruction so that the links between concepts are clearly depicted and repeatedly emphasized. Following a brief review of research on children's understanding of area and volume, this chapter describes a curriculum based on such an approach, exploring how understanding area and volume facilitates comprehension of related science concepts and vice versa. The chapter concludes with a discussion of implications for education.

RESEARCH ON CHILDREN'S UNDERSTANDING OF AREA AND VOLUME

Piaget and his colleagues (Piaget & Inhelder, 1956; Piaget, Inhelder, & Szeminska, 1960) studied children's understanding of qualitative and quantitative measurements of one-, two-, and three-dimensional entities as well as the development of reasoning about situations involving conservation. Among their more controversial conclusions is that the ability to understand the mathematical relationships between the dimensions of a figure and its area or volume is not fully developed until early adolescence. Although Piaget's work offers a number of implications for improving area and volume instruction, particularly at the middle-school level, the focus of derivative research is not primarily educational. Rather, much of the subsequent literature on units and measurement consists of affirmations and refutations of the findings of Piaget and his colleagues (Carpenter, 1976; Clements & Battista, 1992; Miller, 1989; Miller & Baillargeon, 1990).

Mathematics researchers often investigate students' understanding of area and volume by examining their solutions to area and volume problems. Such studies have found that students often have difficulty deciding when rules that apply to rectangles can or cannot be applied to nonrectangular figures (Lindquist & Kouba, 1989b). Similarly, a significant number of middle-school students do not recognize that the area of a figure is equal to the sum of the areas of all of its parts (Hart, 1981; Lindquist & Kouba, 1989a). Burger and Shaughnessy (1986) and Mayberry (1983) examined the geometric reasoning of middle-school students and found that many are not able to analyze figures in terms of component attributes and do not understand common geometric terms. They conclude that students, especially ninth graders, tend to take an algorithmic approach to geometry tasks, and to view mathematics as a subject requiring factual and procedural recall.

Fuys, Geddes, and Tischler (1988) reviewed the presentation of geometry concepts in three K–8 mathematics textbook series. All three series introduce area formulas in Grade 5, teaching students to calculate area by counting squares and rapidly proceeding from "four rows of five" to the formula length times width. The textbooks focus predominantly on rote application of formulas and, for the most part, ignore underlying reasons, principles, and interrelationships. Consequently, "having learned or memorized rules relating to perimeter, area, and volume of figures, many students were totally confused about these concepts and the units of measure (e.g., 'cubic square inches')" (pp. 137–138).

Science research on children's understanding of area and volume has focused largely on tasks involving conservation or differentiation. In conservation tasks, students observe some sort of transformation and are asked whether a property changes or remains the same. Students might, for example, be shown a ball of clay and be asked whether the volume changes when the ball is flattened. Conservation tasks have been utilized by such researchers as Brainerd (1971), Brainerd and Allen (1971), Hooper (1969), Howe and Butts (1970), Lovell and Ogilvie (1961), and Northman and Gruen (1970). Differentiation tasks test the ability to distinguish between two related concepts such as area and perimeter, volume and surface area, volume and mass, or volume and weight. Students are presented with two objects that are equal in one quantity and unequal in another and are asked which factor causes a specific outcome. For example, two objects of equal volume but unequal weight can be used to elicit ideas about water displacement. Students' responses indicate whether they are able to distinguish between the two properties as well as which they think is responsible for the outcome (Halford, Brown, & Thompson, 1986; Linn & Pulos, 1983; Lynch & Dick, 1980).

Much of this research evolved out of interest in devising strategies for dealing with students' typical misinterpretations and alternative frameworks. Hewson (1986), for example, investigated cognitive conflict as an intervention mechanism, and Smith, Carey, and Wiser (1985) developed science instruction to facilitate conceptual change by helping students differentiate between weight and density (Smith, Snir, & Grosslight, 1992; Snir, Smith, & Grosslight, 1993).

Klopfer, Champagne, and Chaiklin (1992) identified specific topics in each subject area in which the "ubiquitous quantities" of volume, mass, and weight figure predominantly. In physics, for example, the pertinent instructional topics include mechanics, buoyant force, Archimedes' principle, and pressure. They pointed out that the ability to reason with these fundamental concepts is essential to comprehension of related concepts. Students cannot gain deep, transferable knowledge unless they are familiar with basic concepts, and the interconnections among concepts are clearly delineated.

In examining children's ideas about area and volume, mathematics and science researchers have identified specific problem areas, including the following:

- Even young children have rudimentary notions about geometric figures and measurement. Unfortunately, the ability to apply those ideas across situations is extremely limited.
- Students have difficulty understanding standard units used in one-, two-, and three-dimensional measurement. Most students do not have a clear understanding of what Piaget et al. (1960) called *logical multiplication*—that adding, subtracting, multiplying, or dividing dimensions is not simply a numeric operation.
- Students are often confused by the fact that conservation of one quantity involves compensation in another quantity, apparently because of the difficulty of simultaneously coordinating interrelated variables.

- Although students become proficient at rote formula application, tasks requiring procedural adaptation reveal their underlying knowledge to be fragile and limited. Students do not grasp the principles behind the formulas, and, as a consequence, have difficulty recognizing that a formula is inappropriate or modifying a procedure to fit a particular situation.
- Most students fail to recognize how area and volume concepts relate to other areas of mathematics and science. Students apparently have not experienced alternative situations in which they can utilize and see the relevance of area and volume concepts, thus learning to apply abstract knowledge to a real-world context "in which empiricism and deduction coexist and reinforce each other" (Schoenfeld, 1986, p. 262).

Several educational implications emerge from this body of research. Before being introduced to area and volume concepts, for example, students must fully understand one-, two-, and three-dimensional space, why units of measurement differ according to dimension, and how a unit of measurement serves as a standard for comparison. As students begin working with formulas, they need frequent reminders that adding and multiplying dimensions is different from adding and multiplying numbers. In addition, students need to be asked to define what they mean whenever they use the term *size*, to specify whether they are referring to height, width, length, perimeter, area, or volume. Students need diverse experiences enabling them see how area and volume relate to other areas of science and mathematics, and they must be encouraged to recognize similarities across seemingly dissimilar tasks. In so doing, their grasp of new material will be swifter and surer, and their understanding of area and volume concepts will be enhanced.

THE MARS APPROACH

The computer-supported Model-based Analysis and Reasoning in Science (MARS) curriculum presents a coherent network of concepts within a model-centered context. In computer tasks, students work with dynamic, visual models of theoretical entities and systems. By manipulating the computer models and observing the results, students are able to link observable events with the laws, regularities, and representations that govern and describe them. The curriculum provides varied instructional experiences including numerous hands-on and problem-solving activities, all of which are intended to explicate and reinforce the computer tasks. Through the model-centered approach, the MARS project seeks to interconnect the factual, procedural, and theoretical aspects of science, coordinating explanatory frameworks and underlying theories with procedural skills, analytical processes, and mathematical abilities. Readers interested in a detailed description of the curriculum are referred to Raghavan and Glaser (1995).

Overview of the Curriculum

The sixth-grade curriculum focuses on the network of concepts involved in balance of forces, beginning with area and volume and concluding with floating and sinking. The nine-unit curriculum comprises three major sections: properties of objects, force concepts, and applications of balance of forces. Each unit builds on concepts learned in prior units, and the connections between concepts are made explicit through the use of computer models. Area and volume are components of many of the concepts students encounter, including mass, weight, forces exerted by fluids, buoyant force, and floating and sinking. By providing considerable practice within a sequential and supportive context extended over a semester, the MARS curriculum affords students the time and opportunity to assimilate, revisit, and build on topics and skills.

The properties of objects section deals with area, volume, and mass, fundamental concepts in science. Students learn to compare and calculate areas and volumes of regular and irregular figures. They are then introduced to mass as a property of an object that depends on both its volume and the material of which it is made. Comparison and conservation are the major themes underlying these basic units.

Building on intuitive notions of push and pull, the force section begins by introducing the two important components of a force: direction and strength. Students learn that an object acted on by two forces of equal strength in opposite directions is in a state of equilibrium. By encountering increasingly complicated situations, students recognize equilibrium as a state in which all the forces acting on an object are balanced. They learn to calculate net force as a single, imaginary force that would produce the same outcome as a combination of forces on an object and to decompose a given net force to produce a variety of component combinations. The unit on weight integrates force concepts with properties of objects. Students apply what they have learned about volume, mass, force, net force, and equilibrium to conduct computer-based experiments involving forces, objects, and springs.

In the third section, students again apply area, volume, weight, net force, and equilibrium to analyze the forces exerted by the surrounding fluid on different surfaces of an immersed object. They discover that the resultant of those forces is always upward, that its strength depends on the volume of the object and the density of the fluid, and that it is called the buoyant force. The curriculum culminates with students applying all of these concepts to investigate the conditions under which immersed objects float and sink.

As Fig. 11.1 illustrates, area and volume are fundamental concepts within the MARS curriculum. The mass of an object depends on its volume and density; weight is the gravitational pull on an object, and its strength varies with the mass and, therefore, the volume of the object; a fluid exerts force on the surface of an immersed object, and the strength of this force is proportional to the area of the surface; the resultant of all the surface forces exerted by a fluid on an object is the buoyant force; the strength of the buoyant force is proportional to the volume of the object; and finally, buoy-

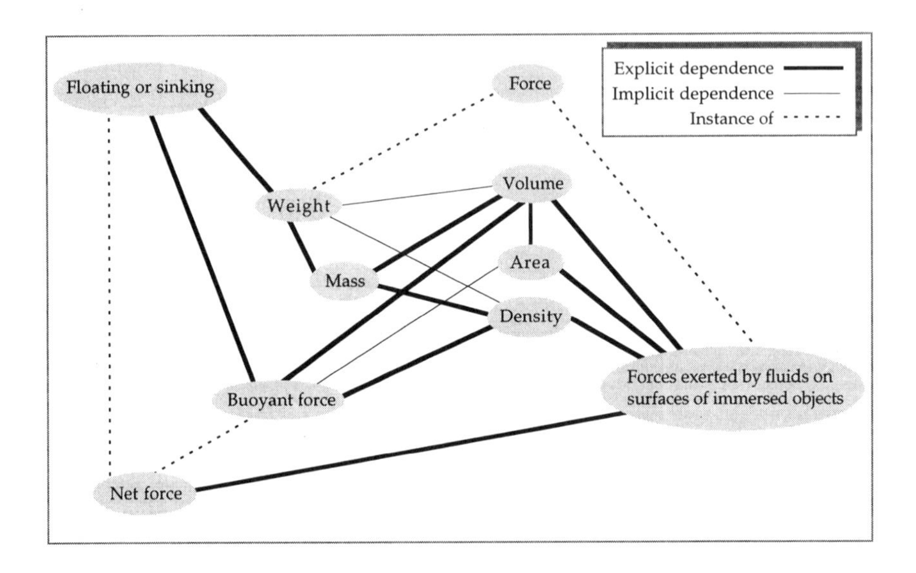

FIG. 11.1 Conceptual structure and area volume in the MARS curriculum.

ant force and weight together determine whether an object will float or sink. A critical feature of the curriculum is the use of computer models to depict clearly all concepts integral to understanding composite concepts, thus empowering students to grasp essential interrelationships.

The model-centered context of the MARS curriculum is designed to nurture the ability to use models to analyze and reason about the concepts they represent. Reflecting the curriculum's conceptual structure, the models in MARS form a nested hierarchy, in which simple models are introduced and subsequently used as components of system models. For example, students begin with the simple model of an area tile, which is subsumed in the volume cube model. Representations for density are then incorporated into the volume model to produce models of mass.

In all MARS interactive software, an option called "View model world" enables students to view specific representations for underlying theoretical constructs. In the volume unit, for example, students learn to use the model-view option to select "volume view," causing the figure on the screen to be segmented into volume units. In all subsequent activities involving volume, the corresponding volume view is included as an option. By encountering the same models—and thus the same concepts—within increasingly complex situations, students are able to examine interrelationships and to practice such skills as synthesis, comparison, and generalization.

The instructional process of the MARS curriculum is learner centered, providing opportunities to students of diverse educational and experiential backgrounds. In addition to interactive computer tasks, students complete hands-on experiments, problem-solving activities, small-group and whole-class discussions, and individual writing assignments. All activities are designed to illustrate the value of models as learning tools that can help students understand, think about, and explain abstract and complicated phenomena. Throughout instruction, the teacher acts as a facilitator, gently challenging assertions, asking students to elaborate explanations and to define the terms they use, recognizing and addressing conceptual difficulties, and offering suggestions as needed.

The context of the MARS instructional environment is depicted in Fig. 11.2. Each unit begins with one or more hands-on activities that introduce and elicit students' ideas about the phenomenon to be investigated in the computer model. At the computer, students move back and forth between two microworlds. The computer real world simulates the scientist's theoretical interpretation of the physical phenomenon. The student depicts his or her initial interpretation of the situation within the computer model world and then compares this model world with the computer real world. By revising and reevaluating the model world until it matches the computer real world, the student eventually produces a validated model, which is then applied in related hands-on activities.

Classroom Implementation

The semester-long MARS curriculum was implemented in five Grade 6 classes in a suburban Pittsburgh middle school. The area and volume units were introduced by a mathematics teacher, and the remaining units were taught by a science teacher. During the previous summer, both teachers worked with MARS project personnel to familiarize themselves with the curriculum and to offer suggestions for improving materials and activities. Both teachers used the curriculum approximately 3 days per week. Each class period was about 40 min. in duration. Throughout implementation, classes were observed and videotaped daily by MARS researchers. All 107 students took written tests at the end of each unit. In addition, 35 students were interviewed at the end of each of the three main sections of the curriculum, and all of their written work, including classwork, homework, and test papers, was photocopied for analysis.

The impact of the area and volume units is detailed in Raghavan, Sartoris, and Glaser (1997). By way of summary, MARS sixth graders

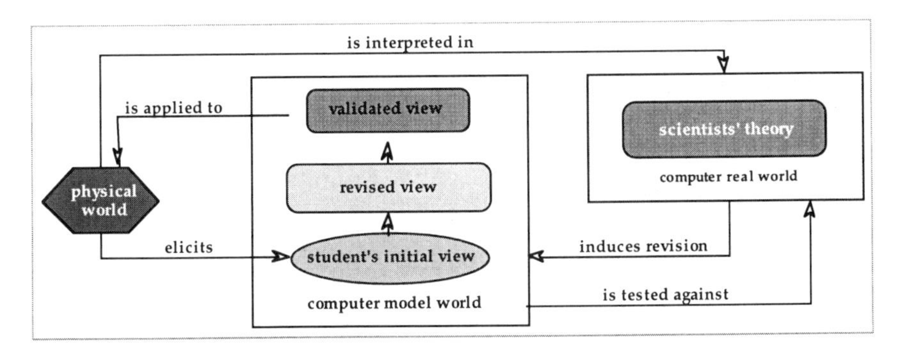

FIG. 11.2 MARS instructional environment: science as model building.

($MARS_6$) performed well on a free-response written test of area and volume concepts administered on completion of the volume unit. In particular, MARS students were successful on problems requiring analytical reasoning and procedural adaptation and demonstrated understanding of conservation and of the relationship between area and volume. In addition, performance on an equivalent test administered to 28 MARS students in December of their eighth-grade school year indicated that students had retained and were able to apply conceptual understanding after 2 years.

The focus here is on whether a firm grasp of area and volume concepts facilitates acquisition of related science concepts. Test performance of MARS students is compared with that of non-MARS students to identify differences in comprehension of mass, weight, and buoyant force. In addition, in-class and interview performances of MARS students are examined to determine their grasp of interrelationships between area and volume, among volume, mass, density, and weight, and among volume, density, and buoyant force.

COMPARATIVE PERFORMANCE

The middle school where MARS was implemented has an integrated science curriculum, in which topics from physics, chemistry, biology, and earth science are introduced and revisited in successive years. To help gauge the effectiveness of the curriculum, MARS tests were given to non-MARS students. Science teachers at all three grade levels volunteered to administer MARS tests to their classes on completion of relevant topics. At the sixth-grade level, one teacher used the MARS curriculum. The other teacher followed the regular school curriculum, which includes units on mass and weight. At the end of each of those units, the teacher administered MARS written tests to his students (non-$MARS_6$). A seventh-grade teacher administered the weight unit test to his five classes (non-$MARS_7$), and an eighth-grade teacher administered the buoyant force unit test to his five classes (non-$MARS_8$). Please note that the comparisons reported here are based on tests designed by MARS researchers and are not intended as an evaluation of non-MARS instruction. Rather, the comparisons investigate whether MARS students were able to do things they would not have been able to do without the solid understanding of basic concepts MARS is designed to promote.

Understanding Mass

At the time they took the mass test, non-$MARS_6$ students had recently studied mass and weight and were just beginning a unit on Newton's laws of motion. The test consisted of one vocabulary question, one problem to solve, and seven items focusing on the relationship among mass, volume, and density. Out of a maximum possible 35 points, the mean score of MARS students was 19.7 ($n = 107$, $SD = 6.1$), and the non-MARS$_6$ mean was 9.5 ($n = 91$, $SD = 4.0$). The difference between the two means is significant ($t = 13.4$, $p[196] < .0001$). A detailed comparison of the test performances of the two groups can be found in Raghavan, Sartoris, and Glaser (1996). The results summarized next lend credence to the hypothesis that a firm grasp of the basic "mathematics" concept of volume provides leverage in learning science concepts—in this case, mass.

Definition of Volume. In the vocabulary item, descriptions of the words *mass*, *matter*, *model*, and *volume* were provided, and students were asked to generate the corresponding scientific term. Although there was no difference between groups in generating the word *mass* to go with "amount of matter in an object" and *matter* for "anything that takes up space," there was a significant difference in supplying the term *volume* for "the amount of space inside an object." Seventy-three percent of the 107 MARS$_6$ students provided the correct term, compared to only 19% of the 91 non-MARS$_6$ students.

The fact that so many non-MARS$_6$ students failed to generate the word *volume* strongly suggests that the term had not been used recently. In fact, an examination of the non-MARS$_6$ textbook revealed a scant five passages referring to mass. Those passages are scattered across two chapters and contain no reference to volume other than a terse statement that size alone does not determine the mass of an object. This is a significant deficiency in light of the considerable research identifying students' inability to differentiate between volume and mass (e.g., Halford et al., 1986). Differentiation between two concepts is not likely to occur unless both concepts are considered simultaneously and with equal emphasis.

Volume as Displacement. One way to determine the volume of an object is by immersing it and measuring the displaced liquid. This procedure is particularly valuable for irregularly shaped objects. Although the displacement technique is not explicitly taught in the MARS curriculum, the Mass unit test includes a question on volume displacement to determine if students recognize the space occupied by an object in a different medium as volume. Student responses to this question provide insight into the depth of their understanding of the concept of volume.

The question involves two objects—a 1-cubic-inch aluminum cube of mass 5 g and a 1-cubic-inch steel cube of mass 50 g. The information on mass is included to detect mass–volume confusion. Students are to draw a picture showing what happens when the steel cube is immersed into an overflow jar filled with water. In addition, they are to determine the amount of overflow when the steel cube is immersed in water, when the aluminum cube is immersed in water, and when the steel cube is immersed in corn oil. In all three cases, the amount of overflow is 1 cubic inch, the volume of the immersed object.

Table 11.1 compares the performance of MARS$_6$ and non-MARS$_6$ students on the four aspects of the problem. The first column displays the number of students who provided an illustration conveying an adequate grasp of the situation. The second column shows the number who said the

TABLE 11.1 PERFORMANCE ON THE VOLUME-AS-DISPLACEMENT QUESTION

Group	Adequate Representation	Correct Displacement	Aluminum vs. Steel (No Difference)	Water vs. Oil (No Difference)
Non-$MARS_6$ ($n = 91$)	40.5 (50%)	10.5 (12%)	14 (15%)	15.5 (17%)
$MARS_6$ ($n = 107$)	90 (84%)	32 (3%)	32.5 (30%)	38 (36%)

cube would displace 1 cubic inch of water. The third and fourth columns respectively reflect students who indicated that the material of the cube and the kind of liquid would not make a difference.

The MARS group's performance was significantly better than that of the non-$MARS_6$ group ($t = 5.49$, $p[196] < .0001$). Students were given partial credit for providing evidence of emergent understanding, such as indicating that the material of the object would not make a difference but specifying that both cubes would displace the same mass. As expected in a curriculum that emphasizes modeling and pictorial representations, MARS students did very well in the representation aspect of the question. In addition, about a third of the MARS students revealed an understanding of volume as the amount of space occupied by an object in any medium and an ability to differentiate volume from mass.

Understanding Weight

Upon completion of the weight unit, the unit test was administered to MARS, non-$MARS_6$, and non-$MARS_7$ students. Not long before taking the test, the non-$MARS_6$ group had studied mass, weight, and Newton's laws of motion and falling bodies. The non-$MARS_7$ students, after reviewing laws of motion, had studied potential and kinetic energy, concepts that build on the foundational concepts of volume, mass, and weight.

The questions, general in nature, involve the interrelationships among volume, mass, weight, force, and balance of forces. The overall performance of the three groups is depicted in Table 11.2. MARS students performed significantly better than both non-$MARS_6$ ($t = 11.5$, $p[202] < .0001$) or non-$MARS_7$ ($t = 10.6$, $p[201] < .0001$) students. There was no significant difference between the performances of non-$MARS_6$ and non-$MARS_7$ students.

Defining Volume, Mass, and Weight. The first question in the weight test required students to define the terms *volume, mass,* and *weight.* In scoring

TABLE 11.2 OVERALL PERFORMANCE ON THE WEIGHT UNIT TEST

Group	n	Mean (Max. = 20)	Standard Deviation
$MARS_6$	108	13.6 (30%)	3.9
Non-$MARS_6$	96	7.5 (37.5%)	3.5
Non-$MARS_7$	95	7.8 (39%)	3.8

the item, formulaic definitions such as "volume is length times width times height" or imprecise definitions such as "weight is the heaviness of an object" were assigned half a point. Complete definitions were worth one point. Table 11.3 lists each group's total score for each definition. Roughly two thirds of the MARS students provided adequate definitions, compared to less than a third of all non-MARS students. Among MARS students, even incorrect responses—defining volume as length times width, for example—reflect a clear association between volume and size. The fact that 20 non-MARS students defined volume as "loudness," suggests that the relationship between volume and weight was not emphasized in their curriculum.

Conservation of Properties. Another question focuses on a drawing in which a ball is dropped from identical heights on the earth and on the moon. Students are to describe similarities and differences between the two situations. They are asked specifically about four items: the volume, mass, and weight of the ball, and the outcome. Each item is worth two points—one for correctly deciding whether the characteristic would stay the same or change, and one for explaining why. Table 11.4 lists each group's total score on each of the eight components. MARS students were better not only at identifying which properties would be conserved but also at providing explanations reflecting knowledge of the concepts, saying, for example, that mass stays the same because "you can't change what's inside the ball." In contrast, many non-MARS students offered such nonexplanations as, "Same. Volume does not change," and one wrote that there is "no volume on the moon."

Both the definition and conservation questions are typical textbook-style questions, probing student understanding of volume, mass, and weight, the building blocks of Newtonian mechanics. Comparative results

TABLE 11.3 PERFORMANCE ON THE DEFINITION QUESTION

Group	*Volume*	*Mass*	*Weight*
$MARS_6$ ($n = 108$)	63 (59%)	74.5 (69%)	60.5 (56%)
$Non\text{-}MARS_6$ ($n = 96$)	5.5 (6%)	39.5 (41%)	29 (30%)
$Non\text{-}MARS_7$ ($n = 95$)	29 (31%)	12.5 (13%)	24.5 (26%)

TABLE 11.4 PERFORMANCE ON THE CONSERVATION-OF-PROPERTIES QUESTION

Group	*Volume*		*Mass*		*Weight*		*Outcome*	
	Same	*Why?*	*Same*	*Why?*	*Different*	*Why?*	*Different*	*Why?*
$MARS_6$ ($n = 108$)	100 (93%)	84 (78%)	93 (86%)	78.5 (73%)	95 (88%)	77 (71%)	87.5 (81%)	62 (57%)
$Non\text{-}MARS_6$ ($n = 96$)	40 (42%)	15 (16%)	64 (67%)	28.5 (30%)	69 (72%)	45 (47%)	71 (74%)	46 (48%)
$Non\text{-}MARS_7$ ($n = 95$)	51 (54%)	24.5 (26%)	53 (56%)	11.5 (12%)	66.5 (70%)	44 (46%)	57 (60%)	40 (42%)

suggest that teaching advanced topics of motion and energy without first establishing the underlying infrastructure will not likely engender conceptual understanding.

Understanding Buoyancy

After the weight unit, MARS students study the forces exerted by fluids on the surfaces of immersed objects. They learn that those forces combine to produce an upward force called the buoyant force. In the final unit, students discover that the resultant of an immersed object's weight and the liquid's buoyant force determines whether the object floats or sinks. On completion of the curriculum, a final free-response test was administered to $MARS_6$ students. The test was also given to a group of non-$MARS_8$ students completing a science curriculum that included water pressure, buoyant force, floating and sinking, Archimedes' principle, and lift on flying objects. The test encompassed forces exerted by liquids, buoyant force, and interrelationships among area, volume, and density.

The mean score of $MARS_6$ students was 11.2 out of a possible 22 points ($n = 107$, $SD = 3.8$), compared to the non-$MARS_8$ mean score of 4.5 ($n = 95$, $SD = 2.6$). The significantly better performance of MARS students ($t = 14.4$, $p[200] < .0001$) is at least partly attributable to their firmer grasp of area and volume concepts. There was almost no mention of area or volume in any of the non-$MARS_8$ responses. MARS students, on the other hand, applied their knowledge of area to compare forces exerted on immersed surfaces and their understanding of volume to calculate the buoyant force on an immersed object. Moreover, asked to explain buoyant force, 67% drew pictures to show an upward force due to the liquid that, together with the gravitational force, determines whether an object will float or sink.

Force Exerted by a Liquid on an Immersed Boundary. In the MARS curriculum, students investigate internal forces in liquids, using the computer model to "draw" boundaries within a container of liquid and to examine the forces exerted on bounded areas. The computer "column model" demonstrates that the downward force on a horizontal bounded area is related to the weight of the column of liquid above the boundary. The strength of the force increases with the area and depth of the boundary and the density of the liquid. To assess understanding of this idea, a test question displays a container of water made up of four differently shaped sections connected at the bottom (see Fig. 11.3). The connected container is a standard textbook illustration used to convey that water pressure varies with depth and is independent of container size. Students are asked to compare the downward forces exerted by the water on four marked boundaries, one in each section. The boundaries are equal in area and at equal depths. To answer this question correctly, students must realize that the important factors to consider are the areas and depths of the boundaries and that the sizes and shapes of the container sections are irrelevant. Although this was the first time they had encountered such a container, 67 of the 107 $MARS_6$ students (63%) explained why the downward forces on the boundaries would be equal. Some even drew columns indicating equal volumes of water above the boundaries to augment their explanations. Figure 11.3 shows illustrative student responses. Only 3 of the 95 non-MARS$_8$ students (3%) responded correctly. Such results indicate that, as a result of the MARS curriculum, students were better equipped to interpret and reason about the kind of diagram frequently encountered in textbooks than they would otherwise have been.

Area, Volume, and Buoyancy. In the question shown in Fig. 11.4, students were asked to calculate the buoyant force exerted on an irregularly

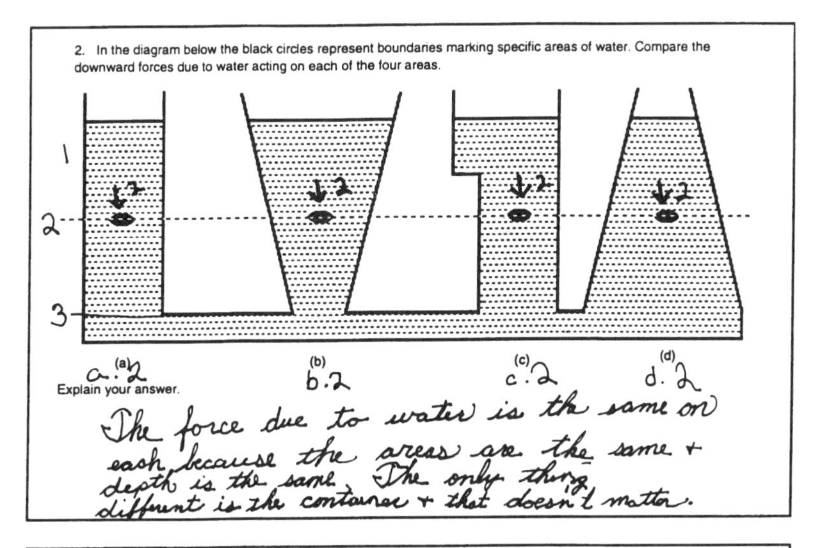

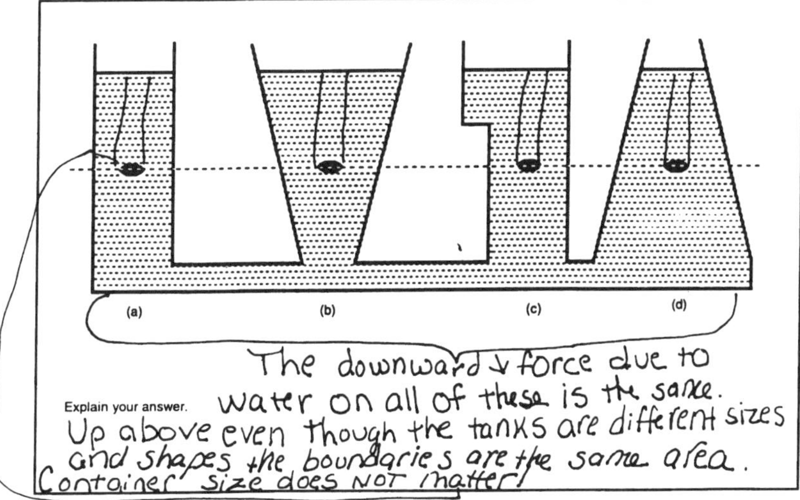

FIG. 11.3 Sample MARS responses to force-exerted-by-liquid test question.

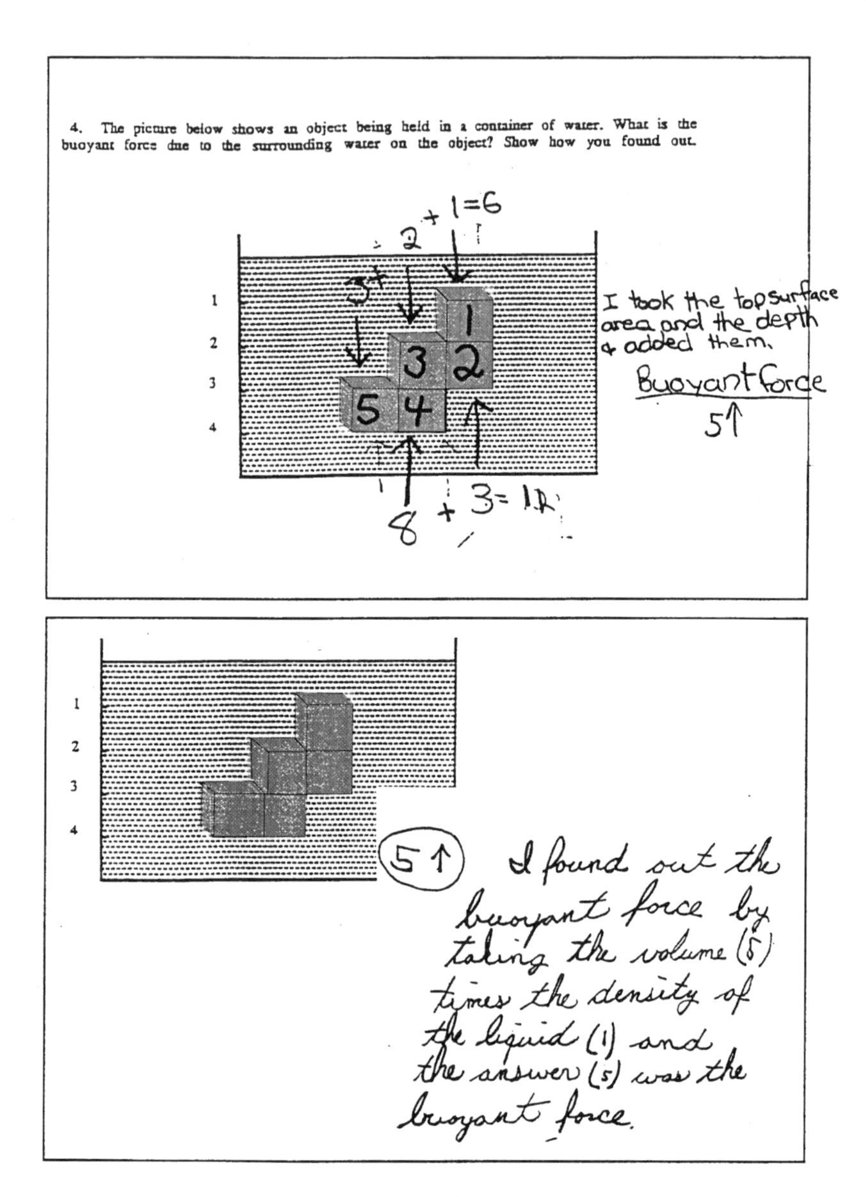

FIG. 11.4 Sample MARS responses to buoyant-force test question.

shaped object immersed in a container of water. As described earlier, MARS students learn to calculate the buoyant force as the resultant of forces exerted on the different surfaces of an immersed object. Some students discovered a "shortcut" strategy for calculating buoyant force—multiply the volume of the object by the density of the liquid. Either strategy could be used to solve this problem. In calculating buoyant force,

$MARS_6$ students had worked exclusively with cubes and rectangular prisms, so the irregularly shaped object was new to them. Still, 83 of the 107 $MARS_6$ students solved at least part of the problem, their most common error being the omission of one or more surface forces. In comparison, only a few of the 95 non-$MARS_8$ students even attempted the problem. The average score for $MARS_6$ students on this item was 53%, compared to 0% for non-$MARS_8$ students. The few non-$MARS_8$ students who tried to answer the question mentioned only the depth of the top or bottom surface. None of them mentioned the areas of those surfaces or the volume of the object. Figure 11.4 shows sample MARS solutions—one using object volume and one using the resultant of surface forces.

In summary, compared to non-MARS sixth, seventh, and eighth graders, MARS sixth-grade students demonstrated a clearer understanding of interconnections among mathematics and science concepts, including area, volume, mass, weight, and buoyant force. Further, they were far more apt appropriately and effectively to apply previously learned concepts and strategies to solve problems and explain phenomena. By gradually building understanding of area and volume through experiences in which those concepts apply and interact with science concepts, MARS students gain a solid foundation for further learning in science and mathematics.

IN-CLASS AND INTERVIEW PERFORMANCE

In this section, selected responses to classwork and interview questions are used to investigate how understanding area and volume facilitates comprehension of such concepts as mass, weight, and buoyancy. As previously mentioned, 35 students were interviewed three times during the course of the curriculum, and their classwork was collected for analysis. These students, roughly one third of the MARS sixth graders, form a representative cross section of the population.

Understanding Interrelationships Between Area and Volume

As Klopfer et al. (1992) pointed out, the word *volume* has several contextual meanings, even within the domain of science. It can refer to the amount of space occupied by an object or substance, the capacity of a container, or the amount of a substance within a container. Yet textbooks commonly provide just one definition, usually the first. Similarly, the method used to measure volume varies with the object or substance being measured. For example, the water displacement method used to measure a rock might not be suitable for a pile of salt.

Specific instructional exercises, test questions, and interview questions are posed throughout the curriculum to determine the breadth of students' conception of volume and their ability to adapt a procedure to fit a particular situation. One such question is the aquarium problem. Students are shown a picture of a partially filled aquarium and asked how they would figure out how much water it contains. The picture includes a bird's-eye view of the base of the aquarium, showing a trapezoid composed of a rectangle adjacent to one leg of a right triangle. To solve this problem, the student must devise a new strategy, adapting familiar procedures to compute the volume of an unfamiliar shape.

Different students came up with different strategies for calculating the volume, but most involved multiplying the area of the base times the height of the water. To determine the area of the base, many students described a decomposition strategy, segmenting the base into a rectangle and a triangle and computing the area of each segment. Others used a composition strategy, doubling the base to create a rectangle, and then halving the area of that rectangle. A few students decomposed the three-dimensional water-filled region into a rectangular and a triangular prism, and then described how to calculate the volume of each three-dimensional segment. The following is an illustrative example of a decomposition strategy:

Student: You could take like the bottom base of the aquarium and just kind of make a line through here. And then you could—like it would be through here. And then you could multiply the width, height, and depth, and then that would be the volume for how much water is in this section, this part.

Interviewer: OK. I am saying this for the tape recorder. She is dividing the base into a triangle and a rectangular piece and she is calculating the volume of?

S: The square.

I: The rectangular piece. OK.

S: And then I would do the same thing for the triangle, and then whenever I have that total, I would divide it in half because it is only half of the square.

I: When you say I would do the same thing for the triangle, are you talking about completing this? [Indicates doubling the triangle to create a rectangle]

S: Yes. And whenever I have that, I would divide it in half because it is only one part of the square.

I: OK. And what is the height you would be considering? Can you point to me?

S: Oh, the height would be from the bottom of the aquarium to the water level.

Although almost every student displayed a qualitative understanding of the problem situation, there were differences in the autonomy and completeness of their procedural explanations. Some students focused on just parts of the problem—the area of the base, for example, or the height of the water—and needed prompting to integrate the various parts into a complete solution. Overall, however, students revealed an ability to flexibly and strategically utilize their knowledge of area and volume concepts in an unfamiliar and somewhat complicated context.

Understanding Interrelationships Among Volume, Mass, Density, and Weight

Immediately following the area and volume units, MARS students are introduced to the concept of *mass*. After hefting two cubes that look identical but are unequal in mass, they are asked to represent the similarities and differences between the cubes without using words. In a follow-up worksheet aimed at eliciting ideas about the interrelationships among volume, density, and mass, students are asked to argue for or against the following two statements:

- Large objects are always heavier than small objects.
- Objects made of heavy material are always heavier than objects made of lighter material.

Perhaps because of their recent experience with the equal-sized cubes, all 35 students disagreed with the first statement. Several generated examples to demonstrate that size alone does not determine mass. The second statement did not provoke such unanimous response. Still, 71% of the students disagreed, explaining that very large objects of light material could be heavier than small objects of heavy material. Again, many substantiated their claims with examples.

Early in the weight unit, a modified version of the same question is posed. Students are asked to react to the following conversation:

Kermit: I have two objects—a thingamajig and a doohickey. I also have a spring. Guess which object will stretch the spring farther.
Fozzie: Which object is bigger?
Kermit: The thingamajig has more volume than the doohickey.
Fozzie: The thingamajig will stretch the spring farther.

Thirty-three of the 35 students (94%) disagreed with Fozzie, explaining that volume information was not sufficient to predict the stretch of the spring. Citing an example from an earlier hands-on activity, one student wrote, "Disagree because the biggest cylinder weighs less than the medium cylinder." Another student wrote, "No, because the doohickey could have greater density but less volume."

The second part of the conversation is as follows:

Kermit: What do you think, Gonzo?
Gonzo: What are the objects made of?
Kermit: The thingamajig is made of plastic. The doohickey is made of glass.
Gonzo: Glass is heavier than plastic, so the doohickey will stretch the spring farther.

The response to this latter exchange paralleled the earlier result, with 25 of the 35 (71%) disagreeing with Gonzo. Explanations ranged from barely coherent—"Disagree because might have more plastic making heavier than glass"—to very explicit—"Disagree because there isn't enough information. You need to also know the size. If you don't, it could be either one." Regardless, however, they all conveyed the idea that relative weight judgments cannot be based solely on density and volume comparisons but require more specific information.

In addition to knowing how volume, mass, weight, and density are related, students must be able to differentiate among the four concepts. During formative testing of the curriculum, many students stated that there was no gravity outside the earth's surface, and some could not distinguish mass from weight. In a worksheet designed to address these notions, students are to complete a data table by determining the missing volumes, masses, and weights of items on earth and a fictional planet, Vinstef. To do so, students must utilize the conservation principle for volume and mass, recognizing that both remain unchanged despite a change in physical location. In addition, students must use the existing data to determine that planet Vinstef has half of the earth's gravitational pull and use that information to calculate the missing weights. All 35 students were able to fill in all of the missing data correctly. In response to an accompanying question about the difference among volume, mass, and weight, 30 students wrote something like, "Volume and mass stay the same, but weight changes when gravity changes," or "Volume and mass are properties, so they don't change. Weight is not a property, so it does change." Thus, in addition to calculating correct answers, the majority of MARS students were able to explicate the rationale underlying their work.

Understanding Interrelationships Among Volume, Density, and Buoyant Force

In the final section of the curriculum, students learn that a liquid exerts a net upward force, the buoyant force, on an immersed object and deduce that this force depends on the volume of the object and the density of the liquid. This is Archimedes' principle for an immersed object. The curriculum does not explicitly link the weight of the displaced liquid with the buoyant force. By providing the "column model" as feedback (see Raghavan & Glaser, 1995), however, the computer model encourages students to make that connection on their own. A modified version of the volume-as-displacement test item, described in the section on comparative performance, was asked during the final interview to determine whether students had indeed made that inference.

Students are shown a list of statements with which they are to agree or disagree, offering explanations and conducting experiments with hands-on materials to support their positions. The available materials include two equal-volume cubes with unequal mass, two unequal-volume cylinders with equal mass, a small block of balsa wood, orange-colored water and green-colored salt water, two overflow jars, several beakers, a spring, a ruler, and a wooden spring stand. Two of the statements involve volume as displacement and are discussed here.

Statement 1. "When an object is placed in a container filled to the brim with a liquid, the amount of liquid that will overflow is equal to the mass of the object." There are at least three ways of proving this statement false with the given materials. Students can immerse the equal-volume/unequal-mass cubes in filled overflow jars and show that volumes of displaced liquid are the same. Alternatively, they can immerse the equal-mass/unequal-volume cylinders and prove that they displace different amounts of liquid. They can also immerse just one object and then compare the mass of that object to the mass of the liquid it displaces. In any event, the question requires that students differentiate volume from mass and realize that displacement is caused by volume and not by mass.

Statement 2. "The buoyant force on an object will be different in different liquids." Although the focus of this statement is on buoyant force, it also elicits ideas about volume and density. Again, there are several ways to demonstrate the truth of this statement. First, students can hang an object from the spring and immerse the object in each of the two liquids to demonstrate a difference in the amount of spring stretch. Second, they can immerse one object in the two different liquids and compare the masses of the displaced liquids. Students who use the latter procedure demonstrate a firm grasp of the fact that, irrespective of the liquid, an object always displaces the same volume. They also recognize that the buoyant force exerted by a liquid depends on its density and that, if one liquid has more mass than an equal volume of another liquid, it also has greater density.

The protocol transcripts of student responses to both statements were analyzed for conceptual knowledge and the ability to coordinate theory with real-world experiments. The responses were sorted into four levels of proficiency: integrated, associated, undifferentiated, and fragmented. The levels are characterized in the sections that follow.

Integrated. Students at this level reveal an understanding of the underlying concepts and theory and are able to map this knowledge into practice through pertinent experiments. In the following excerpt, for example, a student proves Statement 1 false with an appropriate experiment and an explanation:

> *Interviewer*: If an object is placed in a container that is filled to the brim with a liquid, the amount of liquid that will overflow is equal to the mass of the object.
>
> *Student*: No. I don't agree.
>
> *I*: OK. Tell me why.
>
> *S*: Because it's equal to the *volume* of the object. And I will show by using these. [Selects the equal-mass/unequal-volume cylinders, drops the larger cylinder into the overflow jar, and collects the displaced water in a beaker. Holds up the smaller cylinder.] Now, I put this in and less [water] comes out. And [the cylinders] are the same mass. So, equal to the *volume* comes out.

Although students at the integrated level may start out with an incorrect opinion, they will reverse themselves during the explanation or based on unexpected experimental results. The example that follows is an excerpt from the protocol of a student who changes her mind while explaining her position:

Interviewer: The amount of liquid that will overflow is equal to the mass of the object. Do you agree or disagree?
Student: I agree.
I: You agree that it is equal to the mass of the object?
S: Yes. Because whatever, like how much, whatever the mass is, it's like taking up that much space. OK. I changed my mind now.
I: OK. You changed your mind.
S: I disagree with that statement. It doesn't really matter how much the mass is, but like how much volume—like how much it is taking up in the water. [Immerses the equal-volume/unequal-mass cubes and demonstrates that they displace the same amount of water.] Well, I could have done all of this with the same mass [cylinders] to show that mass doesn't really matter.

The following are two examples of integrated-level responses to Statement 2. One student agrees that the buoyant force on an object is different in different liquids. Asked to prove it, she suggests putting the same object in each of the two liquids. She explains that the amount of overflow will be the same and that if the beakers containing the displaced liquids were placed on opposite sides of a balance, the salt-water side would go down because it had greater density. To illustrate, she draws two partially filled beakers, using dots to represent the difference in densities.

The second student explains why he agrees with the statement, transferring his experience with the computer model to support his argument. In the following excerpt, the student refers to the blue and yellow liquids depicted on the computer. (The yellow liquid is three times denser than the blue liquid.)

Student: I am going to say that I agree with that.
Interviewer: And why do you agree?
S: Because, when I saw the computer models, like if you tried it on there, and you used it in the yellow liquid, it would be three times.
I: OK. The buoyant force is three times for the yellow liquid. Now, can you think of an experiment to prove that?
S: [Points to the salt water.] This is more dense. This would be better to use for it. [Picks up a cube.] Maybe if you could put this one into that liquid. Say if you put—this would be one volume unit. OK. This would be one volume unit. [Points to the top and bottom surfaces of the cube.] Top area, the surface area would be the same thing.
I: So, you are taking the cube, one of the cubes, and you are putting it in and . . .
S: [Calculating the relative downward and upward forces acting on the top and bottom faces of the cube, the net force being the buoyant force.] Let's say that the depth is 3 and 4 again. It

would be, right here it would be 3. And this would be 4. But in the other liquid, this would be 12 and 9. So it would be up 3.

I: OK. So, the buoyant force would be up 3, if this were the computer model and this were the yellow liquid.

S: Right. Up 3.

I: Which is a good way, OK. Once again, that is a real good way to explain what you are thinking. And it's a good theoretical way to justify your thinking. But still, the computer model is theory. What I want is a hands-on, real-life experiment that proves what you are saying is true.

S: [Hangs the cube from the spring and immerses it in both liquids to demonstrate that salt water pushes the cube up farther than water because of its greater density.]

This student clearly demonstrates a firm grasp of the relationship between volume, density, and buoyant force, and this understanding enables him to coordinate the theory modeled by the computer program with practical experimentation.

Associated. Students at this level may exhibit sound theoretical knowledge but lack the procedural skill to generate relevant experiments. Conversely, they may design appropriate experiments but have difficulty linking results to underlying theory. The following segment illustrates such a response:

Interviewer: When an object is placed in a container filled to the brim with a liquid, the amount of liquid that will overflow is equal to the mass of the object.

Student: Disagree. Because the object—the amount that will overflow will be the *volume* of the object.

I: The volume of the object. OK. Do you see anything that you could use to set up an experiment that will prove what you say is true?

[The student asks for a cube, and the interviewer asks her to specify which cube.]

S: It doesn't matter. Either one. Because they are the same volume. [Immerses the cube and collects the displaced water. Claims—but is unable to prove—that the volume of the displaced water is equal to the volume of the cube.] Because it wasn't the mass that made it overflow, it was the volume that makes it overflow.

Asked how she could convince a fifth grader that the overflow depends on volume and not mass, she concedes that she does not know.

Responding to Statement 2, some students were able to explain theoretically why the buoyant force is different in different liquids by describing their experiences with the computer program. However, they were not sure how to use the available materials as proof, or they were unable to provide a clear interpretation after conducting experiments. Several stated that the displaced volumes will be the same when an object is immersed in two liquids, but they did not relate the difference in densities to the difference in buoyant force. For instance, one student who agreed with Statement 2 immersed the large cylinder in the green salt water and collected the displaced liquid. She then repeated the process with the orange water. After several moments of silence, she said, "The only thing is that they are different colors. That is really it then, the same amount came out." The student seemed unable to proceed, and the interviewer decided to offer support:

Interviewer: What if we had a balance here? If we had a really good sensitive balance that could tell the difference?
Student: Can I pick these up?
I: Yes. How do they feel?
S: They feel almost the same. This might, this seems a little bit heavier.
I: The green seems just a tiny bit heavier. Say that we put them on a real sensitive balance. Do you think that they would balance? Or do you think that one side might go down?
S: I think the green might go down a little bit.
I: You think the green might go down? So what does that tell you?
S: That the buoyant force is greater for green.

Unfortunately, the interviewer did not press the student to explain why the denser liquid would have stronger buoyant force. However, even if the student did not yet fully understand the connection between liquid density and buoyant force, she clearly understood the difference between volume and mass well enough to support such a connection.

Undifferentiated. Responses at this level exhibit evidence of confusion between two related concepts. For example, some students are unable to distinguish volume from mass, and their interpretations of experimental outcomes reflect this confusion. In the following excerpt, the student initially disagrees with Statement 1. When asked why, he is not able to offer an explanation. The interviewer asks if he could do an experiment to prove the statement false. The student immerses the large cylinder and collects the water that overflows:

Interviewer: OK. And now what do you want to do?
Student: Well, this—that isn't—that isn't—the water that came isn't the same mass as that. [Points to the large cylinder.]
I: O.K. How can you prove that is not the same mass? Or how do you know that is not the same mass?
S: Ah, because, ah, because this, uh, I don't know. Because if you fill that—that tube up with water, it wouldn't be as much. If you filled this up. [Picks up the large cylinder.] Let me see. Ah, well, this, uh, water isn't the same, like. If we could just fill this up, it wouldn't fill to the top.
I: I'm sorry—say that again?
S: Like, it would still weigh the same, but it was, like, hollow. But even if it was the same and we could pour stuff in it, it wouldn't go to the top.
I: So the amount of water that came in here [points to the displaced water and then to the large cylinder], if this were a hol-

low container. If we could open it up and pour this water in here—the orange water that came out into this container—would not be enough to fill up this container? If it were a container?
[The student agrees with the interviewer's interpretation of his words.]

This student does not appear able to differentiate between mass and volume. To prove that the displaced water is not equal to the mass of the cylinder, he focuses on volume and argues that there is not enough water to "fill up" the cylinder.

Another student said that Statement 2 is true "because, on the computer model, we used, ah, there is two different liquids. And if you used the yellow liquid—compared to the water—there would be a different buoyant force." Asked *why* the buoyant force is different in different liquids, he replied, "Because different liquids have different masses. So that—that would matter." When asked to prove this with an experiment, he immersed an object in each liquid and collected the overflow. Seeing that the displaced amounts were the same, the student hesitated, then said the buoyant forces were the same. When the interviewer asked if there was anything else to take into account to interpret the results of the experiment, he decided to hang the object from the spring and compare the spring stretches in each of the liquids. Unable to see the slight difference in spring stretch, he concluded that the "different experiments made me feel that [the buoyant forces] are the same." This student had previously stated that the mass of the liquid matters for buoyant force. Faced with a real-world incarnation of that idea, however, he does not recognize that the equal volumes of displaced liquid would have different masses and would therefore exert different buoyant forces.

Fragmented. Students at this level appear to have acquired several pieces of knowledge, but there is no evidence of interconnections linking those pieces into a coherent whole. Such students generally offer ambiguous or inconsistent explanations and have difficulty designing experiments to test specific ideas or interpreting experimental outcomes. They appear to have a difficult time deciding which concepts are relevant to a particular situation. For example, one student agreed with Statement 1. When asked why, she said she was not really sure and asked to do an experiment. Picking up a cube, she explained, "The amount that is spilled out is the mass of this block." She then immersed the cube and collected the displaced water. She is holding the cube in one hand and the beaker of displaced water in the other when the interviewer asks the following:

Interviewer: You are hefting them in order to try to get a sense whether they are the same or different?

Student: Yeah. I don't know how to tell if they are the same or different. But, well, if you drop this in a really big container and fill it up to the brim, it will spill out a little. But in this case, there is only a little, and this spilled out a lot. [She performs several experiments and concludes that both container size and object size determine the amount of displacement. When asked why the size of the object matters, she experiments with the equal-mass/unequal-volume cylinders, collecting the overflow from both objects in separate containers.] Oh, why did I do that?

I: Tell me what you are comparing now.
S: The volume of it. The two different cylinders' volumes.
I: Yeah.
S: Some of the water fell out, so I'd say that the volume does not matter.
I: The volume does or does not matter?
S: It does not because some of the water fell out.
I: So you are telling me that objects with different volumes will cause the same amount of water to come out or a different amount of water to come out?
S: The same.

After struggling to identify relevant concepts, this student designs and runs an experiment but fails to recognize that the results support the hypothesis she was testing. Instead, she adopts a conclusion that contradicts the experimental results, which she apparently discounts because "some of the water fell out."

Table 11.5 displays the distribution of proficiency-level scores for each of the two statements. Students at both the integrated and associated levels demonstrate a sound understanding of volume as it relates to mass and to displacement in liquids. Improvement over the course of the curriculum can be ascertained by comparing the first and second columns of Table 11.5 with the third and fourth columns of Table 11.1, which reflect parallel questions from the mass unit test. Both sets of questions examine the interrelationships among volume, density, and buoyant force. Because students must design, perform, and interpret real-world experiments to substantiate abstract ideas, however, the interview task is far more complex than the written test items. Still, 69% of the students successfully demonstrated that the amount of liquid an object displaces is equal to its volume. Only 30% were able to explain this at the end of the mass unit. In addition, 63% of the students showed that the buoyant force is different in different liquids, although some required support in interpreting experimental results. In contrast, only 36% of the students recognized this at the end of the mass unit.

It is particularly encouraging to note that most students who conducted experiments were clear about the purpose and were able to offer convincing explanations that recognized conservation of volume before

TABLE 11.5 LEVELS OF PERFORMANCE BASED ON INTERVIEW RESPONSES OF $MARS_6$ STUDENTS ($n = 35$)

Concept	*Integrated*	*Associated*	*Undifferentiated*	*Fragmented*
Volume as displacement	21 (60%)	3 (9%)	8 (23%)	3 (9%)
Buoyant force in two liquids	13 (37%)	9 (26%)	10 (29%)	3 (9%)

and after an object is immersed. One student, for example, explaining why the amount of displaced liquid depends on the volume of the object, said, "Because it is flowing out, this is taking the place of the liquid. So, if it is taking the place of the liquid, then the same amount is going to come out, since it is taking the place. It is like if a family of four moves into another house and another family moves out, [the house] is not changing, it is just taking the place."

CONCLUSIONS AND IMPLICATIONS

A mutually supportive relationship inextricably binds science and mathematics: "Science provides mathematics with interesting problems to investigate, and mathematics provides science with powerful tools to use in analyzing data.... They are part of the same endeavor" (Project 2061, 1990, pp. 16–17). In order for students to benefit from this interrelation, however, they must encounter scientific contexts in which the application of mathematical knowledge provides leverage and vice versa. Chi, Feltovich, and Glaser (1981) identified the presence or lack of interlinked, principle-based organization of knowledge as an important factor differentiating the problem-solving abilities of experts and novices. The computer-supported MARS curriculum provides a learning environment supportive of extended chains of inference and conducive to model-based reasoning and conceptual understanding. To accomplish this, two essential features are included in the design of the curriculum: cohesive content structure, and continuous support for conceptual integration through computer models that coordinate mathematical and scientific representation.

Overall, students' classwork, test performance, and interview responses provide evidence that such cohesive support facilitates growth in understanding of the foundational area and volume concepts and of the science concepts to which they are related. MARS sixth graders appear better equipped to analyze and explain phenomena involving area and volume than they would have been without the curriculum, acquiring not only interconnected declarative knowledge but also procedural skills enabling them to apply that knowledge. By the end of the curriculum, students routinely used models to help explain their ideas and to depict and solve complicated problems. Moreover, several students learned to use models as reasoning tools. Some, like the student who used the computer model to reason about buoyant force, spontaneously used models to think about concepts and phenomena. Others were able to do so when prompted.

Although the overall results are encouraging, there are some problem areas. For instance, some students seem to have a narrow understanding of volume, leading to a misunderstanding of mass. Such students interpret volume only as empty space inside an object and misinterpret mass as filled space in an object. Students also appear to have difficulty assimilating the full implication of the concept of density. Many were able to use such terms as "denser" appropriately and to relate them to substance, but it is unclear whether they genuinely understood the concept, especially in the case of liquids and gases. Throughout the forces-exerted-by-fluids section of the curriculum, students repeatedly saw the column model. In the final interview, many students clearly demonstrated an association between the concepts of density and buoyant force, but only a few explicitly described the connection between the weight of the displaced liquid and buoyant force.

Implications for the MARS Curriculum

At present, the volume unit deals only with solids and provides too few hands-on and qualitative activities. Students need extensive experiences actually measuring volumes of solids and liquids. In particular, they need to measure internal capacity, external space occupied, and filled capacity and to recognize all of these as volume.

The curriculum also needs to provide increased opportunities to explore the concept of density of liquids as well as of solids and to compare and measure the masses of different liquids. Students' qualitative and relational conceptions need to be strengthened by making appropriate connections to mathematical concepts. One planned modification is to link ratio and proportion, concepts that students normally learn in middle-school mathematics, with the density concept. This would be a mutually beneficial way to master all of these important concepts.

To provide students continual but varied contacts with fundamental ideas and processes, the curriculum needs to be extended to other grade levels. A natural progression would be to integrate science topics such as motion and energy with linear functions, slopes, and straight-line graphs, mathematics concepts currently taught in middle school.

Implications for Instruction

Accumulation of facts and development of computational skills appear to be the main objectives of science and mathematics instruction. Different subject areas are compartmentalized with few connections between disciplines. As a consequence, most students minimally process information learned in one discipline and find it difficult to elicit meaning when that information is encountered in another discipline. This is critical in the case of fundamental mathematics concepts such as area and volume, as the comparative results given earlier indicate. The gradual but steady progress of MARS students demonstrates that understanding is not instantaneous but develops only through continuous effort to apply and extend partially understood concepts. Instruction on such integral concepts as area and volume must be coordinated with instruction on related concepts in order to enable students to generate interconnections and teachers to draw on, extend, and exploit those links.

The MARS approach is to integrate instruction on area and volume into the science curriculum and to offer recurring encounters with both concepts throughout the curriculum. Even in the absence of integrated curriculum materials, there is a need for coordination of topics in mathematics and science instruction and for cooperation among teachers to provide complementary learning experiences. Such an approach might help students and teachers alike appreciate how mathematics is so admirably adapted to reality.

ACKNOWLEDGMENTS

Development of this curriculum was sponsored by the Andrew W. Mellon Foundation as part of the Varieties of Reasoning in the Social and Natural Sciences program at the Learning Research and Development Center of the University of Pittsburgh. Thanks to Ann Davidson for help with data collection and to Bill Moore for help in software development. We are extremely appreciative of our partners in education Dr. Alice Hirsch (curriculum coordinator), Al Herrle (science teacher), and Judy Kielman (mathematics teacher) of the Montour School District for their support in implementing the curriculum. This is LRDC Research Report No. MARS-004.

REFERENCES

Berlin, D. F., & White, A. L. (1991). *A network for integrated science and mathematics teaching and learning* (Monograph Series No. 2). Columbus, OH: National Center for Science Teaching and Learning.

Brainerd, C. J. (1971). The development of the proportionality scheme in children and adolescents. *Developmental Psychology, 5*, 171–197.

Brainerd, C. J., & Allen, T. W. (1971). Training and generalization of density conservation: Effects of feedback and consecutive similar stimuli. *Child Development, 42*, 693–704.

Burger, W. F., & Shaughnessy, J. M. (1986). Characterizing the van Hiele levels of development in geometry. *Journal for Research in Mathematics Education, 17*(1), 31–48.

Carpenter, T. P. (1976). Analysis and synthesis of existing research on measurement. In R. A. Lesh & D. A. Bradbard (Eds.), *Number and measurement: Papers from a research workshop* (pp. 47–83). Athens, GA: Georgia Center for the Study of Learning and Teaching Mathematics.

Chi, M. T. H., Feltovich, P. J., & Glaser, R. (1981). Categorization and representation of physics problems by experts and novices. *Cognitive Science, 5*(2), 121–152.

Clements, D. H., & Battista, M. T. (1992). Geometry and spatial reasoning. In D. Grouws (Ed.), *Handbook of research on mathematics teaching and learning* (pp. 420–464). New York: Macmillan.

Fuys, D., Geddes, D., & Tischler, R. (1988). The van Hiele model of thinking in geometry among adolescents. *Journal for Research in Mathematics Education*, Monograph No. 3.

Gallagher, J. (1979). Basic skills common to science and mathematics. *School Science and Mathematics, 79*, 555–565.

Halford, G. S., Brown, C. A., & Thompson, R. M. (1986). Children's concepts of volume and flotation. *Developmental Psychology, 22*(2), 218–222.

Hart, K. (1981). Measurement. In K. Hart (Ed.), *Children"s understanding of mathematics: 11–16* (pp. 9–22). London: John Murray.
Hewson, M. G. (1986). The acquisition of scientific knowledge: Analysis and representation of student conceptions concerning density. *Science Education, 70*(2), 159–170.
Hooper, F. (1969). Piaget conservation tasks: The logical and developmental priority of identity conservation. *Journal of Experimental Child Psychology, 8*, 234–249.
Howe, A. C., & Butts, D. P. (1970). The effect of instruction on the acquisition of conservation of volume. *Journal of Research in Science Teaching, 7*(4), 371–375.
Klopfer, L. E., Champagne, A. B., & Chaiklin, S. D. (1992). The ubiquitous quantities: Explorations that inform the design of instruction on the physical properties of matter. *Science Education, 76*(6), 597–614.
Kouba, V.L. (1989). Common and uncommon ground in mathematics and science terminology. *School Science and Mathematics, 89*, 598–606.
Lindquist, M. M., & Kouba, V. L. (1989a). Geometry. In M. M. Lindquist (Ed.), *Results from the fourth mathematics assessment of the National Assessment of Educational Progress* (pp. 44–54). Reston, VA: National Council of Teachers of Mathematics.
Lindquist, M. M., & Kouba, V. L. (1989b). Measurement. In M. M. Lindquist (Ed.), *Results from the fourth mathematics assessment of the National Assessment of Educational Progress* (pp. 35–43). Reston, VA: National Council of Teachers of Mathematics.
Linn, M. C., & Pulos, S. (1983). Male-female differences in predicting displaced volume: Strategy usage, aptitude relationships, and experience influences. *Journal of Educational Psychology, 75*(1), 86–96.
Lovell, K., & Ogilvie, E. (1961). The growth of the concept of volume in junior school children. *Journal of Child Psychology and Psychiatry, 2*, 118–126.
Lynch, P. P., & Dick, W. (1980). The relationship between high IQ estimate and the recognition of science concept definitions. *Journal of Research in Science Teaching, 17*(5), 401–406.
Mayberry, J. (1983). The van Hiele levels of geometric thought in undergraduate preservice teachers. *Journal of Research in Mathematics Education, 14*(1), 58–69.
Miller, K. F. (1989). Measurement as a tool for thought: The role of measuring procedures in children's understanding of quantitative invariance. *Developmental Psychology, 25*(4), 589–600.
Miller, K. F., & Baillargeon, R. (1990). Length and distance: Do preschoolers think that occlusion brings things together? *Developmental Psychology, 26*(1), 103–114.
National Research Council. (1994). *National science education standards*. Washington, DC: National Academy Press.
Northman, J. C., & Gruen, G. F. (1970). Relationship between identity and equivalence conservation. *Developmental Psychology, 2*, 311–322.
Piaget, J., & Inhelder, B. (1956). *The child's conception of space*. London: Routledge & Kegan Paul.
Piaget, J., Inhelder, B., & Szeminska, A. (1960). *The child's conception of geometry*. New York: Basic Books.
Project 2061. (1990). *Science for all Americans*. New York: Oxford University Press and American Association for the Advancement of Science.
Raghavan, K., & Glaser, R. (1995). Model-based analysis and reasoning in science: The MARS curriculum. *Science Education, 79*(1), 37–61.
Raghavan, K., Sartoris, M. L., & Glaser, R. (1996). *The impact of model-centered instruction on student learning: The mass unit*. Manuscript submitted for publication.
Raghavan, K., Sartoris, M. L., & Glaser, R. (1997). The impact of model-centered instruction on student learning: The area and volume units. *Journal of Computers in Mathematics and Science Teaching,16*(2/3), 363–404.
Schoenfeld, A. H. (1986). On having and using geometric knowledge. In J. Hiebert (Ed.), *Conceptual and procedural knowledge: The case of mathematics* (pp. 225–264). Hillsdale, NJ: Lawrence Erlbaum Associates.

Smith, C., Carey, S., & Wiser, M. (1985). On differentiation: A case study of the development of the concepts of size, weight, and density. *Cognition, 21*(3), 177–237.

Smith, C., Snir, J., & Grosslight, L. (1992). Using conceptual models to facilitate conceptual change: The case of weight-density differentiation. *Cognition and Instruction, 9*(3), 221–283.

Snir, J., Smith, C., & Grosslight, L. (1993). Conceptually enhanced simulations: A computer tool for science teaching. *Journal of Science Education and Technology, 2*(2), 373–388.

12

Geometric Curve-Drawing Devices as an Alternative Approach to Analytic Geometry: An Analysis of the Methods, Voice, and Epistemology of a High-School Senior

David Dennis
University of Texas–El Paso

Jere Confrey
Cornell University

When the concept of analytic geometry evolved in the mathematics of 17th-century Europe, the meaning of the term was quite different from our modern notion. The main conceptual difference was that curves were thought of as having a primary existence apart from any analysis of their numeric or algebraic properties. Equations did not create curves; curves gave rise to equations. When Descartes published his *Geometry* in 1638 (trans. 1952), he derived for the first time the algebraic equations of many curves, but never once did he create a curve by plotting points from an equation. Geometrical methods for drawing each curve were always given first, and then, by analyzing the geometrical actions involved in a physical curve-drawing apparatus, he would arrive at an equation that related pairs of coordinates (Dennis, 1995; Dennis & Confrey, 1995). Descartes used equations to create a taxonomy of curves (Lenoir, 1979). This tradition of seeing curves as the result of geometrical actions continued in the work of Roberval, Pascal, Newton, and Leibniz. As analytic geometry evolved toward calculus, a mathematics developed that involved going back and forth between curves and equations. Operating within an epistemology of multiple representations entailed a constant checking back and forth between curve-generating geometrical actions and algebraic language (Confrey & Smith, 1991). Mechanical devices for drawing curves played a fundamental, coequal role in creating new symbolic languages (e.g., calculus) and establishing their viability. The tangents, areas, and arc lengths associated with many curves were known before any algebraic equations were written. Critical experiments using curves allowed for the coordination of algebraic representations with independently established results from geometry (Dennis, 1995). What we present here is a description of one student's investigation of two curve-drawing devices. The usual approach to analytic geometry in which a student studies the graphs of equations has been reversed, in that the student primarily confronted curves created without any preexisting coordinate system. This student first physically established certain properties of and interrelationships between curves and only afterward came to represent these beliefs using the language of symbolic algebra. This student's actions are interpreted within Confrey's (1993, 1994) framework, which views mathematics as dialogue between "grounded activity" and "systematic inquiry." In this study, we provided curve-drawing devices and posed problems that allowed the student an opportunity to voice both sides of this dialogue.

THE STRUCTURE OF THE INTERVIEWS

The purpose of the investigation was first to create a set of physical tools and then to use that environment to ask a series of questions in a setting where direct physical experiments with curves could shape a student's initial beliefs. The student was asked about how (if possible) each device could be set up to reproduce curves drawn by the other, and how he could be sure that the curves were the same. He was also asked how the action of each device might give rise to an equation of the curve. At no time, however, did the interviewer suggest the use of any particular coordinate system, origin, axes, or unit of measure.

The student was asked to justify his assertions in any way that seemed convincing to him and with as much detail as possible. From his own hypotheses, formed directly from his experience with primary curve-drawing actions, he moved, in different ways, to represent geometric actions with algebraic relations. The ways in which he described his sense of the geometric actions strongly shaped the kind of algebraic language that he employed.

The student was given no prior instruction in the historical, cultural, or mathematical significance of the devices with which he worked. We taught the student rudimentary operations with each device and then posed questions about what kinds of curves each device could draw, the possible situations where different devices might or might not draw the same curve, and how the action of each device might give rise to an algebraic representation. The student justified his answers in any way that seemed appropriate.

The student in this study investigated several curve-drawing devices, but we discuss here only his investigation of the two elliptic devices shown in Figs. 12.1, 12.2, and 12.3. These figures are taken from a popular 17th-century text by Franz van Schooten (1657), who wrote extensive commentaries on Descartes. Figure 12.1 shows a well-known device where a loop of string is placed over two tacks. Figures 12.2 and 12.3 show what is known as a "trammel" device, where two fixed points on a stick (the trammel) move along a pair of perpendicular lines and a curve is traced by any point on the trammel, either between the two pins, as in Fig. 12.2, or outside them, as in Fig. 12.3. We built easily adjustable versions of these devices for use by students. Our string device involved a 3 ft × 4 ft paper-covered sheet of soft plywood into which tacks could be inserted. An adjustable loop of string could then be placed over the tacks and

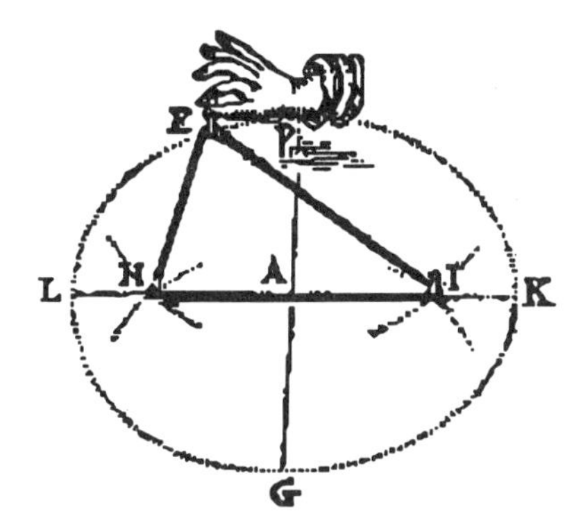

FIG. 12.1 Loop-of-string device.

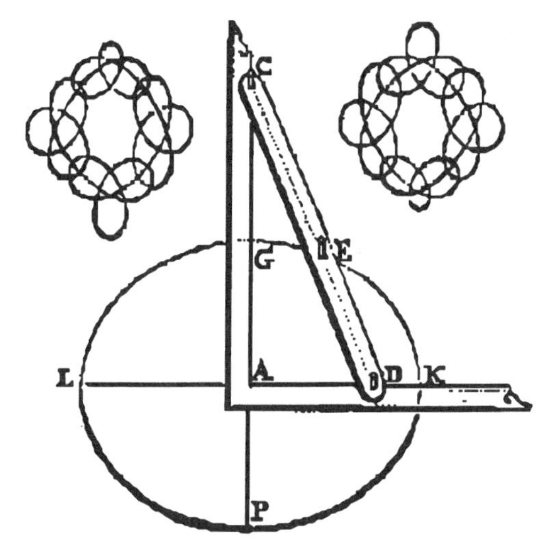

FIG. 12.2 Trammel device (curve traced between pins).

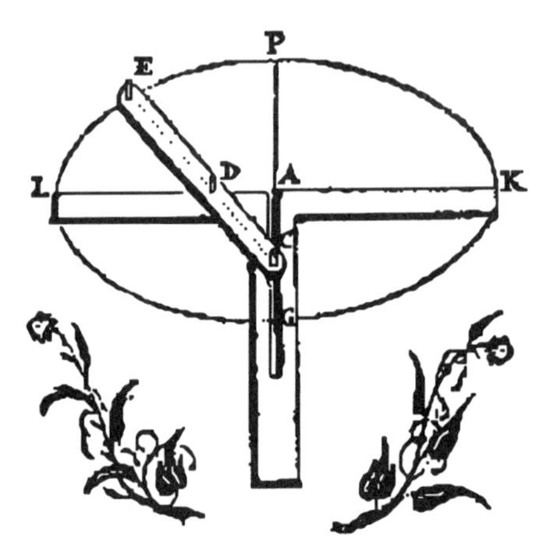

FIG. 12.3 Trammel device (curve traced outside pins).

drawn taut with a colored pencil. The string was tough, braided nylon that would not stretch, and the length of the loop could be quickly and easily adjusted by a spring-locked slide, such as those found on the drawstrings of coats.

Our trammel device worked on a 3 ft × 4 ft sheet of Plexiglas into which two narrow grooves (1/8 in.) had been carved at right angles to each other and bisecting both dimensions of the sheet. The trammel itself (see Fig. 12.4) was made from a 25-in. slotted wooden stick with a fixed pin at one end, which slid in one of the grooves in the Plexiglas. An identical pin, which slid in the other groove, protruded from a small aluminum holder that fit into the slotted stick and could be locked with a thumbscrew at any chosen distance (2 in. to 25 in.) from the fixed pin, thus creating a trammel of adjustable length. The path of motion of any point on the trammel could be traced on the Plexiglas by a pen fitted in one of the adjustable penholders. Aluminum pieces, drilled as penholders for dry-erase colored marking pens, were fitted into the slot of the trammel. These penholders could be locked at any position on the trammel with a thumbscrew. The pen fit tightly into the penholder and did not need to be held by hand during the drawing motion of the trammel. One

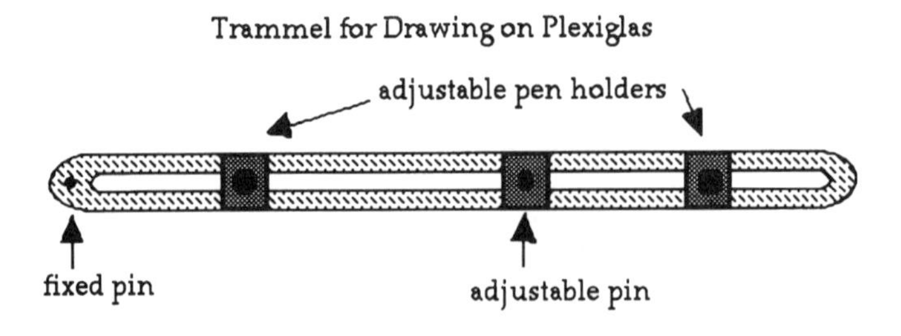

FIG. 12.4 Trammel for drawing on Plexiglas.

penholder was placed between the two pins to form a trammel device as in Fig. 12.2, and the other was beyond the second pin for drawing curves as in Fig. 12.3.

We chose to build these two particular devices for use in student interviews for the following reasons. Ellipses are common in visual experience, science, and art; they are aesthetically pleasing; and, although the curve is not the graph of a function in a narrow sense, it has an equation that is fairly simple and familiar to students. These devices are all relatively simple to build, demonstrate, and experiment with. The actions involved in them can be felt directly and intuitively. Although either device can be used to draw the entire family of ellipses, they feel quite different, and the adjustments for changing the elliptic parameters work in quite different ways; hence, any algebraic relations that emerge directly from the actions will, at first, have different forms. There is an immediate physical element of surprise when such different actions produce curves that look the same. As the interviews show, this first impression provided a strong motivation for the student to search for a coordination of his physical experience with symbolic mathematical representation.

The high-school student interviewed had heard about the loop-of-string device and had seen his teacher derive an elliptic equation from the constant sum of the two focal distances implied by action of the device. Although the student had not personally experimented with such a device and, hence, had little practical instinct for exactly how variations in the length of the loop and the distance between the tacks would affect the curve, this device provided the student an initial sense of familiarity. The student had never seen a trammel device used to draw curves. Making connections between the loop-of-string and the trammel provided a rich problem-solving experience grounded in an immediately tangible situation. No matter how this situation is approached, it involves some kind of geometrical or algebraic transformation because the former device starts by establishing the foci whereas the latter device starts by establishing the lengths of the axes, giving no immediate indication as to focal position. Algebraic representation of the loop-of-string device tends to utilize the distance formula, whereas the trammel lends itself more toward similarity and proportion.

The subject of these interviews ("Jim") was a senior chosen from a class of New York State Regents' Course Four Mathematics at Ithaca High School in Ithaca, New York, in spring 1994. The student was not chosen for any special background or ability. Course Four Mathematics in New York State is a high-school precalculus course taken by seniors as part of the regular Regents' sequence beginning in ninth grade with Course One. Roughly half the students at Ithaca High School eventually take Course Four Mathematics.

Several students were interviewed concerning a variety of curve-drawing devices, but what we present here focuses only on Jim and his work with the two devices just described. Several times during the interviews, Jim described himself laughingly as a "terrible student" and said that if his teacher "saw these videotapes he would probably be horribly embarrassed." Jim's teacher described him as a fair-to-average student who had to struggle hard to keep up with his work. Jim's teacher also found him to be very helpful and cooperative in class. He was a very open, friendly, talkative person, which made it easy to interview him. He talked almost constantly about what he was doing and thinking, with little or no prodding. He seemed to have no inhibitions about being videotaped.

Two individual interviews, about 2 hr each, were videotaped. The second interview occurred 1 week after the first. Jim was asked not to discuss the project with others until after the completion of all of the interviews. During the week between his two interviews, although he did not have the devices, he was free to work by himself on any unresolved questions and consult any mathematical source material that he thought might be helpful. Jim was not provided with any references or background material until after the completion of the interviews.

The interviews were structured around the following questions:

1. Are these devices capable of drawing the same curves?
2. Is there any curve that one device can draw but that the other cannot?
3. How exactly do you go about setting up one device so as to reproduce a curve drawn with the other device?
4. Is there any way to find an equation of a curve directly from the actions involved in the device used to draw that curve?
5. What convinces you of your claims, and how would you go about justifying them?

Jim expressed a strong preference for geometry over algebra and preferred physical experimentation. He said that he really enjoyed "fooling around with stuff" and wished that there could be far more geometry discussed in high school. Jim obviously enjoyed experimenting with the curve-drawing devices and rapidly generated and rejected a whole series of conjectures about how they might relate to each other. Although many of his guesses seem, at first, a bit wild and random, the interviews show that his overall pattern of refining his experiments displays an astute sense of geometric proportion and invariance. He voiced many guesses based on things that were visually and physically suggestive to him.

Jim openly admitted that he easily got lost in algebra and that he found it very boring. He said that he wished that his algebra skills were better, and he thought that this was something he would "have to work on." During his second interview, Jim eventually expressed algebraically the proportions that he found from the geometry of the trammel device, but when asked if these equations were equivalent to the one that he wrote from the loop-of-string device, he paled at the thought of having to attempt an algebraic reduction. His usual cheerful demeanor seemed to darken abruptly. When told that he did not have to do this and reminded that he was free to end the interview whenever he wished, Jim said that it would give him real satisfaction to see the algebra "come out." He asked the interviewer to watch the algebra carefully because Jim knew that he would make mistakes. Sometimes Jim even predicted exactly what type of algebra mistakes he was prone to make, and then several minutes later confirmed his predictions.

Jim was asked how important it was for him to see the algebra "come out" in order to believe that the devices were drawing the same curves. He replied that he had made a big jump in his belief, based on his procedures for reproducing the curves visually, and that the algebraic confirmation was just one more little step. He gestured geometrically with his hands, showing the big jump and the little step. He then estimated the proportions in his gesture at around 8:2 and laughed. Jim had very little confidence in his own algebraic skills, and this seemed to transfer over to his confidence about algebra in general, yet he still wanted to see the algebra confirm what he had learned from his physical experiments. When he became frustrated, he asked the interviewer directly for some algebraic advice and was offered a few procedural hints (e.g., "Try squaring both sides"). Once he had corrected and completed his algebra, he had no trouble at all in interpreting these results in terms of the physical reality of the curve-drawing devices because (as we show) that was the primary source of his beliefs.

JIM'S INTERVIEWS

In this section, we give a general description of the major cognitive steps that Jim made during his encounters with two curve-drawing devices. We broadly classify Jim's investigations into the following seven stages listed in the order in which they occur:

1. Physical exploration of the loop-of-string device and its inherent control parameters.
2. Physical exploration of the trammel device and its inherent control parameters.
3. Development of a systematic method for trammel duplication of curves first drawn with the loop-of-string device.
4. Representation of the action of the loop-of-string device with an algebraic equation.
5. Development of a systematic method for loop-of-string duplication of curves first drawn with the trammel device.
6. Representation of the action of the trammel device with an algebraic equation.
7. Epistemic statements concerning the relations between physical geometry and algebraic representation.

The first three stages of investigations took place during the first interview and the last four take place a week later during the second interview.

First Interview

Exploration of the Loop-of-String Device. Jim begins by drawing some curves with the loop-of-string device and experimenting with the various possibilities that the device allows. Although he is familiar from his mathematics class with the concept of drawing ellipses in this way, he has never personally used such a device to draw curves. He quickly becomes aware that two parameters are involved in this device: the distance between the two tacks, and the length of the loop of string. Jim confirms his expectation that the device will produce ellipses and then states that with a fixed loop of string he can obtain "more eccentric" ellipses by moving the tacks farther apart. His concept of eccentricity is based on a visual geometric sense of curves being stretched away from a circle, and, although he remembers that there is some way to numerically measure eccentricity, he cannot remember how that is done.

Jim is not entirely sure that this device will draw only ellipses. He experiments. He begins to pull at the string in various ways and then wants to try using a third tack in an attempt to use the loop of string to draw a hyperbolic curve. The interviewer asks him to restrict his attention for the moment to what can happen using a loop of string placed over only two tacks. Jim's experiments produce only ellipses, and he hesitantly decides that those might be the only the curves that he can produce in this way:

J: I don't really see how you could draw a hyperbola from this arrangement. Maybe you can... I'm just probably not looking ... I don't see it.

Jim is then asked about how the action of the loop-of-string device might lead to equations of the curves being drawn. Jim says that he has seen this done in his class but he cannot remember how to reconstruct such an algebraic equation. He is, however, convinced from his physical experiments that the loop of string holds the "perimeter" of a shifting triangle fixed and that this fixed perimeter along with the fixed distance between the two tacks will completely determine any equation of the curve:

Interviewer: Would those two measurements be enough to determine an equation, or would you need more information?

J: It seems to me that that should be enough, because all that we're using are these two things... By varying these two distances we can vary the shape of these drawings, so as far as writing an equation, I would think that these two distances would be the only pertinent information... Yeah, I'm pretty certain, because it seems those are the only two things that are interacting on this system right now.

We see here how the results of Jim's physical experiments form the foundation of his beliefs about what is controlling the shape of the curves that can be drawn. We also see how these beliefs shape his expectations about the possible form of any algebraic representation of those curves. Despite his distaste for algebra, Jim will become more and more determined to continue his investigations until his algebraic expressions are reconciled with his geometric experience.

Exploration of the Trammel Device. Jim next turns his attention to the trammel device. Before drawing any curves, Jim makes some guesses as to what might result. If the pen is outside the two pins, as in Fig. 12.3, Jim predicts that the device will draw ellipses. This belief is based on a desktop toy with a similar motion. When the pen is placed between the pins, Jim's first prediction is that the device will produce a cusped star, as in Fig. 12.5. Jim is bit surprised to find that both positions of the pen in the device draw curves that appear elliptic. He wonders if this device will ever produce another type of curve and, after some experiments, decides that it will not.

Jim discusses proportions that he sees in the device that appear to him to be critical to an understanding of its motion. First are the relative rates of motion of the two pins in the tracks. Jim describes qualitatively how, when one pin is near the junction of the tracks, the rate of motion of the other pin is very small and hence the motion of the pen becomes entirely vertical or horizontal, depending on which of the tracking pins has the greater rate of motion. Jim describes how the pen's horizontal (or vertical) motion will always be some particular "fraction" of the horizontal (or vertical) tracking pins. This sense of an invariant proportion in the trammel device stays with Jim and becomes stronger and stronger until he is able to represent this proportionality in an algebraic form that confirms his physical sense of proportional rates of motion. As we see later, his final algebraic expressions are not so much a confirmation of the elliptic motion of the device, but rather a confirmation of the reliability of algebraic expressions to represent physical geometrical action.

Another issue for Jim in his initial trammel explorations is the question of how to get the device to "blow up" a curve (i.e., to create a similar but larger version of a given curve). In particular, Jim is concerned with finding situations where the trammel will draw circles. He first looks at cases where the pins are very close together and the drawing pen is outside the pins as far away as possible. These look quite circular, and he sees these curves as corresponding to the curves drawn with a large loop of string and two tacks very close together. He does not at first notice that

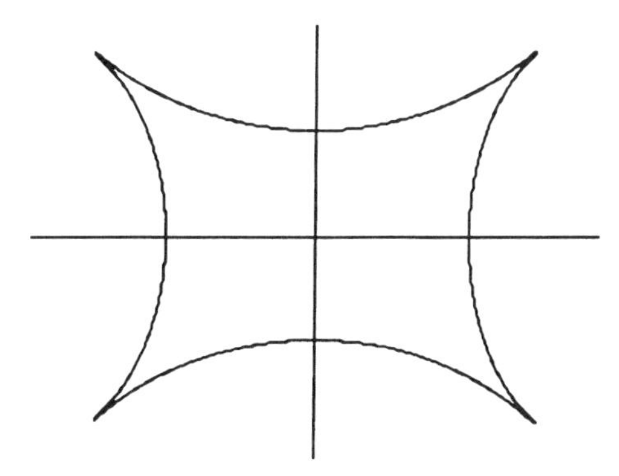

FIG. 12.5 Jim's first prediction of the trammel's motion.

placing the pen at the midpoint between the two pins will produce a circle. When he later discovers this, it is crucial to his seeing how to systematically duplicate the loop-drawn curves with the trammel. Jim has a clear physical concept of enlargement and dilation, which guides much of his later investigations.

Duplication of Loop-Drawn Curves with the Trammel Device. One main focus of this investigation was to create an environment where curve-drawing actions could be coordinated prior to any algebraic representation. To this end, Jim is asked to try to duplicate with the trammel any specific curve drawn with the loop of string and vice versa. He is asked to be as specific as possible about any system of procedures that he might use to accomplish this task in general.

Jim's initial experiments have convinced him that these devices should be able to produce some curves that are the same. He begins by choosing a specific length of string and tack distance (both measured in even inches) and then trying to duplicate this curve with the trammel. His first attempt is to match the distance between the pins to the distance between the tacks and then to set the penholder at a distance that matches half the string. Several times Jim refers to the pins on the trammel as "foci," even though they move during the drawing of the curve. Jim is committed to some correspondence of this type because of his observation that the trammel draws curves that are quite circular when the pins are close together and the pen is far away from them. He sees this as analogous to a large loop of string and two closely spaced tacks.

After various adjustments to this scheme Jim is unable to duplicate the loop-drawn curve, although he does draw a curve that he feels might be a "blow up" of the loop-drawn curve. He then begins experimenting with trammel setups where the pen is between the two pins. Trying to match various measurements from the loop-of-string arrangement to the trammel, Jim eventually sets the pen halfway between the two pins. Before he begins drawing the curve he exclaims:

> *J*: Oh!... This is obviously going to be like a circle. I should have seen this before.

Jim draws the curve, gets what he expects, and then explains how the distances of the pins from the pen holder determine where the curve will cross the horizontal and vertical tracks, which he now calls the x and y axes. In this case those two distances are both equal, and he says that is "a characteristic of a circle." This is an important moment for Jim because he realizes that axis lengths rather than focal distance are inherent in the setup of trammel device:

> *J*: This is pretty much as close as we're going to get to a perfect circle. That's my prediction.
>
> *Interviewer*: Do you think that this is a perfect circle? Or as close as you can get with this device?

J: Theoretically, yeah, it probably is a perfect circle, because this distance here and this distance here [indicates the half-axes] are supposed to be exactly the same... It looks circular to me.

Jim continues experimenting with the trammel device and eventually sees that the distances of the pen from the pins have to match the length of the half-axes on the loop-drawn curve. He easily sees the length of the semimajor axis of the loop-drawn curve and looks for the length of the semiminor axis on the loop-drawn curve. After some initial mismeasurments, Jim accurately reproduces the curve with the trammel, using the semimajor and semiminor axes as the distances from the pen to each of the pins:

J: Looks reasonably close.
Interviewer: Do you have a system at this point for copying any curve over there [string loop] with this thing here [trammel]?
J: I should be able to.

Jim explains that he will use the midpoint between the tacks as a "center" or "origin," measure the half-axes, then set up the trammel accordingly. The interviewer asks him if he can calculate these trammel settings from the tack distance and the length of the loop of string. He tells the interviewer that the semimajor axis is "$L-(1/2)X$" (where L is half the loop, and X is the distance between the tacks; see Fig. 12.6). Jim then explains that by dividing in half the isosceles triangle formed by the loop of string positioned at the end of the minor axis, he will get two equal right triangles each having a hypotenuse of $L-(1/2)X$, and a leg of $(1/2)X$. Using the Pythagorean theorem, he can then find the semiminor axis that he needs to set up the trammel.

Through personal physical experience with curve-drawing devices, Jim came to several important realizations in his first interview. He saw that radically different mechanical actions with very different relative rates of motion could trace the same overall curves. At first Jim expected

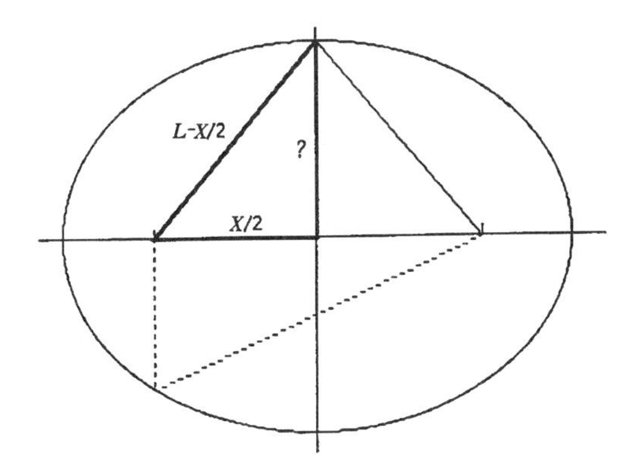

FIG. 12.6 Jim's study of the loop-of-string device.

to find a way to set up the focal distance on the trammel because that was how the loop-of-string device worked. He then saw that the parameters of control inherent in different devices could be different. The realization that the trammel device gave direct control over the lengths of the axes eventually led Jim to a very different approach when he came to an algebraic analysis of the trammel device. Similarity and invariant proportionality played a much more central role in his vision of the trammel's action, and, in a second interview, he saw how his vision could be directly translated into an algebraic equation that both confirmed and validated his physical and geometrical beliefs. His eventual algebraic confirmation of his well-grounded physical beliefs provided Jim with a much more profound belief in the possibility of mathematics to allow for consistency across multiple representations.

Second Interview

An Equation from the Loop-of-String Device. When Jim returns a week later for his second interview, he begins by telling the interviewer that he has looked over some of his notes on conic sections and that he has thought about what is "important" in the loop-of-string device. He puts two tacks in the board and says that the distance between them is "important." He then uses the loop of string to draw an ellipse and chooses a point on the curve and labels it (x,y). After some discussion of the symmetries of the ellipse, Jim explains precisely how the parameters that control the loop-of-string device can be used to write an equation of the curve drawn by the device. His explanation differs somewhat from the exposition given in his textbook in that he makes far more direct appeal to the physical possibilities of the device. After defining the constants a and c on the device (see Fig. 12.7), Jim writes the equation of the curve as:

$$\sqrt{(x + c)^2 + y^2} + \sqrt{(x - c)^2 + y^2} = 2a$$

He then states that this equation can be algebraically reduced to:

$$\frac{x^2}{a^2} + \frac{y^2}{a^2 - c^2} = 1$$

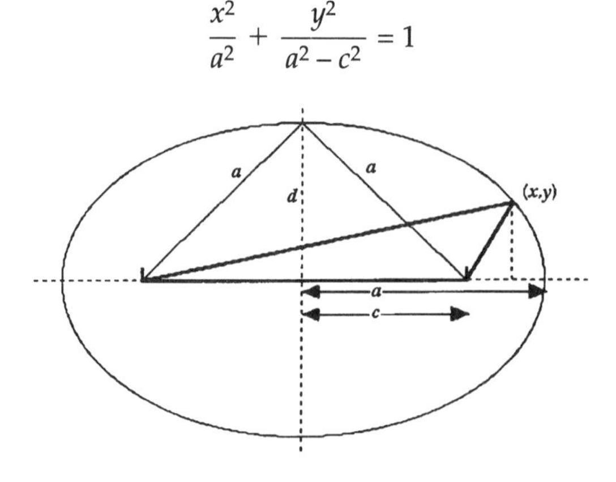

FIG. 12.7 Using the loop-of-string device: Jim's attempt at writing an equation of the curve.

He has seen his teacher make this reduction, and he feels that he could probably reproduce it, but that does not interest him.

Duplication of Trammel-Drawn Curves with the Loop-of-String Device. Jim is quite convinced at this point that the two devices draw the same set of curves. He is then asked how he would go about setting up the loop-of-string device in order to copy a curve first drawn with the trammel device. This procedure is the reverse of the duplication method that he developed during the first interview. After reviewing that method, he proceeds by labeling the semiminor axis d and observing $d^2 + c^2 = a^2$. Jim clearly demonstrates that a and d are the parameters inherent in the setup of the trammel, whereas a and c are the parameters inherent in the setup of the loop-of-string device. He now faces the problem of how to determine the focal distance (tack distance) from a given trammel setup.

Jim sees that he can use the Pythagorean relationship between a, c, and d to calculate c, and then use a and c to set up the loop-of-string device. He tries this and is reasonably satisfied with the results of this first duplication, but he continues searching for more compelling evidence that the curves being produced are, in fact, the same. He returns for a while to his claim that placing the trammel pen halfway between the two pins will produce a circle.

While examining various aspects of the motion of the trammel, Jim discovers a second way to find the foci of any trammel-drawn curve. This method involves using the trammel itself as a compass. After using the trammel to draw an ellipse with axes a and d, Jim takes the trammel out of the tracks, places one pin at the top of the semi-minor axis, swings the pen in an arc of radius a, and marks the places where this arc intersects the major axis (see Fig. 12.8). Because a and d are the lengths on the trammel, this process involves no readjustment of the trammel settings. The pen in the trammel works quite well as a compass, and this direct physical method locates the foci accurately without any calculation or numerical measurement. With the help of the interviewer, Jim holds tacks on the marked foci, places the loop of string on the Plexiglas sheet, and physically traces over the curve that he has just drawn with the trammel. The trace very accu-

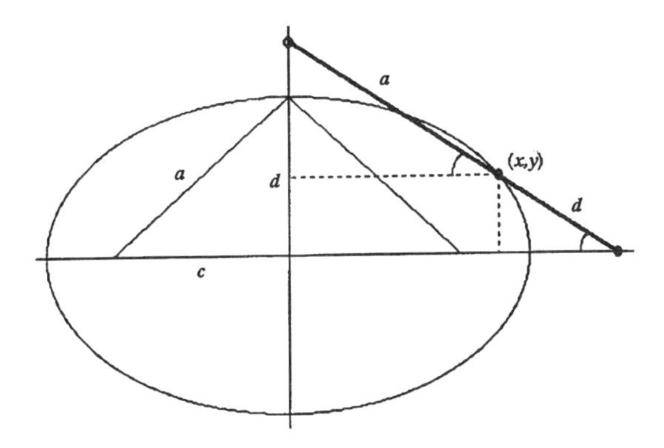

FIG. 12.8 Jim's first analysis of the trammel device.

rately matches the trammel-drawn curve. Again, Jim has ingeniously used the physical geometry of his tools to accomplish accurately mathematical tasks without the use of algebraic notation or calculation.

An Equation from the Trammel Device. Jim next attempts to confirm his physical, geometric experiences by algebraically representing the trammel-drawn curves:

Interviewer: Is there any other kind of argument that could really nail this down?

J: I Well, I'm guessing that the equation is going to be the same for both, since we have the equivalent pieces.

Interviewer: Is there a way to get an equation out of this device [the trammel] that talks about the geometry of this device?

Jim carefully studies the motion of the trammel over one quarter of the curve. He watches as the trammel moves toward its vertical position and observes that the pen and the pin move toward the vertical track "in constant ratio." He labels the pen as (x,y) and draws two dotted lines on the Plexiglas (see Fig. 12.8). Jim observes that the two right triangles with hypotenuses a and d are always similar for any trammel position. He discusses this idea in various ways and says that this "constant ratio" is what he sees as the most essential feature of the trammel's motion.

Jim is not at all sure how to express this idea in an algebraic statement. He spends quite a while physically pointing to lengths that he knows are proportional, using phrases like "this distance here" or "the base of that triangle." Jim paces around and looks at the figure and the curve from various perspectives. He takes off his glasses and appears deep in thought. He mutters repeatedly about points that move "in a fixed ratio." Eventually Jim copies his drawing onto a sheet of paper and begins labeling the lengths in the similar triangles (see Fig. 12.9). It takes a while before he puts in the two square-root expressions.

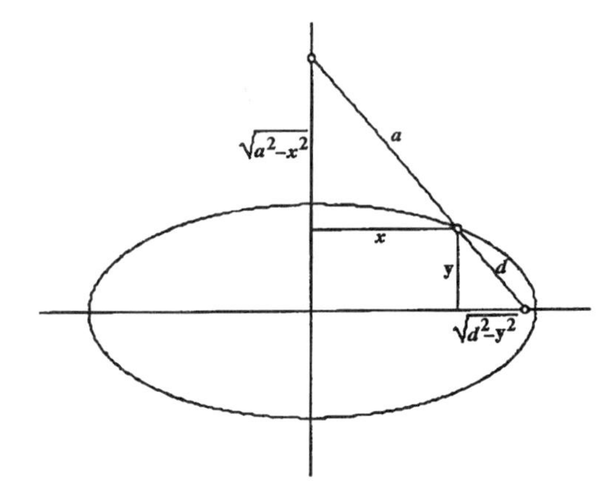

FIG. 12.9 Deriving the equation of a trammel-drawn curve: Jim's use of similar triangles.

Eventually Jim writes down a proportion from his figure:

$$\frac{a}{d} + \frac{x}{\sqrt{d^2 - y^2}}$$

At this point, the interviewer asks Jim what he has written:

Interviewer: Is that an equation for this curve?
J: I don't know.
Interviewer: Looks like an equation.
J: It's an equation, that's for sure [laughs]...but what's it saying...It's giving you...uhhhh...I don't see why not. I mean it's giving you this distance x and y, given an a and a d, which we can get from those things [points at trammel].
Interviewer: OK, so it's an equation that talks about this curve. Is it the same equation that we got over there with that string device? Is this equation equivalent to those two over there, or is it different?
J: It's got the same look to it as far as the ratios go...things like that...you know...the relation of the...but the thing is that there are no squares besides down here [indicates y^2 under the radical but no square on x or a]. Whereas on the other side over there [indicates: $\frac{x^2}{a^2} + \frac{y^2}{a^2 - c^2} = 1$, from loop of string] there are no square roots, there are just squared numbers.

Jim is hesitant to believe that this can be an equation for this curve because it looks very different from the reduced elliptic equation that he knows and because, as he notes later, it has been too easy to obtain. He is expecting some complicated use of the distance formula such as he has seen in class for the loop-of-string device. Using a proportion from similar triangles seems too easy to him. Jim is also very hesitant to perform any kind of algebraic manipulation. He says he is very "bad at algebra," and the thought of having to do algebra makes him very anxious. He mutters to himself in a foreboding tone, "Here come the rules." He stares at his new equation for a while, trying to decide what to do.

With a procedural hint from the interviewer (i.e., "Square both sides"), Jim's final derivation proceeds slowly as follows:

$$\frac{a}{d} + \frac{x}{\sqrt{d^2 - y^2}}$$

$$dx = a\sqrt{d^2 - y^2}$$

$$\frac{d^2x^2}{a^2} = d^2 - y^2$$

$$\frac{x^2}{a^2} = \frac{d^2 - y^2}{d^2} = \frac{d^2}{d^2} - \frac{y^2}{d^2}$$

Jim knows what he is trying to do. Once the radical is gone he immediately tries to obtain the term x^2/a^2 because it appeared in the other equation. Once he has that, he continues on and is pleased when the equation eventually appears as:

$$\frac{x^2}{a^2} + \frac{y^2}{d^2} = 1$$

He looks over at the loop-of-string equation:

$$\frac{x^2}{a^2} + \frac{y^2}{a^2 - c^2} = 1$$

and smiles. When asked about the difference between the two equations, Jim knows immediately that d^2 and $a^2 - c^2$ are the same. That, after all, is the geometric relation that he demonstrated so well when using the trammel as a compass:

J: I'm happy.

Jim's Epistemology. After verifying that the same equation could be derived from the two different devices, Jim discusses the sources of his beliefs and his views on the relations between geometry and algebra. For him the algebraic equations are not so much a proof that the curves are the same, but more a demonstration that the algebraic representation of curves is consistent with physical geometric experience, as illustrated in the following exchanges:

Interviewer: Does this convince you that the curves are the same?

J: [with resignation] Well, if they have the same equation, I guess I *should* be convinced.

Interviewer: But equations, deep down, don't seem to convince you very much. Is that what you're trying to tell me? Do I detect a skeptical note?

J: No, I'm happy. I mean seeing the equation the same makes me happy, but I was more convinced the first time I saw the similar...uhh...graph...or drawing...or whatever you want to call it...Well, I can't say I was more convinced...I was quite certain...I mean I took a large step when I saw the relationship between the drawing tool we had here [trammel] and the string over here, and getting those two to draw the same thing; I immediately thought, OK, they're doing the same operation. They're making the same kind of picture. Therefore, they're doing the same thing. They're operating in the same way. And they probably do have a similar equation. And getting that equation to work out, you know, confirms it...but it's not like it's a great shock...It's something I already knew...You know, I kind of assumed that it was like that.

Interviewer: So the physical experience was really a more convincing experience to you than an algebraic experience?

J: Well, not to belittle the power of the algebra to show you, without a doubt that it's like that, but I mean I was relatively certain... If you can look at my steps of certainty... I took a large step from here to here [gestures about 1 ft on the table] when I first saw it drawn out and I could get it to do the same thing... and from here to here [gestures about 2 in.] when I saw it [the algebra]... well, yeah OK... This being my total amount of certainty.

Interviewer: [laughing] I see... if you had to put it on a 1 to 10 scale? I'm looking at a ratio on your fingers there of ahhh . . . looks like maybe . . .

J: 8 to 2.

Interviewer: Eighty percent confident with the experimentation, and the algebra give you another 20% on top of that? Something like that?

J: [laughs and nods] Yeah... Actually it surprised me that I was able to get it so easily... But it worked out nicely. I guess doing it with similar triangles was a good idea I mean... it looked right.

Interviewer: That was what jumped out to your eye: these two triangles?

J: Yeah, I mean I saw them. When I approach something, I try to draw in everything that I can, so I can get an overall sense of what it's going to look like, and then look at each piece of it with the greatest amount of... uhhh... greatest degree of... uhhh... I want to have all the detail... So then I can look at the overall thing, and then look at each piece and how it relates to the overall drawing, rather than getting caught up in algebra [voice drops]. Algebra for me, it helps to make something certain and to give it a great deal of shape, but *the actual thought of how something's going to work out happens in the geometry* [student's emphasis].

Interviewer: I see. Geometry is somehow more deeply convincing to you?

J: [nods] Also much easier to understand the way that things interact with each other. Watching this piece move along like this [moves trammel along curve], and watching this decrease as this decreases [indicates two horizontal distances, one from the pen to the vertical axis and the other from the horizontally moving pin to the center], I can see that those are in a fixed ratio from watching this thing move.

Jim here reiterated exactly how he saw two points moving in a "fixed ratio." Even before Jim mentioned the pair of similar triangles in Fig. 12.9, his videotaped gestures clearly indicated that he saw pairs of points moving proportionally toward lines. The static figure with the similar triangles did not really convey how Jim experienced and "saw" this invariant relation. Mechanical dynamics was crucial to Jim's understanding of how these curves were being generated.

When Jim arrived at an algebraic equation, he was immediately aware that the equation represented a general ellipse and that this curve was consistent with his geometric experiments. Jim's personal confidence about the curves being the same was not based on achieving an algebraic result, but this confirmation of his experiments in another representation enhanced both his beliefs about the viability of his geometric methods, and (especially) his beliefs about the ability of algebraic expressions to coordinate with these geometrical methods. He very much wanted to see a clear symbolic confirmation of what he already believed. Far more than his beliefs about the curves being the same, the derivation of the equations greatly enhanced Jim's confidence in the language of algebra:

> *J*: It makes me feel good to get that!

We have quoted Jim at length here because he so articulately expressed the sources of his beliefs and how they related to each other. Jim here clearly took a view of mathematics that was at the heart of scientific revolution of the 17th century when one of the main issues was whether algebra could consistently represent kinetic mechanical geometry (Dennis, 1995). We invite the reader to compare Jim's epistemological and psychological statements with some from René Descartes' *Rules for the Direction of the Mind* (written in 1625, about 10 years before he would publish his famous *Geometry*, in which he analyzed many curve-drawing devices):

> Rule 13: If we understand a problem perfectly, it should be considered apart from all superfluous concepts, reduced to its simplest form, and divided by enumeration into the smallest possible parts.
>
> Rule 14: The same problem should be understood as relating to the actual extension of bodies and at the same time should be completely represented by diagrams to the imagination, for thus will it be much more distinctly perceived by the intellect.
>
> Rule 15: It is usually helpful, also to draw these diagrams and observe them through the external senses, so that by this means our thought can more easily remain attentive.
>
> (Descartes, 1625/1961)

CONCLUSIONS

Skills and habits of observation and investigation such as Jim's are unlikely to be engaged by our traditional mathematics curriculum. The ability to play and tinker and hypothesize in a physical setting is not often called for in mathematics classes. Even Jim's refined visual sense of ratio helped him only on few occasions, due to the paucity of physical geometry in our secondary curriculum. What passes for "context" in classrooms is most often sets of "word problems," which may describe a situation but rarely involve designing or physically experiencing a particular "context." Most "contextual problems" in mathematics curriculum are, in fact, decontextualized.

For example, the trammel involves the same action as a ladder sliding down a wall, a common rate problem in calculus, yet few teachers of mathematics know that the motion of any point on that sliding ladder is elliptical. We have asked many experienced calculus teachers this question, and although they were all familiar with the common rate problem, they were all very surprised that the motion of points on a sliding ladder is elliptical. The most common first guess is that the motion of points on the sliding ladder is hyperbolic (i.e., something like the graph of $y = 1/x$), followed by guesses that it resembled some kind of cusped curve, like Jim's star (see Fig. 12.5). Calculus teachers (ourselves included) have taught this "contextual rate problem" for years, without ever physically examining the action involved. Traditional mathematics curricula do not tend to develop strong instincts for motion, although the genesis of calculus was based on such experience (Dennis, 1995). A student like Jim is far more creative and inspired when given a physical action to control and observe. Jim could clearly see rates and "constant ratios" long before he could express them in algebra. For him, algebra had to emerge from geometrical motion before it became a viable form of expression. Historical records show that this process was also true for many mathematicians, particularly those (e.g., Descartes, van Schooten, Roberval, Pascal, and Newton) who were involved in the original creation of analytic geometry and calculus (Dennis, 1995).

The recent educational reform emphasis on "visualization" is still locked into an epistemic hierarchy where equations create curves, but rarely vice versa. Graphs are mostly seen as secondary facilitators that help students visualize an equation or numerical data (see, e.g., any of the articles in Romberg, Fennema, & Carpenter, 1993). Although such reform efforts contribute many important educational insights, they do not give truly independent status to different representations, and the approach to analytic geometry taken by Descartes and other 17th-century mathematicians is almost entirely absent. Even most "reformed" curricula fail to complete a cognitive feedback loop where multiple representations, including physical dynamic geometry, are given fully equal status.

Jim showed no strong inclinations toward algebra or traditional functional notation. He was much happier using statements about changing rates and the invariance of ratios that directly expressed his geometric vision. Jim preferred to see the ratios inherent in the dynamic system and to study their operation physically. Although he did not use functional language, his notion of the physical parameters that control and prescribe the motion of a device was astute, and his sense of how an equation "talked about" what is happening with respect to the motion of a device was well expressed.

Many students, like Jim, express a longing to return to geometry, as the piece of mathematics that they most love. We think that some experience, much earlier in the curriculum, with curve drawing and dynamic geometry would help to inspire them and give voice to their perceptions. In such a context, they might find a way to engage more profoundly their gift for seeing ratios. This experience could go a long way toward changing their attitudes about algebra and mathematics in general. Jim's beliefs were formed mainly by physical and geometric experience, as he so directly expressed (80%). For algebra to be meaningful to Jim, it had to be a careful and precise confirmation of what he had physically experienced. We feel that there are many more students like Jim for whom a belief in the viability of algebra would best evolve through coordination of symbolic expression with physical geometric experience.

Although Jim disliked algebra, the experience of connecting and confirming geometric experience with algebraic expression was both engaging and satisfying. Interviews with Jim point up the need to bring about a more balanced dialogue both between geometry and algebra and between physical experience and theoretical language. Jim could have benefited greatly from experiences with curve drawing long before he reached his senior year of high school. Curve drawing could have been introduced in middle school before the equations of curves were even mentioned. It could have been connected to many other empirical activities where curves are directly generated (e.g., sundials). Having a base of such grounded activity would have been beneficial in many ways. Although we have presented here only two curve-drawing devices, this approach to curves is quite general, in that all algebraic curves can be drawn with mechanical linkages, a theorem little known in the United States, even to professional mathematicians (Artobolevskii, 1964; Dennis, 1995).

Such a base might give many students an entirely different feeling about algebra. If they see algebra as a systematic language developed to allow for the expression of physical and mechanical visions, they are much less likely to come to see algebra as boring or fearsome. Even after having developed certain debilitating attitudes, Jim was still able to work clearly and precisely within a problem-solving situation where his visual skills were clearly valuable and connected to the problem. He did not want to avoid algebra at all costs. He wanted to see how it could express what he saw and validate what he experienced. By reversing the usual epistemic hierarchy where curves are defined from algebra, the curve-drawing devices gave him the stamina to work on a difficult problem. His physical certainty as to what would "come out" gave him the determination to try to confirm his beliefs within an algebraic representation.

If mathematical language is to become comprehensible to a broader audience, it must display early on its capacity for expressing a wide variety of situations. Most often in our curriculum, the linguistic form of mathematics (usually algebra) dictates in advance both the forms of classroom discourse and the allowable span of activities. Physical activities and "contextual problems" are introduced as examples or applications of preestablished linguistic skills and concepts. The language and symbolisms are not generated in response to student activity, but vice versa. Because symbolism usually dictates in advance the content of mathematics curriculum, students are only allowed to discuss activities that fit those forms, and often even simple "activities" are only discussed hypothetically and never materially explored. Even more disturbing is the way algebraic simplicity and convenience dictate which curves are admitted for discussion in mathematics classrooms. Algebraic simplicity and mechanical simplicity are not the same. Very simple devices are quite capable of drawing fourth and eighth degree curves (Dennis, 1995). The artificial curtailment of the objects allowed for classroom discussion creates the false sense that algebra is always the best way to go.

Such a situation severely disadvantages students like Jim. Their skills, thoughts, and epistemic inventions remain largely unengaged. Jim did not really hate algebra; what he hated was the way that linguistic rules had come to dominate the content of his mathematics courses. When language flowed from physical experience, Jim was quite ready to push very hard to coordinate and reconcile language with experience. As he said, "the thinking happens in geometry." Jim had a vision of what he expected of geometry, but that vision remained out of touch with school mathematics. Jim's vision was largely a 17th-century mechanical geometric vision, like that of Descartes and Pascal, which involved architecture, civil engineering, and mechanical devices. For example, Jim was disappointed that the geometry that he learned in high school never helped him even to begin to analyze the motion of the mechanical apparatus that reset pins in a bowling alley where he had worked.

Jim clearly benefited from his experience with these curve-drawing devices. His engagement with the curve-drawing devices was profound: They satisfied in him a longing for what he saw as the geometry of the world. We learned a great deal from watching and listening to Jim. Jim's phrase "these move in a fixed ratio," combined with his hand gestures, remain with us. They have already become part of our thinking about the learning and teaching of dynamic geometry.

If our curriculum is allowed to confront the uncertainties and ambiguities of how language interacts with the physical world—if mathematical language, symbols, and notations are allowed to grow directly from experiences and be shaped by them—then this fully circular feedback loop could evolve into a powerful epistemological model based on the coordination of multiple representations (von Glasersfeld, 1978). The algebra of equations and functions would then be more than merely just what Jim despairingly referred to as "the rules." More students would then be able to say genuinely, as Jim did at the end of his derivation, "It makes me feel good to get that."

REFERENCES

Artobolevskii, I. I. (1964). *Mechanisms for the generation of plane curves*. New York: Macmillan.

Confrey, J. (1993). The role of technology in reconceptualizing functions and algebra. In J. R. Becker & B. J. Pence (Eds.), *Proceedings of the 15th annual meeting of the North American Chapter of the International Group for the Psychology of Mathematics Education* (Vol. 1, pp. 47–74). San Jose, CA: Center for Mathematics and Computer Science Education, San Jose State University.

Confrey, J. (1994). A theory of intellectual development. *For the Learning of Mathematics, 14*(3), 2–8; *15*(1), 38–48; *15*(2).

Confrey, J., & Smith, E. (1991). A framework for functions: Prototypes, multiple representations, and transformations. In R. Underhill & C. Brown (Eds.), *Proceedings of the 13th annual meeting of the North American Chapter of the International Group for the Psychology of Mathematics Education* (pp. 57–63). Blacksburg, VA: Virginia Polytechnic Institute and Christianbury Printing Company.

Dennis, D. (1995). *Historical perspectives for the reform of mathematics curriculum: Geometric curve-drawing devices and their role in the transition to an algebraic description of functions*. Unpublished doctoral dissertation, Cornell University, Ithaca, NY.

Dennis, D. & Confrey, J. (1995). Functions of a curve: Leibniz's original notion of functions and its meaning for the parabola. *The College Mathematics Journal, 26*(2), 124–130.

Descartes, R. (1952). *Geometry.* (D. E. Smith & M. L. Latham, Trans.). LaSalle, IL : Open Court. (Original work published 1638)

Descartes, R. (1961). *Rules for the direction of the mind.* (L. J. Lafleur, Trans.). New York: Bobbs Merrill. (Original work published 1625)

Lenoir, T. (1979). Descartes and the geometrization of thought: The methodological background of Descartes' geometry. *Historia Mathematica, 6,* 355–379.

Romberg, T., Fennema, E., & Carpenter, T. (Eds.). (1993). *Integrating research on the graphical representation of functions.* Hillsdale, NJ: Lawrence Erlbaum Associates.

van Schooten, F. (1657). *Exercitationum Mathematicorum, Liber IV, Organica Coniccarum Sectionum in Plano Descriptione.* Leiden. (Original edition in the Koch Rare Books Collection, Cornell University, Ithaca, NY)

von Glasersfeld, E. (1978). Cybernetics, experience and the concept of self. In M. Oser (Ed.), *Toward a more human use of human beings* (pp. 109–122). Boulder, CO: Westview.

13

Conjecturing and Argumentation in High-School Geometry Students

Kenneth R. Koedinger
Carnegie Mellon University

There is a tendency to think of individuals who can discover new ideas or develop convincing arguments as having special "talent" or superior "intelligence." This view of conjecturing and argumentation abilities as fixed traits suggests that instruction directed toward such abilities is pointless for all but the most gifted of students. In direct contrast to that view, this chapter argues that successful conjecturing and argumentation performances are the consequence of particular skills and knowledge. In an appropriately structured learning environment, such skills can be acquired by anyone.

One reason for doubt regarding the instructability of conjecturing skills is our limited understanding of what these skills are. Developing a model of these skills is a key step toward creating effective learning environments for conjecturing. This model can then provide design guidance in creating elements of a learning environment: conjecturing tasks that appropriately challenge students and forms of assistance (including manipulatives, facilitative talk, and computer software) that support student learning. The learning approach being advocated here has important similarities with the Vygotskian notion of assisted performance (Vygotsky, 1978) and more recent variations like cognitive apprenticeship (Collins, Brown, & Newman, 1989).

This chapter presents a cognitive analysis of student conjecturing that includes a task analysis and an initial model of conjecturing, observations of student performance in a dynamic assessment, and modifications to the proposed model as guided by the results of this assessment. This cognitive analysis is then used to suggest forms of student assistance, including computer software tools, activities that draw out conjecturing skills, and facilitative "talking points" (hints or prompts), to help students through the most difficult terrain on their ways to conjecturing skill.

A TASK ANALYSIS AND PRELIMINARY MODEL

The cognitive analysis of conjecturing focuses on a representative conjecturing task called the Kite task, an adaptation of one that appears at the end of *Discovering Geometry* (Serra, 1989):

Kite Task—Part 1.

> A "kite" is a special kind of quadrilateral whose four sides form two pairs of congruent adjacent segments. In other words, a kite is a quadrilateral ABCD with AB congruent to CB and AD congruent to CD.
>
> Investigate these figures called kites using whatever tools you would like and discover and write down what must be true of every kite.

Kite Task—Part 2.

> Here is a statement that a student in another class made: "The diagonals of every kite bisect each other." Do you believe it? Why or why not?

The Kite task was designed as a dynamic assessment whereby students are given a difficult open-ended problem in the presence of an interviewer who plays the combined roles of assessor and tutor. The interviewer was prepared with a set of prompts or hints that could be cautiously given to students when they showed clear signs of being stuck. These prompts were organized in a hierarchy of goals created through a rational task analysis of the demands of the conjecturing task. The Kite interviewer's *only-as-needed* presentation of prompts ensures that we can see some signs of success from all students.

A Preliminary Task Analysis of the Goals of Conjecturing

As a first step toward understanding the goals of conjecturing and argumentation more generally, we performed a task analysis (Newell & Simon, 1972) to identify the major goals involved in performing the Kite task:

1. Draw an example of a kite.
2. Draw an example that is not overly specific (not a rhombus).
3. Make a conjecture.
4. Prove a conjecture.
5. Find a counterexample to reject a false conjecture.

We considered what knowledge was needed to achieve each goal and, in particular, in what ways students were likely to have trouble. The output of this analysis was a list of hints and prompts within each goal. These hints were the basis for our dynamic assessment. The initial hints within each goal are vague; subsequent hints get successively more specific. This approach has proven successful in our previous work on intelligent computer-based tutors (Anderson, Corbett, Koedinger, & Pelletier, 1995).

These goals and hints reflect our initial hypotheses about the nature of students' conjecturing and argumentation skills. One hypothesis was that getting started on this open-ended problem would be difficult for students who are used to 1-min. problems with a single right answer (Schoenfeld, 1989). Thus, the hints for the first goal ("Draw an example of a kite") prompt the student toward getting started: "Think about what a kite looks like" and then "Why don't you draw an example of a kite?"

A second hypothesis was that students were likely to create and investigate overly specific instances of kites, in particular, drawing a rhombus rather than a kite. Hints suggesting that students "Draw some more kites" and, more explicitly, that "Your kites all have four congruent angles" address this concern.

Third, we thought that focusing students' attention on individual geometric objects and on measuring these objects would help students make conjectures. Thus, within the third goal are hints like "Look at the angles in a kite" and "Use your protractor."

Fourth, we hypothesized that students might not spontaneously provide evidence for their conjectures and thus were prepared with the prompts "How do you know the things you wrote down are always true?" and "How would you convince someone else?" Further, we thought students' arguments might progress in sophistication from statements of self-evidence, like "I just know," through reference to a single supporting example, then to multiple examples, and finally to a deductive argument.

Fifth, based on the results of Senk (1983) we knew that even for students who were successfully writing geometry proofs in class, formulating a proof problem from a conjecture was likely to cause considerable difficulty. Thus, we included the subgoals "4.2.1 Identify a reference diagram," "4.2.2 Identify the goal statement," and "4.2.3 Identify the given statements" and provided a substantial number of prompts within each like "What do you know about kites? That should be your given."

Sixth, we thought that a substantial number of students would not generate a false conjecture and therefore not need spontaneously to engage in conjecture-defeating argumentation. Thus, we added a conjecture evaluation task (Part 2) to address the goal "Find a counterexample to reject a false conjecture."

THE QUALITATIVE STUDY

Method

About 60 students were interviewed at two sites in the Pittsburgh Public Schools. One school was using the *Discovering Geometry* textbook (Serra, 1989) with its emphasis on introducing students to geometry through inductive investigations often done in collaborative groups. The other school was using a traditional geometry textbook with its emphasis on introducing definitions and postulates, proving and applying theorems. About half the students at each site got a variation of the Kite task where we changed the term "kite" to "tike." Given the everyday meaning of the term kite, which is suggestive of a particular form, we decided to see what effect the term had on student behavior by comparing against a neutral term "tike."

Each of two interviewers performed two interviews during a 44-min. class period. Interview time was about 20 min. Students were randomly selected from the geometry classes and came to a separate, quiet office for the interview. The interviews were performed in late May at the end of a year of geometry instruction. During the interview, students were provided with the following tools: compass, ruler, protractor, pencil, and as much paper as they needed. Students were asked to think out loud, and we prompted them to "keep talking" whenever they fell silent (Ericsson & Simon, 1984).

Procedure. The interviewer began by reading the statement:

> [Student name], we are interested in how geometry students think about problems. I am going to give you a problem to work on and while you are working on the problem, I'd like you to think out loud. In other words, I'd like you to say what you are thinking as you are thinking it. You don't have to worry about whether everything you say is right or wrong, just tell me what you are thinking. Does that make sense? [Pause and answer any student question.] If you don't mind, we are going to record this. Of course, this data will be confidential and we will not use your name when we refer to it. Do you mind if we record this? [Pause and answer any questions.]
>
> Occasionally I may say "keep talking" if you become silent. Please talk loudly and clearly. [Start the tape.] For the tape, this is student ID number [student's ID number]. Here's the problem [give student the problem] and there's extra paper if you need it. Start by reading the problem out loud and then keep talking. [Start timing]

The student then read the problem statement, "Draw an example of a kite," and was asked to think aloud as he or she worked. Our goal was for students to do as much independent work as possible. However, when the interviewer judged that student was at an impasse, he or she would give one hint as guided by the interviewer form. Signs of a student impasse included long pauses, statements of confusion or frustration, or off-task behavior.

The interviewer checked off hints as they were given and, if necessary, indicated additional hints in the blanks provided. As much as possible, hints were chosen sequentially from the earliest goal not yet achieved and, within each goal, from the earliest hint not yet checked off. Students, however, would sometimes begin to pursue a "later goal" before completing an "earlier one," for example, beginning to make conjectures (goal 3) before having drawn a kite that is not overly specific (goal 2). If hints were needed in such cases, the interviewers would follow through on the student's goal if it made sense (e.g., if the student's difficulty was in how to achieve this goal). If, instead, the student wasn't sure what to do next, the interviewer would choose a prompt from the earliest goal that the student had not yet achieved. In general, the interviewers put priority on coherent interaction over strict adherence to the interview form.

The initial hints within each goal were vague, but further hints were progressively more detailed and directive. The interviewer treated each goal independently. A detailed hint might be provided on one goal, but on the next goal the student was left on his or her own unless the student showed clear evidence of another impasse.

The original plan was for interviewers to pace their hints such that students achieved goal 3 within 10 min. and goal 4 within 15 min. After achieving goal 4, the student was to be given Part 2 of the Kite Task, the conjecture evaluation question. We found this schedule difficult to maintain. Further, we soon recognized that most students were making false conjectures spontaneously; that is, goal 5 was being addressed in the context of the other four goals. Thus, the pacing designed to ensure an opportunity for students to evaluate a false conjecture proved unnecessary. We relaxed the timing constraints and, unless a student was going particularly fast, left off Part 2 in favor of giving students more time on the earlier goals.

Overall Results

Curriculum Comparison. The comparison between students using the *Discovering Geometry* text and those using a traditional text showed no substantial difference. Although it remains possible that a more complete and detailed quantitative analysis might yield some differences, we did not see the kind of qualitative differences that might be expected from the nontraditional approach of *Discovering Geometry.* This result should not be interpreted as critical of this particular text, as at least three mediating factors reduced the likelihood of an effect: (a) greater teacher experience using the traditional text, (b) great variability in the way different teachers implement the *Discovering Geometry* curriculum, and (c) high variability and generally poor preparation of this urban student population. This result *should* be considered as evidence for substantial difficulties in implementing curriculum reform in a way that yields substantial student achievement gains. It takes much more than a textbook.

Kite–Tike Manipulation. The Kite versus Tike manipulation showed a large difference in the initial example drawing (goals 1 and 2). Although both groups were equally likely to draw overly specific diagrams (discussed later), the kinds of diagrams drawn were distinctly different in the two groups. Figure 13.1 shows examples of the initial diagrams typical of the Kite and Tike groups. The Kite group (see Fig. 13.1a) drew figures that looked like rhombi with slanted sides such that the diagonals, if drawn, would be horizontal and vertical. The Tike group (see Fig. 13.1b), on the other hand, drew figures that looked like squares (sometimes rectangles

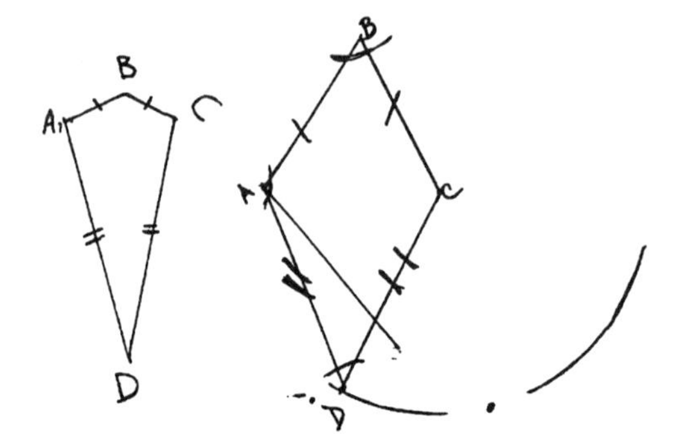

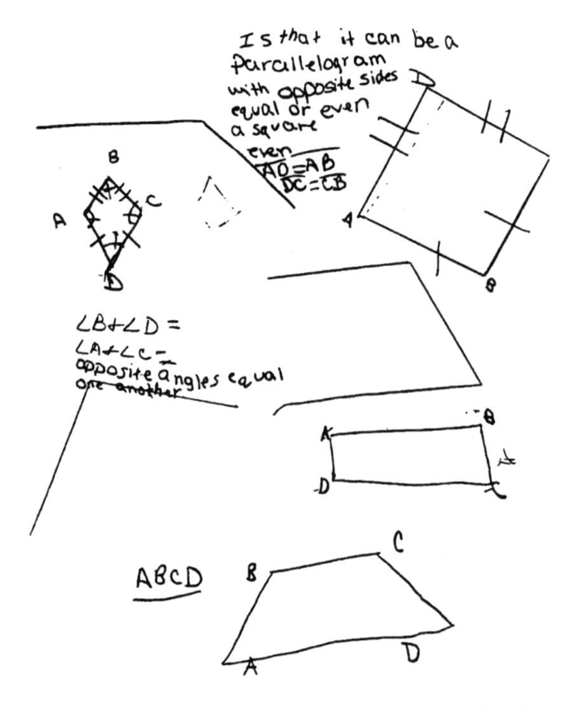

FIG. 13.1 (a) A sample of student work on the Kite task. (b) A sample of student work on the "Tike" version of the Kite task.

initially) with the sides drawn horizontally and vertically and the diagonals slanted.

Although quite strong, these differences were limited to the way the figures were drawn and had little bearing on students' success. The Kite condition might be expected to yield more fluent conjecturing because the specialized world knowledge cued by "kite" might help generate ideas. However, this expectation was not borne out. Similarly, there were no signs of an effect of the Kite–Tike manipulation on conjecture evaluation (goal 4).

Qualitative Analysis

We observed a wide range of performance in these urban high-school students. At one extreme, we saw that there was plenty of room for improvement in conjecturing skill—many students had difficulty right from the start in simply creating an initial example to investigate (drawing an example kite). At the other extreme, we saw a few students (2 or 3 out of 60) exhibit a full range of conjecturing skills as they successfully performed the task with little or no assistance, for example, by conjecturing and proving that the diagonals of a kite are perpendicular. In between these extremes, the dynamic assessment method allowed us to observe a wide variety of skill levels across students and across goals. Because we provided assistance only as needed, some students lacking skill in one aspect of conjecturing (e.g., example creation) were able to exhibit skill in another (e.g., conjecture induction). Although most students would have gotten practically nowhere on their own, almost all students, with the support of occasional prompts, were actively engaged in this task. Each student made reasonable progress, and many showed glimmers of talk and reasoning characteristic of mathematicians.

What follows is a descriptive analysis of the variety of student performances within each of the major goals. The focus is to provide a qualitative characterization of what conjecturing skills were or were not exhibited by students.

Goals 1 and 2: Drawing an Example That Is Not Overly Specific. As we anticipated, many students had difficulty getting started. Twenty-five percent of the students needed to be prompted before they began drawing an example kite. Students had even more difficulty in drawing a kite that was not overly specific. Only 15% did so without prompting, another 35% were able to do so on their own after some prompting, and a full 50% needed to be shown how to do so. Many of students' overly specific initial diagrams were the result of the same common pattern of behavior. Such students would first draw two sides of equal length that shared a vertex (Fig. 13.2a). When drawing a third side, they would invariably begin to draw it parallel to an existing side (Fig. 13.2b). The consequence is that when the student tried to make a fourth side equal to the third side, they were forced to create a rhombus. (If $1 = 2$ and $2 \parallel 3$ and $3 = 4$, then the figure is a rhombus.) Students' tendency to draw a rhombus appeared to re-

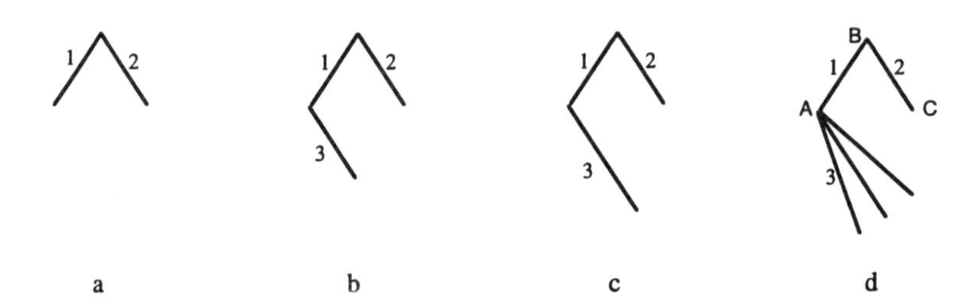

FIG. 13.2 A typical student's attempt at constructing a kite. (a) Sides 1 and 2 are drawn equal in length. (b) Side 3 tends to be drawn parallel to side 2 and thus, to make sides 3 and 4 equal in length, both sides 3 and 4 end up equal in length to sides 1 and 2. (c) With prompting to make sides 3 and 4 different lengths from those of 1 and 2, the student reaches an impasse. (d) The impasse is resolved by swinging the ruler to find a point where sides 3 and 4 intersect and are equal in length to each other, but not to sides 1 and 2.

sult less from a preconceived notion of what the complete figure should look like, than from a bias to draw parallel lines.

When prompted to draw a kite where sides 1 and 2 were different from sides 3 and 4, the following behavior of one student provides an illustration of the way many students struggled. This student picked 2 in. as the length of sides 1 and 2, and then decided that sides 3 and 4 should be 3 in. long. When he drew a longer side 3 (again parallel to side 2), he reached an impasse (see Fig. 13.2c). Putting his ruler between the unattached end points of sides 2 and 3, he saw that side 4 would not be 3 in.

At this point many students decided it wasn't possible to make sides 3 and 4 different from sides 1 and 2. But this student continued to experiment trying different positions for side 3 (Fig. 13.2d). Some found the right spot with such discrete experiments, but this student decided that he could swing his ruler, fixed at the intersection with side 1, to mark off all the possible points 3 inches from point A. Then he performed the same ruler-sweeping procedure from point C and so found the point at which sides 3 and 4 would each be 3 in. long. Indicative of how limited to particular temporal and situational contexts knowledge can sometimes be, this student reinvented the standard compass construction procedure (using his ruler) without recalling he had been taught this procedure the prior fall *and* even though there was a compass in the tool set in front of him.

Goal 3: Making a Conjecture. Students did not have much trouble in coming up with *some* conjecture, that is, in saying something they thought would be true of every kite. However, many of these conjectures were trivial, essentially repetitions of the givens (e.g., a kite "has four sides" or "has AB = BC and CD = AD"). Most students were able to make a meaningful, nontrivial conjecture, either on their own or with some prompting. However, 25% were unable to do so within 20 min. These students typically had much difficulty in the drawing phase and were weak in their knowledge of basic geometric concepts and notations (e.g., angles and how to label them).

Although we expected some students to draw overspecialized figures and, thus, be in a position to make false conjectures, we were surprised to see how common such conjectures were. The most frequent false conjecture was "$\angle B = \angle D$." Students who drew rhombi correctly observed that these angles looked equal in their overspecialized version of a kite. Some other false conjectures were also consequences of drawing a rhombus (e.g., "All the 'little triangles' are congruent" from a student who had drawn the diagonals).

Other conjectures, like "$\angle EAD = 60°$," seemed to have their source in the particular examples the student drew or as variation of a geometry theorem the student recalled (e.g., an equilateral triangle has 60° angles). This second kind of memory-based influence on conjecture generation is notable. It is not a form of inductive reasoning in the usual sense of generalizing from instances, nor is it deductive, deriving logically from existing postulates or theorems.

Goal 4: Prove a Conjecture. Almost all students seemed satisfied to stop after making one or a few conjectures from the example(s) they had drawn. Only a rare few showed any unprompted signs of thinking that further evidence was necessary or desirable (goal 4.1), and, in at least one such case, the student appeared to provide a proof more out of adherence to classroom habit than out of a self-motivated desire to validate his conjecture. More often, students seemed confused when asked for further evidence, as in the first hint for goal 4.1: "How do you know the things you wrote down are always true?" The next prompt, "How would you convince someone else?," was considerably more effective in eliciting an attempt to provide some justification.

Many students passed through the progression of first saying the example they had drawn was enough evidence and then, with prompting, that many positive examples are sufficient evidence. Most students had to be asked explicitly to write a proof "like you do in class" before they made such an attempt.

Students had clear difficulty in setting up proofs for their conjectures (goal 4.2). Even though they had been exposed to many proofs during the school year, they had difficulty expressing a conjecture in terms of the "given statements" and the "goal statement." Students' conjectures were usually stated just as a conclusion (e.g., "Angles A and C are always equal") without stating the premise (e.g., "In kite ABCD, angles A and C are always equal"). Accordingly, students found it easier to specify the goal of the proof problem, but establishing the givens, that ABCD is a quadrilateral with $AB = BC$ and $AD = DC$, was more difficult. Few students spontaneously referred to or drew and labeled specific figures as reference diagrams for the proofs.

Few students (about 10%) successfully formulated proof problems and began to work on them. One of the big difficulties in coming up with a proof (goal 4.3) was adding the segment BD (a "construction") to create congruent triangles. Even when told to draw this segment, a number of students needed quite specific hints in order to finish the proof (e.g., hints 4 and 5 in goal 4.3.1: "What methods do you know for proving [congruent triangles]?" and "Do you remember [Side–Side–Side]?").

At the other extreme, two students with little or no prompting were able to conjecture and prove that the diagonals of a kite are perpendicular. Interestingly, while working toward a proof of this conjecture, one student deduced a new conjecture she hadn't previously thought of. Although the discovery of a conjecture is usually the product of induction from examples, this student's work illustrates that proof itself can also serve as a discovery tool.

Summary of Observations

We made the following observations of students' performance on this conjecturing task:

1. Experimenting with a class of figures by constructing and examining examples is a difficult but significant skill for students.
2. Example construction is strongly biased by subtle perceptual (e.g., parallel-lines bias) and linguistic (e.g., analogical suggestiveness of the label *kite*) influences. It is difficult for students to break out of the "set" caused by these influences.
3. Generating conjectures is not difficult per se; however, many conjectures students generated were either trivial repetitions of the problem conditions (e.g., AB = CB), not particular to kites (e.g., has 4 sides), or false inferences from overspecialized examples (e.g., $\angle B = \angle D$).
4. Students have difficulty differentiating claims (conjectures) and evidence for those claims (argument). When an argument is elicited, it is much more likely to be in the form of failure to falsify, "Every time I tried it, it worked," than in the form of a deductive proof.
5. Formally stating conjectures is difficult.
6. Deductive proof can lead to new conjectures.

TOWARD A COGNITIVE MODEL OF CONJECTURING

Framework for a Conjecture Model

Figure 13.3 shows the hypothesized components of a cognitive model of conjecturing, organized in a goal-structure diagram. As indicated by the top goal, conjecturing skills are relevant not only for discovery, but also for problem solving and recall. The two major subgoals are Generate Conjecture and Argue For Or Against. This goal-structure depiction indicates the major component processes in an approximate ordering. The ordering, however, is only suggestive. These processes or strategies are not executed sequentially by students, by mathematicians, or by scientists; instead, they are opportunistically applied based on specific problem demands. It is also not the case that each subgoal has a unique process for achieving it. A particular subgoal may have multiple processes or strategies for achieving it. For example, to Generate Conjecture one can Analogize, Investigate, or Deduce. Conversely, a particular strategy may be useful for multiple subgoals; for example, Investigate can be used to Generate Conjecture or to Find Examples For or Find Counterexamples Against.

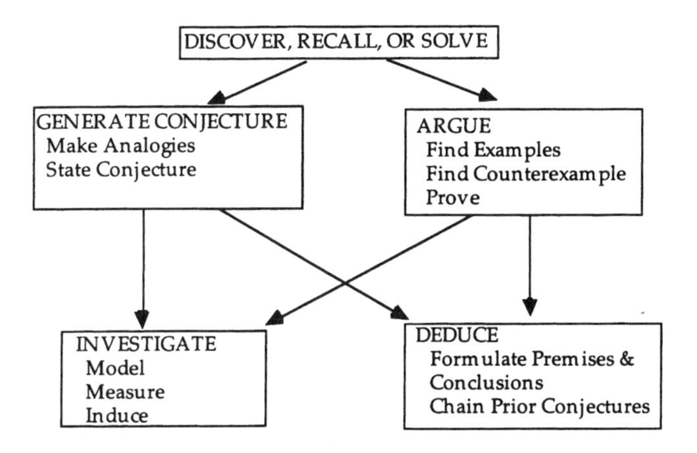

FIG. 13.3 The goal structure for conjecturing and argumentation skills.

There is a temptation to associate empirical investigation with conjecture generation and deduction with conjecture evaluation; however, evidence presented here and elsewhere suggests otherwise. Using terminology more common to science, Klahr and Dunbar (1988) presented a model of scientific discovery in which "experimentation" (a kind of Investigate) is relevant for both generating hypotheses (Conjectures) and for testing hypotheses (Examples For and Counterexamples Against). Chazan (1993) documented "student preference for empirical arguments over deductive arguments when presented with mathematical problems." In Chazan's interviews with high-school students, he found students were more likely to view empirical evidence as "proof" than they were to see deductive proof as such.

The role of deduction in conjecture generation was not recognized in these prior works. The possibility of deductively generated conjectures is clear from rational grounds. Postulates and theorems can be used to chain forward from the given premises of the investigation (e.g., In kite ABCD, AB = BC and AD = CD) toward new conclusions (ones not previously generated by other means). When these processes are used to aid conjecture recall or rediscovery, deduction provides an important avenue to reconstruction of forgotten conjectures. The opportunistic and sometimes redundant application of alternative strategies to achieve a variety of goals is a crucial aspect of mathematical understanding (Koedinger & Anderson, 1991; Tabachneck, Koedinger, & Nathan, 1994).

Conjecture Generation

Conjecture generation includes generating an idea for a conjecture, making sure it is "interesting," and making sure it is at least consistent with the examples that are being investigated. The simplest filter on the interest of a conjecture is that it at least say something new, something beyond repeating the premise. A number of students in the Kite study made trivial conjectures ("Kites are quadrilaterals" or "AB = BC and AD = BD") that simply repeated an aspect of the definition. It may be that students are unable to distinguish between such trivial conjectures and potentially more interesting ones, but it seems more likely that students were unable to access more sophisticated strategies for conjecture generation than simply repeating statements from the text.

A slightly more sophisticated strategy, although still quite weak, is to reason by analogy, using similar prior knowledge. We saw this approach in the Kite study when students conjectured the conclusion "$\angle ADC = 60°$" by analogy to equilateral triangles. Klahr and Dunbar (1988) also found analogy to be an important source for hypothesis generation in scientific discovery. The analysis here, however, focuses on two more-systematic approaches for conjecture generation, Investigate and Deduce.

Investigation Skills

One method for generating a conjecture involves investigating examples and inducing any relationships that appear invariant. In the Kite task, students drew examples of kites, inspected and sometimes measured them, and noticed relationships (e.g., the two sets of opposite angles look congruent, or the diagonals look perpendicular). More generally, to Investigate, one must Model, Measure, and Induce (see Fig. 13.3).

Model Creation. One way scientists begin an investigation is to explore examples or models in some general area of interest, like falling objects (to study gravity) or the *Escherichia coli* bacteria (to study germ propagation). An initial model can be created using that part of the conjecture that comes from the research question, the premise in a premise-driven investigation, or the conclusion in a conclusion-driven investigation. To investigate "If a figure is a kite, then X," we create diagrammatic models of kites and explore their characteristics. To investigate "If X, then success in college is more likely," we find models of college success (college graduates) and explore their characteristics. In both cases, it is also important to explore nonexamples (quadrilaterals that are not kites or students that didn't graduate) to avoid conjectures that apply more generally ("Kites have four sides" or "Students who can read are more likely to complete college").

An important general heuristic for modeling is to create or find models that are not overspecialized (i.e., having more features than is necessary). For instance, we don't use a quadrilateral with all sides equal to explore kites; we don't use doctoral students to explore characteristics of undergraduate success. Related to this "general model" heuristic is an "extreme model" heuristic, which suggests investigating extreme cases, like a concave kite (where D appears "inside" the figure). Most students in the Kite study did not invoke these heuristics. Many of their models were either overspecialized (e.g., all sides equal, parallel sides, or right angles) or prototypical (e.g., horizontal–vertical orientation, convex, average "thickness").

In many domains, like geometry, it is typical for researchers to construct models to investigate. To solve all but the simplest construction problems requires the following problem-solving steps: (a) decompose the problem into solvable subproblems, (b) compose and integrate the solutions to these subproblems, and (c) manage the inevitable interactions between subproblem solutions. For instance, in the Kite task where the construction problem is to create a quadrilateral ABCD with AB = BC and AD = CD, successful students decomposed the task into simpler problems of constructing segments. The subproblem of constructing segment AB equal to segment BC is solved with little trouble. The typical student used a ruler to draw two equal-length segments, say 2 in., that shared point B. The third and fourth segments, AD and CD, present an analogous subproblem and students typically solved it in the same way, which resulted in segments with symmetric orientations and the same lengths (i.e. , AD and CD are also 2 in. long). Applying the general-model heuristic adds the constraint that the pairs of segments should not be equal. Thus, the new subproblem is to make AD and CD, say, 3 in. long. By itself, this subproblem is as easy as the first (make AB and BC each 2 in. long); however, managing the interaction between the two subproblems, that points A and C must be shared, is difficult (as was illustrated earlier in this chapter).

Model construction is sometimes taught in science classes as part of instruction on the "scientific method" or, more specifically but less commonly, instruction on "experimental design." In geometry instruction, the methodology for model construction is typically compass and straight-edge construction. Traditional instruction has so isolated and overrefined this method, however, that its function in empirical investigation is lost on most students and teachers alike. Despite weeks of instruction on compass and straight-edge constructions, students in the Kite study did not employ this method when it was appropriate. However, students are capable of less formal approaches to model construction, including freehand drawing and the use of other measurement and drawing tools. Most students in the Kite study were able to create adequate diagrammatic models, but did so using the ruler and protractor.

By not presenting construction within the context of its use in investigations, traditional geometry instruction leads to student knowledge structures that tie construction methods only to teacher-imposed goals. Thus, students are not likely to access construction knowledge when it is really needed to aid recall or help solve a problem.

Measurement. In premise-driven investigation, generating a conjecture amounts to finding a conclusion that follows from the given premises. The focus of measurement is on possible conclusions. The student can measure such things as segment lengths, angles, and areas. Looking for constants or identities among such measures is a good way to find many interesting results. Although there are results that are difficult to rediscover empirically (e.g., the Pythagorean theorem), many conjectures in high-school geometry are "rediscoverable" through straightforward measurement strategies.

Measurement strategies aid in the generation of conjectures that are at least consistent with the examples that have been constructed. In the Kite investigation, many students made the conjecture that angles A and C were equal as the result of focusing their attention on these angles and noticing that their measures looked the same. Having started with an overly specific model (a rhombus), many students also made the consistent but false conjecture that angles B and D are equal. The investigation process of noticing equivalence in measurements is the same for both conjectures.

Induction. Koedinger and Anderson (1990) presented a detailed cognitive model of inductive competence in geometry as a key component of a model proof planning. We found that despite years of contact with geometric theorems, high-school geometry experts do not plan proofs by using heuristics to search iteratively through the space of possible theorem applications, as was previously thought (e.g., Anderson, Greeno, Kline, & Neves, 1981; Gelernter, 1963). Instead, they initially make an abstract proof plan by using perceptually based knowledge to lay out the key proof steps, leaving theorem application to fill in the details. They employ a type of inductive reasoning to do so.

A key component of this planning ability is also critical to conjecturing, namely, the use of perceptual knowledge of prototypical diagram configurations to "parse" diagrams into useful parts (e.g., triangle pairs, triangles, angles, and segments). This perceptual configuration knowledge provides a source of guidance in conjecture generation. It cues what parts of a diagram to look at, compare, and measure. Once a student sees a diagram in terms of relevant parts, how they overlap and interconnect, it is not difficult to go the next step and inquire what parts might have invariant properties alone or relative to other parts.

Using "Model and Measure" in Problem Solving. Investigation skills are relevant not only for conjecturing, but also for problem solving. A "model and measure" strategy can be used to solve problems without recourse to the deductive application of theorems. Thus, it is possible to use this strategy to solve problems prior to having learned the normative technique for doing so. Consider a high-school geometry student who does not know trigonometry and is faced with the following problem. He works for a small company building customized stereo speaker boxes and needs a method to figure out the length of boards needed to create a slanted roof on the top of the box. He knows the angle of the slant (20°) and the horizontal distance (10 in.). If he knew trigonometry, he could find the length of the top by solving "$\cos 20° = 10/x$" for x. Alternatively, the model-and-measure strategy can be used (see Fig. 13.4). Draw a scale model of the situation (1 cm = 1 in.) so that it fits the givens of the problem: that is, the width of the box is 10 cm and the angle of the slant (CAB) is 20°. Measure the desired quantity, in this case segment AC, which is 10.6 cm, and apply the scale factor. The top should be about 10.6 in.

Although this strategy sacrifices some precision and takes more time to perform than the deductive application of a theorem, it is quite suffi-

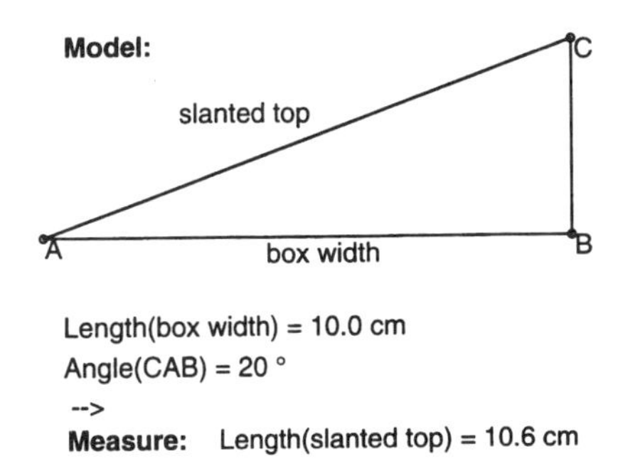

FIG. 13.4 Applying the "model and measure" strategy in problem solving. The goal is to find the length of the slanted top (AC) of a speaker box given the box width (AB = 10) and the angle of the slant ($\angle CAB = 20°$). The strategy does not require trigonometry: draw a scale model, measure the desired length, and apply the scale factor.

cient for many problem situations. More importantly, the model-and-measure strategy has three advantages over theorem application. First, it is more general—a student doesn't need to have learned a theorem to apply it. It can be applied, for example, to "trigonometry" problems prior to having learned trigonometry theorems. Second, it is easier to learn and apply than most theorems. For example, the model-and-measure alternative to the Pythagorean theorem does not require algebraic computations. Third and most importantly, the model-and-measure strategy can play a sense-making role in helping students learn new theorems, recall old theorems, and check solutions generated using those theorems.

Stating Conjectures

Articulating generalizations is difficult. Students can reason generally about unknowns prior to being able to articulate the generalization with which they are working (Koedinger & Anderson, 1996; Nathan, Koedinger, & Tabachneck, 1996). To do so requires self-reflection and extra cognitive resources that come only from practice. In the Kite task, students' initial conjectures tended to be stated simply as conclusions, with the premise presumed. For example, they didn't say "In kite ABCD, $\angle ADC = 60°$" but simply "$\angle ADC = 60°$"; not "Kites have four equal sides," but "four equal sides"; not "Opposite angles of kites are equal," but "opposite angles are equal." By assuming the premise and essentially forgetting about it, students are better able to focus their cognitive resources on what they need to measure and generate. However, once a likely conclusion has been discovered, the conjecture does need to be stated in full.

The preceding examples illustrate two different ways of stating conjectures: one, more symbolic, referring to point labels, and one, more natural. The use of natural language tends to express better the generality of the conjecture, whereas the symbolic form may contribute to some students' difficulty in distinguishing evidence and claim, namely, in thinking that the conjecture is about a particular figure rather than about kites in general (see Chazan, 1993). On the other hand, the natural-language form can become cumbersome and difficult to interpret. For instance, attempts to find a natural-language form for "In kite ABCD with AB = BC and AD = CD, angle A is congruent to angle C" can lead to jargon-laden statements like "The opposite angles formed by noncongruent adjacent sides of a kite are congruent."

Argumentation

Argumentation involves making empirical or deductive arguments for or against a conjecture. Mathematicians commonly convince themselves of a conjecture by searching for a counterexample—if they do not find one after significant time and effort, this itself is reasonable evidence for the truth of the conjecture, particularly when widely varying positive examples can be demonstrated. Extended failure to find a counterexample has been the key source of argument for the conjecture referred to as Fermat's Last "Theorem." There is a recent well-substantiated claim of a proof of this conjecture, but over the years the effort to find a proof has been less justified by a need to convince us of its truth and more justified as a way to facilitate the generation of new mathematical ideas and conjectures. Mathematicians also use deduction to aid conjecture generation. In reasoning outside of mathematics (and perhaps even within it), empirical arguments are much more common than deductive ones (Kuhn, 1991).

In the Kite task, students rarely provided an argument for their conjectures without some prompting. As Kuhn (1991) found, many students appear to lack epistemic knowledge that distinguishes between claims and evidence and, further, lack an appreciation for the value of evidence in convincing others. A key step in the development of argumentation skill is the recognition of the importance of multiple examples as providing better evidence than a single example. Testing a conjecture with multiple examples involves the same Investigate skills—modeling and measuring—discussed previously. In the Kite task, students often referred back to the instances they used in conjecture generation. When pushed for a convincing argument, it was typical for them to invoke measurement (Chazan, 1993).

Few students spontaneously attempted a deductive proof. When prompted to do so, most had trouble formulating their conjectures as proof problems. The skills required for proof formulation are similar to conjecture-stating skills discussed earlier. Skills for finding a proof were more fully addressed in Koedinger and Anderson (1990), in which we presented a cognitive model and computer simulation of expertise in high-school geometry proofs.

In addition to needing skills for performing investigation and deduction in the service of argument, students need knowledge to make conclusions from these strategies, decide between them, and check their conclusions:

Drawing conclusions:

- If I find a counterexample, the conjecture is false.
- If I find a proof, the conjecture is true.

Switching strategies:

- If many attempts to find a counterexample fail, perhaps the conjecture is true, and I should try to prove it.
- If I can't find a proof, perhaps the conjecture is false, and I should look for a counterexample.

Checking for errors:

- If I've found a counterexample, I should still check that I cannot find a proof.
- If I've found a proof, I should still check extreme examples to make sure there is no counterexample.
- If I seem to have a proof and a counterexample, perhaps I've formulated the proof problem wrong or made incorrect measurements in my counterexample.

This higher-order knowledge about the nature of argument is lacking in many students.

The model of conjecturing proposed here is intended to fall somewhere between a descriptive model of existing student conjecturing skills and a normative model of what these conjecturing skills should be. The attempt is to characterize the edge of high-school student competence, discover which skills are present and which are lacking, and provide a direction for instructional design efforts. In the following section, we discuss some instructional implications of the model and, in particular, the role of modern software in learning conjecturing.

MODEL-BASED DESIGN OF CONJECTURING ACTIVITIES AND SOFTWARE

This section provides suggestions for instructional activities and software directed at helping students acquire better conjecturing and argumentation skills. Most of the software suggestions concern ways to use existing tools, but I also propose the design of new tools to address aspects of conjecturing not well addressed by existing software.

Activities and Software to Enhance Investigation

This section illustrates suggestions for software-related activities for Investigation using the Kite activity as an example. Such an activity could also be pursued with paper and pencil. I discuss how the activities differ and what advantages and disadvantages the technology brings.

Model Construction and Measurement Activities. I argued earlier that instruction in traditional compass and straight-edge construction is too isolated and overemphasized in typical high-school curricula. Other means of construction can get students more quickly involved with activities directed toward more general and powerful conjecturing and argumentation skills. The common placement of construction at the beginning of courses does not allow students to acquire the goal structures that frame construction techniques as useful knowledge. Instead students acquire construction skills within an arbitrary school-imposed goal structure and, as we saw in the Kite study, are not able to access these skills when they are truly needed. By starting with simpler approaches to construction, students can learn, first, when and where construction skills are needed. More specialized approaches to construction can emerge later in the curriculum with "pull" activities, in which there is a felt need for more efficient methods or more precise diagrams.

I use *diagram drawing* to refer to the simplest approach to diagrammatic modeling, in which figures are drawn by any method or tool set, and *diagram construction* to refer to the use of Euclidean methods and a restricted tool set. Examples of paper-based diagram drawing, from freehand to marked-ruler use, were given earlier. Students in the Kite study showed a substantial preference for diagram drawing over construction. When construction occurred, it was often prompted by other needs (e.g., to create a nonrhomboid kite).

Computer-Based Diagrammatic Modeling. The distinction between diagram drawing and diagram construction in computer-based tools is illustrated in Figs. 13.5 and 13.6. The kite in Fig. 13.5a was drawn, whereas the one in Fig. 13.6a was constructed. The steps to create Fig. 13.5a were (a) use the segment tool to draw four connected segments, (b) use the distance-measure tool to measure these segments, and (c) use the selection/move arrow to adjust the points until the segments AB and BC are equal, and AD and CD are equal. The steps to create Fig. 13.6a are analogous to a Euclidean construction: (a) draw a circle with center B that will determine the lengths of segments AB and BC, (b) draw a second circle with center D that intersects the first and determines the lengths of segments AD and CD, and (c) draw segments from the circle centers, B and D, to the two points where the circles intersect (points A and C).

Tools like Geometer's Sketchpad (Jackiw, 1991) make it easier to perform lower-level drawing, construction, and measurement steps. Thus, students can better focus on conjecturing and argumentation. Another advantage of these tools is that diagrams are "dynamic," making it easy for students to perform numerous geometry experiments. The diagrammatic models in Figs. 13.5 and 13.6 are dynamic in two ways: (a) the diagrams can be moved with their structural features maintained, and (b) the measurements of parts are dynamically updated. I illustrate two simple experiments with the drawn and constructed kites in Figs. 13.5 and 13.6.

Diagram drawing, besides being easier to do, has the advantage of facilitating the identification of "trivial conjectures" described earlier. Figure 13.5b shows the result of moving point A to modify the diagram in Fig.

a

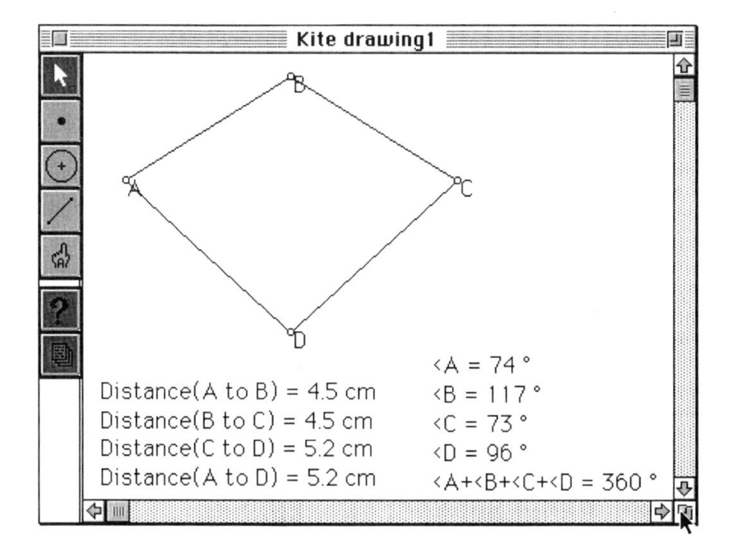

b

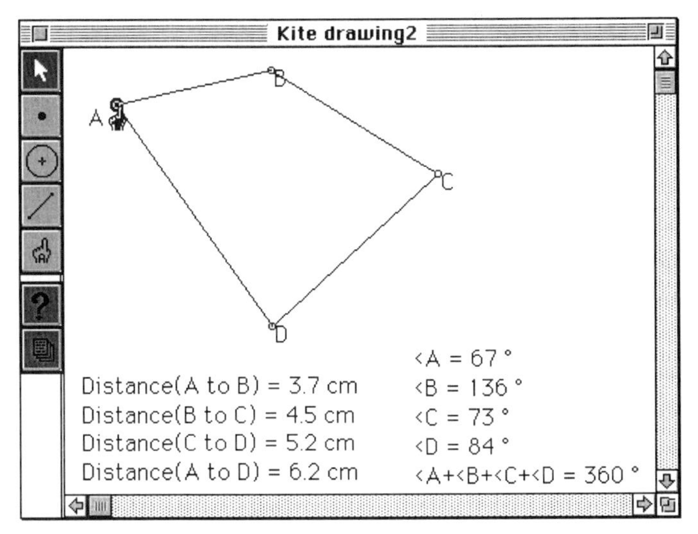

FIG. 13.5 Kite drawing (student work). (a) The points have been adjusted so that AB = BC and CD = AD. This example is consistent with the conjecture "The angles of a kite add to 360°." (b) Moving point A creates a figure that is no longer a kite. The angles still add to 360°, indicating that this property is not particular to kites.

13.5a. The diagram is no longer a kite, but one of the conjectured conclusions about kites, that the angles sum to 360°, still appears to be true. Thus, the experiment shows that this conclusion is not a characteristic particular to kites.

The disadvantage of diagram drawing is that it is difficult to get the premises exactly right. (In some cases it is impossible; e.g., an equilateral

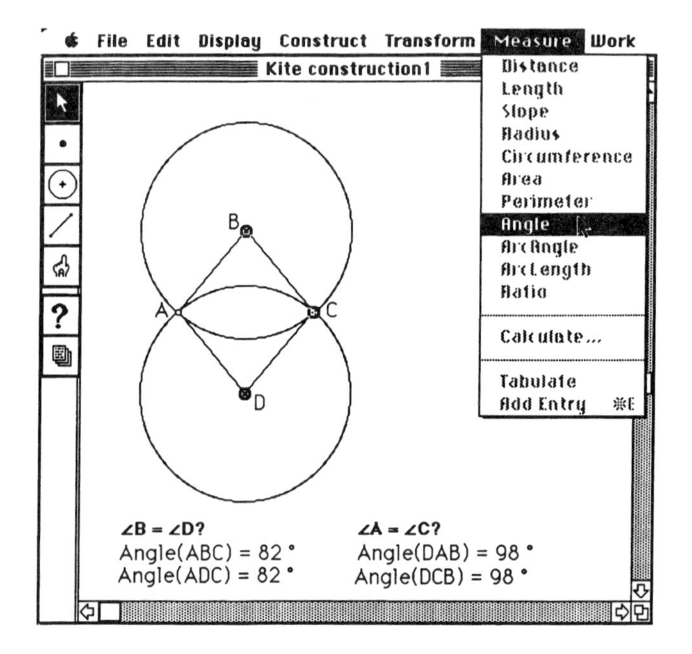

FIG. 13.6 The construction and measurement of a kite in Geometer's Sketchpad. (a) The Measure menu is illustrated, showing that when three points are selected (B, C, D in order) the angle formed by them can be measured. The result is recorded on the screen (e.g., "Angle(DCB) = 98°" at the bottom right). (b) The result of experimenting by moving point D. Sketchpad updates the measurements: Now angles B and D are no longer equal, but angles A and C still are. The student has hidden the circles used in the construction to better focus on the figure itself.

triangle cannot by drawn precisely on the pixel grid of a computer screen). A student must carefully move points and watch the measurements change until the conditions of the premises are met. Further, each new experiment usually requires that the same painstaking process is performed again. For instance, to get Fig. 13.5b back to being a kite, say a nonconvex one, would require moving points A and D until both AB = BC and AD = CD again.

Diagram construction like that in Fig. 13.6a facilitates experimentation because the diagram not only *looks* like a kite, but it *behaves* like one too. Sketchpad animates the movement of the diagram while maintaining the constraints of the construction: Although other things may change, segments AB and AC always stay equal because they are radii of the same circle—similarly for segments BD and CD. Sketchpad also dynamically updates the measurements as the diagram moves. Thus, in Fig. 13.6b we see the result of moving point D and the updated measurement values. Note that the student has "hidden" the circles used in the construction (see Fig. 13.6a). The figure still behaves in the same way, but this "hide" feature of Sketchpad allows the student to better focus on the figure itself. As a result of this experimentation, students may begin to modify their initial conjecture ideas and decide that although two of the opposite angles ($\angle A$ and $\angle C$) are congruent, the other two ($\angle B$ and $\angle D$) are not.

One can see from this example that Sketchpad simplifies the process of investigation in that a single construction can serve as a generator for multiple experiments. This kind of functionality is a general feature of many modern computational tools (e.g., spreadsheets, computer-aided design), in which setting up a system of constraints gives the user the power to automatically address "what if?" questions. Such dynamic models are a clear advantage over paper-based tools where exploring multiple models is considerably more time-consuming.

A Prototype Intelligent Tutor Agent for Investigation. We created a prototype cognitive Tutor Agent for diagram construction in Sketchpad (Ritter & Koedinger, 1995). Figure 13.7 illustrates the prototype system on the first exercise in chapter 3 of *Discovering Geometry:* copying a segment. The window at the top is Sketchpad, and the Messages window below is provided by the Tutor Agent.

Instruction in investigation, whether from teacher, text, peer, or computer tutor, needs to address two general classes of student errors: those involving the use of the tool, and those involving the mathematical content. The Tutor Agent helps with both types of errors. Providing automated remediation of low-level errors frees both students and teachers to focus on higher level content issues. Students may also get stuck in problem solving. As illustrated in Fig. 13.7, students can request a hint from the Tutor Agent, which can provide successively more specific hints, much like those in the Kite Interview Form, but specialized to each student's particular approach to a problem.

To summarize, computer-based tools for diagrammatic modeling such as Geometer's Sketchpad offer some increased benefits and reduced costs over paper-based tools. The key benefit is that computer diagram-

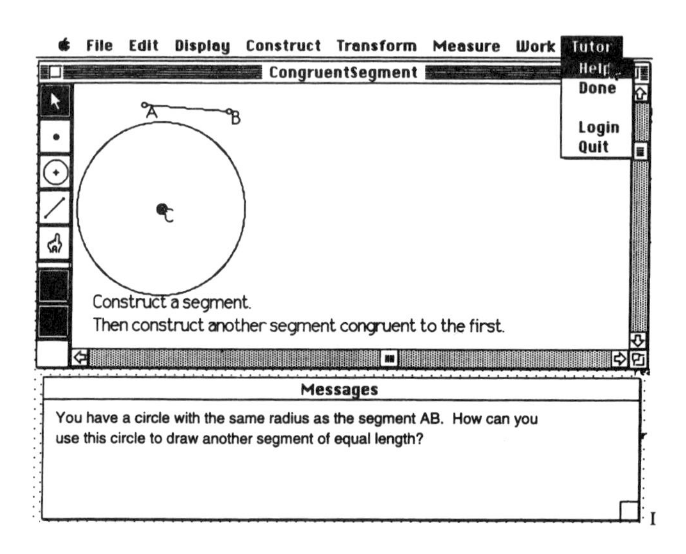

FIG. 13.7 Geometer's Sketchpad and an intelligent Tutor Agent for geometric construction. The rightmost Tutor menu was added to Sketchpad and allows the student to request hints from the Tutor Agent. Hints then appear in the Messages window.

ming allows for the creation of a dynamic model through which investigation is greatly facilitated by making it easy for students to modify the diagrammatic model. On the cost side, many low-level concerns are reduced or eliminated by providing automated drawing, measurement, and construction capabilities. Errors due to compass slippage or protractor misalignment are eliminated. Computer diagramming does have its own unique tool errors, though. Such difficulties can be intimidating to students and often even more so to teachers, who may encounter conflicts with their paper-based construction methods. The creation of intelligent Tutor Agents can help reduce these barriers for both students and teachers.

Activities and Software to Enhance Conjecturing

Although software tools for the other skill components are available, there is nothing that directly supports conjecturing. What is needed is a "Conjecture Editor" tool that aids students in formulating conjectures and recording them for future reference. By analogy to structured editors in computer programming environments, a Conjecture Editor could provide support through displaying a list of context-sensitive conjecture templates. The need to support students in conjecture formulation is evidenced by the approach in the *Discovering Geometry* textbook, in which students are given nearly completed conjectures to complete. For example, in the first discovery, which investigates the properties of the angles formed by crossing lines, students are prompted to make a conjecture that starts: "If two angles are vertical angles, then the angles are____." Teachers whom I have worked with at the urban schools report that when students are not given such prompts, they have great difficulty in expressing their conjectures. However, such prompts have negative cognitive consequences. Students are not engaged in the target skill of conjecturing, but in a fill-in-the-blank task. Further, students lose the experience of creation and discovery that conjecturing activities are intended to achieve. A Conjecture Editor could preserve the desired role of student as creator yet provide subtle support to make the task within reach of students.

Activities and Software to Enhance Argumentation

Following students' greater readiness for empirical argument over deductive proof, initial activities should emphasize empirical argument. Such arguments are a natural extension of conjecturing activities, but students need to be pressed to provide multiple examples in support of their conjectures. Further, the importance of argument processes will not be recognized by students unless they have numerous experiences with conjectures that seem true but are false. We have used what we call "Truth Judgment" activities to engage students in argument processes by asking them to evaluate conjectures that may or may not be true: for instance, "Is the following conjecture always true: 'The diagonals of a quadrilateral cross.' " The conjecture-evaluation skills needed for Truth Judgment problems are skills scientists use in reviewing a paper or listening to a talk. Related skills are used in evaluating claims in a newspaper article or a political speech.

Both the Investigate and Deduce strategies are relevant to Truth Judgment problems, but students are more likely to engage initially in investigation. A typical student or class might begin by arguing the conjecture just given is true by drawing a diagrammatic model of a quadrilateral, likely a rectangle, with crossing diagonals. When pressed "Do the diagonals always cross?," students will draw multiple quadrilaterals. Because of a bias toward convex quadrilaterals, it is likely that many students will become convinced the conjecture is true. Teachers need to wait and fight their urge to help, giving only vague hints if necessary (e.g., "Have you tried all kinds of examples, including weird ones?"). The more time it takes for a student to find a counterexample and convince others, the deeper will be the lesson. Activities like this, where intuitions are misleading, enable students to experience the utility of the argument process. In traditional geometry classes, students get few such opportunities because the predominant activity is to prove conjectures that are already known to be true.

Earlier we illustrated the use of Geometer's Sketchpad to support the Investigation process to help both generate and test conjectures. In previous work we built a software learning environment (ANGLE), which provides a graphical interface for discovering geometry proofs and online intelligent help (Koedinger & Anderson, 1993a, 1993b). Here we illustrate how ANGLE supports the Deduce process, not only for the usual purpose of verifying a conjecture, but also in finding evidence against a conjecture and in facilitating the discovery of a new conjecture. Figure 13.8a shows

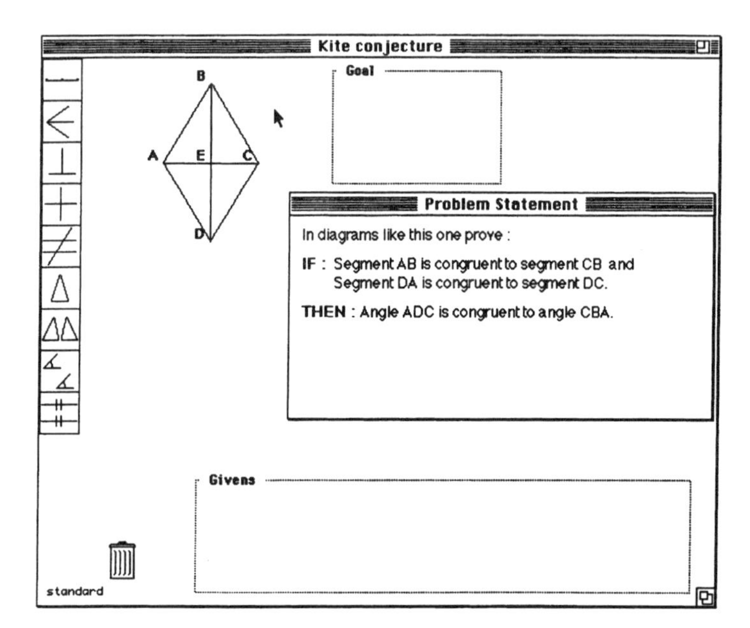

a

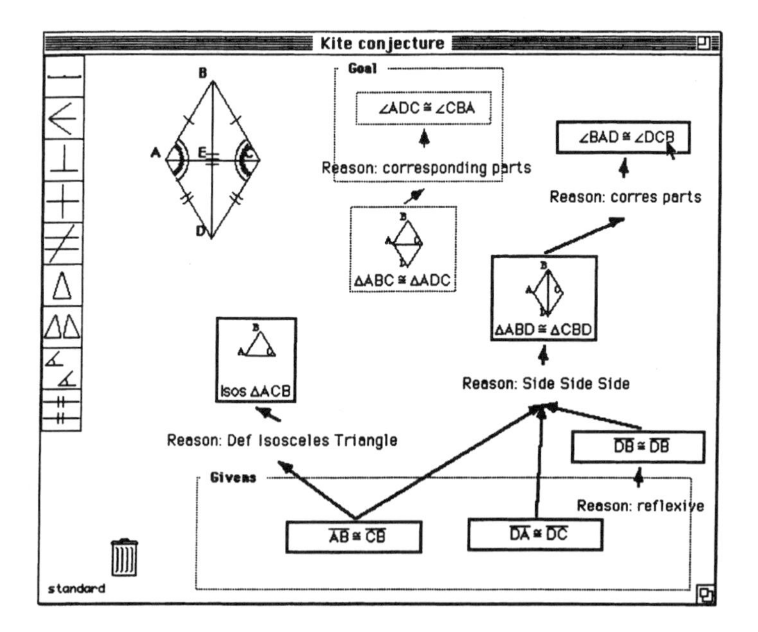

b

FIG. 13.8 Student work in ANGLE, an editor and intelligent tutor for geometry proof. (a) The student has entered a conjecture. (b) The student has been unable to find a proof for the entered conjecture ($\angle ADC \cong \angle CBA$), but the attempt leads to a discovery (and proof) of a related conjecture ($\angle BAD \cong \angle DCB$).

the ANGLE screen with a student-entered conjecture for the Kite problem—the most frequent false conjecture from the Kite study. Figure 13.8b shows the screen some time later. After attempts to work forward from the given (e.g., proving triangle ACB is isosceles) and backward from the goal (e.g., posing the subgoal to prove triangles ABC and ADC congruent), the student can find no way to prove her conjecture. In the process, however, she deduces a new conjecture by finding she can prove triangles ABD and CBD congruent and, in turn, angles BAD and DCB congruent.

Students can learn from independent use of ANGLE, but ANGLE is more effective when well integrated into a classroom curriculum. In a laboratory study where students worked independently with ANGLE for a total of 8 hr, we found their proof-writing achievement increased dramatically (by 60%) from pretest to posttest (Koedinger & Anderson, 1993a). In a classroom study, students using ANGLE and Truth Judgment problems with an experienced teacher performed one standard deviation better than students in classes with the same teacher but without ANGLE (Koedinger & Anderson, 1993b).

The Conjecture Editor, described earlier, could be combined with ANGLE to create a "Conjecture Manager" tool. This tool not only would provide students with the capabilities to enter conjectures and proofs, but also would provide a way of recording new discoveries and building on old ones. As students progress through the curriculum, they should essentially "build a book" (see Healy, 1993) by recording conjectures, labeling some conjectures as postulates, and proving other conjectures using these postulates. Figure 13.9 shows a mock-up menu for the Conjecture Manager. The Conjecture menu would allow students to "enter" new conjectures, "see" previous conjectures, promote some as "postulates," and "prove" others as theorems.

Choosing the Prove menu item and selecting a conjecture would initiate ANGLE, in which students would draw a reference diagram, enter the conjecture with reference to this diagram as Givens and a Goal, and then search for a proof. The National Council of Teachers of Mathematics (NCTM) Standards emphasize developing "short sequences of proofs" that show how one conjecture can support another (NCTM, 1989). A Conjecture Manager would facilitate a build-a-book approach (Healy, 1993) in which students could create their own sequencing of definitions, postulates, and theorems in geometry.

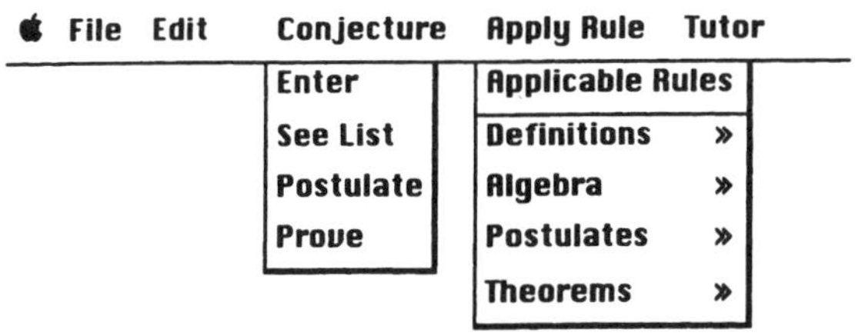

File	Edit	Conjecture	Apply Rule	Tutor
		Enter	Applicable Rules	
		See List	Definitions »	
		Postulate	Algebra »	
		Prove	Postulates »	
			Theorems »	

FIG. 13.9 Menu items in the proposed Conjecture Manager.

CONCLUSION

I have presented a model of conjecturing and argumentation skills based on the analysis of student problem solving in an open-ended conjecturing task. Conjecturing is the process of generating generalizations about a class of phenomena, whereas argumentation is the process of finding support for a generalization. Both processes are supported by two complementary reasoning strategies: model-based investigation, and rule-based deduction. Conjectures can be generated by investigating models of the phenomena and inducing new conjectures, or by chaining together prior "rules" (well-supported conjectures) and deducing new conjectures. Similarly, arguments for a conjecture can cite inductive support in positive examples (models) of the phenomenon, or deductive support in the chaining together of prior rules that show the logical link between the premise and conclusion of the conjecture.

This model of conjecturing and argumentation has similarities to cognitive models of scientific discovery (e.g., Klahr & Dunbar, 1988; Langley, Simon, Bradshaw, & Zytkow, 1987) and argumentation (e.g., Cavalli-Sforza, Lesgold, & Weiner, 1992). It is distinguished from these previous models in its integration of both empirical and deductive methods for discovery and argumentation.

In this chapter, I have attempted to provide evidence that the ability to discover new ideas and develop convincing arguments is not talent, but the consequence of particular skills and knowledge. The importance of such skills goes beyond mathematical discovery; conjecturing and argumentation skills are also relevant for nonroutine problem solving and for aiding recall. The model-and-measure strategy (see Fig. 13.4) illustrates the application of conjecturing skills in problem solving. Recall can be aided by both conjecturing and argumentation skills because forgotten facts can be rediscovered or rederived.

Not only are conjecturing and argumentation skills relevant for tasks besides discovery, they are also relevant in other domains. Further, there is evidence that such skills are lacking in the adult population. Many of the reasoning difficulties we observed in the Kite study show up in the reasoning patterns of adults arguing for claims outside of mathematics. Kuhn (1991) studied the argumentation skills of 160 U.S. adults who were asked to defend claims about familiar social policy issues (e.g., causes of school failure or unemployment). She found that more than half have poor reasoning abilities; in particular, they fail to make arguments that rely on evidence or accepted generalizations in a sound way. Like many students in the Kite study, adults in Kuhn's study tended to cite single examples (often personal experiences) as sufficient support for their claims. Both groups showed signs of confusion about the difference between claims and evidence. In the Kite study, many students had difficulty seeing conjectures as more than statements about a particular diagram, that is, as general claims about all kites. In the Kuhn study, many adults talked about claims and examples as one and the same (e.g., I believe school failure results from poor teaching because when I was in school I had a lot of bad teachers). Kuhn

(1991) commented that for many people "the evidence is not sufficiently differentiated from the theory itself" (p. 285). She suggested that "to progress beyond the fusion of theory and evidence to the full differentiation and coordination of the two requires... thinking about theories, rather than with them, and thinking about evidence and its bearing on a theory, rather than merely being influenced by it" (Kuhn, 1991, p. 285).

If students who pass geometry tend to acquire better abilities to differentiate and coordinate claims and evidence, then Pelavin (1990) may be correct in his claim that high-school geometry provides a unique opportunity to learn fundamental reasoning skills that are important for college success. This claim is consistent, though not uniquely so, with the results of his College Board study (Pelavin, 1990) where he found that among a number of other potential factors, passing high-school geometry is the best correlate with college success. An alternative explanation for this correlation is that high-school geometry courses are uniquely difficult and thus have served as a selection filter through which flow only those students who enter the course with the reasoning and learning skills needed for success in this difficult course and in college. Between these extremes is the likely reality. Some general reasoning and learning skills are probably improved through successful experiences in geometry. We need further empirical and theoretical research (cognitive models) to identify clearly what these skills are and the extent to which they can be acquired generally (i.e., not tied to the domain of instruction). Attempts to teach general reasoning skills have often failed, but successes at "transfer" have had in common a clear identification of the targeted general skills, often in the form of a cognitive model, and this model has been the basis for careful instructional design (e.g., Klahr & Carver, 1988; Lehrer, Randle, & Sancilio, 1989; Schoenfeld, 1985; Singley & Anderson, 1989).

One example of a class of skills that appears relevant both in geometry and more generally is skills for "set breaking" that can be used to avoid rigid thinking and lead to more flexible problem solving (Luchins & Luchins, 1959). Geometry activities can help students address their own biases toward prototypical figures and "perceptual set." How well general set-breaking skills and heuristics can be acquired in geometry is not clear, but it is clear that such skills cannot be acquired without opportunities that illustrate the need for them. The Kite problem is one such activity. It provides a good context for instruction on heuristics like the "general model heuristic": If your goal is to construct and investigate an example or model of a phenomenon (e.g., a kite), then construct as general an example as possible—avoid adding properties or constraints that are not required (e.g., making all four sides equal).

Designing instruction to help students learn discovery skills should be distinguished from a discovery learning approach. Just giving students discovery tasks and little support can lead to interesting results, but is too slow and frustrating for many students (Healy, 1993). Instead, students should learn discovery skills "by doing" in a supportive environment deliberately designed to achieve the skills outlined in the cognitive model. The elements of this environment should include well-chosen activities that elicit student thinking and classroom debate, well-timed instructor-facilitated introduction of better conjecturing and argumentation techniques, and the use of modern software tools that facilitate the conjecturing process and that include, where possible, the additional just-in-time support of a computer-based Tutor Agent.

ACKNOWLEDGMENTS

This research was supported by a minigrant from the National Center for Research in the Mathematical Sciences Education at the University of Wisconsin–Madison. Thanks to Jeremy Resnick for his major contributions to the Kite study and to Rich Lehrer for extensive comments on earlier drafts.

REFERENCES

Anderson, J. R., Corbett, A. T., Koedinger, K. R., & Pelletier, R. (1995). Cognitive tutors: Lessons learned. *The Journal of the Learning Sciences, 4*(2), 167–207.

Anderson, J. R., Greeno, J. G., Kline, P. J., & Neves, D. M. (1981). Acquisition of problem-solving skill. In J. R. Anderson (Ed.), *Cognitive skills and their acquisition.* (pp. 191–230). Hillsdale, NJ: Lawrence Erlbaum Associates.

Cavalli-Sforza, V., Lesgold, A., & Wiener, A. (1992). Strategies for contributing to collaborative arguments. In *Proceedings of the fourteenth annual conference of the Cognitive Science Society* (pp. 755–760). Hillsdale, NJ: Lawrence Erlbaum Associates.

Chazan, D. (1993). High school geometry students' justifications for their views of empirical evidence and mathematical proof. *Educational Studies in Mathematics, 24*(4), 359–387.

Collins, A., Brown, J. S., & Newman, S. E. (1989). Cognitive apprenticeship: Teaching the crafts of reading, writing, and mathematics. In L. B. Resnick (Ed.), *Knowing, learning, and instruction: Essays in honor of Robert Glaser.* Hillsdale, NJ: Lawrence Erlbaum Associates.

Ericsson, K. A., & Simon, H. A. (1984). *Protocol analysis: Verbal reports as data.* Cambridge, MA: MIT Press.

Gelernter, H. (1963). Realization of a geometry theorem proving machine. In E. A. Feigenbaum & J. Feldman (Eds.), *Computers and thought* (pp. 134–152). New York: McGraw-Hill.

Healy, C. (1993). *Creating miracles: A story of student discovery.* Berkeley, CA: Key Curriculum Press.

Jackiw, N. (1991). *The Geometer's Sketchpad* [Computer software]. Berkeley, CA: Key Curriculum Press.

Klahr, D., & Carver, S. M. (1988). Cognitive objectives in a LOGO debugging curriculum: Instruction, learning, and transfer. *Cognitive Psychology, 20,* 362–404.

Klahr, D., & Dunbar, K. (1988). Dual space search during scientific reasoning. *Cognitive Science, 12,* 1–48.

Koedinger, K. R., & Anderson, J. R. (1990). Abstract planning and perceptual chunks: Elements of expertise in geometry. *Cognitive Science, 14,* 511–550.

Koedinger, K. R., & Anderson, J. R. (1991). Interaction of deductive and inductive reasoning strategies in geometry novices. In *Proceedings of the thirteenth annual conference of the Cognitive Science Society.* Hillsdale, NJ: Lawrence Erlbaum Associates.

Koedinger, K. R., & Anderson, J. R. (1993a). Reifying implicit planning in geometry: Guidelines for model-based intelligent tutoring system design. In S. Derry & S. Lajoie (Eds.), *Computers as cognitive tools* (pp. 15–45). Hillsdale, NJ: Lawrence Erlbaum Associates.

Koedinger, K. R., & Anderson, J. R. (1993b). Effective use of intelligent software in high school math classrooms. In *Proceedings of the world conference on artificial intelligence in education* (pp. 241–248). Charlottesville, VA: American Association for Computers in Education.

Koedinger, K. R., & Anderson, J. R. (in press). Illustrating principled design: The early evolution of a cognitive tutor for algebra symbolization. *Interactive Learning Environment.*

Kuhn, D. (1991). *The skills of argument.* Cambridge: Cambridge University Press.

Langley, P., Simon, H. A., Bradshaw, G. L., & Zytkow, J. M. (1987). *Scientific discovery: Computational explorations of the creative processes.* Cambridge, MA: MIT Press.

Lehrer, R., Randle, L., & Sancilio, L. (1989). Learning pre-proof geometry with LOGO. *Cognition and Instruction, 6,* 159–184.

Luchins, A. S., & Luchins, E. H. (1959). *Rigidity of behavior: A variational approach to the effects of Einstellung.* Eugene: University of Oregon Books.

Nathan, M. J., Koedinger, K. R., & Tabachneck, H. J. M. (1996, April). *Difficulty factors in arithmetic and algebra: The disparity of teachers' beliefs and students' performances.* Paper presented at the annual meeting of the American Educational Research Association, New York.

National Council of Teachers of Mathematics. (1989). *Curriculum and evaluation standards for school mathematics.* Reston, VA: Author.

Newell, A., & Simon, H. A. (1972). *Human problem solving.* Englewood Cliffs, NJ: Prentice Hall.

Pelavin, S. (1990). *Changing the odds: Factors increasing access to college* (Report No. 003969). New York: College Board Publications.

Ritter, S., & Koedinger, K. R. (1995). Toward lightweight tutoring agents. In *Proceedings of the 7th world conference on artificial intelligence in education* (pp. 91–98). Charlottesville, VA: Association for the Advancement of Computing in Education.

Schoenfeld, A. H. (1985). Mathematical problem solving. Orlando, FL: Academic Press.

Schoenfeld, A. H. (1989). Explorations of students' mathematical beliefs and behavior. *Journal for Research in Mathematics Education, 20*(1), 338–355.

Senk, S. L. (1983). *Proof-writing achievement and van Hiele levels among secondary school geometry students.* Unpublished doctoral dissertation, University of Chicago.

Serra, M. (1989). *Discovering geometry: An inductive approach.* Berkeley, CA: Key Curriculum Press.

Singley, M. K., & Anderson, J. R. (1989). *Transfer of cognitive skill.* Cambridge, MA: Harvard University Press.

Tabachneck, H. J. M., Koedinger, K. R., & Nathan, M. J. (1994). Toward a theoretical account of strategy use and sense-making in mathematics problem solving. In *Proceedings of the sixteenth annual conference of the Cognitive Science Society* (pp. 836–841). Hillsdale, NJ: Lawrence Erlbaum Associates.

Vygotsky, L. S. (1978). *Mind in society.* Cambridge, MA: Harvard University Press.

Part III

Defining A New Semantics of Space: Computers, Software, and the Electronic World

Computer tools (software, computer-based video toolkits, electronic discussion, and news groups) have the potential not only to define new approaches to geometry education, but also to define a new semantics of space, a potential explored by the contributors to this final section of the volume.

Goldenberg and Cuoco start off by surveying issues raised by dynamic geometry, a geometry that, although derived from Euclid, has a number of distinctive features with no parallels in Euclidean geometry. The authors examine the effects on teaching and learning of one key feature of dynamic geometry—dragging—and look in particular at the way students perceive figures (and geometry) because of the defaults built into the "drag" mode of operation of these electronic tools. They look at the notion of continuum as applied to two-dimensional figures, at the possibility of two-dimensional "monsters" (e.g., self-intersecting bow-tie quadrilaterals), and examine the implications of "stretchy" line segments, which preserve ratios between (non-Euclidean) "movable" points rather than the particular, measured distances between (Euclidean) fixed points.

De Villiers, in chapter 15, also notes the effect of the dynamic geometry on the teaching and learning of geometry. Raising the issue of the traditional geometric ritual of proof, he examines student conceptions of proof, suggesting that dynamic geometry pushes proof away from verification toward discovery and deductive explanation. This theme is revisited by Olive (in chap. 16), who also focuses on yet another aspect of dynamic geometry raised by Goldenberg and Cuoco, the potential fusion of geometry and algebra. Olive describes a broad spectrum of student work with dynamic geometry, ranging from transformations and symmetry to functions and conic sections.

The next two chapters explore a different dynamic, that of geometry as applied mathematics. Watt describes a prototype fifth-grade geometry unit and child-appropriate computer-aided design (CAD) program. Fifth-grade students used KidCAD to model furniture rearrangment in their classroom and to design a new school space. Watt characterizes this activity as negotiation between the model space and the actual space of the world. Noting the simplicity of this software in comparison to dynamic geometry software, he reminds us that the successful use of software to develop students' mathematical power depends not only on the software itself, but on the way that software supports students' exploration of mathematical concepts.

Zech and her colleagues at the Cognition and Technology Group at Vanderbilt University developed computer-based videos and visual toolkits to help middle-school students explore activities such as designing playgrounds. Like other "Jasper" adventures, the mathematical concepts are anchored to video stories that make the motivation for and nature of the mathematical application clear. The authors paint a portrait of classrooms where student understanding of geometry and measurement emerges as students consider situated problems and use electronic toolkits that not only aid visualization, but also support reflection and revision.

In chapter 19, Renninger, Weimar, and Klotz suggest that mere memorization in geometry has the same educational value as memorizing a page from the city directory. However much this insight seems obvious, they suggest that its implications demand a restructuring and reorganization of professional development and the learning environment. They describe their efforts to create and sustain an electronic Math Forum as a learning environment for both students and teachers. Consideration of the authors' efforts to create and sustain this electronic community suggests one direction for the future of geometry education.

Taken as whole, these chapters suggest a wide range of electronic environments and applications and question our assumptions about the rituals of and approaches to geometry education. The authors, in their eagerness to rethink, redesign, and reexamine computer-based learning environments, provoke us, as do many of the authors in previous sections, into stretching the limits of geometry and reconsidering the nature of teaching and learning mathematics.

14

What Is Dynamic Geometry?

E. Paul Goldenberg and Albert A. Cuoco
Education Development Center, Inc.

Only a few years ago, computer- and then calculator-aided graphing captured the attention and imagination of mathematics educators. These technologies quickly came to exert a major and continuing influence over thinking about content, curriculum materials, and pedagogy in a substantial area of secondary and postsecondary mathematics, and made a remarkably rapid entry into schools.

In much the same way, a new kind of software, which has come to be referred to collectively as dynamic geometry (DG), is now engendering such great interest and enthusiasm that it, too, is making a quick entry into schools. The software provides certain primitive objects (e.g., points, lines, circles), basic tools (e.g., perpendicular to line l through point P) for assembling these into composite objects, and several possible transformations, including, for example, reflection through a point or line. It also allows the user to measure certain parts of the drawing, and, typically, to trace the path of points, segments, or circles as dynamic transformations are applied.

The term *dynamic geometry,* originally coined by Nick Jackiw and Steve Rasmussen, has quickly entered the literature as a generic term because of its aptness at characterizing the feature that distinguishes DG from other geometry software: the continuous real-time transformation often called "dragging." This feature allows users, after a construction is made, to move certain elements of a drawing freely and to observe other elements respond dynamically to the altered conditions. As these elements are moved smoothly over the continuous domain in which they exist, the software maintains all relationships that were specified as essential constraints of the original construction, and all relationships that are mathematical consequences of these. Current implementations of DG include Geometer's Sketchpad (Jackiw, 1991, 1995), Cabri (Baulac, Bellemain, & J. M. Laborde, 1992, 1994), Geometry Inventor (Brock, Cappo, Dromi, Rosin, & Shenkerman, 1994), and, in a partial way, SuperSupposer (Schwartz & Yerushalmy, 1992).

This chapter focuses on the learning issues raised by the dragging feature and its direct consequences and suggests the kinds of questions that remain to be answered by further study. Although what you are now reading appears as a chapter in a book, it must be considered a working paper—the first installment of what will become a history and analysis of the ideas behind "dynamic geometry."

SOME ISSUES RAISED BY DYNAMIC GEOMETRY

How Do Students Perceive the Moving Displays?

How a displayed movement is perceived may depend on the context in which it is set. For example, on the basis of introspective analysis alone (as yet without the needed experiments to confirm it with students), we hypothesize that when an endpoint of a stretchy segment is moved, and the segment is the only object present, the user perceives the movement as a *translation* of the point. That is, dragging A to A′ (see Fig. 14.1) may feel psychologically like a translation. The display may also tend to be seen more as a mapping of A (in its various positions) to C than as a mapping of *A* and C to A′ and C′, respectively.

But other situations may lead to very different perceptions. For example, consider the same construction with a perpendicular to $\overline{AB}$ at B (see Fig. 14.2). A comparable movement of *A* now appears to rotate the system; the sense that *A* is being translated is considerably diminished.

Even though we may have perceived our movement of A as a translation, we now see a rotational aspect of it: The newly constructed perpendicular rotates unchanged around B, drawing our attention to the rotation of $\overline{AB}$ around B. When this is coupled with the movement of C, the various movements of A may now begin to feel more like a composition $\delta \circ \rho$ of (possibly null) rotation ρ about B and (possibly null) dilation δ centered at B. What students make out of this we don't yet fully know.

In pedagogy designed around highly focused or "leading" questions, the exact nature of students' perceptions may not matter much. However, in designing curricula that make significant use of experimentation and investigation, we must understand what students glean from their experiments, and that clearly depends on what features of the experiment they notice (cf. the illusions in perceiving graphs described in Goldenberg, 1988, 1991).

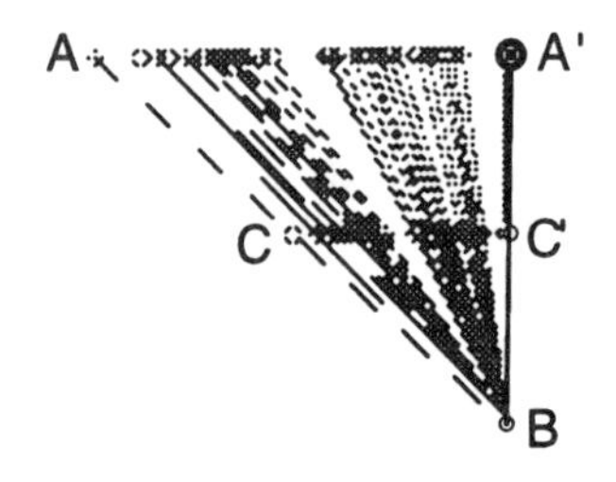

FIG. 14.1 Movement of A suggests translation.

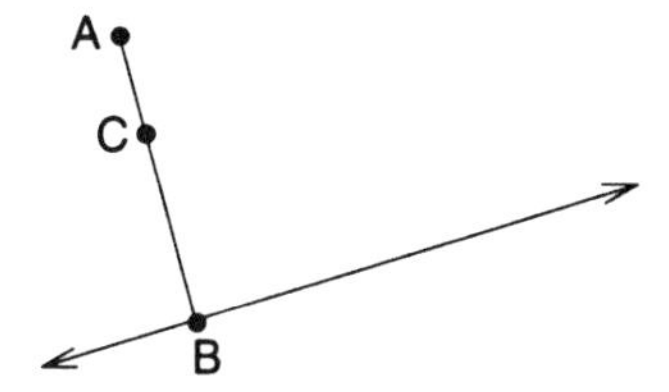

FIG. 14.2 If the line through B always remains perpendicular to segment $\overline{AB}$, then the movement of A can suggest rotation.

How Do Students Interpret What They See?

There are also other aspects of interpretation that must be better understood. For example, asked to investigate proportions in similar triangles, a student builds ΔABC (see Fig. 14.3), constructs $\overline{DE}$ parallel to $\overline{AC}$, and measures a few ratios, including BD/BE and BD/BA.

As D is moved along $\overline{AB}$, $\overline{DE}$ stretches and shrinks to preserve the only property it was given: the fact that it is parallel to $\overline{AC}$ and terminated by the sides of the triangle. The value of BD/BE remains fixed. To find out whether or not this is an accident of the particular shape of ΔABC, the student moves to create a different triangle. Because dynamic geometry allows segments like $\overline{AB}$ to be stretched by moving an endpoint, the implementors must decide the fate, under such a transformation, of a point like D, placed arbitrarily on the segment. The default behavior of such an arbitrarily placed point is that its movement preserves the ratio of the two portions of the line segment it partitions.[1] So as the student adjusts the shape of the triangle by moving A, the value of BD/BA remains constant (see Fig. 14.4).

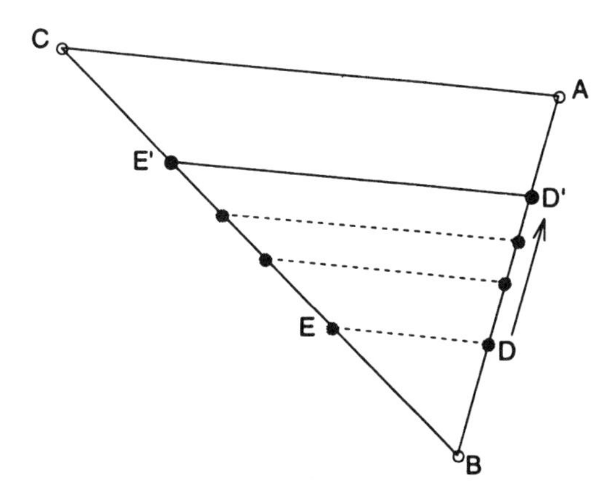

FIG. 14.3 Investigating proportions in similar triangles.

[1] Experimental versions exist that do not share this design feature, and randomization of the location of arbitrarily placed points can be achieved, for example, as an advanced feature in Sketchpad 3.

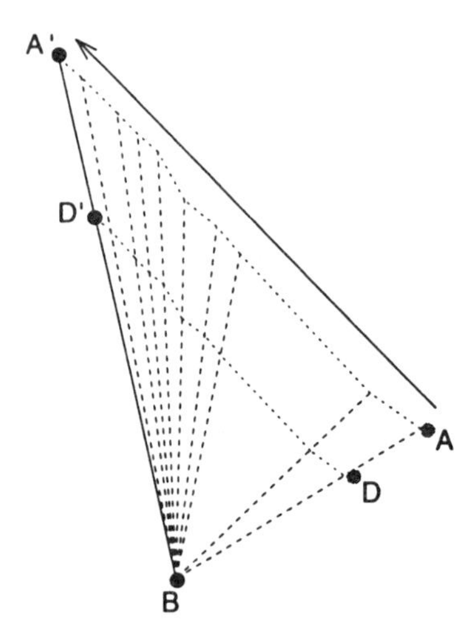

FIG. 14.4 Movement of A preserves BD/BA.

What does a student make of these two observations? One is an important *theorem* of Euclidean geometry. The other is an artifact of the design of usable software, or, thought about another way, a postulate of (a particular) geometry (that is not being taught and hasn't even been quite defined). But a student may well treat both of these discoveries equivalently, as geometric "facts." Optimal curriculum design must be aware of places like this, where reasoning may be confounded by behaviors that are built into DG tools but are not part of the (conscious) postulate set of the students. Research must also clarify what students do, in fact, notice, and how they do interpret what they see. (Of course, there is something of a chicken-and-egg problem here: What they notice is affected by what they learn in their curriculum.)

What Is That Picture on Paper or Screen?

When we look at pictures like the ones in Fig. 14.5, we tend to interpret one as a triangle, ignoring its generality as a polygon, and the other as an arbitrary polygon, ignoring its particularity as a heptagon.

In fact, we *always* interpret when viewing a drawing, treating some features as inessential (adding some generality), and preserving others as essential (not adding too much generality). Thus, the triangle is generalized to an arbitrary triangle (not just the one depicted) but is not generalized all the way to "polygon." Likewise, the heptagon is generalized to "polygon," but not all the way to "curve." Minds tend to interpret symbols at something of a "middle level" of abstraction—neither too narrowly nor too broadly—unless special knowledge (gained formally or informally) dictates a different course of action. This phenomenon applies to linguistic symbols (see, e.g., Pinker, 1995) as well as to pictorial ones and appears to be a general cognitive principle (Rosch, 1978).

Corresponding roughly to our use of *particular* and *general*, respectively, is the distinction that Parzysz (1988) highlighted when he introduced the terms *drawing* and *figure* to the research literature in geometric learning. A drawing was one picture; a figure was a class (often infinite) of drawings related by some underlying commonality. This distinction was refined into a tripartite taxonomy by Colette Laborde and Bernard Capponi (1994), who referred to (a) geometric objects (abstract mathematical objects defined by a set of relationships); (b) drawings, which necessarily contain both too much information (e.g., details such as size and orientation are often irrelevant) and too little information (e.g., the degree of generalization intended); and (c) figures (the set of psychological features, interpretations, and extra baggage [cf. Vinner, 1983] relating the mathematical object to the particular drawing).

If our primary focus is on making optimal use of the new opportunities afforded by nice geometry software, why should we be worrying about tripartite taxonomies and other such theories of mind? For one thing, the DG environments are built on the same principles: The user specifies to the computer the underlying relationships (the mathematical object), and the computer must preserve that object while leaving surface features (the drawing) completely malleable. To craft curricula that "work" with students, we must know what the student makes of the result (the figure) and the connection between the display and the commands the student used to create it.

Reexamining Definitions: What Is a Quadrilateral?

Parzysz's (1988) terms, clarifying a distinction apparently long made in geometry teaching in France, are useful contributions to a language for talking about how students interpret what they see in static pictures like those in Fig. 14.5, but there may yet be more for us to understand about students' perceptions in an environment where drawings are not static, single instances of the figure. The work of C. Laborde and Capponi (1994) represents one extension to the theory, but it is also useful to look at the situations, new with DG, that might force us to reexamine the theory.

Nonstatic drawings can confound our thinking about how far to generalize in interpreting a picture. Consider, for example, students in-

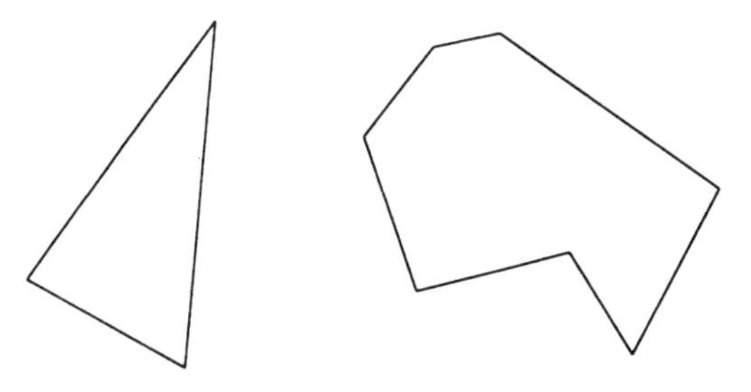

FIG. 14.5 Interpreting figures: Which is general and which is particular?

vestigating a construction in which the midpoints of the four sides of a quadrilateral (ABCD) are connected in order. The resulting figure appears to be a parallelogram. Two obvious questions are, will that be true for all quadrilaterals or just particular ones? and why?

Using a DG tool, the student might begin by drawing a quadrilateral ABCD. Although the original drawing is necessarily particular, DG treats it as a stand-in for a class of objects; suitable "dragging" allows it to be transformed to any other member of that class. Figure 14.6a attempts to suggest the motion with five frames, but the direct manipulation of the quadrilateral gives the impression of a continuously changing figure. (See Fig. 14.6b for another static attempt at conveying this image.)

Slight changes in the location of one point (*D* in Fig. 14.6a) may, depending on the direction of change, deform the quadrilateral into a kite, a trapezoid, or some other convex quadrilateral. More drastic changes can create a concave quadrilateral. But some possibly unanticipated things can also happen. Sufficient movement of that vertex can create a self-intersecting figure (the "bow tie") that the student most likely does not classify as a quadrilateral (see the case of "monster" posed by de Villiers, chap. 15, this volume). And in moving from the original quadrilateral to the bow tie, the figure inevitably passes through a degenerate case that is, in fact, a triangle (the middle frame in Fig. 14.6a). Yet the observation (not yet a theorem until students show it to be) about the connected midpoints holds even for this figure.

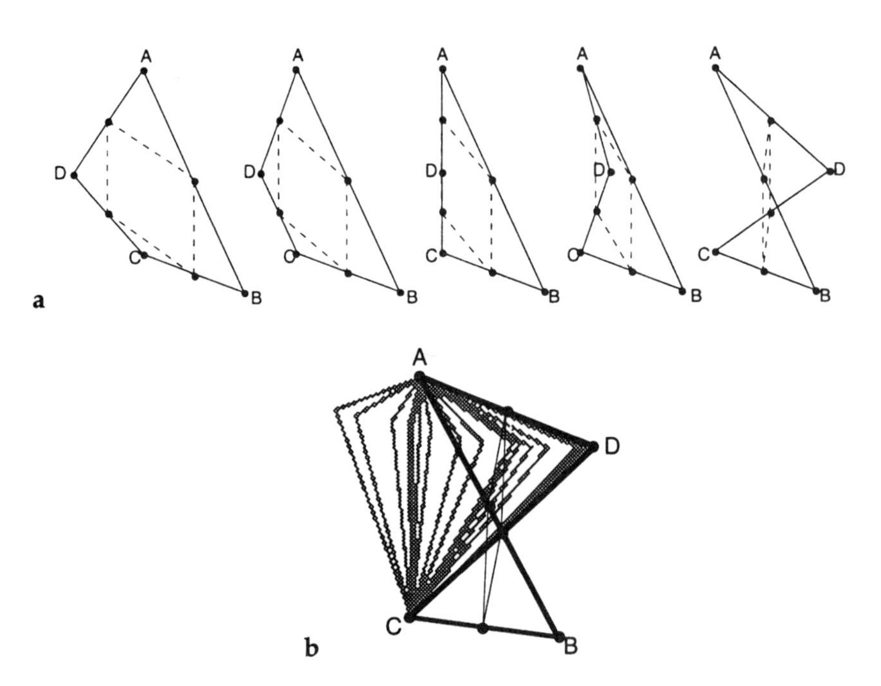

FIG. 14.6 Transforming one quadrilateral into another. (a) Five frames in the life of a quadrilateral. (b) A "continuum" of quadrilaterals.

How do students interpret the effect of moving a point? Do they see ABCD as one thing being deformed in various ways, or as many different things, each one happening to consist of four connected points? Do they construct the notion of a continuum of quadrilaterals? How do they handle the fact that moving a single point can also create "monster quadrilaterals," such as the triangular configuration or the self-intersecting bow tie? Do they seem to fail to notice these cases altogether, or ignore them as if they do not exist or are a kind of irrelevancy? Do they think of these as interesting but separate by-products of a set of observations about quadrilaterals? Or do students experience this as conflicting with their previous notions of "quadrilateral"? If so, do they extend the definition? Define exceptional cases? Do they resolve the conflict by what Lakatos (1976) called "monster-barring"—deliberate and careful reworking of all relevant definitions (in this case, of *quadrilateral*) for the purpose of rejecting aberrant or troublesome cases? This same problem domain can also give us insight into students' conceptions of *vertex* (does D remain a vertex when it is collinear with A and C?) and *point* (if D is not a "vertex" when it is collinear with A and C, how, if at all, is its status different from any other point on the figure?; cf. results in Goldenberg, 1988; Goldenberg & Kliman, 1990).

Questions of this kind do not so naturally occur when software presents discrete cases preselected—pre-monster-barred—by restrictive definitions of quadrilaterals built into the software. By contrast, dynamic geometry offers an interesting arena within which to watch students construct or reconstruct definitions of categories of geometric objects, because it allows the students to transgress their own tacit category boundaries without intending to do so, creating a kind of disequilibrium, which they must somehow resolve. Confusions can be beneficial or destructive. Understanding how students resolve such conflicts will help us devise educationally better uses of the software, ones that maximize the opportunities and minimize the risks of the confusions created by such transgression of accepted definitions.

Interestingly, the opportunities are not for researchers alone. The making of definitions is central to mathematics. To learn the importance and purpose of careful definition, students must be afforded explicit opportunities to participate in such definition-making themselves (de Villiers, 1994). Calling explicit attention in class to issues like the ones just described may help students both to examine and to refine their definitions of these specific mathematical objects and even to think more deeply about the very nature of definition.

Thinking About Functions, Variables, and Experiments

The principal contribution of DG to the nature of experimentation in geometry is the dynamic dependency of its display on the position of some movable object. This suggests a kind of function, and yet another perspective on the figure–drawing distinction.

Recall Parzysz's (1988) "figure" as a collection of drawings. If the *figure* is Triangle, then each *drawing* represents a single triangle. We may want to equip this set with a topology, so we'll call the drawings *points* of a figure *space*. In fact, we'll often want to regard certain points in that space as identical (identifying, for example, points that represent congruent or even similar triangles), so the figure space may actually be a quotient of the set of drawings by some relation. A *geometric experiment*, in the most ordinary sense of the term, could then be seen as a function defined on a figure space together with a method for selecting drawings on which to apply the function.

The kinds of functions that students usually define (e.g., on the space of triangles) are specified by a construction, some composition of allowed actions on these triangles. The function value is some observation that the experimenter chooses to make—often a measurement, possibly combined with calculation. We'll call this function the *procedure* of the experiment.

From the figure space, students can select points (that is, drawings) on which to apply a procedure in several ways. For example, they can simply pick a new drawing at "random," or they can isolate some aspect of the drawing that seems to be important and try another point that has the same property (e.g., carry on the experiment only with isosceles triangles). This amounts to defining another function, a "selector" function, that produces data on which the experiment will be conducted.

Different computational environments allow for different kinds of experiments. With the Geometric Supposers (Schwartz & Yerushalmy, 1983–1991, 1992), pioneers of geometry experimentation environments, the figure space is essentially predetermined by the choice of software (triangles, quadrilaterals, circles, etc.), and the student typically starts by selecting a member of that class of geometric objects (e.g., triangles)—a single point in that figure space. In DG environments, depending on the student's experimental intent, he or she may have to define the figure space through a construction. This can be easy if the figure space is quite general (quadrilaterals), and can entail more work if the figure space is quite specific (squares).

The second step is to perform some construction on the starting drawing. The construction methods for such procedures in the menu-driven and direct-manipulation environments are quite similar. In both environments, the student builds composites of primitive constructions—taking midpoints, drawing parallels or perpendiculars, joining points, and so on. The definition of the procedure stays constant over the course of the experiment (e.g., midpoints stay midpoints regardless of the fate of the point in the figure space).

Students then make observations about the result and may conjecture about how their observations might be affected if the same construction were performed on another starting triangle. Again, the Supposer and DG environments are similar.

The greatest difference is apparent when the students come to test their conjecture. On the Supposer, the "Repeat Construction" menu item allows students to specify a new triangle by particular features, or to choose a class of triangles (or apply no restriction) and have the machine select randomly within that class. The inputs to both of these selector functions are statements, with support from menus. The output is a new triangle (or other object). Step by step, at a pace controlled by the student, the computer then reexecutes the construction on the new object. The student can see, at the end, whether the original observation remains valid in this new trial.

DG developers chose a different selector mechanism. The mouse interface allows users to see other instances of a construction by "clicking and dragging" primitive elements (e.g., vertices) of the picture (a triangle, in our example) on which the construction was built. (Sketchpad has made it a design principle not to restrict the draggable objects only to the most primitive elements.) As the design on the screen is altered, users see what seems like a continuum of intermediate states, so that the original instance of the construction appears to deform continuously into the final instance, giving the impression that the user is checking an infinite number of cases "on the way" to the new selection. The selector function's output, as with the Supposer, is an object (e.g., triangle) on which a construction and some observations are based, but the input—the independent variable—is a feature (e.g., a vertex or a side) of the currently selected object, and the method of obtaining an output is by mousing on that feature and moving it to another part of the plane.

This whole selection method suggests a topology in the figure space in which two points (recall that each point in figure space refers to an entire drawing on the Euclidean plane) are close if one can be perturbed slightly to obtain the other. This can, of course, be made precise (e.g., by defining the distance between two such points to be the maximum distance between pairs of corresponding vertices), but the basic idea is that small changes in a feature of a drawing produce small changes in the drawing.

The difference in emphasis may affect how students perceive the procedures in the two environments. In a DG experiment, the sense of "closeness" may invoke a sense of continuity and dynamic dependency; in the Supposer environment, attention is drawn to function as algorithm.

This distinction fits in quite well with the taxonomy, described in Cuoco (1995), of certain functions (defined using **R** as an auxiliary set). In the Supposer experiment, the student's attention is drawn to the constructive algorithm: Each new experiment recapitulates the construction. Even if the student is interested solely in the outcome, the activity on the screen draws attention, at least partly, to the steps along the way. In DG software, once the construction is complete, the only activity on the screen is the movement of the varied element—attached to the user's hand—and the consequent deformation of the dependent parts of the construction. The user can ignore the steps of the construction and regard it, in its entirety, as a single, responding thing.

It is instructive to depart momentarily from geometry and compare this to some representations of polynomial functions. In the BBN Function Machines language (Feurzeig, 1993), elementary functions are composed into larger ones by "piping" data from the output of one to the input of another. An $x\ (x + 3)$ machine (see Fig. 14.7) is thus created from a plus and a times machine by piping the input x to both, and establishing a constant 3 in the plus machine. Each experiment with this $x\ (x + 3)$ machine re-

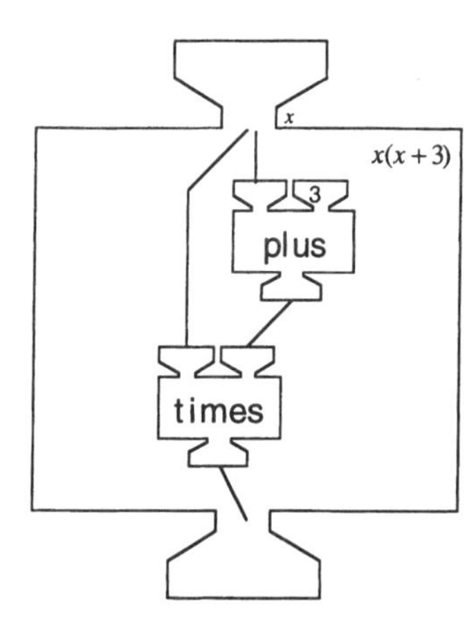

FIG. 14.7 An example of a BBN Function Machine.

quires placing a specific number, in this case 1.5, in the machine's hopper, watching it flow to the two internal machines, watching the 4.5, $(x + 3)$, flow from the output of the plus to the input of the times, and watching the product 6.75, $x\ (x + 3)$, flow from the output of the times through the output of the $x\ (x + 3)$ machine.

This is directly analogous to the Supposer's "Repeat Construction" command. By contrast, the imagery provided by the EDC DynaGraph more closely resembles the DG situation (Fig. 14.8).

With a suitable set of tools, the user again constructs the function $f(x) = x\ (x + 3)$, but having completed it, the user then drags the wedge representing x to the left or right along the x number line and watches as the other wedge, sitting on the upper number line and representing the value $f(x) = x\ (x + 3)$, responds dynamically.

Each system brings certain properties of functions to the fore, at the sacrifice of other properties. When using the EDC DynaGraph, the user is almost inescapably drawn to the function behavior—for example, if we move x at a constant "velocity," the velocity of $f(x)$ is not constant. In fact, when scale information is suppressed (as in Fig. 14.8) there is little else to look at but the dynamic dependency of $f(x)$ on x. A consequence is that students working with this tend to develop vocabulary about change and transformation (Goldenberg, Lewis, & O'Keefe, 1992). At least one of the sacrifices is that the user is no longer necessarily aware of what function's behavior he or she is observing. In fact, the functions can be supplied from a menu of "canned" behaviors (for the sake of examining some kinds of behavior, or, at a later stage of learning, designing functions that have sim-

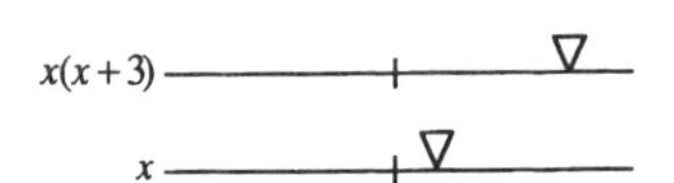

FIG. 14.8 An example of an EDC DynaGraph: As x moves, $x\ (x + 3)$ moves accordingly.

ilar behaviors) without knowing anything about what steps might be involved in generating such a behavior. In contrast, when working with Function Machines, users cannot see the behavior of this function without working through examples and constructing that behavior for themselves, but the algorithmic structure of the function is readily apparent and available for study and modification, as is the notion of composing complex structures out of simpler ones, encapsulating these as machines, and then using them as building blocks for still more complex structures (Cuoco, 1993).

These two interpretations of the concept of mathematical function—function-as-dynamic-dependency (the kinds studied in physics and analysis), and function-as-algorithm (the kinds, like Euclid's GCD algorithm, that are a mainstay of algebra)—run throughout mathematics. Together, the two kinds of tools may help develop these complementary images of function. It seems reasonable to assume a similar kind of complementarity between drag and non-drag geometric experiments. If that assumption stands up to scrutiny, it would have much to say about the development of curriculum materials and pedagogy that make use of geometric software.

WHAT IS THE MATHEMATICS OF DG?

A mathematical analysis of dynamic geometry has a practical role in the development of pedagogy. Although adults trained on Euclid can recognize familiar ideas in a demonstration of, for example, the concurrence of the medians of a triangle in Cabri, there is no reason at all to assume that students who are first encountering geometry through this medium will construct the same ideas that we bring to its interpretation. If we start, as we do, with the notion that students construct their own mathematical ideas based on the phenomena that they experience, then it becomes important to ask not only about how they experience the display—matters touched on in the previous section—but also about what the displayed phenomena are, and how they might fit into a coherent mathematical picture. In effect, this question asks what mathematics we should *expect* students to construct.

Nick Jackiw, in a November 7, 1994, posting to the Swarthmore Geometry Forum, said "at its heart, 'dynamic geometry' is not a well-formulated mathematical model of change, but rather a set of heuristic solutions provided by software developers and human-interface designers to the question 'how would people like geometry to behave in a dynamic universe?' "

The truth of this becomes increasingly apparent as each new version of DG software introduces features whose interactions with each other and with older features is fully determined algorithmically (the stuff must work, after all!) but whose consequences are neither fully predicted in advance nor part of a predesigned mathematical theory. On the other hand, it is also true that each implementation of DG software has a "behavior," and students' experiences with dynamic geometry are based on their observations and interpretations of that behavior.

Thus, to understand students' (mathematical) experiences, it is important to characterize DG's behavior (in a mathematical way). Such an analysis might be treated as a definition of (existing) DG and should naturally precede empirical work with students: It is the preparation after which we can ask which phenomena students notice, and how they perceive and interpret those phenomena.

We know of no such theoretical analysis of a geometry with the properties of DG, but we can anticipate at least two directions that such an analysis might go: giving dynamic geometry its own axiomatic foundation, or explaining it in terms of existing theory.

Fitting Existing Theory

The DG experiments with which we are primarily concerned involve properties that remain invariant under a particular kind of continuous deformation of a geometric construction, or the locus of some object as another is transformed in a continuous way. Such constructions represent certain kinds of continuous functions, suggesting strong links with analytic methods not native to classical Euclidean arguments.

For example, imagine starting with ΔQRU as shown in Fig. 14.9, and dragging U so that the triangle looks like the tallest one in that figure. (Some of the "intermediate" triangles are shaded in gray.)

An existing mathematical construction explains the seeming continuum of triangles: Imagine a function ψ, defined and continuous on the unit square and taking values in $\mathbf{R}^2$, so that the image of the bottom edge of the square is the original ΔQRU, the image of the top edge of the square is the final instance of ΔQRU, and the image of any segment parallel to these edges is an "intermediate" triangle with base $\overline{QR}$ In topology, such a function is called a *homotopy* between these two instantiations of ΔQRU. Intuitively, we can think of a homotopy as a continuous deformation of the image of the bottom edge of the square to the image of the top edge.

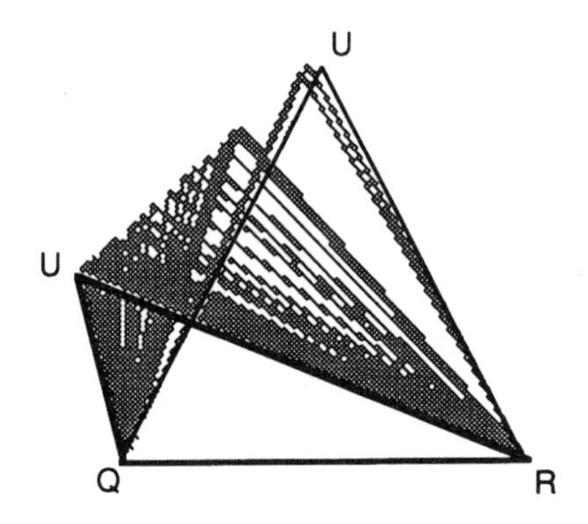

FIG. 14.9 Dragging U to deform one triangle into another.

Homotopy theory is a rich and widely used part of mathematics and finds elementary applications (converting Celsius to Fahrenheit scales or deforming a circle into an ellipse can be viewed as homotopies) as well as uses in research in algebraic topology. Fig. 14.10 is meant to suggest that a DG deformation can be viewed as a picture of a homotopy. As a segment parallel to the base of the square rises from the bottom to the top, it is transformed into a sequence of triangles that "interpolate" between the original and final positions of ΔQRU.

In dragging a triangle, the final image need not be displayed until the mouse button is released. What DG does, therefore, reflects a deliberate choice of the developers, and is a departure from the many environments that display the effects of transformations on the plane by showing only the pre-image and the image, even when a homotopy between these two figures exists.

New Foundations

Although proof of a dynamically derived conjecture may invoke classical Euclidean methods, a reliance on Euclidean methods alone ignores a key aspect of the experiment, thereby risking the integrity of the mathematics

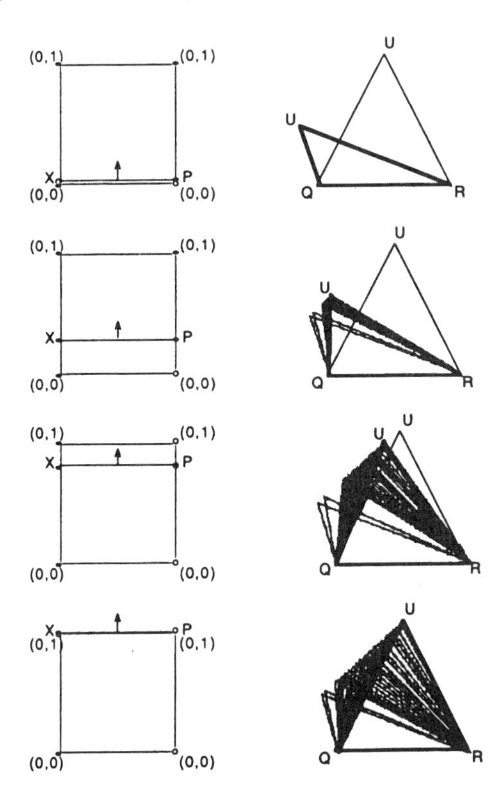

FIG. 14.10 Viewing a DG deformation as a homotopy between triangles.

and—as is true with any other geometry—failing to take special advantage of that geometry's unique power.

Can an axiomatic foundation be given to dynamic geometry, in very much the same way as we lay down the axioms for projective geometry? This approach has been taken to describe the mathematics underlying other computational media. For example, in Abelson and diSessa (1980), the "turtle geometry" underlying Logo's turtle graphics has been analyzed as a system that is quite close to differential geometry.

Such an analysis would begin by identifying the objects and *assumptions* of a DG environment. For example, the "stretchy" line segment is a "new" object in this environment, and its ratio-preserving property (e.g., if P is a point placed arbitrarily on $\overline{AB}$, and if we stretch or shrink $\overline{AB}$ by mousing on *B*, *P* moves in a way that preserves PA/PB) is an assumption (a "postulate" of DG but not of Euclid). Another assumption—very important but almost unnoticeable because it seems so natural—is that dragging a vertex of a triangle has no effect on the opposite side. The set of such specialized objects and assumptions has not yet been fully described.

The mathematical implications of the transformation (dragging) must also be better understood. Dragging is clearly not a transformation on the entire plane. What *is* a clear mathematical characterization of this transformation?

Finally, we must look past postulated invariants, such as the ratio-preserving property, to the theorems of invariance that follow from the assumed properties and transformations. Does DG, in fact, specify a different mathematical system, largely compatible with that of Euclid, but differing in important respects? If we are to discover the real power of this (hypothetically new) geometry, we need to examine and describe the relationship between these "new" theorems (and postulated invariants) and the theorems of Euclidean geometry.

WHERE DOES THIS LEAVE US?

Although research still leaves many unanswered questions about how students use static visualizations, DG raises the ante and requires us to understand how students glean geometric ideas from pictures that involve motion, often quite complex. How do students develop a sense of where to look, what objects to track, what questions to ask, what experiments to perform? In fact, with so many degrees of freedom, how do they learn what experiments *not* to perform? Preliminary observations among college mathematics majors suggest that students who are deforming a figure tend not to let go of the point that they are dragging when the result would be a "monster" (like a cross-quadrilateral) but, instead, return points to tame positions before letting go. If so, are they truly ignoring the continuity of the change and treating the screen data as discrete tame cases that just happen to be connected? And, if so, might they not be better off with the discrete experiments in the first place?

In fact, the shock and delight that students often express at some unexpected behavior seems a pretty good indicator that they are *not* ignoring the monster cases. The avoidance of *stopping* at these cases might then be interpreted as evidence that students *are* attending to the variables and degrees of freedom, and trying to manage them while they come to understand the geometry and the display better.

Dynamic geometry is seen (e.g., C. Laborde & Capponi, 1994) as an environment to help students develop the notion of "figure"—attending to underlying relationships rather than to the particulars of a specific drawing. The fact that students need such help is well enough known, but, to the best of our knowledge, the level of generality or particularity that students apply in their interpretation of drawings is not as well known. Neither is the flexibility of their definitions, especially in the face of movable points.

What is quite clear is that geometry on a computer is different from geometry on paper. C. Laborde (1992) remarked that geometric tools like Logo, DG environments, and the Supposers differ from drawing tools like MacPaint in that "the process of construction of the drawing [with the latter] involves only action and does not require a description" (p. 129). What she is referring to is the fact that the geometric tools require the user to *describe* intended relationships: Two lines are parallel, or two segments congruent, because they are declared to be, not just because they happened to be drawn to look that way. The details of such specification vary greatly among tools like Logo, DG environments, and the Supposers. Those details influence how the user interacts with the tool, and plausibly influence how a student learns with it. But the common feature is that the user is interacting both with the visual elements and with a declarative description of the visual.

In fact, except for what students do in their heads, paper-and-pencil geometry also "involves only action and does not require a description." Part of what students learn in geometry is, as Poincaré put it, the art of applying good reasoning to bad drawings—adding the descriptions that specify which features or relationships in the drawings are intended, which are incidental, and which are to be totally ignored as they attempt to draw inferences about the figure depicted.

It is also clear that a "plastic" geometry is quite different from a geometry that is fixed (see, e.g., the growth of children's ideas about morphing in Lehrer, Jacobson, et al., chapter 7, this volume). Movable points cause segments (and circles and conics and so on) to stretch and deform, giving rides—sometimes very surprising rides—to points that sit on them. This cascade of consequences of movable points creates new mathematical raw materials for students to observe. This chapter describes just a few of the new mathematical and pedagogical issues that a dynamic geometry raises.

Finally, if the potential hazards and pitfalls seem larger and more numerous than we might have imagined, it is also clear that the opportunities to make new and important mathematical connections between classical geometric content and big ideas from other mathematical areas are too intriguing to ignore (see, e.g., Cuoco & Goldenberg, in press). So is the evidence in students' eyes and on their faces. We must come to understand the new terrain well, so that its roughness becomes a part of, and not a detraction from, its beauty.

ACKNOWLEDGMENTS

The perspective, rationale, planning, and writing of this chapter were supported in part by development funds from Education Development Center, Inc. (EDC), with additional support from the National Center for Research in Mathematical Sciences Education (NCRMSE) and EDC's *Connected Geometry* project, funded by the National Science Foundation (grant MDR-9252952). Major funding for the interviews with Jean-Marie and Colette Laborde and others at Institut d'Informatique et Mathématiques Appliquées (IMAG) was provided by NCRMSE, with additional support from *Connected Geometry*. Interviews with Nick Jackiw, Eugene Klotz, Judah Schwartz, and Michal Yerushalmy (whose significant intellectual contributions are, as yet, only partially reflected in this writing) and the final preparation of this chapter were supported in part by National Science Foundation grant RED-9453864. We are particularly grateful to Jean-Marie and Colette for the time they spent not only during our initial discussions, but also in reviewing the manuscript. We also acknowledge the contributions of Nicolas Balacheff, Bernard Capponi, and James King. Opinions (and errors) are ours, and do not necessarily reflect the views of any of the funders or contributors.

REFERENCES

Abelson, H., & diSessa, A. (1980). *Turtle geometry,* Cambridge, MA: MIT Press.

Baulac, Y., Bellemain, F., & Laborde, J. M. (Designers). (1992). *Cabri: The interactive geometry notebook* (Cabri Géomètre). Pacific Grove, CA: Brooks-Cole.

Baulac, Y., Bellemain, F., & Laborde, J. M. (Designers). (1994). *Cabri II.* Dallas, TX: Texas Instruments.

Brock, C. F., Cappo, M., Dromi, D., Rosin, M., & Shenkerman, E. (Designers). (1994). *Tangible math: Geometry Inventor.* Cambridge, MA: Logal Educational Software and Systems.

Cuoco, A. (1993). Action to process: Constructing functions from algebra word problems. *Intelligent Tutoring Media, 4*(3–4), 117–127.

Cuoco, A. (1995). Computational media to support the learning and use of functions. In A. diSessa, C. Hoyles, & R. Noss, with L. Edwards (Eds.), *Proceedings of the Advanced NATO Workshop: Computational media to support exploratory learning* (pp. 79–107). New York: Springer-Verlag.

Cuoco, A., & Goldenberg, E. P. (1996). Dynamic geometry as a bridge from Euclidean geometry to analysis. In D. Schattschneider & J. King (Eds.), *Geometry turned on: Dynamic software in learning, teaching and research* (MAA Notes, Vol. 41). Washington, DC: Mathematical Association of America.

de Villiers, M. (1994). The role and function of a hierarchical classification of quadrilaterals. *For the Learning of Mathematics, 14*(1), 11–18.

Feurzeig, W. (1993). Explaining function machines. *Intelligent Tutoring Media, 4*(3–4), 97–108.

Goldenberg. E. P. (1988). Mathematics, metaphors, and human factors. *Journal of Mathematical Behavior, 7,* 135–173.

Goldenberg. E. P. (1991). The difference between graphing software and educational graphing software. In W. Zimmerman & S. Cunningham (Eds.), *Visualization in teaching and learning mathematics* (MAA Notes, No. 19, pp. 77–86). Washington, DC: Mathematical Association of America.

Goldenberg, E. P., & Kliman, M. (1990). *Metaphors for understanding graphs: What you see is what you see* (Center for Learning, Teaching, and Technology Tech. Rep. No. 90-4). Newton, MA: Education Development Center.

Goldenberg, E. P., Lewis, P. G., & O'Keefe, J. (1992). Dynamic representation and the development of an understanding of functions. In. G. Harel & E. Dubinsky (Eds.), *The concept of functions: Aspects of epistemology and pedagogy* (MAA Notes, No. 25, pp. 235–260). Washington, DC: Mathematical Association of America.

Jackiw, N. (Designer). (1991). *The Geometer's Sketchpad.* Berkeley, CA: Key Curriculum Press.

Jackiw, N. (Designer). (1995). *The Geometer's Sketchpad, v3.0.* Berkeley, CA: Key Curriculum Press.

Laborde, C. (1992). Solving problems in computer-based geometry environment: The influence of the features of the software. *Zentralblatt fur didaktik der mathematik, 4,* 128–135.

Laborde, C., & Capponi, B. (1994). Cabri-Géomètre constituant d'un milieu pour l'apprentissage de la notion de figure géométrique. *DidaTech Seminar, 150,* 175–218.

Lakatos, I. (1976). *Proofs and refutations.* Cambridge, England: Cambridge University Press.

Parzysz, B. (1988). "Knowing" vs. "seeing": Problems of the plane representation of space geometry figures. *Educational Studies in Mathematics, 19*(1), 79–92.

Pinker, S. (1995). *The language instinct.* New York: Harper.

Rosch, E. (1978). Principles of categorization. In E. Rosch & B. Lloyd (Eds.), *Cognition and categorization.* Hillsdale, NJ: Lawrence Erlbaum Associates.

Schwartz, J., & Yerushalmy, M. (Designers). (1983–1991). *The Geometric Supposers,* Pleasantville, NY: Sunburst Communications.

Schwartz, J., & Yerushalmy, M. (Designers). (1992). *The Geometric SuperSupposer.* Pleasantville, NY: Sunburst Communications.

Vinner, S. (1983). Concept definition, concept image and the notion of function. *International Journal of Mathematics Education in Science and Technology, 14*(3), 293–305.

15

An Alternative Approach to Proof in Dynamic Geometry

Michael de Villiers
University of Durban-Westville

> [There is an underlying] formalist dogma that the only function of proof is that of verification and that there can be no conviction without deductive proof. If this philosophical dogma goes unchallenged, the critic of the traditional approach to the teaching of proof in school geometry appears to be advocating a compromise in quality: he is a sort of pedagogic opportunist, who wants to offer the student less than the 'real thing.' The issue then, is not, what is the best way to teach proof, but what are the different roles and functions of proof in mathematics. (adapted from Hersh, 1979, p. 33)

The problems that students have with perceiving a need for proof are well known to all high-school teachers and have been identified without exception in all educational research as a major problem in the teaching of proof. Who has not yet experienced frustration when confronted by students asking "Why do we have to prove this?" Gonobolin (1954/1975) noted that "the students ... do not ... recognize the necessity of the logical proof of geometric theorems, especially when these proofs are of a visually obvious character or can easily be established empirically" (p. 61).

The recent development of powerful new technologies such as Cabri-Geometre and Geometer's Sketchpad with drag-mode capability has made possible the continuous variation of geometric configurations and allows students to quickly and easily investigate the truth of particular conjectures. What implications does the development of this new kind of software have for the teaching of proof? How can we still make proof meaningful to students?

In this chapter, a brief outline of the traditional approach to the teaching of proof in geometry is critiqued from a philosophical as well as a psychological point of view, and in its place an alternative approach to the teaching of proof (in a dynamic geometry environment) is proposed.

THE TRADITIONAL APPROACH TO THE TEACHING OF PROOF IN GEOMETRY

Underlying Philosophy

Philosophically, the traditional view of proof has been and still is largely determined by a kind of philosophical rationalism, namely, that the formalist view that mathematics in general (and proof in particular) is absolutely precise, rigorous, and certain. Mathematics from this perspective is seen as the science of rigorous proof. Although this rationalistic view has been strongly challenged in recent years by the quasi-empirical (or fallibilist) views of, for example, Lakatos (1976), Davis and Hersh (1983, 1986), Chazan (1990), and Ernest (1990), it is probably still held by the vast majority of mathematics teachers and mathematicians.

In an extreme version of this view, the *function* (or purpose) of proof is seen as only that of the verification (conviction or justification) of the correctness of mathematical statements. In other words, proof is narrowly seen merely as a means to remove personal doubt and/or that of skeptics, an idea that has one-sidedly dominated teaching practice and most discussions and research on the teaching of proof (even by those who profess to oppose a formalist philosophy). Consider the following quotes, which only emphasize the verification function of proof:

> "A proof is only meaningful when it answers the student's *doubts*, when it proves what is not obvious" (emphasis added; Kline, 1973, p. 151).
>
> "The necessity, the functionality, of proof can only surface in situations in which the students meet *uncertainty* about the truth of mathematical propositions" (emphasis added; Alibert, 1988, p. 31).
>
> "A proof is an argument needed to *validate* a statement, an argument that may assume several different forms as long as it is *convincing*" (emphasis added; Hanna, 1989, p. 20).
>
> "Why do we bother to prove theorems? I make the claim here that the answer is: so that we may *convince people* (including ourselves) . ... we may regard a *proof as an argument sufficient to convince a reasonable skeptic*" (emphasis added; Volmink, 1990, pp. 8, 10).

Arguing from the viewpoint that the results of all inductive or quasi-empirical investigation are unsafe, proof is seen (basically) as a prerequisite for conviction—therefore, proof is required as the absolute guarantee of their truth. In other words, the only purpose of proof is to give the final stamp of approval:

> Reasoning by induction and analogy calls for recourse to observation and even experiment to obtain the facts on which to base each argument. But the senses are limited and inaccurate. Moreover, even if the facts gathered for the purposes of induction and analogy are sound, these methods do not yield unquestionable conclusions. ... To avoid these sources of error, the mathematician utilizes another method of

reasoning. ... In deductive reasoning the conclusion is a logically inescapable consequence of the known facts. (Kline, 1984, pp. 11–12)

Traditional Teaching Approach

Because proof is seen only as an instrument for the removal of doubt, the typical approach is to create doubts in the minds of students, thereby attempting to motivate a need for proof. Traditional approaches in motivating proof in geometry can be classified into three main types, namely, pattern failure, optical illusion, and false conclusion.

Type 1: Pattern Failure. Pattern failure is often the "easiest" (most visual) way to create doubts:

- $n^2 - n + 41$ gives prime numbers for $n = 1, 2, 3, \ldots$ but breaks down when $n = 41$.
- n points on a circle, when all are connected, divide it into 2^{n-1} regions. This holds for $n = 1$ to $n = 5$, but breaks down when $n = 6$ (see Fig. 15.1).

There are also famous historical examples where inductive generalizations eventually turned out false. About 500 BC, Chinese mathematicians (and much later also Leibniz) conjectured that if $2^n - 2$ is divisible by n, then n must be prime. It turns out that the empirical investigation supports the conjecture up to $2^{340} - 2$. In all these cases $2^n - 2$ is divisible by n when n is prime, and not divisible by n when it is composite. However, in 1819 it was discovered that $2^{341} - 2$ is divisible by 341, even though 341 is composite ($341 = 11 \times 31$).

In 1984 Odlyzko and Te Riele showed that a conjecture by Franz Mertens (a contemporary of Riemann) over 100 years ago was actually false, despite computer support that showed that it was true up to $n = 10^7$.

After giving some such examples to students, teachers are usually satisfied that a sufficiently critical attitude has been cultivated and proceed to introduce the proofs of geometrical results, as a means of verifying that those results are correct.

Type 2: Optical Illusion. Another traditional approach is to provide students with optical illusions in order to caution them against putting too

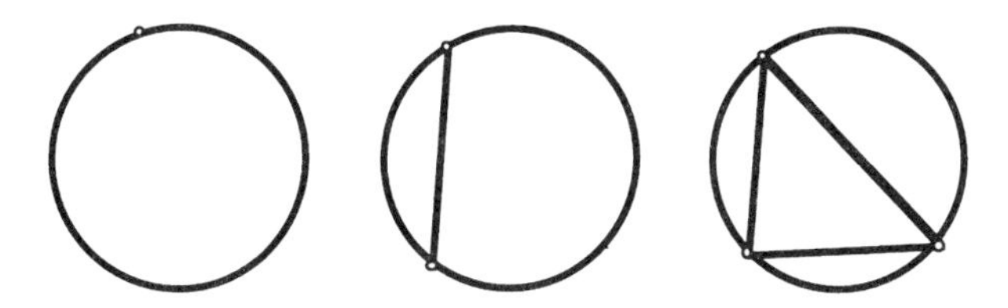

FIG. 15.1 Pattern failure approach of proof. Given n points on a circle, all connected, 2^{n-1} = number of regions created. Pattern holds if $1 \leq n \leq 5$ but breaks at $n = 6$.

much faith in the way a figure looks. The intention is to show the "superiority of reasoning over experience." Consider the two examples in Fig. 15.2a. In both figures, although AB appears to be shorter than CD, AB and CD are, in fact, of equal length. Another often used example (see Fig. 15.2b) is an 8 × 8 square divided into four pieces, cut out and rearranged to form a 5 × 13 "rectangle" whose area is now suddenly 1 unit "greater" than that of the original square. What happens here is that actually a small narrow parallelogram with area of 1 unit is formed in the middle of the figure on the right, and the four pieces only appear to form a rectangle.

After some such examples, students are assumed to be convinced about the dangers of visual observation, and proof is then introduced as the "safe and sure" means of validating geometric statements that students have already confirmed experimentally.

Type 3: False Conclusion. Another approach is to give students a diagram like that in Fig. 15.3a for any arbitrary triangle ABC, followed by a "proof" that it then follows that CA = CB (i.e., that any triangle is isosceles). Children are then told that the obviously false conclusion arises because of the inaccuracy of the diagram (one of the points D and F always falls inside and the other outside, as shown in Fig. 15.3b), and are cautioned to be careful about how a figure looks or is drawn—that, in fact, all sketches are essentially unreliable and we should only rely on our power of reasoning.

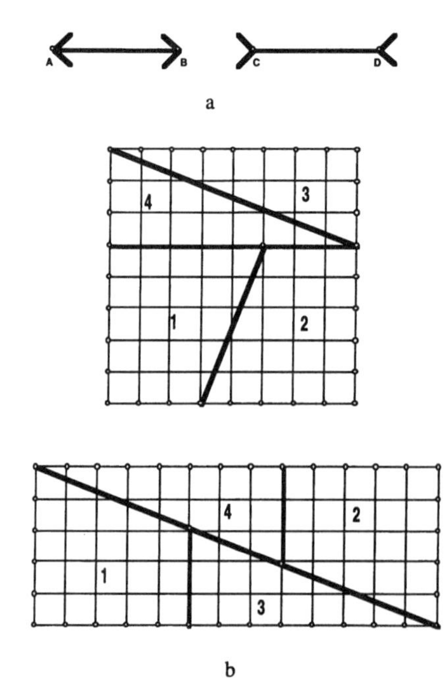

FIG. 15.2 Optical illusion approach to proof. (a) AB = CD. (b) Greater area.

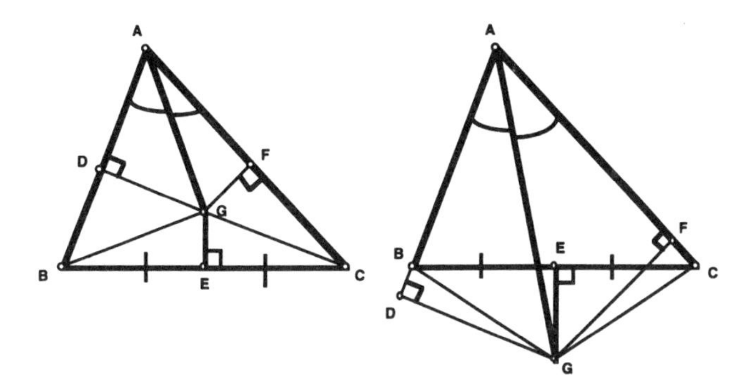

FIG. 15.3 False conclusion approach to proof.

A CRITICISM OF THE TRADITIONAL APPROACH TO THE TEACHING OF PROOF IN GEOMETRY

In what follows the author first criticizes the three types of examples given in the traditional teaching approach just given, before giving a more general critique of the underlying philosophy.

A Criticism of the Traditional Teaching Approach

The problem with the Type 1 examples given is that they are all actually from number theory, and not geometry at all (note that the second example is merely disguised as geometry). Although such examples are very appropriate in motivating proof within the context of number theory, their potential for motivating proof specifically within geometry is highly questionable. The author has yet to see a geometric configuration that has an invariant property for a very large number of cases (e.g., $n = 10^{10}$), but then suddenly breaks down for $n + 1$ cases!

It is furthermore important to note that there are subtle qualitative differences between such number-theoretic examples and the results of elementary geometry. First, the variables involved in the former are discrete, but the variables are continuous in the latter (e.g., angles, lengths). This is particularly the strong feature of drag-mode software like Cabri (Laborde & Bellemain, 1994) and Sketchpad (Jackiw, 1994), namely, that they allow for the *continuous* arbitrary variation and evaluation of geometric configurations. (Strictly speaking, these drag-mode transformations are only extremely good simulations of continuous variation, because the computer can calculate only discretely. More correctly, these variations are *near-continuous*.)

Second, many of the elementary geometry theorems are far more self-evident than these number-theoretic examples and can even be explained informally. Although students may not be able to articulate these subtle differences, many do sense it and are not necessarily convinced by such number-theoretic examples of the need for proof (as a means of verification) in geometry. (Many children of course quickly learn to play the teacher's game and start subscribing to this view only to please the teacher. After many years of "indoctrination" of this type, it sometimes takes a lot of deep probing to uncover children's actual personal views in this regard.)

The Type 2 examples (optical illusions) are deceitful: They do not encourage a need for deductive proof at all. For example, in the first two cases children are told to actually *measure* the lengths of AB and CD, finding them in fact to be equal. These examples therefore encourage measurement as the appropriate means of conviction/verification.

One of the famous French mathematicians once said, "Geometry is the art of drawing correct conclusions from incorrectly drawn sketches." But the false conclusion in the Type 3 example shows how easily a (correct) logical argument can lead to a fallacy because of a construction error or a mistaken assumption in a sketch. Instead of motivating a need for proof, such examples actually emphasize the importance of quasi-empirical testing (i.e., the *accurate* construction of some examples).

These strategies of attempting to raise doubts in order to create a need for proof are even less likely to be successful when geometric conjectures have been thoroughly investigated through their continuous variation with drag-mode software like Cabri or Sketchpad (see Olive, chap. 16, this volume). When students are able to produce numerous corresponding configurations easily and rapidly then they simply have no (or very little) need for further conviction/verification. The problem is further intensified by a facility on Cabri that enables students to check whether certain features of configurations such as concurrency, collinearity, parallelness, perpendicularity, and equality of lengths are true in general. If not true in general, this facility produces a counterexample shown on screen. The computer, functioning as a "proof machine," reduces (in effect, eliminates) the students' need for self-generated proof (verification).

A Criticism of the Underlying Philosophy

As pointed out by Bell (1976), the traditional view of verification/conviction being the main (or only) function of proof "avoids consideration of the real nature of proof [because conviction in mathematics is often obtained] by quite other means than that of following a logical proof" (p. 24). Research mathematicians, for instance, do not often scrutinize published proofs in detail, but are rather led by the established authority of the author, the testing of special cases, and an informal evaluation whether "the methods and result fit in, seem reasonable" (Davis & Hersh, 1986, p. 67).

With very few exceptions, teachers of mathematics seem to believe that a proof, for the mathematician, provides absolute certainty and that it is therefore the absolute authority in the establishment of the validity of a conjecture. They seem to hold the naive view described by Davis and Hersh (1986) that behind each theorem in the mathematical literature there stands a sequence of logical transformations moving from hypothesis to conclusion, absolutely comprehensible, and irrefutably guaranteeing truth. This view, however, is false. Proof is not necessarily a prerequisite for conviction—to the contrary, conviction is probably far more frequently a prerequisite for the finding of a proof.

A mathematician simply does not think: "Hmm ... this result looks very doubtful and suspicious; therefore, let's try to prove it." For what other weird or obscure reasons, would we then sometimes spend months or years to prove certain conjectures, if we weren't already reasonably convinced of their truth?

Polya (1954) wrote, for example, that

> having verified the theorem in several particular cases, we gathered strong inductive evidence for it. The inductive phase overcame our initial *suspicion* and gave us a strong *confidence* in the theorem. Without such *confidence* we would have scarcely found the courage to undertake the proof which did not look at all a routine job. When you have satisfied yourself that the theorem is *true*, you start *proving* it. (emphasis added, pp. 83–84)

In situations where conviction provides the motivation for looking for a proof, the function of an eventual proof for the mathematician clearly cannot be that of verification/conviction, but has to be looked for in terms of explanation, discovery, communication, systematization, self-realization, and so forth.

Absolute certainty also does not exist in real mathematical research, and personal conviction usually depends on a combination of intuition, quasi-empirical verification, and the existence of a logical (but not necessarily rigorous) proof. In fact, a very high level of conviction may sometimes be reached even in the absence of a proof. For instance, in their discussion of the "heuristic evidence" in support of the still unproved twin prime pair theorem and the famous Riemann hypothesis, Davis and Hersh (1983) concluded that this evidence is "so strong that it carries conviction even without rigorous proof" (p. 369).

That conviction for mathematicians is not reached by proof alone is also strikingly borne out by the remark of a previous editor of the *Mathematical Reviews* that approximately one half of the proofs published in it were incomplete and/or contained errors, although the theorems they were purported to prove were essentially true (Hanna, 1983, p. 71). It therefore seems that the reasonableness of results often enjoys priority over the existence of a completely rigorous proof. It is furthermore a commonly held view among today's mathematicians that there is no such thing as a rigorously complete proof (see Hanna, 1983, 1989; Kline, 1982). First, there is the problem that no absolute standards exist for the evaluation of the logical correctness of a proof nor for its acceptance by the mathematical community as a whole. Second, as Davis and Hersh (1986) pointed out, mathematicians usually only publish those parts of their arguments that they deem important for the sake of conviction, thus leaving out all routine calculations and manipulations, which can be done by the reader. Therefore a "complete proof simply means proof in sufficient detail to convince the intended audience" (Davis & Hersh, 1986, p. 73).

In addition, attempts to construct rigorously complete proofs lead to such long, complicated proofs that an evaluative overview becomes impossible and at the same time the probability of errors becomes dangerously high. For example, Manin (1981, p. 105) estimated that rigorous proofs of the two Burnside conjectures would run to about 500 pages each, and a complete proof for Ramanujan's conjecture would run to about 2000 pages. Even the proof for the well-known, but relatively simple, theorem of Pythagoras would take up at least 80 pages, according to Renz (1981, p. 85).

Limitative theorems by Gödel, Tarski, and others during the early part of this century have highlighted the inadequacy of the axiomatic method in general (and deductive proof in particular) for establishing firm foundations for the whole of mathematics. Lakatos (1976, 1978) also argued, from an epistemological analysis of examples from the history of mathematics, that proof can be *fallible,* and that it does not necessarily provide an absolute guarantee for the attainment of certainty:

> There have been considerable and partly successful efforts to simplify Russell's *Principia* and similar logistic systems. But while the results were mathematically interesting and important they could not retrieve the lost philosophical position. The *grandes logiques* cannot be proved true—nor even consistent; they can only be proved false—or even inconsistent. (Lakatos, 1978, p. 31)

When investigating the validity of an unknown conjecture, mathematicians normally not only look for proofs, but also try to construct counterexamples at the same time by means of quasi-empirical testing because such testing may expose hidden contradictions, errors, or unstated assumptions. Frequently, the discovery/construction of counterexamples necessitates a reconsideration of old proofs and the construction of new ones. Personal certainty consequently also depends on the continued absence of counterexamples in the face of quasi-empirical evaluation. More generally, the attainment of personal conviction depends on positive justification and/or negative falsification (see Fig. 15.4).

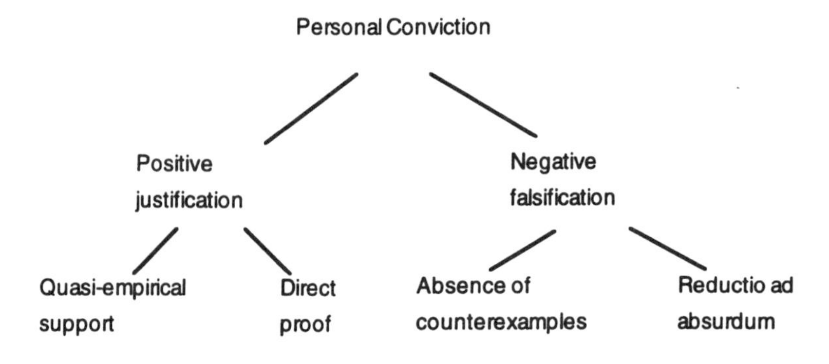

FIG. 15.4 Underpinnings of personal conviction.

THE COMPUTER AS A MEANS OF EXPLORATION AND VERIFICATION

Recent years have seen an explosion in the use of computers as a means of exploration and verification in many areas of mathematics:

> We find ourselves examining on the machine a collection of special cases which is too large for humans to handle by conventional means. The computer is encouraging us to practice unashamedly and in broad daylight, certain customs in which we indulge only in the privacy of our offices, and which we never admitted to students: experimentation. To a degree which never appears in the courses we teach, mathematics is an experimental science.... The computer has become the main vehicle for the experimental side of mathematics. (Pollak, 1984, p. 12)

But typically, the question arises: "How do we know that the computer has not made a mistake?" As Appel and Haken (1984), however, pointed out:

> When proofs are long and highly computational, it may be argued that even when hand checking is possible, the probability of human error is considerably higher than that of machine error; moreover, if the computations are sufficiently routine, the validity of programs themselves is easier to verify than the correctness of hand computations. (p. 172)

Grünbaum (1993), in talking about his computer proof of a result from Euclidean geometry, presented an interesting argument based on work by Davis (1977) that the probability of his (Grünbaum's) findings being false are, for all practical purposes, zero:

> The question arises what is the character of the facts I have been discussing? Do we start trusting numerical evidence (or other evidence produced by computers) as proofs of mathematics theorems? ... Is there any consequence to be drawn from the fact that in example after example numerical evidence establishes the homothety of Q and Q2? ... If we have no doubt—do we call it a theorem? ... I do think that my assertions about quadrangles and pentagons are theorems.... the mathematical community needs to come to grips with the possibilities of new modes of investigation that have been opened up by computers. (p. 8)

This, of course, raises a serious question: With the increasing use of the computer as a means of verification, is there still a need for a deductive proof? Of course, if we see the function of a deductive argument as only that of verification, we might as well now start saying, as Horgan (1993) did, that "proof is dead or dying" and bury it. However, as discussed in the next section, a deductive proof is useful for other reasons.

PROOF AS A MEANS OF EXPLANATION AND DISCOVERY

Although it is possible to achieve confidence in the general validity of a conjecture through the use of dynamic geometry features like property checkers and the drag-mode effect of continuous transformation across the screen, such features offer no satisfactory explanation why a given conjecture may be true. The software merely confirms that the conjecture is true, and even though the consideration of more and more examples may fortify a student's confidence, it gives no psychologically satisfactory sense of illumination—no insight or understanding into how a given conjecture is the consequence of other familiar results. Despite the convincing heuristic evidence in support of the earlier mentioned Riemann hypothesis, for example, we may have a need for explanation:

> It is interesting to ask, in a context such as this, why we still feel the need for a proof.... It seems clear that we want a proof because ... if something is true and we can't deduce it in this way, this is a sign of a lack of understanding on our part. We believe, in other words, that a proof would be a way of understanding *why* the Riemann conjecture is true, which is something more than just knowing from convincing heuristic reasoning that it is true. (Davis & Hersh, 1983, p. 368)

Gale (1990) also clearly emphasized, with reference to Feigenbaum's experimental discoveries in fractal geometry, that the function of the eventual proofs of these discoveries was that of explanation, not of verification:

> Lanford and other mathematicians were not trying to *validate* Feigenbaum's results any more than, say, Newton was trying to validate the discoveries of Kepler on the planetary orbits. In both cases the validity of the results was never in question. What was missing was the *explanation.* Why were the orbits ellipses? Why did they satisfy these particular relations?... There's a world of difference between validating and explaining. (p. 4, emphasis added)

Proof, then, when the results concerned are intuitively self-evident and/or supported by convincing quasi-empirical or computer evidence, is not concerned with "making sure," but rather with "explaining why."

Furthermore, for most mathematicians, the clarification/explanation aspect of a proof is probably of *greater importance* than the aspect of verification. Halmos (quoted in Albers, 1982) noted that although the computer-assisted proof of the four-color theorem by Appel and Haken convinced him that the theorem was true, he would still personally have preferred a proof which also gave an "understanding":

> I am much less likely now, after their work, to go looking for a counterexample to the four-color conjecture than I was before. To that extent, what has happened convinced me that the four-color theorem is true. I have a religious belief that some day soon, maybe six months from now, maybe sixty years from now, somebody will write a proof

> of the four-color theorem that will take up sixty pages in the *Pacific Journal of Mathematics.* Soon after that, perhaps six months or sixty years later, somebody will write a four-page proof, based on the concepts that in the meantime we will have developed and studied and understood. The result will belong to the grand, glorious, architectural structure of mathematics... mathematics isn't in a hurry. Efficiency is meaningless. Understanding is what counts. (pp. 239–240)

Also to Manin (1981) and Bell (1976), explanation is a criterion for a "good" proof when stating respectively, that it is "one which makes us wiser" (Manin, 1981, p. 107) and that it is expected "to convey an insight into *why* the proposition is true" (Bell, 1976, p. 24).

Critics of the amount of deductive rigor at school level often note that deduction in general (and proof in particular) is not a particularly useful heuristic device in the actual discovery of new mathematical results. This view, however, is false. There are numerous examples in the history of mathematics where new results were discovered/invented in a purely deductive manner; in fact, it is completely unlikely that some results (e.g., the non-Euclidean geometries) could ever have been chanced on merely by intuition and/or only using quasi-empirical methods. A proof that explains a result can often lead to unanticipated generalizations. To the working mathematician, therefore, proof is not merely a means of a posteriori verification, but often also a means of exploration, analysis, discovery, and invention (e.g., compare De Jager, 1990; Schoenfeld, 1986), as well as a means of systematization or communication (see de Villiers, 1990; van Asch, 1993).

AN ALTERNATIVE APPROACH TO PROOF IN GEOMETRY

Although most students who have extensively explored geometric conjectures in dynamic geometry environments usually have no further need for conviction (cf. Chazan, 1993), the author has found it relatively easy to solicit further curiosity by asking students why they think a particular result is true—to challenge them to explain it (see also Schumann & de Villiers, 1993). Students quickly admit that inductive verification merely confirms; it gives no satisfactory sense of illumination. They find it quite satisfactory, therefore, to view a deductive argument as an attempt at explanation rather than at verification.

Particularly effective as a first introduction to deductive proof appears to be to present students early on with results where the provision of proofs enables surprising further generalizations. In what follows, four examples of introductory activities that the author has used with his own mathematics education students (prospective senior primary/junior secondary teachers) are briefly discussed, and worksheets for these activites are provided in the Appendix. Some of these student teachers also have tried out similar ideas with their students with a fair amount of success in microteaching contexts or in interview situations with individual students. The author has also conducted a number of workshops with secondary teachers around these ideas, and preliminary feedback seems to indicate that such an approach to proof in dynamic geometry is meaningful (see also Koedinger, chapter 13, this volume).

Working with a Kite

Purpose. The aims of this worksheet (see Worksheet 1 in the Appendix) are (a) to allow students to discover and formulate a conjecture and (b) to guide them toward an explanation that illustrates the discovery function of proof.

Formulation. The line segments consecutively connecting the midpoints of the adjacent sides of a kite form a rectangle (see Fig. 15.5).

Deductive Explanation. A deductive analysis shows that the inscribed quadrilateral is always a rectangle, because of the *perpendicularity* of the diagonals of a kite. For example, according to an earlier discussed property of triangles, we have $EF \parallel AC$ in triangle ABC and $HG \parallel AC$ in triangle ADC (see Fig. 15.5a). Therefore, $EF \parallel HG$. Similarly, $EH \parallel BD \parallel FG$, and therefore EFGH is a parallelogram. Because $BD \perp AC$ (property of kite) we also have, for instance, $EF \perp EH$, which implies that EFGH is a rectangle (a parallelogram with a right angle).

Looking Back. Notice that the property of equal adjacent sides (or an axis of symmetry through one pair of opposite angles) was not used at all. In other words, we can immediately generalize the result to a "perpendicular" quad as shown in Fig. 15.5b. Furthermore, note that the general result was not suggested by the purely empirical verification of the original conjecture. Even a systematic empirical investigation of various types of quadrilaterals would probably not have helped to discover the general case because most people would probably have restricted their investigation to the more familiar quadrilaterals such as parallelograms, rectangles, rhombuses, squares, and rectangles. (Note that from the preceding explanation we can also see that EFGH will always be a parallelogram in any quadrilateral.)

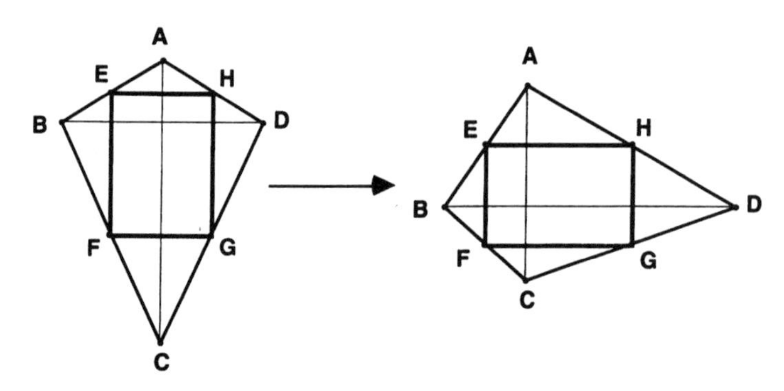

FIG. 15.5 Explaining and generalizing to perpendicular quad.

Working with a Triangle

Purpose. The aims of this worksheet (see Worksheet 2 in the Appendix) are (a) to allow students to discover and formulate a conjecture and (b) to guide them toward an explanation that illustrates the discovery function of proof.

Formulation. The medians of a triangle are concurrent (see Fig. 15.6).

Deductive Explanation. CD and BF are medians intersecting at O. Join A with O and extend to E on BC. We now have to show that E is the midpoint of BC. If we denote the areas of the various triangles by the following notation, area $\Delta ABC \leftrightarrow (ABC)$, we have

$$\frac{BE}{EC} = \frac{(ABE)}{(AEC)} = \frac{(OBE)}{(OCE)} = \frac{(ABE) - (OBE)}{(AEC) - (OCE)} = \frac{(ABO)}{(ACO)}$$

Similarly, we find

$$\frac{CF}{FA} = \frac{(BCO)}{(ABO)} \quad \text{and} \quad \frac{AD}{DB} = \frac{(ACO)}{(BCO)}$$

But AD = DB and CF = FA. Therefore, (ACO) = (BCO) and (BCO) = (ABO), which implies (ACO) = (ABO). But the areas of these two triangles are proportional to BE and EC as shown by the first equation. Thus, BE/EC = 1 implies BE = EC.

Looking Back. Looking back at the first part of the explanation, it is interesting to note that the product of the given ratios is always equal to 1, irrespective of whether D, E, and F are midpoints. For example,

$$\frac{BE}{EC} \times \frac{CF}{FA} \times \frac{AD}{DB} = \frac{(ABO)}{(ACO)} \times \frac{(BCO)}{(ABO)} \times \frac{(ACO)}{(BCO)} = 1$$

This immediately implies the following general result: If three line segments AE, BF, and CD of ΔABC are concurrent, then

$$\frac{BE}{EC} \times \frac{CF}{FA} \times \frac{AD}{DB} = 1$$

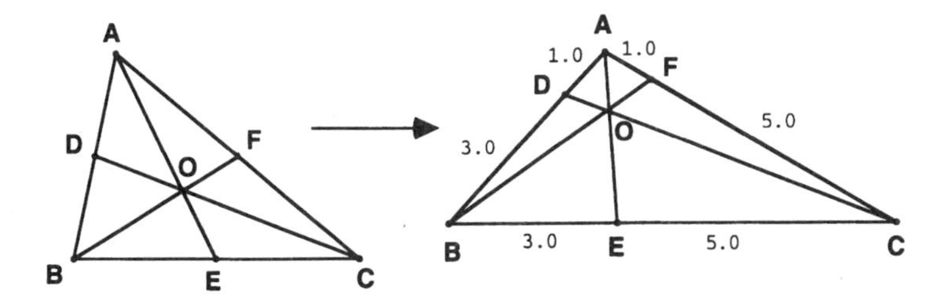

FIG. 15.6 Explaining and generalizing to Ceva's theorem.

This interesting result, Ceva's theorem, was named after the Italian mathematician who published it in 1678. In his honor, the line segments AE, BF, and CD joining the vertices of a triangle to any given points on the opposite sides are called cevians. (Note that apart from the medians, the altitudes and angle bisectors of a triangle can be considered as cevians, if extended to meet the opposite sides.) Although it is not known exactly how he discovered this result, it is likely that he discovered it in a similar fashion as just outlined, and not merely by using construction and measurement.

Working with Equilateral Triangles

Purpose. The aims of this worksheet (see Worksheet 3 in the Appendix) are (a) to allow students to discover and formulate a conjecture and (b) to guide them toward an explanation that illustrates the discovery function of proof.

Formulation. If equilateral triangles DAB, EBC, and FCA are constructed on the sides of a right triangle ABC with $\angle B = 90°$, then DC, EA, and FB are concurrent (see Fig. 15.7). If we call the observed point of concurrence O, then it looks as if the six angles formed at O are each equal to 60°. By measurement and transformation on Cabri or Sketchpad, this can easily be confirmed (see Fig. 15.7a). In other words, quadrilaterals ADBO, BECO, and CFAO must be cyclic because the exterior angles are equal to the opposite interior angles (60°). We can now use this observation to produce the following explanation. (Note that this illustrates another function of quasi-empirical testing and exploration, namely, assistance in the discovery/invention of a deductive explanation.)

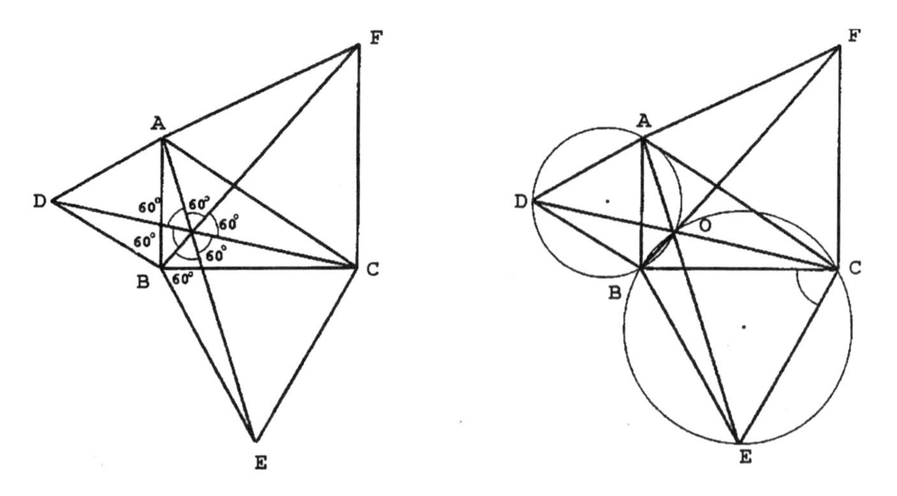

FIG. 15.7 Explaining and generalizing to any triangle.

Deductive Explanation. Construct circumcircles ADB and BEC to intersect in B and O (see Fig. 15.7b). Connect O with A, B, C, D, E, and F. Then $\angle BOE = \angle BCE = 60°$ (inscribed angles on the same chord). But $\angle BOA = 120°$ because ADBO is cyclic. Therefore, AOE is a straight line. Similarly, DOC is a straight line. Also $\angle AOC = 360° - (\angle BOA + \angle BOC) = 360° - 240° = 120°$. Therefore, CFAO is also cyclic, and, as before, it follows that BOF is a straight line.

Looking Back. Because we did not use the property that $\angle B = 90°$, it follows that this result is true for *any* triangle ABC. Again we see that the insight obtained from constructing a deductive explanation enables a further generalization. (It should be pointed out that the point O is normally called the Fermat point of a triangle.)

Working with a Quadrilateral

Purpose. The aims of this worksheet (see Worksheet 4 in the Appendix) are (a) to allow students to discover and formulate a conjecture, (b) to guide them toward two different explanations, and (c) to provide them with a Lakatosian experience by confronting them with a crossed quadrilateral shortly after they've discovered the interior angle sum of simple closed quadrilaterals.

Primitive Formulation. The sum of the angles of any quadrilateral is equal to 360° (see Fig. 15.8).

Deductive Explanation. Consider the convex and concave quadrilaterals ABCD shown in the tessellations in Fig. 15.9. In both cases, drawing diagonal BD would divide quadrilateral ABCD into two triangles ABD and BCD. (Note that the *reflexive* angle at D is the interior angle of concave quad ABCD and is therefore $360° - 119° = 241°$). Because the sum of the angles of a triangle is equal to 180°, the sum of the angles of a quadrilateral is $2 \times 180° = 360°$.

Alternative Deductive Explanation. Consider the convex and concave quadrilaterals ABCD shown in Fig. 15.9. Imagine that you are a bug crawling from A to B, and at B you turn through the angle p as indicated to face in the direction of C. Continue crawling around the perimeter, turning as indicated through angles q, r, and s at C, D, and A, respectively, until at A you are facing in the same direction as you started. Because you are now facing in the same direction as when you started, you must have completed one revolution, 360°.

Therefore, $p + q + r + s = 360°$. The sum of the interior angles of ABCD is given by $(180° - p) + (180° - q) + (180° - r) + (180° - s) = 720° - (p + q + r + s) = 360°$.

(Note that in the concave case, angle r is negative in relation to the other angles because it has an opposite direction of rotation. The size of the interior angle at D is therefore $180° - r = 180° + |r|$.)

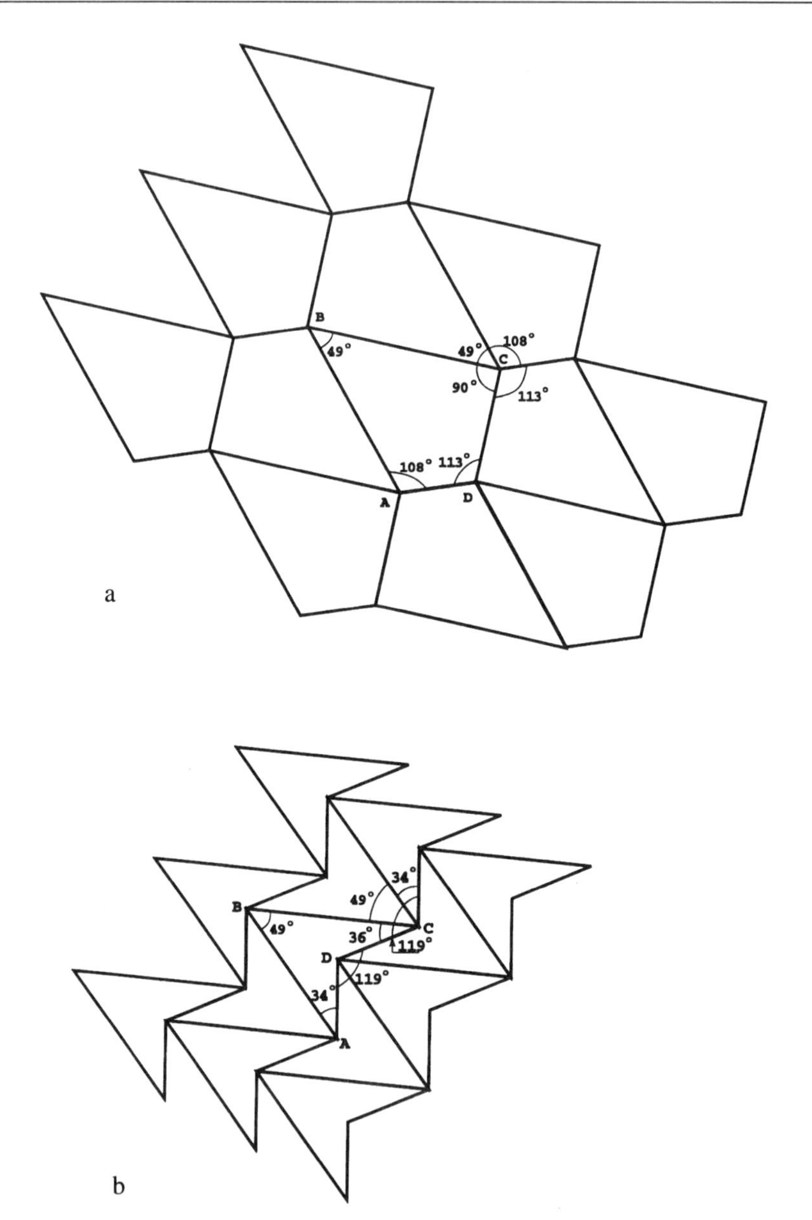

FIG. 15.8 Discovering interior angle sum of quadrilaterals.

A Counterexample?

Construct a quadrilateral ABCD and measure its angles. Drag vertex D over side AB to obtain a figure similar to the one shown in Fig. 15.10. (On several occasions I've actually observed students using dynamic geometry software accidentally dragging polygons into crossed configurations

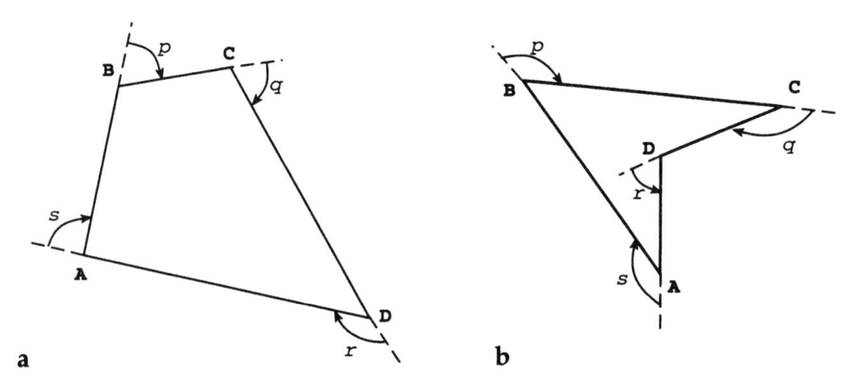

FIG. 15.9 Explaining exterior angle sum of convex (a) and concave (b) quadrilaterals.

such as this, something that is clearly not likely to arise or even be considered in standard paper-and-pencil work.) Is the sum of its interior angles equal to 360°? Is the figure ABCD a "quadrilateral"? How does this relate to the result formulated and explained earlier? What do we mean by "interior" angles?

Most people's first reaction to such a counterexample is one of "monster-barring," in support of the theorem that the sum of the interior angles of all quadrilaterals is 360°. We might therefore try to define a quadrilateral in such a way that figures like these are excluded. Lakatos

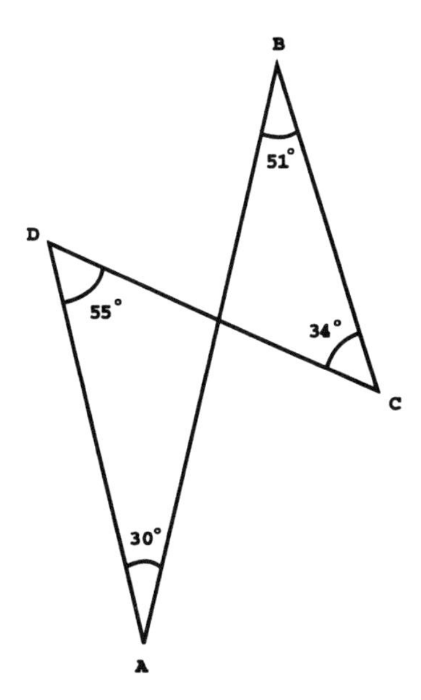

FIG. 15.10 The crossed quadrilateral.

(1976) describes a similar situation after the discovery of a counterexample to the Euler–Descartes theorem for polyhedra by the characters in his book:

Delta: But why accept the counter-example? We proved our conjecture—now it is a theorem. I admit that it clashes with this so-called "counter-example." One of them has to give way. But why should the theorem give way, when it has been proved? It is the "criticism" that should retreat. It is fake criticism. This pair of nested cubes is not a polyhedron at all. It is a *monster*, a pathological case, not a counter-example.

Gamma: Why not? *A polyhedron is a solid whose surface consists of polygonal faces*. And my counter-example is a solid bounded by polygonal faces.

Delta: Your definition is incorrect. A polyhedron must be a *surface:* it has faces, edges, vertices, it can be deformed, stretched out on a blackboard, and has nothing to do with the concept of "solid." *A polyhedron is a surface consisting of a system of polygons*. (p. 16)

From the preceding extract, we also see that refutation by counterexample usually depends on the meaning of the terms involved, and consequently definitions are frequently proposed and argued about. How can we define quadrilaterals? What do we mean by "interior" angles? How can we "save" the preceding theorem?

Defining

The intuitive essential meaning of a quadrilateral is that it has four sides (or line segments) and four vertices (or points). In other words, we could define a quadrilateral ABCD as a plane figure consisting of four (non-collinear) points A, B, C, and D, connected by four line segments AB, BC, CD, and DA.

According to this definition, the figure shown in Fig. 15.10 is a quadrilateral, and we refer to it as a *crossed* quadrilateral. (It also makes good sense to consider it a quadrilateral because the midpoints of its sides also form a parallelogram—see Worksheet 1 in the Appendix.) We can also refer to convex and concave quadrilaterals as simple quadrilaterals because they do not have any sides crossing each other.

Typically, students do not want to accept Fig. 15.10 as a quadrilateral. The following are responses obtained during individual interviews:

A (Grade 11): But the way I look at it is that these sides [AC & BD] haven't been put in. This is the one side, and this is the other... AB and DC are just diagonals. If you add those sides, then it'll be a complete quadrilateral, and you have four angles, and it'll equal 360°.

B (Grade 12): One can't say it's a quad if the angles are not 360°. It is not 360°, therefore it is not a quadrilateral....

C (Grade 11): Can't you make this into a quadrilateral? . . . Put D where B is, and C where D is . . . if it lies on another type of plane, one can see that it is merely twisted.

D (Grade 11): I can't give a reason, but it is not a quadrilateral . . . I don't know why . . . If I add these two angles [indicates angles AOD and BOC, where O is the intersection of AB and CD], then it will give you 360°.

Students then also spontaneously tried to define a quadrilateral in such a way as to exclude crossed quadrilaterals:

A (Grade 11): We should say that two sides may not cross . . . they can't intersect. Yes, that would be the best thing, then you can't draw something like that.

D (Grade 11): It must be consecutive points on a circle. How can I put it? If one goes clockwise, then the points must be consecutive, for example, A, B, C, and D.

One way to extend the notion of "internal" angles to crossed quadrilaterals is by first analyzing and defining the notion of internal angles for convex and concave quadrilaterals and then consistently applying that definition to crossed quadrilaterals. (This is a strategy often used in mathematics when extending certain concepts beyond their original domain.)

Suppose we walk clockwise from A to B, B to C, and so on, around the perimeter of the convex quadrilateral shown in Fig. 15.11. The internal angle at each vertex will then be the angle through which the next side must be rotated clockwise (with the vertex as rotation center) to coincide with the previous side. In the same way we can now determine the internal angles of the crossed quadrilateral by walking around the perimeter as shown. This leads us to the surprising conclusion that the two reflexive angles indicated at A (360° – ∠BAD) and D (360° – ∠ADC) are the *"internal"* angles of the crossed quadrilateral ABCD.

We can now also calculate the sum of the interior angles of a crossed quadrilateral as follows: ∠ABC + ∠BCD = acute ∠BAD + acute∠CDA; therefore, ∠ABC + ∠BCD + (360° – ∠BAD) + (360° – ∠CDA) = 720°.

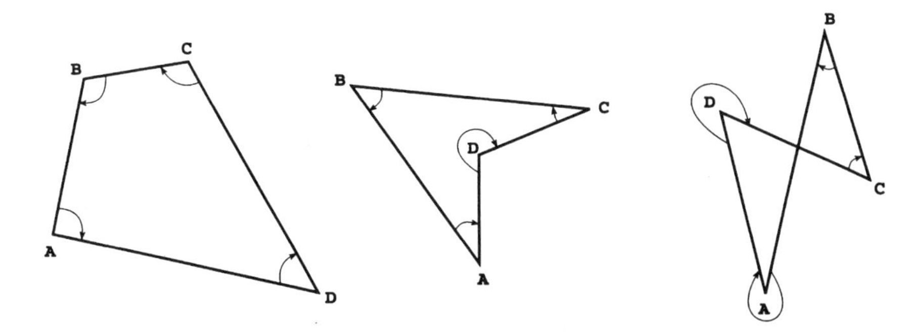

FIG. 15.11 Defining internal angles in quadrilaterals.

Reformulation

1. The sum of the interior angles of any simple quadrilateral is 360°. (Note that the first explanation assumes that at least one of the diagonals falls inside the quadrilateral, which makes the explanation invalid for crossed quadrilaterals. The second explanation is invalid for crossed quadrilaterals because the total turning $p + q + r + s$ is not 360°, but 0°—the two clockwise turns at B and C are canceled out by the two anticlockwise turns at D and A).

2. The sum of the interior angles of any crossed quadrilateral is 720° (see preceeding explanation).

DISCUSSION

"Proof" was not used anywhere in the preceding activities or in their worksheets in the Appendix. Instead, the word *explanation* was used precisely to emphasize the intended function of the given deductive arguments. The problem is that the word *proof* in everyday language carries with it predominantly the idea of verification or conviction, and to use it in an introductory context would implicitly convey this meaning, even if the intended meaning is that of explanation. Tentative results with worksheets like those in the Appendix indicate that the presentation of proof in dynamic geometry as a means of explanation appears to be a viable alternative to the traditional approach.

The teacher's language is particularly crucial in this introductory phase. Instead of saying the usual, "We cannot be sure that this result is true for all possible variations, and we therefore have to (deductively) prove it to make absolutely sure," students find it much more meaningful if the teacher says: "We now know this result to be true from our extensive experimental investigation. Let us however now see if we can EXPLAIN WHY it is true in terms of other well-known geometric results, in other words, how it is a logical consequence of these other results."

It is necessary to discuss in some detail what is meant by an "explanation." For example, regular observation that the sun rises every morning clearly does not constitute an explanation; it only reconfirms the validity of the observation. To explain something, we have to explain it in terms of something else (e.g., the rotation of the earth around the polar axis). Students may need to be guided to appropriate explanations (proofs), the production of alternative explanations, and their comparison. Lack of initial participation in the actual activity of explaining (proving) has also been reported by teachers who have tried out some of these ideas at school level, and it appears that, in our experience, only after considerable concerted exposure to work of this kind do students become proficient in constructing their own explanations and critically comparing them. What is significant, however, is that, when proof is seen as explanation, substantial improvement in students' *attitudes* toward proof appears to occur.

The activities in this chapter, of course, assume that students have already, over a period of, say, 2 to 3 years, accumulated quite a large body of geometric knowledge by experimental exploration with dynamic geometry software. For example, students should already know various properties of quadrilaterals and that the line segment connecting the midpoints of two sides of a triangle is parallel to and equal to half the third side. They should also already know the area formula for triangles, properties of cyclic quadrilaterals, and the sum of the angles of a triangle.

Later we can reason "backward" to arrive at the basic axioms and definitions of geometry. This process of a posteriori axiomatization is typically used in real mathematical research: Axioms are usually not the beginning, but the end of such research.

The last section (working with a counterexample) is intended to recreate a typical Lakatosian situation where a counterexample to a result is presented after its deductive explanation (proof). To convince the students that it might be possible to consider Fig. 15.11 as a quadrilateral, it is useful to remind them of the exploration of the work with the kite and the consequent explanation, showing them that if we consecutively connect the midpoints of its sides, we also obtain a parallelogram. Only after much discussion is it possible to introduce and clarify an acceptable definition for quadrilaterals in general, making distinctions between simple closed quadrilaterals and crossed quadrilaterals, and appropriately reformulating the result. Another example that could be used as a follow-up is to consider the following and its explanation: "The opposite angles of a cyclic quadrilateral are supplementary." Again a crossed cyclic quad is the counterexample. It also should be noted that although all simple quadrilaterals tessellate, crossed quadrilaterals cannot tessellate because they overlap.

It should perhaps be pointed out that I fully agree with Hanna (1995) that Lakatos's model of heuristic refutation ought not to be taken as an all-encompassing model for the philosophy of mathematics, nor for curriculum development and design in general, as there are many historical counterexamples to the process Lakatos describes. However, I believe that the Lakatosian (fallibilist) view compliments the Platonist and Formalist views of mathematics and that we should ensure that students are given activities and experiences that reflect each of these. As Davis and Hersh (1983) pointed out, none of these views are "correct" as each one is incomplete and one-sided if taken only by itself (p. 359).

SOME CONCLUDING COMMENTS

Traditionally the role and function of proof in the geometry classroom have either been completely ignored, or proof has been presented as a means of obtaining certainty (i.e., within the context of verification/conviction). However, as pointed out in this chapter, mathematicians often construct proofs for reasons other than that of verification/conviction (cf. Hersh, 1993). The popular formalistic idea of many contemporary mathematics teachers, that conviction is a one-to-one mapping of deductive proof, should therefore be completely abandoned; conviction is not gained exclusively from proof alone, nor is the only function of proof that of verification/conviction. Not only does such an approach in a dynamic geometry environment represent intellectual dishonesty, but it does not make sense to students.

Rather than one-sidedly focusing on proof as a means of verification in geometry, the more fundamental function of explanation and discovery ought to be used to present proof as a meaningful activity. At the same time, attention should be given to the communicative aspects thereof by actually negotiating with students the criteria for acceptable evidence, explanations, and arguments. Furthermore, in mathematics, as anyone with a bit of experience will testify, the systematization function of proof comes to the fore only at a very advanced stage and should, therefore, be withheld in an introductory course to proof.

REFERENCES

Albers, D. J. (1982). Paul Halmos: Maverick mathologist. *Two-Year College Mathematics Journal, 13*(4), 234–241.

Alibert, D. (1988). Towards new customs in the classroom. *For the Learning of Mathematics, 8*(2), 31–35, 43.

Appel, K., & Haken, W. (1984). The four-color problem. In D. M. Campbell & J. C. Higgens (Eds.), *Mathematics: People, problems, results* (Vol. 2, pp. 154–173). Belmont, CA: Wadsworth.

Bell, A. W. (1976). A study of students' proof-explanations in mathematical situations. *Educational Studies in Mathematics, 7,* 23–40.

Chazan, D. (1990). Quasi-empirical views of mathematics and mathematics teaching. In G. Hanna & I. Winchester (Eds.), *Creativity, thought and mathematical proof.* Toronto: Ontario Institute for Studies in Education.

Chazan, D. (1993). High school students' justification for their views of empirical evidence and mathematical proof. *Educational Studies in Mathematics, 24,* 359–387.

Davis, P. J. (1977). Proof, completeness, transcendentals and sampling. *Journal of the Association of Computing Machines, 24,* 298–310.

Davis, P. J. & Hersh, R. (1983). *The mathematical experience.* London: Pelican.

Davis, P. J. & Hersh, R. (1986). *Descartes' dream.* New York: Harcourt, Brace, & Jovanovich.

De Jager, C. J. (1990). When should we use pattern? *Pythagoras, 23,* 11–14.

de Villiers, M. D. (1990). The role and function of proof in mathematics. *Pythagoras, 24,* 17–24.

Ernest, P. (1991). *The philosophy of mathematics education.* Bristol: Falmer.

Gale, D. (1990). Proof as explanation. *Mathematical Intelligencer, 12*(1), 4.

Gonobolin, F. N. (1975). Students' comprehension of geometric proofs. In J. W. Wilson (Ed.), *Soviet studies in the psychology of learning and teaching mathematics* (Vol. 12), *Problems of instruction.* Chicago: University of Chicago Press. (Original work published 1954)

Grünbaum, B. (1993). Quadrangles, pentagons, and computers. *Geombinatorics, 3,* 4–9.

Hanna, G. (1983). *Rigorous proof in mathematics education.* Toronto: Ontario Institute for Studies in Education.

Hanna, G. (1989). More than formal proof. *For the Learning of Mathematics, 9*(1), 20–23.

Hanna, G. (1995). Challenges to the importance of proof. *For the Learning of Mathematics, 15*(3), 42–49.

Hersh, R. (1979). Some proposals for reviving the philosophy of mathematics. *Advances in Mathematics, 31,* 31–50.

Hersh, R. (1993). Proving is convincing and explaining. *Educational Studies in Mathematics, 24,* 389–399.
Horgan, J. (1993, October). The death of proof. *Scientific American,* 74–82.
Jackiw, N. The Geometer's Sketchpad [Computer software]. Berkeley, CA: Key Curriculum Press.
Kline, M. (1973). *Why Johnny can't add: The failure of the new math.* New York: St. Martin's.
Kline, M. (1982). *Mathematics: The loss of certainty.* Oxford: Oxford University Press.
Kline, M. (1984). The meaning of mathematics. In D. M. Campbell & J. C. Higgens (Eds.), *Mathematics: People, problems, results* (Vol. 2, pp. 11–18). Belmont, CA: Wadsworth.
Laborde, J.-M., & Bellemain, F. (1994). Cabri-geometry II [Computer software]. Dallas, TX: Texas Instruments.
Lakatos, I. (1976). *Proofs and refutations.* Cambridge: Cambridge University Press.
Lakatos, I. (1978). *Mathematics, science and epistemology.* Cambridge: Cambridge University Press.
Manin, Y. I. (1981). A digression on proof. *Two–Year College Mathematics Journal, 12*(2), 104–107.
Pollak, H. O. (1984). The effects of technology on the mathematics curriculum. In A. I. Olivier (Ed.), *The Australian experience: Impressions of ICME 5.* Centrahil, South Africa: Association for Mathematics Education of South Africa.
Polya, G. (1954). *Mathematics and plausible reasoning: Induction and analogy in mathematics* (Vol. 1). Princeton, NJ: Princeton University Press.
Renz, P. (1981). Mathematical proof: What it is and what it ought to be. *Two-Year College Mathematics Journal, 12*(2), 83–103.
Schoenfeld, A. H. (1986). On having and using geometric knowledge. In J. Hiebert (Ed.), *Conceptual and procedural knowledge: The case of mathematics* (pp. 225–264). Mahwah: NJ: Lawrence Erlbaum Associates.
Schumann, H., & de Villiers, M. D. (1993). Continuous variation of geometric figures: Interactive theorem finding and problems in proving. *Pythagoras, 31,* 9–20.
van Asch, A. G. (1993). To prove, why and how? *International Journal of Mathematical Education in Science and Technology, 24*(2), 301–313.
Volmink, J. D. (1990). The nature and role of proof in mathematics education. *Pythagoras, 23,* 7–10.

APPENDIX

Worksheet 1

(a) Construct a dynamic kite using the properties of kites explored and discussed in our previous lessons.
(b) Check to ensure that you have a dynamic kite, i.e., does it always remain a kite no matter how you transform the figure? Compare your construction(s) with those of your neighbors—is it the same or different?
(c) Next construct the midpoints of the sides and connect the midpoints of adjacent sides to form an inscribed quadrilateral.
(d) What do you notice about the inscribed quadrilateral formed in this way?
(e) State your conjecture.

(f) Grab any vertex of your kite and drag it to a new position. Does it confirm your conjecture? If not, can you modify your conjecture?
(g) Repeat the previous step a number of times.
(h) Is your conjecture also true when your kite is concave?
(i) Use the property checker of Cabri to check whether your conjecture is true in general.
(j) State your final conclusion. Compare with your neighbors—is it the same or different?
(k) Can you explain why it is true? (Try to explain it in terms of other well-known geometric results. Hint: construct the diagonals of your kite. What do you notice?)
(l) Compare your explanation(s) with those of your neighbors. Do you agree or disagree with their explanations? Why? Which explanation is the most satisfactory? Why?

Worksheet 2

(a) First construct a triangle. Next construct the midpoints of the sides.
(b) Now connect the midpoint of each side with the opposite vertex of your triangle. These line segments are called medians. What do you notice about these medians?
(c) State your conjecture.
(d) Grab any vertex of your triangle and drag it to a new position. Does it confirm your conjecture? If not, can you modify your conjecture?
(e) Repeat the previous step a number of times.
(f) Is your conjecture also true when your triangle is obtuse, scalene, or right-angled?
(h) Use the property checker of Cabri to check whether your conjecture is true in general.
(i) State your final conclusion. Compare with your neighbors—is it the same or different?
(j) Can you explain why it is true? (Try to explain it in terms of other well-known geometric results. Hint: Consider the ratios between the areas of triangles ABO and ACO, BCO and ABO, and ACO and BCO.)
(k) Compare your explanation(s) with those of your neighbors. Do you agree or disagree with their explanations? Why? Which explanation is the most satisfactory? Why?

Worksheet 3

(a) First construct a dynamic right triangle ABC with $\angle B = 90°$.
(b) Using the macro-construction facility of Cabri (or the script facility of Sketchpad), outwardly construct equilateral triangles DAB, EBC and FCA on the sides of your right triangle.

(c) Draw DC, EA and FB. What do you notice?
(d) State your conjecture.
(e) Grab any vertex of your right triangle and drag it to a new position. Does it confirm your conjecture? If not, can you modify your conjecture?
(e) Repeat the previous step a number of times.
(f) Is your conjecture also true for any shape of your right triangle?
(h) Use the property checker of Cabri to check whether your conjecture is true in general.
(i) State your final conclusion. Compare with your neighbours—is it the same or different?
(j) Can you explain why it is true? (Try to explain it in terms of other well-known geometric results. Hint: Consider the six angles at the point of intersection.)
(k) Compare your explanation(s) with those of your neighbours. Do you agree or disagree with their explanations? Why? Which explanation is the most satisfactory? Why?

Worksheet 4

(a) Construct a dynamic quadrilateral ABCD. Rotate the quadrilateral through around the midpoints of all its sides (give it half-turns). (Time-saving hint: In Cabri first define a macro-construction for half-turning a quadrilateral around the midpoint of one of its sides, and then use it for [a] and [b].)
(b) Give each of the newly formed quadrilaterals a half-turn around the midpoint of its sides.
(c) Measure the four angles of your quadrilateral ABCD, as well as the four angles around vertex C.
(d) Carefully compare the angles around vertex C with the angles of ABCD. What do you notice?
(e) What can you say about the sum of all the angles around vertex C? What does this say about the sum of the angles of quadrilateral ABCD?
(f) Grab any vertex of your quadrilateral ABCD and drag it to a new position. What can you say about the sum of all the angles around vertex C? What does this say about the sum of the angles of quadrilateral ABCD?
(g) Repeat the previous step a number of times.
(h) Is your observation also true if ABCD is concave?
(i) State your final conclusion. Compare with your neighbors—is it the same or different?
(j) Can you explain why it is true? (Try to explain it in terms of other well-known geometric results. Hint: Draw a diagonal of the quadrilateral ABCD.)
(k) Compare your explanation(s) with those of your neighbors. Do you agree or disagree with their explanations? Why? Which explanation is the most satisfactory? Why?

16

Opportunities to Explore and Integrate Mathematics with the Geometer's Sketchpad

John Olive
University of Georgia

A PROBLEM WITH GEOMETRY LEARNING IN SCHOOLS

The *Curriculum and Evaluation Standards for School Mathematics* (National Council of Teachers of Mathematics [NCTM], 1989) recommends that students in elementary and middle-school grades study geometric properties and relationships through practical experiences with shapes and patterns, integration of mathematical experiences with their art work, and the use of geometric computer programs. The *Standards* emphasize that "insights and intuitions about two- and three-dimensional shapes and their characteristics, the interrelationships of shapes, and the effects of changes to shapes are important aspects of spatial sense" (p. 48). These geometric experiences are critical for our young students. Research based on the van Hiele levels of geometric thinking (Fuys, Geddes, & Tischler, 1988; Scally, 1990; van Hiele, 1986) has indicated that students entering high school have very little knowledge or experience of geometric properties and relationships; most are operating at the visual level of geometric thought in terms of the van Hiele model. These students can do little more than recognize different geometric shapes. They do not realize that a square is also a rectangle and a rhombus, or that all three are also parallelograms. Many do not realize that a square must have four right angles or that all four sides are congruent. Most have never heard of a "line of symmetry," let alone understand why a square has four lines of symmetry whereas a non-square rectangle has only two. It is not surprising, then, that results of the Fourth National Assessment of Educational Progress (NAEP) indicate that fewer than a third of Grade 7 students (and only a quarter of the Grade 11 students not taking geometry) could identify perpendicular lines, and that for the same set of students, responses to items involving properties of simple figures (such as triangles and circles) were very poor (ranging from 8 to 42% correct; Lindquist, 1989). It is no wonder then that a decade ago fewer than one third of our high-school students made it through high-school geometry (National Commission on Excellence in Education, 1983).

A COMPUTER TOOL FOR TEACHING AND LEARNING GEOMETRY

Several new computer tools, such as Cabri Geometer (Baulac, Bellemain, Laborde, 1992, 1994), The Geometer's Sketchpad (Jackiw, 1991), and Geometry Inventor (Brock, Cappo, Dromi, Rosin, & Shenkerman, 1994) have been developed during the last 5 years. These new tools provide teachers and students with an opportunity to explore geometric properties and relationships both intuitively and inductively (see Goldenberg, Cuoco, & Mark, chap. 1, this volume). The dynamic nature of these tools makes them both exciting and accessible even to elementary students. Although these three software tools share similar dynamic qualities, they have developed in slightly different ways, each providing unique capabilities not shared by the others. This chapter focuses on teachers' and students' use of the Geometer's Sketchpad.

It is difficult to describe this software in the medium of static print or even with the aid of pictures. Its uniqueness and its power lie in its dynamic qualities. It is a dynamic geometric construction kit, which takes full advantage of the mouse interface of the computer (see Goldenberg and Cuoco, chap. 14, this volume, for an in-depth discussion on the nature of dynamic geometry). Simply by clicking on a tool icon the user can choose to construct points, lines, line segments, rays, or circles; these geometric objects are then constructed on the screen simply by clicking and dragging the mouse. Geometric figures can be constructed by connecting their components. A triangle, for example, can be constructed by connecting three line segments. This triangle, however, is not a single, static instance of a triangle, which would be the result of drawing three line segments on paper, but in essence a prototype for all possible triangles. By dragging a vertex of this triangle with the mouse, the length and orientation of the two sides of the triangle meeting at that vertex will change as the vertex moves. The mathematical implications of even this most simple of operations was brought home to me when my 7-year-old son was "playing" with the software. As he moved the vertex around the screen he asked me if the shape was still a triangle. I asked him what he thought. After turning his head and looking at the figure from different orientations he declared that it was. I asked him why, and he replied that it still had three sides. He continued to make triangles, which varied from squat fat ones to long skinny ones (his terms), that stretched from one corner of the screen to another (with no one side horizontal). The real surprise came when he moved one vertex onto the opposite side of the triangle, creating the appearance of a single line segment. He again asked me if this was still a triangle. I again threw the question back to him, and his reply was: "Yes.

It's a triangle lying on its side." I contend that this 7-year-old child had constructed for himself during that 5 minutes of exploration with Sketchpad a fuller concept of "triangle" than most high-school students ever achieve. His last comment also indicates intuitions about plane figures that few adults ever acquire: that such figures have no thickness and that they may be oriented perpendicular to the viewing plane. Such intuitions are the result of what Goldenberg et al. (chap. 1, this volume) refer to as "visual thinking."

Sketchpad provides the student with tools to construct midpoints of segments, perpendicular lines, parallel lines, angle bisectors; to construct segments, rays, or lines, given two points; and to construct circles given two points or a fixed radius (which can be used to duplicate segments in the same way that we use a compass). Sketchpad also allows the student to perform geometric transformations (translations, reflections, rotations, and dilations) on any selected group of objects. The transformed images are live images in the sense that any change in the original objects will cause a corresponding change in the transformed image. This facility allows dynamic investigations of the properties of transformations, which are just not possible in any other medium. As students change the shape or position of the original object (using the mouse), they are able to observe simultaneously the corresponding changes in the transformed image. My 7-year-old son again demonstrated for me the mathematical power that such a dynamic facility provides. After creating a double reflection of a triangle in perpendicular "mirrors" (with my assistance), he then experimented by dragging a vertex of the original triangle around the screen (see Fig. 16.1). He was fascinated by the movements of the corresponding vertices of the three image triangles. He was soon challenging himself to predict the path of a particular vertex, given a movement of an original vertex. At one point he went to the chalkboard, sketched the axes and triangles, and indicated with an arrow where he thought an image vertex would move. He then carried out the movement of the original vertex on the screen and was delighted to find his prediction correct. Goldenberg and Cuoco (chap.14, this volume) challenge us to think seriously

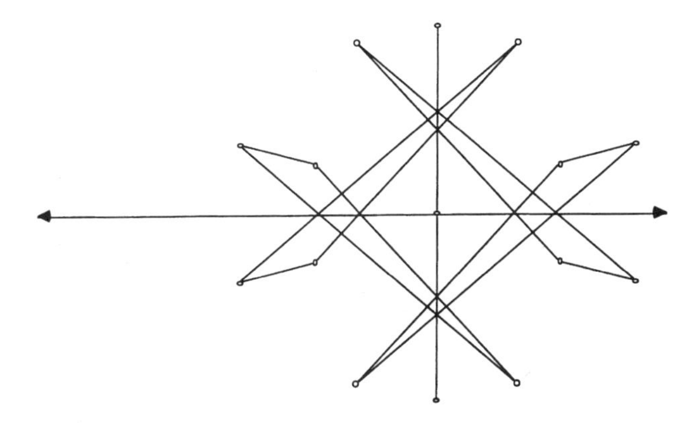

FIG. 16.1 Double reflection of a triangle from Geometer's Sketchpad.

about the educational consequences for children working in an environment in which such mental reasoning with spatial relationships can be provoked.

Sketchpad also provides the student with options to measure lengths of segments, distance between points, perimeter and area of polygons and circles, angles, radius, arc angles, and arc length of circles and to perform calculations using any combination of measurements. The measurements and calculations can be placed anywhere on the screen and are "live" in the sense that they will change as the measured objects are changed by mouse movements. Thus, for example, students can test conjectures about the ratio of the area of a quadrilateral to the area of the quadrilateral formed by joining the midpoints of its sides simply by constructing the figure, measuring both areas, asking the computer to compute and display the ratio, and then changing the original quadrilateral in any way they wish, using the mouse and watching the effect of this change on the areas and their ratio (see Fig. 16.2).

Sketchpad also has the capability to animate points on segments or circles and to trace the loci of any object. These capabilities offer effective and easy ways to investigate loci problems and to generate geometric curves based on their defining loci. For example, a parabolic curve can be generated as the locus of a point that moves so that it is equidistant from a given line and a fixed point not on the line. Investigations of such curves can come within the reach of middle-school students with the use of Sketchpad (see Fig. 16.3).

Perhaps one of the most pedagogically useful aspects of Sketchpad is the ability to record a student's constructions in a readable script, which can be played back on a different set of initial objects (the minimum set of givens for the construction). This script is generated by simply turning on a simulated tape recorder on the screen. The script is written in simple mathematical statements that describe the particular construction (e.g.,

Area of outer quadrilateral = 2.82 square inches
Area of inner quadrilateral = 1.41 square inches
Area of outer quadrilateral/Area of inner quadrilateral = 2.00

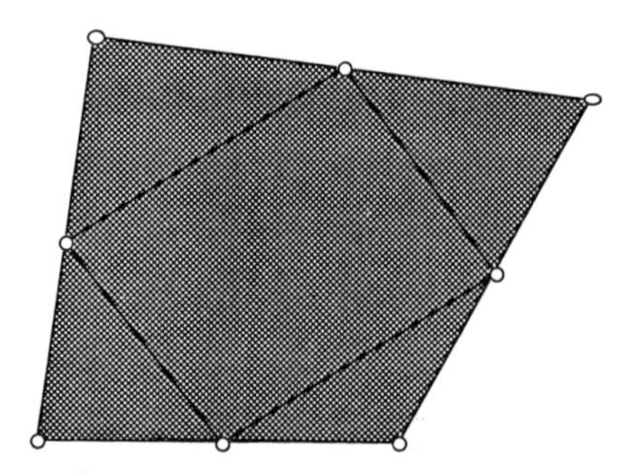

FIG. 16.2 Ratio of the areas of two quadrilaterals (the second formed by joining the midpoints of the sides of the first) from Geometer's Sketchpad.

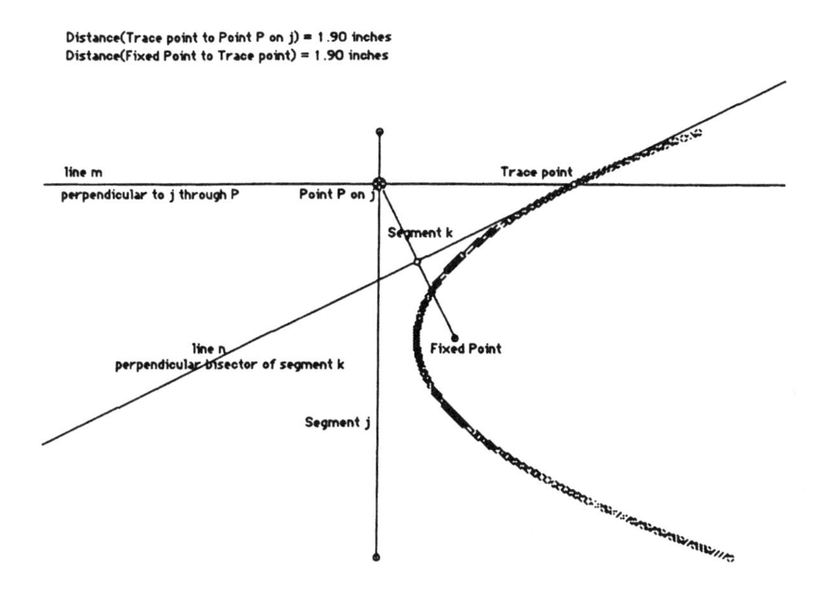

FIG. 16.3 Parabola tracer from Geometer's Sketchpad. Note: As point P moves on segment j, the trace point traces out the parabola.

"Let [D] = midpoint of segment [k]"). The recorder can be stopped at any time simply by clicking on the stop button. If a student forgets to turn on a recorder or wants to experiment without recording, a script can be generated after the fact by simply selecting the finished construction, then selecting "make a script" from the WORK menu. Scripts and finished sketches can both be saved to files for later use. Students working individually or in small groups on a construction can turn in a script and sketch on disk or paper to the teacher for evaluation purposes or for sharing with other students.

EDUCATIONAL IMPLICATIONS

The educational implications of dynamic construction tools for the learning of geometry are far reaching (see Goldenberg and Cuoco, chap. 14, this volume). The traditional approach to geometry instruction based on definitions, theorems, and their proofs makes little sense in an environment in which real-time phenomena can be explored. An inductive approach based on experimentation, observation, data recording, and conjecturing, in which (as deVilliers proposes in this volume, chap. 15) proof takes on the role of explanation rather than verification, would be much more appropriate. Such an approach would give students the opportunity to engage in mathematics as mathematicians, not merely as passive recipients of someone else's mathematical knowledge. From a constructivist point of view, this is the only way children can learn mathematics—by constructing mathematical relationships themselves (Steffe & Cobb, 1988).

Learning in a computer environment also has implications for instruction. The computer environment provides a representational medium different from concrete materials, diagrams or written symbols on paper, or mental representations (see Goldenberg et al., chap. 1, this volume). If used in conjunction with practical, physical experiences (such as ruler and compass constructions on paper) the computer construction tool can provide a link between the physical experiences and the mental representations. Kaput (1991) saw this interplay of representational systems as a powerful aspect of computer-based learning.

The nature of Sketchpad also makes it conducive to collaborative problem solving among small or even large groups of students. Several students gathered around a single computer are easily caught up in the conjecturing process as they watch the changes taking place on the computer screen and are quick to offer suggestions for further experimentation. I have used Sketchpad effectively with whole-class demonstrations to stimulate class discussion of the phenomena. The class can contribute to the demonstration by making suggestions for changes and further constructions. In a networked computer lab setting, Sketchpad provides a built-in "Demonstrate To" facility that allows students to demonstrate and share their work in progress with one another and with the instructor. This nonhierarchical networking makes possible collaboration in real time among students working on different machines and also provides a means for the teacher to monitor the work of all the students in a lab situation from one large-screen monitor or display device. Multiple student demonstrations can also be shown simultaneously on the same classroom display screen. In an even wider context, sketches and scripts with accompanying problems and challenges can be shared among classes, schools, and different learning communities, either by mail, through publications, or through online electronic communications networks. The Math Forum at Swarthmore College has established a newsgroup and World Wide Web server that includes discussions on the use of dynamic geometry software and archives for the electronic sharing of Sketchpad and Cabri files (see Renninger, Weimar, & Klotz, chap. 19, this volume).

CLASSROOM IMPLEMENTATION OF SKETCHPAD

Sketchpad has been used very creatively in elementary classrooms. Teachers in Project LITMUS (Leadership Infusion of Technology in Mathematics and Its Uses in Society)[1] in Grades 2 through 5 used Sketchpad with

[1] Project LITMUS (Leadership Infusion of Technology in Mathematics and Its Uses in Society) was a five-year teacher enhancement project supported in part by a grant from the National Science Foundation (Grant No. TPE-8954793). The goal of the project was to transform the teaching of mathematics K–12 throughout two rural school systems in Georgia through the effective use of computing technology by teachers and students. Leader teachers in each school were prepared to mentor their partner teachers in the effective use of technology in their classrooms. The project aimed to have an impact on every teacher of mathematics (K–12) in the two school systems by summer 1996. Geometer's Sketchpad was used extensively throughout the project.

their students to dynamically investigate reflections, translations, and rotations of constructed figures in very creative ways. The children created stick figures with line segments and circles. They then translated these stick figures several times to create a line of figures. By moving the limbs of their original figures the children made their lines of "paper dolls" dance in unison. This line dancing was fascinating for the children, and the teachers used it as an opportunity to explore the notion of angles and length. Similarly, when the stick figures were reflected about a vertical line segment, properties of reflection were investigated as the children made their "paper dolls" dance with one another (see Fig. 16.4).

These same teachers also linked Sketchpad with art activities such as "curve stitching." Children created a "stitched curve" in Sketchpad simply by connecting series of points on adjacent line segments in reverse order (see Fig. 16.5). A line segment connecting the two endpoints of the figure was then used as a line of reflection for the whole figure in order to

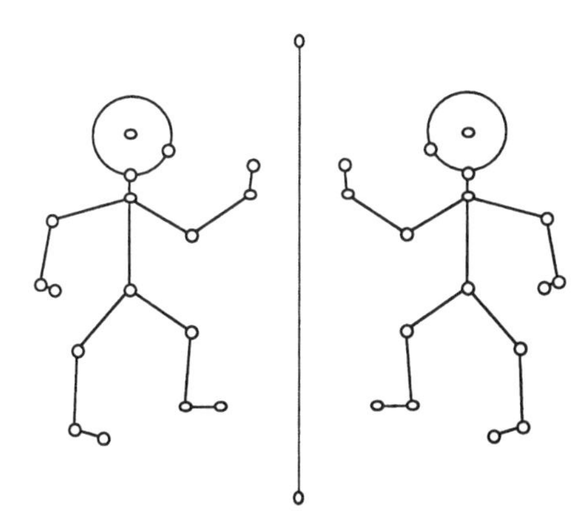

FIG. 16.4 Paper-doll reflection.

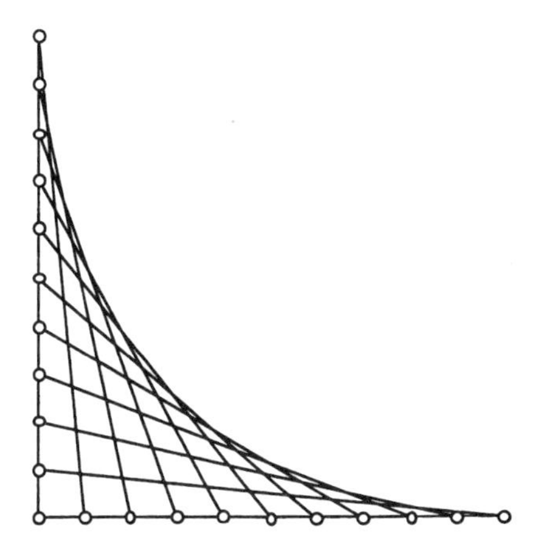

FIG. 16.5 Curve stitching with Sketchpad.

create a petal shape. This petal shape was then used to create a flower structure through a series of reflections about different line segments. The children found through exploration that they could vary the number of petals in their flower by changing the angle between the two original segments (see Fig. 16.6); thus, this "art" activity became an exploration of angle rotation and reflection.

Innovative uses of Sketchpad are also occurring at the secondary level; however, in situations where teachers have had very little exposure to Sketchpad and may feel constrained by the traditional geometry curriculum, the content remains unchanged and the potential of Sketchpad (alluded to earlier) is seldom tapped. Teachers in such situations tend to use Sketchpad mainly as an electronic chalkboard to present diagrams for their students. Their students, in turn, tend to use Sketchpad merely as a drawing tool rather than as a construction tool (Finzer & Bennett, 1995). In one such high-school situation, students were to "construct" right triangles, rhombi, kites, and midpoint quadrilaterals. Although most students did use the CONSTRUCT menu to create the midpoints of the sides of a quadrilateral, their other figures were mostly drawn with the freehand tools, and segments that needed to be parallel or perpendicular were merely aligned so that they *appeared* to be in the correct relationship; a movement of any part of the figure would, however, destroy the desired appearance. Only about one fourth of the 50 students from two classes used construction techniques for any of their figures. None of the students were able to successfully *construct* a rhombus or a kite. One student did extend the construction of an equilateral triangle using congruent circles to produce a daisy design (see Fig. 16.7).

The teacher in this situation had very little experience with Sketchpad herself and had not had any in-service instruction in using it with her students. As with most other educational innovations, just "putting it out

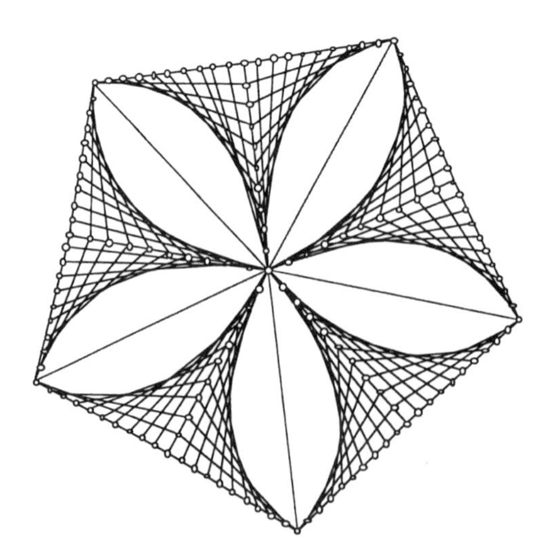

FIG. 16.6 Creating a flower through successive reflections over different line segments (Sketchpad).

FIG. 16.7 Using congruent circles to extend the construction of an equilateral triangle.

there" is not enough. The elementary- and middle-school teachers in Project LITMUS had 6 weeks of in-service, during which they were able to learn how to use Sketchpad (and other software) and explore ways in which they could use Sketchpad with their students. They were also supported by class visits from project staff during the academic year following their summer in-service. Teachers needed this level of support to make fundamental changes in the way they teach and to use technology as a tool for their teaching and students' learning (Pagnucco, 1994).

Some teachers have used Sketchpad to present conjectures as fill-in-the blanks on electronic worksheets. Such an approach forces conjectures on the students. Other teachers have used Sketchpad-prepared sketches as starting points for their students' explorations and student-generated conjectures. For example, Karen (a teacher of Grades 7 through 9) prepared sketches for every activity on her worksheets. The student's role was to explore the prepared sketch rather than to construct a figure. The students were to make measurements, drag components of the figure, and make observations about specific properties of the figure listed in the worksheet. Although these investigations were very teacher directed, they did make effective use of the dynamic nature of Sketchpad. I found the explorations and questions the teacher asked to be interesting.

Three students in Karen's eighth-grade integrated algebra and geometry class (Andrew, Jeff, and Rick) used Sketchpad to solve a classic problem illustrated in Fig 16.8. The students' approach reflected the use of measurement variation emphasized in the teacher-directed activities. They constructed congruent circles sharing a common radius (AB) and placed free points (C and D) on each circle to construct the three congruent sides (AC, BD, AB) of the figure. They then constructed the diagonals (AD, BC) and the fourth side (CD) and manipulated the free points (C and D) until the measures of these three segments (AD, BC, CD) were equal. The angle measures then gave the solution (see Fig. 16.9). Rick did add a line through C parallel to AB to help locate the point D on circle 2; how-

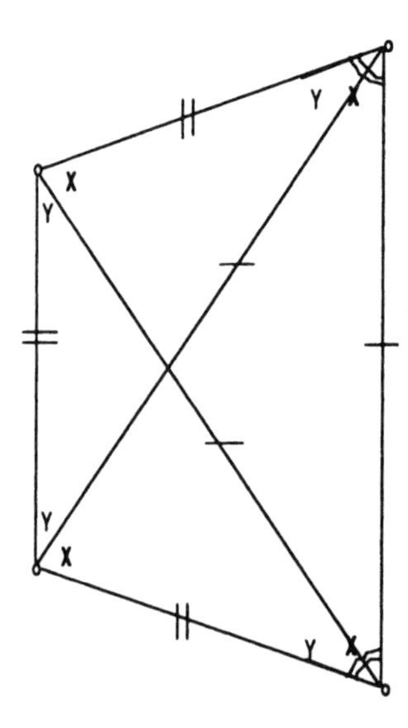

FIG. 16.8 Andrew's problem: Find the measures of angles x and y.

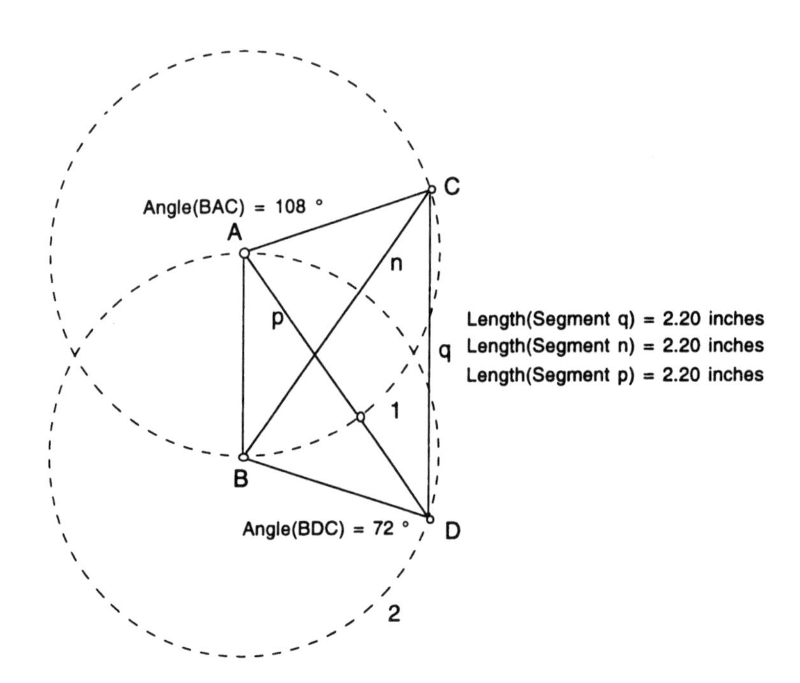

FIG. 16.9 Construction of congruent circles to find the measures of unknown angles (solution to Andrew's problem, Fig. 16.8).

ever, he still used free points on each circle for the segment DC. He did not appear to realize that he could construct point D as the intersection of his parallel line and circle 2.

Dynamic Measurement and Proof

This "measurement" approach to a classic algebraic question raises an issue for me: Should we encourage students to rely on Sketchpad measurements as solutions to problems that can be solved exactly (algebraically) using the geometric properties of the figure? There is an inherent danger in the "measurement" approach: Students may take an approximate measure to be an exact solution. It is important that the teachers in situations like the one just described suggest to students that they substitute the measured angles back into the original diagram to check for consistency. (Do the angle-sums of all the triangles equal 180°?) In checking this essential property, the students may eventually "see" the algebraic solution to this problem.

Sketchpad can be a very powerful aid in the solution of such classic geometric problems. It enables the investigator to discover the constraints of the problem situation through construction of the figure and through possible manipulations of the figure. These capabilities raise important questions as to what constitutes "proof" when working in a dynamic medium. Students in situations similar to those described here have turned in electronic sketches that use dynamic measurement as evidence for proving a conjecture (see Fig. 16.10).

The last statement in the student's argument shown in Figure 16.10 suggests a confusion between the notion of "demonstration" and "proof." de Villiers (this volume, chap. 15) discusses the notion of "proof" in a "drag-mode" geometry environment and suggests that "proof" in mathematics serves many purposes besides "verification" of the truth of a statement. He sees demonstration as both an important component and an important role of proof. He suggests that we focus more on explanation and discovery as functions of proof that are meaningful for students. In the above example, the students should be encouraged to generate an argument for *why* the sum of the measures of the four angles of a quadrilateral is 360 degrees. The dynamic Sketchpad investigation has provided the convincing discovery for the student, but not the explanation. Hofstadter (1997) also suggested that when working in a dynamic medium like Sketchpad some relations can be *seen* to be true through their constancy under variation, and that this kind of "direct contact with the phenomenon is *even more convincing than a proof*" [italics added] (p. 10). He believed that a "*why of insight*" could come from such direct experiences in mathematics, and that such insights could lead to deep understanding in mathematics. Rasmussen (1995) suggested that working with Sketchpad can be an incentive for students to make sense out of the theorems they encounter or discover for themselves, and that this "sense making" is a more productive route to proof in geometry than the traditional deductive argument.

C-6
The sum of the measures of the four angles of a quadrilateral is 360 degrees.

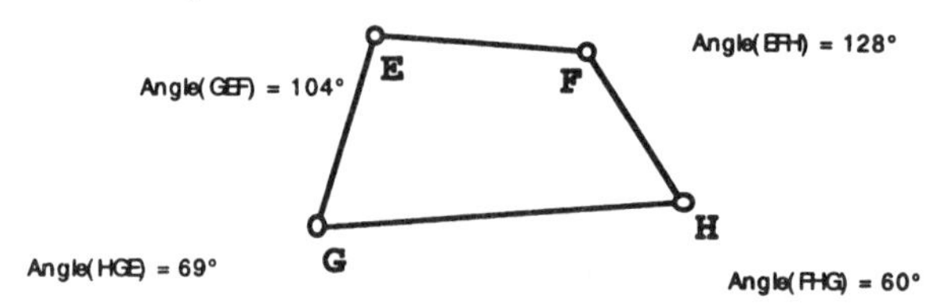

Angle(GEF)	Angle(HGE)	Angle(EFH)	Angle(FHG)	
139.24	55.27	79.26	86.23	=360 degrees
103.53	68.68	127.87	59.91	=360 degrees

In order to prove this conjecture, I randomly made a quadrilateral. I measured the angles and tabulated them. I then moved the points of the quadrilateral to make the angles measure different amounts. I also tabulated those results. I then added each of the sets of angles both of them were equal to 360 degrees.

Thus, proving that the sum of the measures of the four angles of a quadrilateral is 360 degrees.

FIG. 16.10 Substituting dynamic measurement for proof of conjecture: a student "proof" using Sketchpad.

Expanding the Geometry Curriculum

Karen's seventh-grade class on symmetry demonstrated the creativity that can be unleashed with a dynamic medium like Sketchpad. The students' symmetric designs were stunning, amusing, and clever. Many students experimented with the animation features of Sketchpad to bring their creations to life with amusing movements. Figure 16.11 illustrates an example of rotational symmetry. As with the elementary teachers in Project LITMUS, Karen encouraged the integration of art topics to expand and enrich her geometry curriculum in the middle grades.

In a situation where students have open access to computers both during and after class, where the teacher and the school community do not feel constrained by outside pressures to conform to rigid curricula goals, where students are encouraged to pose and explore their own problems, and where the teacher has become very familiar with the capabilities of the Geometer's Sketchpad, Sketchpad can be a catalyst for expanding the scope and depth of the geometric learning taking place at the high-school level. Paul taught calculus and advanced geometry. The geometric topics went beyond the normal high-school curriculum, and the challenges that Paul presented in his Sketches are worthy of a college geometry class. For example, the sketch in Fig. 16.12 presents a challenge to demonstrate the measure of a particular angle in an isosceles triangle con-

FIG. 16.11 Rotational symmetry: integrating art and geometry with Sketchpad.

struction. A similar problem has stumped in-service mathematics teachers that I have worked with at the university.

Paul included an exploration of Napoleon's theorem in his honors geometry class and had students working on original mathematical explorations. For instance, two students were exploring noncongruent triangles with equal area and perimeter. The students found four such triangles and were in competition with a local mathematician who was also working on the problem.

It is interesting to note that students in the regular geometry class at this same school (and presumably with all of the same advantages and resources, but with a different teacher) tended to use Sketchpad merely as a

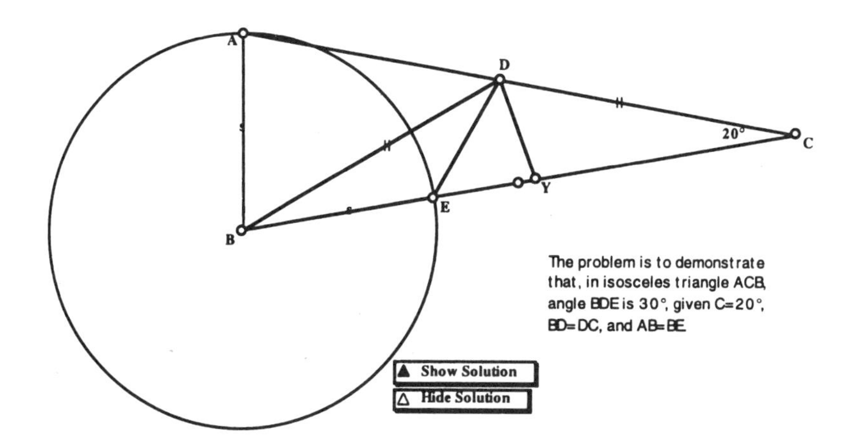

FIG. 16.12 Challenge problem from Paul's class.

drawing tool rather than as a construction tool. The implication of this note might be that the available resources and lack of school constraints do not guarantee innovative work by students—the goals and teaching style of the teacher, the teacher's experience with the software, and possibly the intellectual capabilities of the students may also be important factors.

In a very different situation in Baltimore's Patapsco High School, a geometry student was stimulated to investigate a possible generalization of Marion Walter's theorem, which states that if the trisection points of the sides of any triangle are connected to the opposite vertices, the resulting hexagon has an area one-tenth the area of the original triangle (Key Curriculum Press, 1995; see Fig. 16.13). The student (Ryan) wondered what would happen if the sides of the triangle were partitioned into a number of equal segments other than three (see Fig. 16.14). He attempted to investigate this situation but was frustrated by the limited capabilities of the software he was using. His teacher introduced Ryan to Sketchpad, and Ryan was able to investigate many such situations with relative ease and form his own conjectures. He became the first high-school student to present a colloquium on his conjectures to the mathematics department at the local university. The mathematicians have encouraged Ryan to pursue his conjectures toward the formulation of a new mathematical theorem.

Ryan's case is obviously exceptional, but it does illustrate that where the school environment supports and encourages independent mathe-

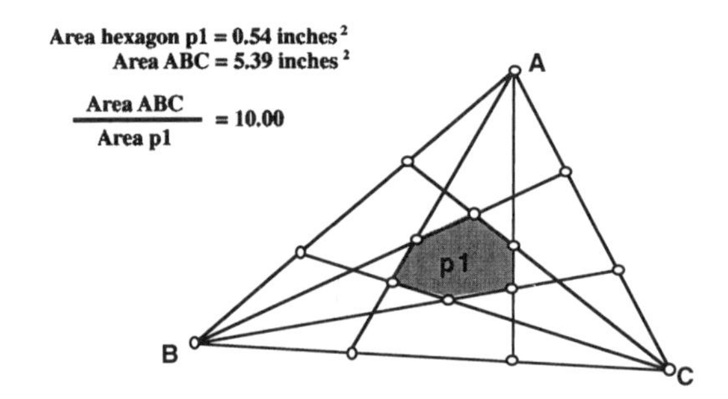

FIG. 16.13 Marion Walter's theorem on Sketchpad.

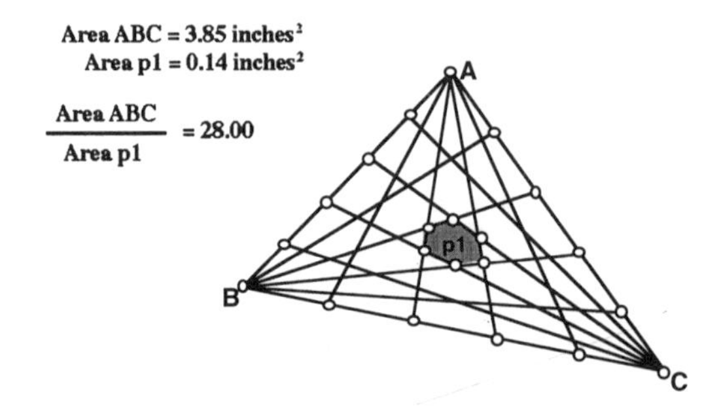

FIG. 16.14 Ryan's extension of Marion Walter's theorem.

matical explorations by students, dynamic geometry software can become a powerful tool for doing and creating new mathematics and thus expanding the current geometry curriculum (see Goldenberg et al., chap. 1, this volume).

GOING BEYOND GEOMETRY

Several teachers, students, educators, and researchers have used dynamic geometry software to investigate topics outside the normal geometry curriculum. Sketchpad has been used as a dynamic medium for exploring conic sections, trigonometry, algebraic functions, and topics in calculus and complex analysis. Outside of mathematics, it has been used to dynamically model and investigate perspective drawing, mechanical linkages, physical trajectories, optics, and data analysis. Goldenberg et al. (chapter 1, this volume) argue that providing the opportunities and dynamic tools for students to make such connections among different topics will promote the "habits of mind" that constitute real "mathematical power."

Exploring Conics

In *Exploring Conic Sections With the Geometer's Sketchpad*, Scher (1995) described a unique construction of an ellipse by a contemporary American high-school student, Danny Viscaino. The construction is dependent on Sketchpad's facility to trace the locus of a constructed point while animating one of its parents. The latest version of Sketchpad now allows the user to actually construct the locus as a solid curve as well as to trace the path of the locus point. Fig. 16.15 shows the construction of Danny's ellipse us-

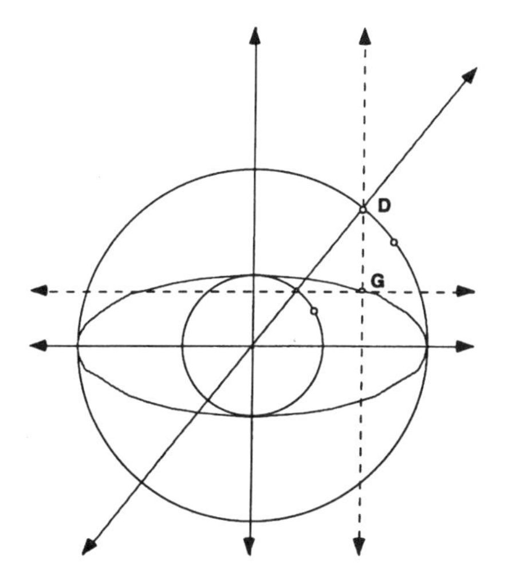

FIG. 16.15 Danny Viscaino's construction of an ellipse.

ing this new feature. The ellipse is the locus of point G as point D moves around the outer circle. Different ellipses can be obtained simply by changing the relative sizes of the two concentric circles.

The generation of conic sections through paper folding has been used by innovative mathematics teachers for many years. The process, however, can become very tedious as many folds are necessary for the conic shape to emerge as the boundary of the envelope of folds. Lin (1994) and Scher (1995) illustrated how students can use Sketchpad to simulate the paper-folding activity dynamically (see Fig. 16.16). In his article, Lin provided descriptions of Sketchpad investigations that progressed from paper folding to the algebraic formula of conics and also explored the definition of conics based on eccentricity. In Fig. 16.17, points A and F1 are the centers of the defining circles for the conics. AK is the radius of one circle and F1R is the radius of the other circle. The ratio AF1/AK defines the eccentricity of the conic. The conic is the locus of point P as the point C moves around the circle with center A. Point P is constructed as the intersection of the perpendicular through F1 to the tangent through point C, and the perpendicular to the ray F1C through the inverse of point C in the circle with center F1 (i.e., the point R′). F1 and F2 are the foci of the ellipse. Line y is one directrix. With this sketch, all three conic curves can be formed simply by changing the relative size or positions of the two circles. When the ratio AF1/AK is less than one, the conic is an ellipse; when it is greater than one it is a hyperbola; when it is equal to one, a parabola is formed.

Lin (1994) claimed that "the eccentricity view of conics integrates the concept of conic curves" (p. 27) and that his dynamic environments, constructed in Sketchpad, made it possible for students to explore the relationship of eccentricity visually and possibly gain some understanding of the laws of planetary movement. The optical properties of the conics can also be explored using Lin's sketches. Figure 16.17 illustrates how a light

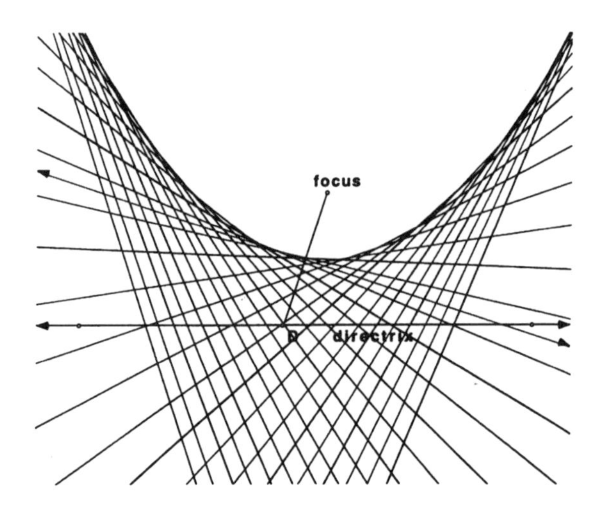

FIG. 16.16 Folding a parabola in Sketchpad.

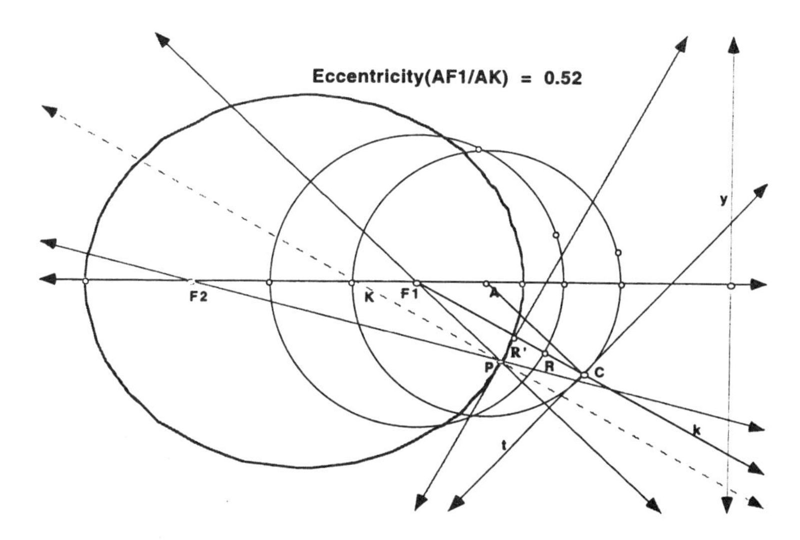

FIG. 16.17 An ellipse based on eccentricity using Sketchpad.

ray emanating from one focus will be reflected by the conic through its other focus. This is true of all three conics. In the case of the hyperbola, the light ray is reflected as if it is emanating from the other focus. In the case of the parabola, the other focus is at infinity; thus light from the focal point produces a parallel set of rays, which is why parabolic reflectors are used for headlight beams and radio-wave collectors (dish antennas).

Exploring Trigonometry

Shaffer (1995) used Geometer's Sketchpad in his teaching of a semester-long course on trigonometry for juniors and seniors in high school. He found that by using dynamic explorations with Sketchpad he could help his students construct the visual intuitions about circles and angles that formed the underpinnings of his own understanding of trigonometry. Figure 16.18 visually illustrates the definitions of the six trigonometric functions based on ratios of segments formed by a point C moving on a unit circle with center at point A. Shaffer (p. 46) used this dynamic diagram to help students define the six functions in terms of proportional lengths of segments found in the diagram.

Shaffer (1995) also used Sketchpad to explore trigonometric identities, such as the ratio of the sine of an angle in a triangle and the length of the side opposite the angle being the same for all three angles (see Fig. 16.19).

Shaffer also generated graphical representations of the trigonometric functions. The animation feature of Sketchpad allows the user to animate a point on a circle and a point on a segment simultaneously. The two points move at approximately the same rate; therefore, by the time the point on the circle has made one circuit, the point on the segment will have moved approximately the length of the circumference of the circle.

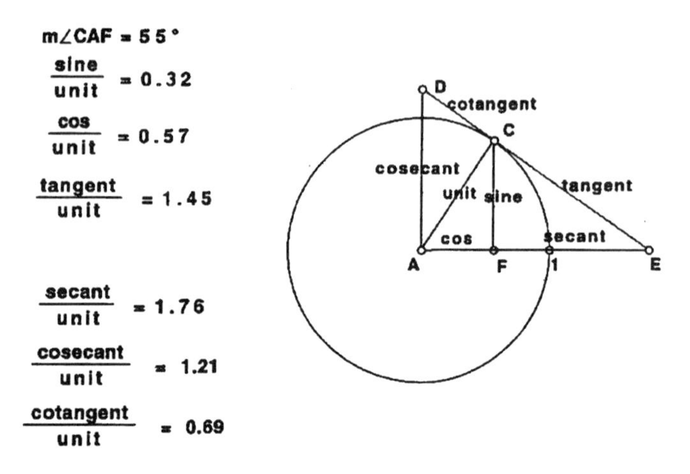

FIG. 16.18 The six trigonometric functions illustrated using Sketchpad.

The two animating points can be simply linked by creating the point of intersection of a line parallel to the segment passing through the point on the circle, and a line perpendicular to the segment passing through the point on the segment. If the segment is part of a line passing through the center of the circle, then the linkage point will trace out the sine or cosine curve as the two points are animated (see Fig. 16.20).

Exploring Coordinate Graphs and Algebraic Functions

I have used Sketchpad to construct a rudimentary version of what Goldenberg et al. (chap. 1, this volume) terms "Dynagraphs," but Lin (Lin & Olive, 1992) was the first to fully exploit the capabilities of Sketchpad as an environment for creating dynamic, linked, multiple representations of algebraic functions. Lin created a coordinate system in the Sketchpad win-

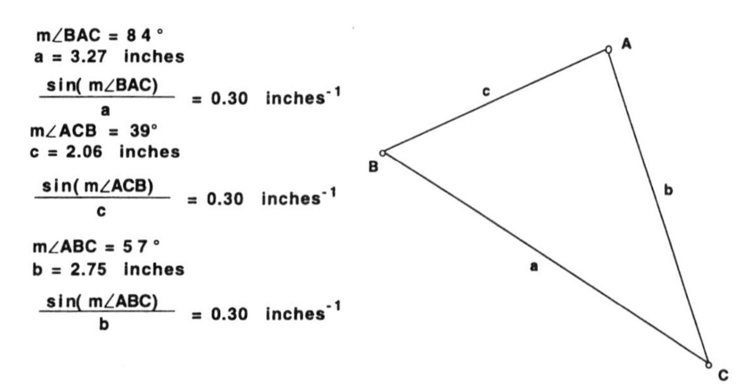

FIG. 16.19 Exploring a trigonometric identity using Sketchpad: the ratio of the sine of an angle to the length of the opposite side.

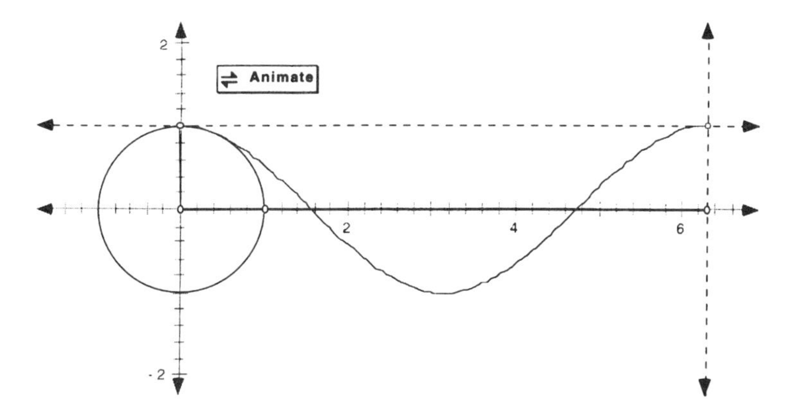

FIG. 16.20 A graph of the cosine function from Sketchpad.

dow and the constructed curves that represented graphs of algebraic functions based on his coordinate system. Because the graphs were constructed geometrically, they could be acted upon as geometric objects in the environment. They could be translated, rotated, reflected, and dilated. Lin generated algebraic expressions for these curves using the dynamic measurement capabilities of Sketchpad. These expressions changed dynamically as the curves were transformed (see Fig. 16.21).

Lin also geometrically constructed function relationships among his graphical objects. In Fig. 16.21, the point P is constructed as the division of

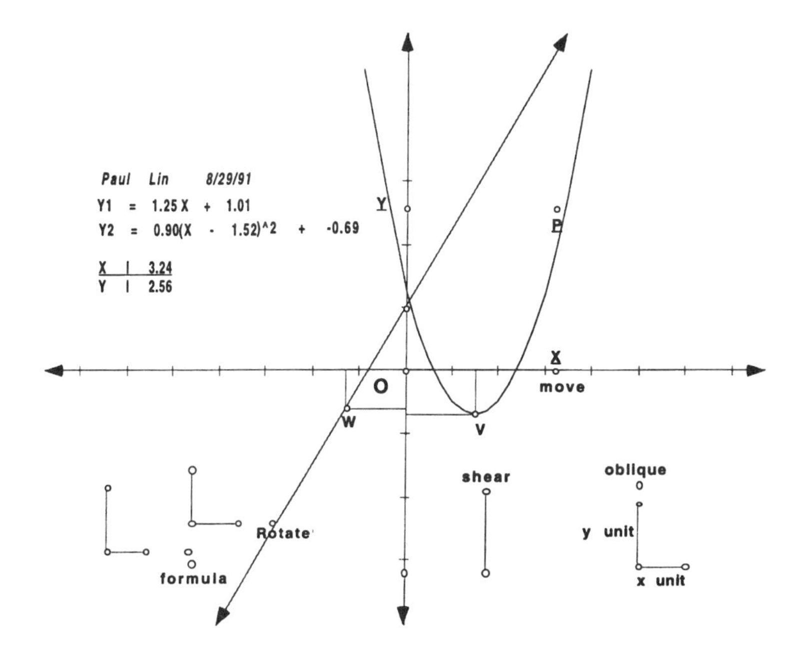

FIG. 16.21 Graphs of linear and quadratic functions in Sketchpad.

the linear function by the quadratic function. The x, y coordinates of P are displayed in the figure. When the point "x move" is moved along the x-axis, the constructed point P traces out the rational function. Figure 16.22 shows the constructed path of point P made using the Construct Locus feature of Sketchpad.

When the graphical objects representing the linear and quadratic functions in Figure 16.22 are moved or altered by the user, the constructed path of point P (the rational function) also changes; thus, the special properties of the rational function can be dynamically observed. For instance, when the parabola is moved above the x-axis, the quadratic function has no zeros, and, consequently, the rational function's discontinuities disappear (see Fig. 16.23).

Lin (1993) used his dynamic graphs with three high-school students to investigate their construction and understanding of transformations on functions. The ability to directly manipulate the graphical objects and also change the coordinate system by simply moving the origin helped these students to think visually about these algebraic relations. Working in these environments encouraged the students to think on three different planes: the physical, the operational, and the representational planes. The experiences with the dynamic environments was most effective for the student who was the least advanced mathematically. He made effective use of the dynamic manipulations to construct his own understanding of translation and scaling of function graphs. The most advanced student found it more difficult to think on the physical and operational planes as she was used to memorizing formulas and applying them on the representational plane. Consequently, her thinking strategies were less creative,

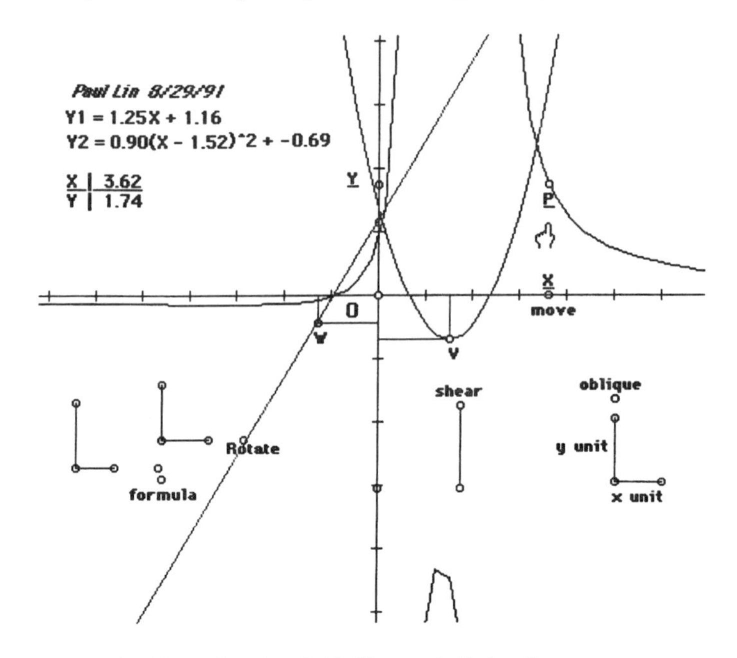

FIG. 16.22 Graph of linear function divided by quadratic function.

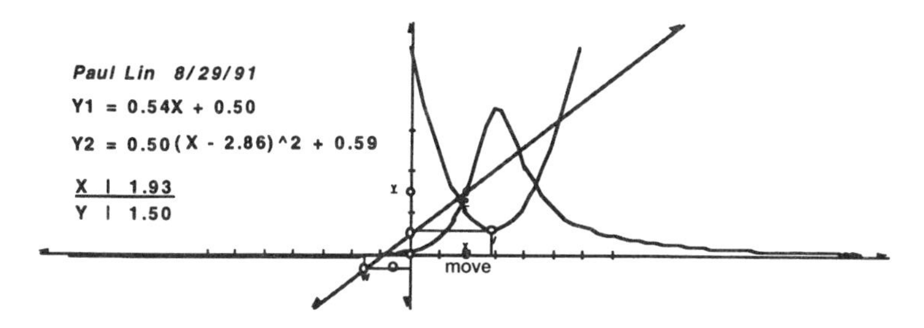

FIG. 16.23 Rational function with no discontinuities.

even though she did use the dynamic manipulations in the environments to check her algebraic results and make sense of them.

Lin and Hsieh (1993) described possible ways to use these dynamic environments for exploring the effects of changing the parameters of linear and quadratic functions, and for solving systems of linear equations.

With the introduction of Sketchpad 3.0, it is now possible to explore algebraic relations through geometric constructions more directly. Both Cartesian and polar coordinate systems can be used in Sketchpad 3.0. The algebraic expressions for lines and circles based on either coordinate system are readily available through the Measure menu. These expressions are dynamic just like any other measurement. As the user moves or changes a line or circle in any way, the expression changes accordingly. The coordinate systems are also dynamic. The location of the origin can be changed simply by moving the origin point. The scale of the axes can be changed simply by moving the unit point relative to the origin point.

Sketchpad 3.0 can function as a full-fledged function plotter: Any algebraic expression can be formed using any measurement with the new calculator functions. For instance, the graph of the function $f(x) = \ln(x + 2)/x$ can be obtained simply by placing a free point on the x-axis, obtaining the x-coordinate of that free point, and using this measurement as input to the calculator to form the desired expression for $f(x)$. The point $[x, f(x)]$ can then be plotted by simply selecting both measurements and using the "Plot point" option under the new GRAPH menu. Once the point $[x, f(x)]$ has been plotted, the locus of this point as the free point moves on the x-axis can be constructed simply by selecting the point and the free point on the x-axis (in that order) and selecting "Construct Locus" under the CONSTRUCT menu. Fig. 16.24 illustrates the result of the above operations.

These powerful new features of Sketchpad 3.0 have yet to be fully explored in classroom situations, but, judging by the diversity of response to the earlier versions of Sketchpad as described in this chapter, I predict a genuine move toward the integration of algebra and geometry, as recommended in the NCTM *Standards* (1989), on the part of teachers and students who have ready access to this latest edition of Sketchpad. There will be a need, however, for ongoing in-service and administrative support for teachers to break away from the traditional separation of algebra and geometry endemic in the high-school mathematics curricula.

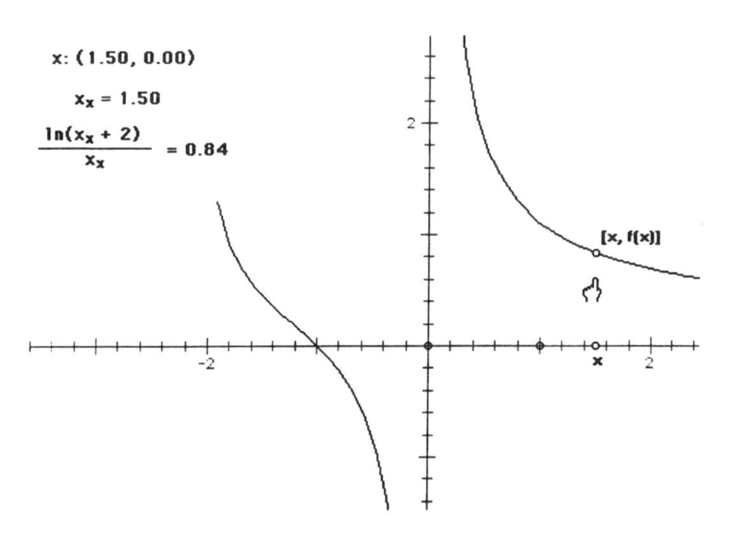

FIG. 16.24 The graph of ln(x + 2) / x constructed in Sketchpad 3.0.

SUMMARY

In this chapter I have attempted to illustrate, by example, the opportunities for exploring and integrating many ideas and topics in mathematics with the Geometer's Sketchpad. Examples were presented from elementary-, middle-, and high-school classrooms where teachers have been using Sketchpad in a variety of ways for diverse purposes. Although most of these examples illustrate the exciting potential of Sketchpad for students' creative explorations, some examples also illustrate potential problems. Some of the problems that emerged from these different implementations of Sketchpad were discussed, such as the difference between "drawing" and "constructing" (see Finzer & Bennett, 1995, for a more in-depth discussion on this topic), demonstration versus proof (see de Villiers, chap. 15, this volume, for a thoughtful discussion of proof in a drag-mode geometry environment), and pedagogical problems associated with a teacher's lack of experience with Sketchpad and the perceived constraints of the "traditional" curriculum.

Examples of resources and ideas for exploring topics outside the normal geometry curriculum were briefly described. These topics included conics, trigonometry, coordinate geometry, and algebraic functions. Goldenberg et al. (chap. 1, this volume) provide other examples. Finally, the potential of Sketchpad for helping teachers and students integrate their developing knowledge of algebra and geometry was illustrated through an exploration of the new locus and graphing features of version 3.0. According to Goldenberg et al. (chap. 1, this volume), such an integration will help students connect with mathematics and build a "habits of mind" perspective that generates mathematical power.

REFERENCES

Baulac, Y., Bellemain, F., & Laborde, J.-M. (1992). Cabri-Géomètre [Computer software]. Pacific Grove, CA: Brooks-Cole.

Baulac, Y., Bellemain, F., & Laborde, J.-M. (1994). Cabri II [Computer software]. Dallas, TX: Texas Instruments.

Brock, C. F., Cappo, M., Dromi, D., Rosin, M., & Shenkerman, E. (1994). Tangible Math: Geometry Inventor [Computer software]. Cambridge, MA: Logal.

Finzer, W. F., & Bennett, D. S. (1995). From drawing to construction with the Geometer's Sketchpad. *Mathematics Teacher, 88*(5), 428–431.

Fuys, D., Geddes, D., & Tischler, R. (1988). The van Hiele model of thinking in geometry among adolescents. *Journal of Research in Mathematical Education Monograph.* Richmond, VA: The National Council of Teachers of Mathematics.

Hofstadter, D. (1997). Discovery and dissection of a geometric gem. In J. King & D. Schattschneider (Eds.), *Geometry turned on!: Dynamic software in learning, teaching, and research.* (pp. 3–14). Washington, DC: Mathematical Association of America.

Jackiw, N. (1991). Geometer's Sketchpad [Computer software]. Berkeley, CA: Key Curriculum Press.

Kaput, J. J. (1991). Notations and representations as mediators of constructive processes. In E. von Glassersfeld (Ed.), *Radical constructivism in mathematics education.* (pp. 53–74). Dordrecht, the Netherlands: Kluwer.

Key Curriculum Press. (1995, April). Baltimore student captures math innovation award. Press release issued at the 73rd Annual Conference of the National Council of Teachers of Mathematics, Boston.

Lin, P. P. (1993). *Learning translation and scaling in dynamic linked, multiple representation environments.* Doctoral dissertation, University of Georgia, Athens.

Lin, P. P. (1994). From paper folding to the algebraic formula of conics. *Mathematics Educator, 5,* 1, 23–28.

Lin, P. P., & Olive, J. (1992). *Dynamic, linked, multiple representation environments using the Geometer's Sketchpad.* Unpublished manuscript.

Lin, P. P., & Hsieh, C. J. (1993). Parameter effects and solving linear equations in dynamic, linked, multiple representation environments. *Mathematics Educator, 4,* 1, 25–33.

Lindquist, M. M. (Ed.). (1989). *Results from the Fourth Mathematics Assessment of the National Assessment of Educational Progress.* Reston, VA: National Council of Teachers of Mathematics.

National Commission on Excellence in Education. (1983). *A nation at risk: The imperative for educational reform.* Washington, DC: U.S. Government Printing Office.

National Council of Teachers of Mathematics. (1989). *Curriculum and evaluation standards for school mathematics.* Reston, VA: Author.

Pagnucco, L. (1994). *Mathematics teaching and learning processes in a problem-oriented classroom.* Doctoral dissertation, University of Georgia, Athens.

Rasmussen, S. (1995, April). *Engaging proofs using the Geometer's Sketchpad.* Paper presented at the 73rd Annual Meeting of the National Council of Teachers of Mathematics, Boston.

Scally, S. P. (1990). *A clinical assessment of the impact of Logo experience on ninth grade students' understanding of angles.* Doctoral dissertation, Emory University, Atlanta, GA.

Scher, D. (1995). *Exploring conic sections with the Geometer's Sketchpad.* Berkeley, CA: Key Curriculum Press.

Shaffer, D. (1995). *Exploring trigonometry with the Geometer's Sketchpad.* Berkeley, CA: Key Curriculum Press.

Steffe, L. P., & Cobb, P. (1988). *Construction of arithmetical meanings and strategies.* New York: Springer-Verlag.

van Hiele, P. M. (1986). *Structure and insight: A theory of mathematics education.* Orlando, FL: Academic Press.

17

Mapping the Classroom Using a CAD Program: Geometry as Applied Mathematics

Daniel Lynn Watt
Education Development Center, Inc.

This chapter describes a prototype fifth-grade geometry curriculum unit and child-appropriate computer-aided design (CAD) program developed as part of the Math and More[1] elementary mathematics project, a long-term research and development project carried out by the Education Development Center (EDC) in collaboration with IBM/EduQuest. Fifth-grade students, using KidCAD[2] and working in collaborative groups, created computer-generated scale drawings of the classroom, rearranged

---

[1] Math and More is a registered trademark of the IBM Corporation. The work described here was supported by IBM and carried out at Education Development Center in Newton, Massachusetts. The author of this chapter was development team leader for the geometry curriculum. This research, however, was a complex effort involving many collaborators, including project directors Myles Gordon and Elizabeth Bjork; curriculum developers Sherry Shanahan and Kristen Bjork; and our program managers at IBM, Elayne Schulman and John Schiener. Wayne Harvey, Kristen Bjork, and I were principally responsible for the prototype software design, which was creatively implemented by Papyrus Software, in an incredibly short time. Our research team, headed by Jan Ellis and including Jill Christiansen and Denise Sergent, gathered and summarized the research data. Our teacher collaborators were Steve Carme and Cynthia Reid of the Bishop School in Arlington, and Peter Escott and Elizabeth Walters of the Lincoln Park School in Somerville. David Nelson and Marlene Nelson (not related) ably captured powerful vignettes in the short video "Plans and Patterns," which has helped keep this work alive both for us and for thousands of viewers at conferences and workshops in the years following the project itself. I also acknowledge the gracious support of my colleagues Sherry Shanahan and Jan Ellis, who read and commented on an earlier draft of this chapter.

[2] About 2 years after the research described in this chapter, Davidson Software released a commercial product called KidCAD, a three-dimensional architectural visualization program that has no relationship to the software described here.

the existing furniture as it appeared on the screen, and then used the program to design innovative classrooms of their own. This research effort represented a significant change in both mathematical content and pedagogy for the teachers and students involved.

The primary purpose of our research with this prototype was to inform our ongoing curriculum and software development efforts to create a computer-rich, inquiry-based mathematical learning environment, not to evaluate the curriculum or the software, or to demonstrate its effectiveness. As curriculum developers we were interested in documenting students' learning, but this was not a major focus of the overall study. We also gathered data on a number of issues that are beyond the scope of this chapter: teachers' and students' comfort with small-group collaboration and the mix of online/off-line activities, the pedagogical changes demanded of teachers, classroom management issues, and the use of the computer as a tool.

THE PREMISES UNDERLYING OUR APPROACH TO CURRICULUM DEVELOPMENT

Educators approach geometry as a subject of study in a variety of ways (e.g., as the traditional Euclidean and non-Euclidean geometry of axioms and theorems concerning the relationships among lines, angles, shapes, and so on; as transformational geometry, with its emphasis on shapes and motions applied to shapes; as topology, with its focus on the properties of shapes on a variety of surfaces; as spatial orientation, focusing on the location and arrangement of objects in space; as spatial visualization, focusing on comparing and transforming mental images of objects; and as spatial measurement, usually taught as a separate, nongeometric strand despite its connections to understandings of geometric concepts). Geometry curriculum for elementary school students, in its most traditional form, is an attempt to familiarize students with the names and properties of plane and three-dimensional shapes in ways that might prepare them for the formal study of Euclidean geometry in high school. In recent years—especially since the publication of the National Council of Teachers of Mathematics (NCTM) *Standards* (1989)—ideas from transformational geometry and the study of patterns and symmetry have been introduced in elementary school as well. In most cases (even considering the NCTM *Standards*), geometry is treated as an abstract study, unrelated to other mathematics, only incidentally related to the world outside the classroom, and typically placed near the end of the textbook—to be dropped from the curriculum if teachers feel pressure to spend more time teaching arithmetic.

In developing our elementary curriculum, we approach geometry as applied mathematics. Rather than trying to "cover" all appropriate geometric knowledge that could be learned in each grade, we began with a few powerful mathematical ideas that we believe students of a particular age can learn. We then worked to identify contexts (of interest or importance to the students themselves), in which they would have to use those ideas to create new objects or to understand something about the natural and the constructed worlds in which they live, and developed software tools that scaffolded the development of students' mathematical thinking. We also forged links between geometry and other areas of elementary mathematics (e.g., number, rational numbers, proportion, arithmetic, measurement).

Developing rich contexts, tools, and challenges, however, is only half the story. We believe that mathematical skills and concepts are not learned effectively in isolation and that powerful mathematical learning must be reflective as well as constructive: Students must be able to identify and clarify concepts, relate them to their prior knowledge, express those concepts in their own words as well as in conventional mathematical terms, and understand how to apply those concepts and related skills in unfamiliar contexts. Because we could not assume that the students (or teachers) would have prior familiarity with both the mathematical ideas and the software tools, the curriculum we designed provides a few preparatory activities that gradually introduce some of the new skills and concepts before students engage in the more challenging collaborative problem-solving activities.

The Project

The goal of the Math and More project was to develop a set of technology-rich mathematics curriculum units for three strands of elementary mathematics: geometry, patterns and relationships, and probability and statistics. The curriculum included the use of video, concrete manipulatives, physical movement, and paper-and-pencil activities in addition to software. Before launching a full-scale development effort for Grades 1–5, we wanted to test the feasibility of our approach and identify some of the issues to be resolved in implementing such a curriculum in typical classrooms.

In fall 1991, the classroom design unit was trial-taught in four Boston-area Grade 5 classes at two different schools. One school was situated in a predominantly white community of working-class and professional families; the other was in an ethnically, culturally, and linguistically diverse community with many low-income families. The four teachers who participated had not previously integrated technology into their mathematics curriculum. Data-gathering methods included classroom observations, focus groups, interviews, and questionnaires. We also collected samples of student work and carried out structured interviews with four students in each classroom at the end of the classroom trials.

From the point of view of teachers and students, the project represented a complex innovation with respect to their prior mathematical learning experience. Five computers (one for every five students) were introduced into the classroom, and students and teachers were asked to learn the use of a new, specialized software tool. During the 4 weeks of the unit, teams of students were responsible for planning the rotation of roles and responsibilities for the activities of complex, extended projects (rather than working on a series of self-contained assignments written in a textbook or on worksheets). Students were frequently involved in different tasks at the same time, including both online and off-line activities. As students carried out specific tasks, they encountered and resolved a range of problems not specifically posed by a teacher. In addition, the subject matter itself was not part of the existing fifth-grade mathematics curriculum and represented novel subject matter for some of the teachers.

The geometric learning activities we created were connected to students' everyday lives and to the use of geometry in cultural artifacts that they saw around them. Students were encouraged to represent their mathematical experiences through drawings, symbols, writing, and oral presentations. They used geometry for constructive and creative purposes, with knowledge socially constructed in the collaborative creation of a scale drawing of a classroom. They were given opportunities to discuss and resolve discrepancies that arose between what they predicted, intended, or expected and what actually happened. In addition to the fifth-grade unit on classroom design (which we discuss in this chapter), we have applied these principles across the elementary grades. In Grade 1, students create simple maps and practice giving and following map directions. In Grade 2, students create geometric quilt designs. In Grade 3, they build a three-dimensional physical model of a small community and make a two-dimensional map using perspective drawings to represent the same community. In Grade 4, they systematically construct two-dimensional tessellations from polygons. In Grade 5, as noted earlier, they map and redesign their classroom.

MAPPING AND REDESIGNING A CLASSROOM—THE UNIT

The specific mathematical content of the classroom design unit was derived from the application: The students learned and used the mathematics needed to carry out the tasks of the unit. The primary mathematical content reflected four interrelated domains: measurement, geometric shapes and transformations, spatial visualization, and scale and proportional reasoning. Secondary mathematical content domains included estimation and arithmetic (with and without a calculator). Mathematical processes as defined by the first four NCTM Standards (problem solving, reasoning, communication, and connections) were also prominent in the unit.

The classroom design unit focused on the application of measurement and geometry skills to the task of measuring, scaling, mapping, and reinventing the interior spaces and furnishings of their classroom in the form of a top-view floor plan similar to those used by architects, engineers, and builders. Each group of four to six students was assigned its own computer, and students worked collaboratively to carry out the tasks involved. The unit ended with a final collaborative task: Design a new space within the school, determine its purpose, design the floor plan and furniture for that space, and justify the decisions made. Teachers supported the teams in their planning of complex activities and in their sharing and rotating of responsibilities and roles involved in carrying out their plans.

Sequence of Activities

The unit involved six major activities, each of which took from two to five class periods to complete. The first three activities were designed to provide preparation and practice for the three extended challenges that followed.

Top-View Representations (Activity 1). Students first examined a blueprint of a school library as an example of a top-view representation, then sketched (by hand) a top view of their own classroom.

Scale Models (Activity 2). Students compared a set of scale models of classroom furniture (e.g., tables, chairs, bookshelves), measured them with scale rulers, and sketched their top-view representations on graph paper.

Scale Drawing (Activity 3). Using KidCAD, students practiced drawing a simple room outline and then created, moved, and rotated furniture on the screen.

Measurement and Mapping (Activity 4). Students measured their classroom and the furniture in it, then made a scale representation using KidCAD. This activity was the central focus of the unit and was carried out in structured phases: Students first measured classroom walls and doors and created the classroom outline on the computer. They then measured the furniture in the classroom and (using KidCAD) created the top view of each different piece of furniture. They then inserted these furniture objects into the appropriate locations within the KidCAD-generated classroom floor plan. As they worked through this task, students made printouts of their work in progress, used scale rulers to check their representations against the real-world objects, and made any needed adjustments.

Classroom Rearrangement (Activity 5). After creating a scale drawing of the classroom, students used KidCAD to rearrange the classroom, talking about and experimenting with space needed for aisles and around chairs and computer workstations. They made large-scale printouts of their design plans, presented them to their classmates, and compared their team's work with the designs of other teams.

Final Design Project (Activity 6). In the culminating project, students designed a new room for their school. They decided on the function for this room (a room the size of their current classroom), created and arranged appropriate furnishings for the space as it would be used, and wrote a team report presenting their design to their classmates.

MAPPING AND REDESIGNING THE CLASSROOM—THE SOFTWARE

We designed the software for fifth-grade students with limited computer experience and mathematical knowledge, recognizing as well that most teachers were neither math nor computer specialists and that few would have time to practice using the software before their students used it in class. This recognition, coupled with the need to design and code the software in a very short time, meant that the prototype software had relatively few tools and options. In retrospect, the simplicity of the software turned out to be one of its strongest features.

Software for Drawing or for Thinking?

The prototype software for this unit was designed to incorporate elements of two different software paradigms. CAD software, as a tool, should allow students to draw objects as quickly and easily as possible, should not make assumptions about the objectives or prior knowledge of the user, and, ideally, should build in features that anticipate possible uses. CAD software, as a mathematical microworld (Papert, 1980), however, should require students to use mathematical ideas explicitly as they create their floor plans and incorporate a limited number of objects and structures, the relationships among which should enhance awareness and understanding of a particular mathematical concept or relationship. These two software paradigms embody subtly different design principles and philosophies of learning. This led to a certain amount of creative tension and intense discussions during the design process. What follows are two illustrations of how we attempted to resolve these tensions, one concerning how mathematical shapes (in this case, rectangles) were defined on the screen and the second concerning how rotational transformations of shapes were performed.

Shape Definition and Properties. The typical point-and-click tool allows the user to draw a rectangle by clicking on a starting point and dragging the cursor until the appropriate rectangle appears on the screen, whereas a microworld rectangle tool might require the user to type a specific length and width to create that rectangle, thus reinforcing through repeated use the numerical properties and relationships, the proportions of different rectangles, and the numerical differences among different rectangles.

To resolve these differences, we used the point-and-click tool, but as the cursor moves, the scale dimensions of the rectangle (length, width, and perimeter) appear in a dialogue box at on the bottom of the screen. Students can draw a rectangle based on the on-screen grid, or they can refer to the dialogue box to decide exactly where to set the second vertex of the rectangle.

Rotational Transformations. We wanted students to use the idea of rotation explicitly so that they can learn that shapes do not change when they are rotated and that copies of the same shape can represent several different objects with the same dimensions, in different orientations. With a point-and-click tool, the user selects an object, then clicks on a rotation icon until the object looks the way it should on the screen. A microworld approach to rotation might require the student to specify the degree of rotation, then correct the position with an additional (or reverse) rotation if necessary. This approach allows students to focus on the angle of rotation, encourages them to estimate the amount of rotation, and familiarizes them with degrees as a measurement of rotation and with the size of frequently used rotations (90°, 180°, and 270°), which can then be used as anchors for estimating rotations that fall between.

To resolve these differences, we used the point-and-click tool and rotation icon, with each click rotating the object 15° clockwise. The degrees of the total rotation appear in a dialogue box at the bottom of the screen. Because students have no need to know the amount of rotation, however, these data can easily be ignored.

The Drawing Screens

KidCAD features two drawing modes with different-sized grids: the Main Drawing Screen (30 × 42 grid array) and the Furniture Drawing Screen (9 × 12 grid array). Rulers along the edges of both screens display the distance (in feet) from the origin.

The Main Drawing Screen is used to create the floor plan of the room. The Line Drawing Tool draws line segments and polygons. Segments that are nearly horizontal or vertical snap into alignment when the mouse button is released. Segments that end close to an existing vertex snap to that vertex. The length of the current line segment and the total length (perimeter) of the current polygon are displayed in a dialogue box at the bottom right of the screen. The name of the function currently selected is displayed in a dialogue box at the bottom left of the screen.

In addition to the Line Drawing Tool, the screen features a Measuring Tool (which displays the distance between any two points), a Make Furniture Tool (which shifts to the Furniture Drawing Screen), a Get Furniture Tool (which displays a list of named furniture objects and places any selected object on the main screen), a Move Object Tool (which allows a selected object to be "picked up" and moved), a Copy Object Tool, and a Rotate Object Tool. Delete and Clear perform the expected functions. The Computer icon brings up a dialogue box that allows for saving and loading files and printing the main screen. The Print function (which allows students to print their floor plans at both 1/4-in. and 1/2-in. scales) requires a custom printer driver.

The Furniture Drawing Screen includes the same Line Drawing Tool, Measuring Tool, Delete, and Clear functions as the Main Drawing Screen and adds the Rectangle Tool (described earlier) and the Circle Tool. (The

---

[3] The Save function did not save the object to the disk, but instead assigned a name to an object as a part of a particular workspace. The entire workspace, floor plan and objects, was saved as a whole by using the Save command from the computer menu.

first click on the Circle Tool defines the center of the circle, and the second click defines its radius, with the lengths of the radius and circumference of the circle displayed in the measurement box.) The misleadingly named[3] Save function allows the current drawing to be assigned a name as a furniture object and adds it to the list of such objects. The Done Tool shifts back to the Main Drawing Screen.

SUPPORTING MATHEMATICAL THINKING BY CONNECTING ONLINE AND OFF-LINE ACTIVITIES

It is the goal of all our curriculum development work to create computer tools that complement, extend, and work in conjunction with other modes of learning activity. KidCAD exemplifies this by providing a direct link between the physical environment of the classroom and the abstract model constructed on the computer.

Because KidCAD scale drawings require accurate lengths (in feet and inches), students are encouraged to measure distance carefully and to check and compare their measurements, moving screen objects as necessary. But students' ability to approximate, interpret, and negotiate also comes into play in creating top-view representations. When students look closely at furniture, they notice that tables often have rounded corners and that chairs might have slightly curved backs. In using a tool that provides only rectangles and circles as preprogrammed shapes, students have to approximate the actual shape of most classroom objects. Making decisions about how and where to approximate is an essential component of model making. Such visual approximation is also typical of the way geometry is used in many engineering and architectural applications—but these ways of thinking are rarely encountered in school curricula.

EXAMPLES FROM THE CLASSROOM

To get a sense of how students approached the task of mapping and designing space, we look at examples of students' work in four stages: students' hand drawings of the classroom floor plan created at the beginning of the unit, students' representations of the classroom using KidCAD, students' plans for rearranging their classroom, and students' designs for new rooms of the same size. In order to show progression of work, all examples are drawn from the same class.

Sketching a Floor Plan

On the first day of the unit, students were shown an architect's drawing of the floor plan of a school library (Fig. 17.1). Students were asked to guess the type of space represented, identify the objects in the drawing, and dis-

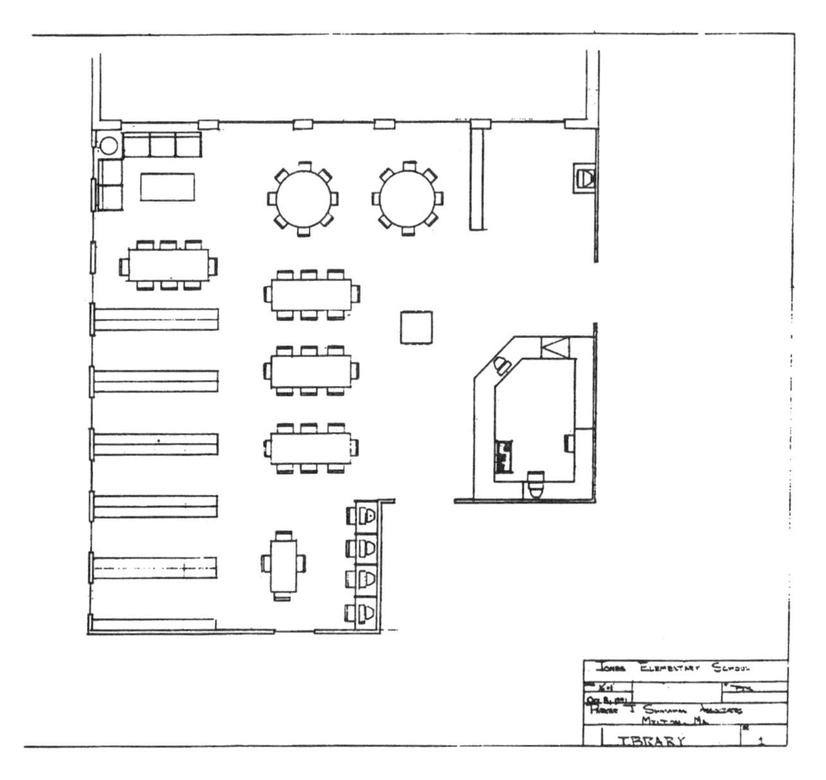

FIG. 17.1 Architect's floor plan used in classroom activity.

cuss both how they were able to identify the objects and the way those objects could have been accurately placed in the floor plan. Following an extensive discussion about what can be seen in a top-view perspective, students were asked to imagine that they were looking down at their classroom from the ceiling and to draw what they saw (to sketch a top view of their classroom). They were not asked to make any measurements—but were not prevented from doing so—and were given the choice of whether or not to use rulers to help them make their sketches. The results give us a sense of students' ability to visualize how their classroom looked from above, what they considered important to include, and their sense of the relative proportions of different pieces of classroom furniture and the open space in the classroom.

The six examples shown in Fig. 17.2 are representative of these classroom sketches. The drawings differ in what students chose to represent and how they represented it. Although all the drawings show five clusters of four desks, they vary considerably in how much space the desks take up, in how closely the desks are spaced to the walls and to each other, and how consistently the students maintained the top-view perspective. Student A, for example, drew desks that almost completely filled the classroom space, whereas the drawings of students B and E show a great deal of space between desks and in other parts of the room. Student E, how-

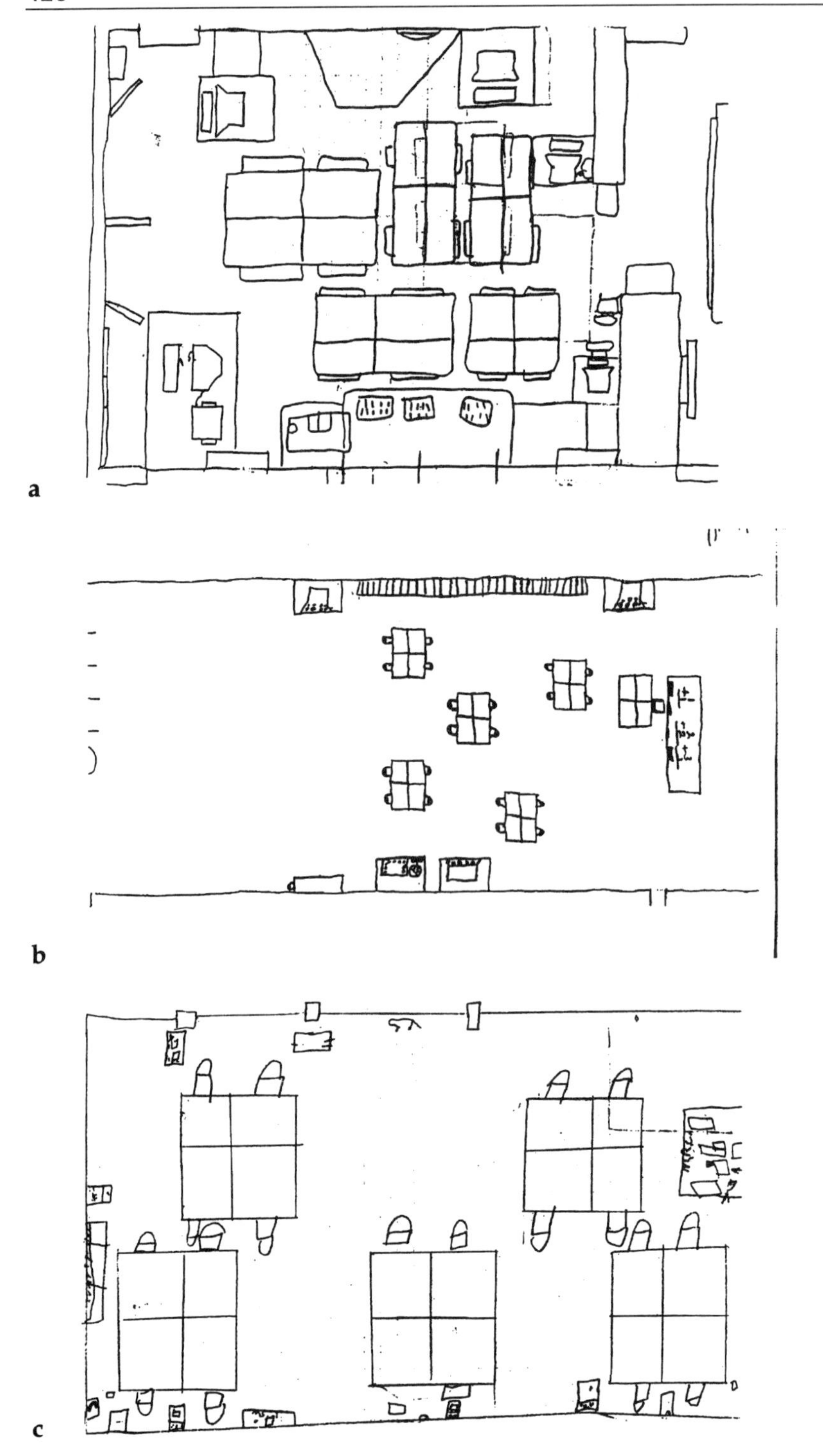

FIG. 17.2 Students' top-view sketches of floor plans.

d

e

f

ever, did not include chairs—but did create images of the children at each table. Students E and C did not include the coat closets at the end of the room (which were represented in some way by most of the other students). A number of students chose to include what could be seen on the walls. Student B chose to draw in the blackboard at the front of the room. Student D represented the front, back, and one side wall as if the room had been cut open at the wall intersections and three of the walls then folded back. Student F (as did many others) drew the chairs in side view.

Scale Drawings Made with the Software

The four scale drawings are very similar. Unlike the hand-drawn sketches in Fig. 17.2, the KidCAD drawings show classroom objects with very similar proportions and relationships to other objects in the classroom. Small details in representing the classroom and its furniture, on the other hand, posed interesting challenges for students and allowed a fair amount of creativity.

Consider the ways chairs are represented in the four drawings shown in Fig. 17.3. Groups B and D represented each chair as a single line—as if the chairs had been pushed all the way into the desks. Group C represented chairs in full top view, with seat and back clearly distinguished. Group A, in contrast, represented chairs as narrow rectangles, almost as if the chairs had been pushed partway into the desks. These differences give very different impressions of the space available in the room. Drawing C, which shows chairs in a full top view, makes the room appear much more crowded than do the other drawings.

Interestingly, all the groups were consistent and accurate in the orientations of the chairs to the main clusters of desks. (Four desks made up a cluster.) Two clusters of desks had chairs facing the front and back of the

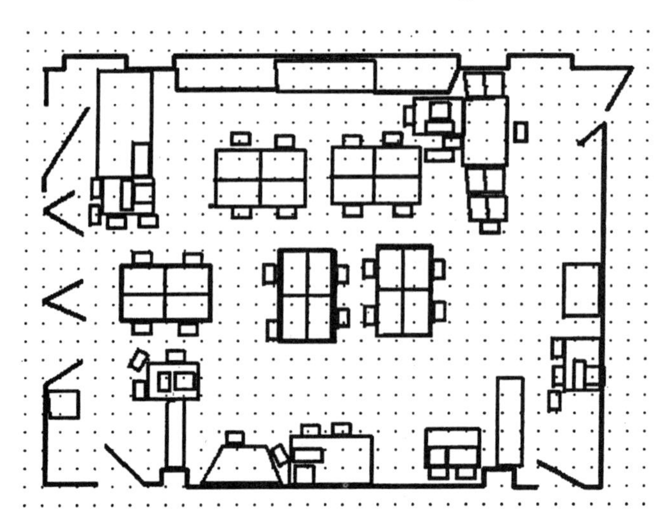

a

FIG. 17.3 Students' scale drawings of classroom (using KidCAD).

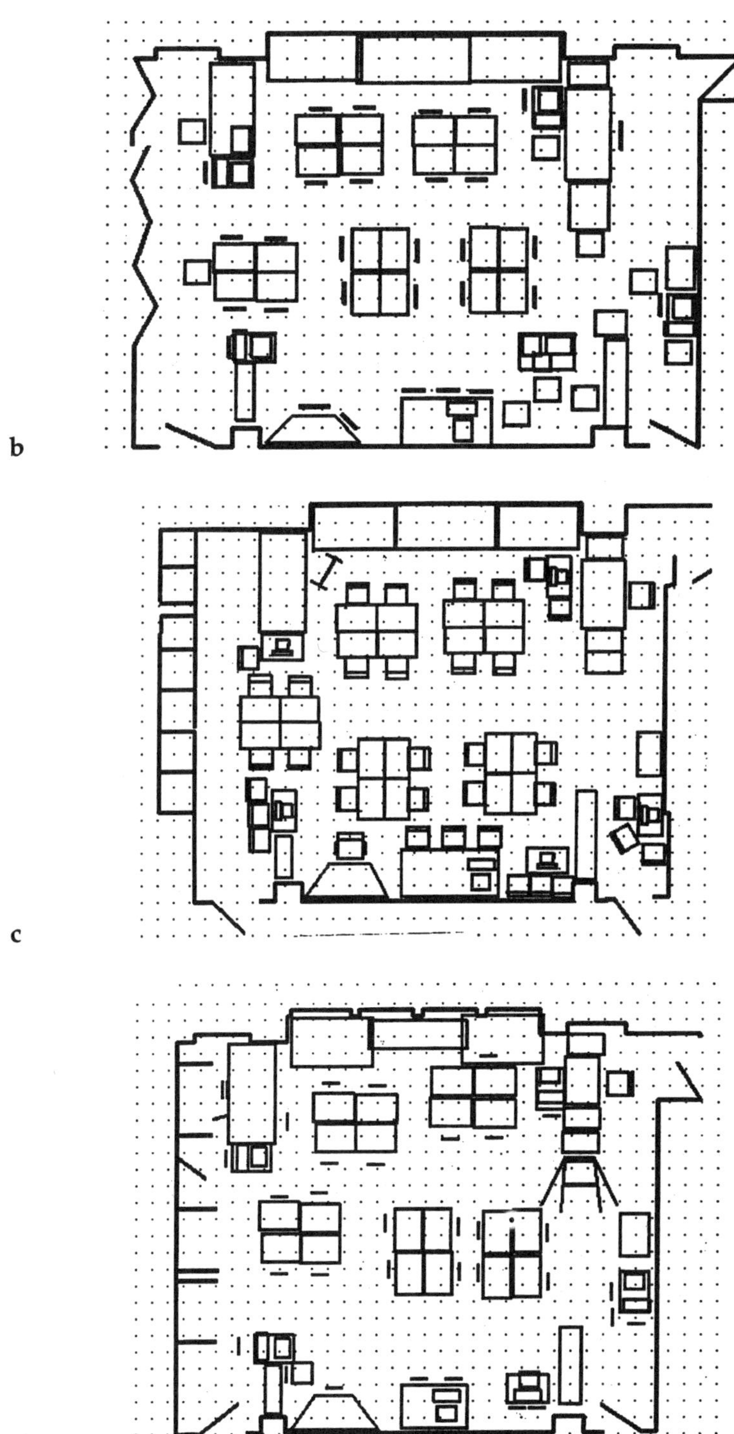
b
c
d

room; the other three clusters had chairs facing the sides. By way of contrast, only one of the hand sketches (see student A, Fig. 17.2), showed all the chairs in the correct orientation.

Drawing a top-view representation of the classroom's portable easel was especially difficult. Groups A and B did not include the easel in their drawings at all. Group C represented the easel as a large "I" shape at the lower right of their drawing. Group D represented the easel (near the left side of the drawing) from a mixed front and side perspective.

Rearranging the Classroom

The examples in Fig. 17.4 show students' success at using KidCAD to represent a classroom that did not previously exist. This rearranged classroom had all the same furniture, but was organized differently. Before creating the floor plan of the rearranged classroom, each group of students had to agree on the changes they wanted to propose and write a rationale for them. After all teams had "rearranged" their classroom, they made a presentation to an adult jury, and after much discussion of all plans, this class and their teacher decided to reorganize the classroom according to one of the plans (Plan A). The three other teachers, however, chose not to carry the activity as far as reorganizing the actual classroom. Our observations and documentation have convinced us that, although the unit was successful without the physical rearrangement of the classroom, this part of the activity was beneficial both to students' motivation and to their actual learning.

Designing a New Space

As a concluding activity, students were challenged to design a new room—to be used for any legitimate school purpose. This task provided students an opportunity to apply what they had learned about using KidCAD, along with their knowledge of measurement, top-view representations, and relationships among objects in space, to create an image of a novel space that existed only in their minds. The examples in Fig. 17.5 show how students were able to represent very different types of furniture and room arrangements to serve a variety of purposes. Students were able to imagine tables of different sizes, plan for occupancy by a particular number of students, and design spaces that could allow for different types of activity occurring at the same time. The drawings show students' awareness of appropriate relationships between tables and chairs, efforts both to design furniture that would fit available space and to balance furniture with open space, and a clear differentiation according to the particular activities and the number of people they envisioned for their alternative classroom.

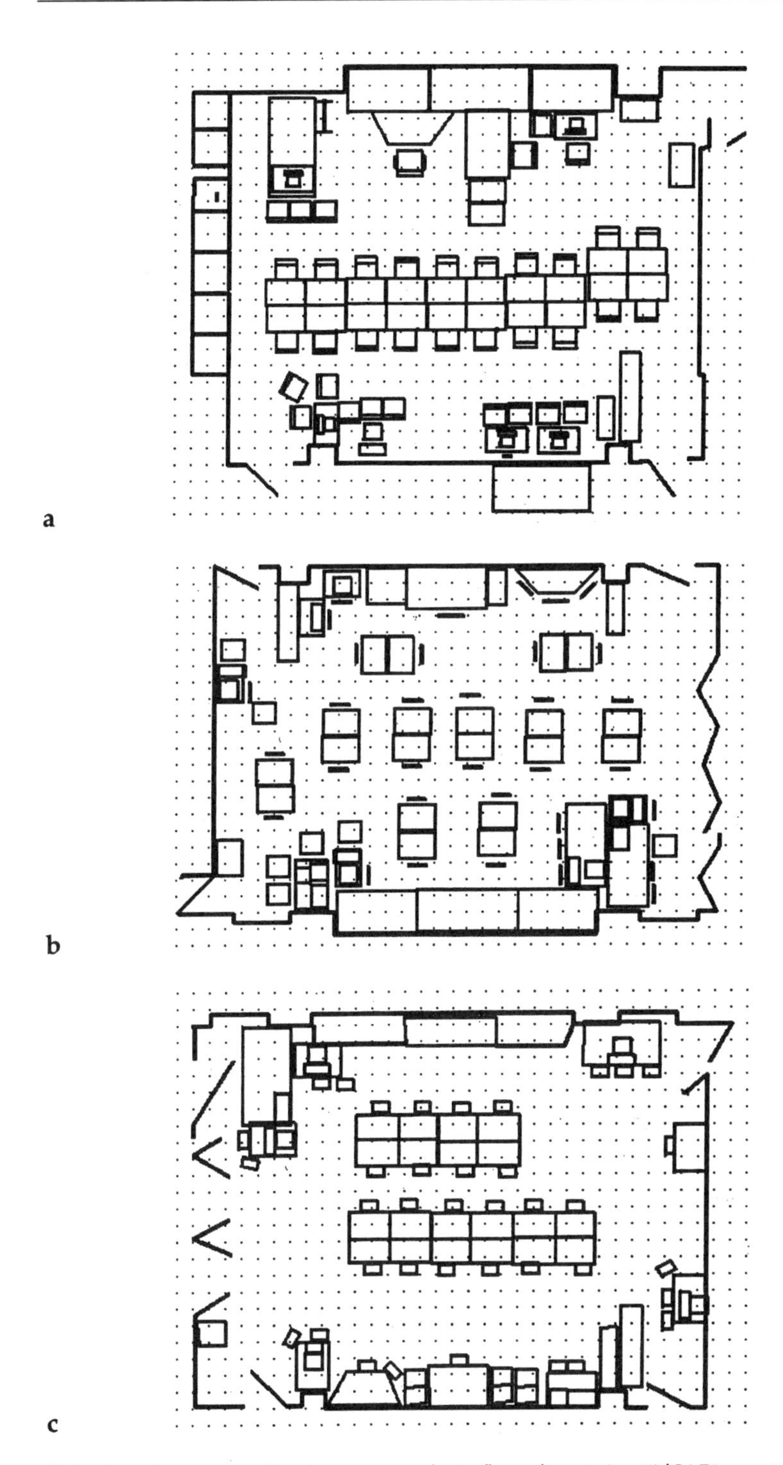

FIG. 17.4 Rearranging the classroom: students' floor plans (using KidCAD).

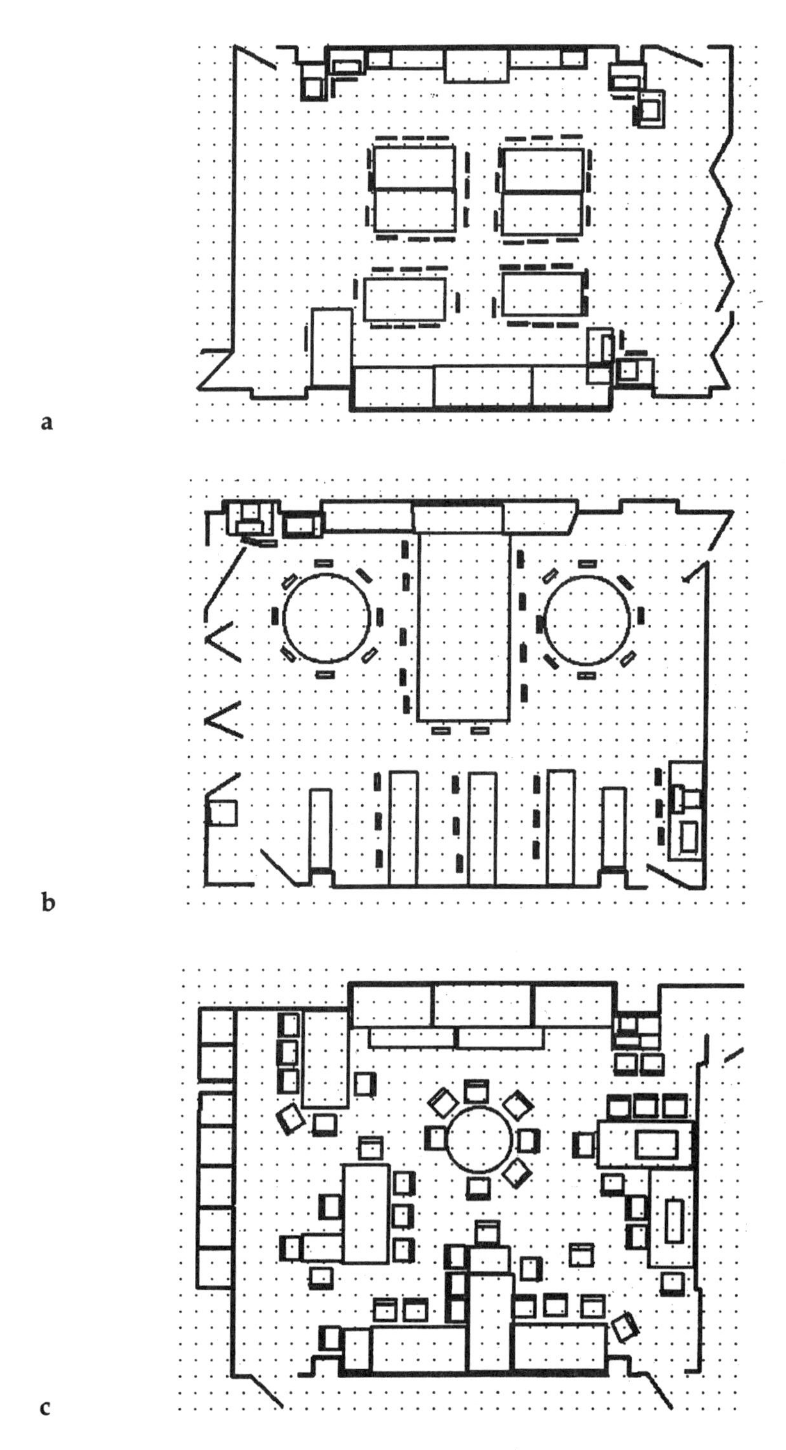

FIG. 17.5 Designing new space: students' floor plans (using KidCAD).

STUDENT LEARNING OUTCOMES

Although, as noted earlier in this chapter, our research was not intended to provide a fine-grained analysis of student learning or of differences among students using the software, we can make some general observations about what students learned in relation to four areas of mathematics content: measurement, geometry, spatial visualization, and scale. What we discuss here is based on classroom observations, examination of student work, and interviews with 16 students (4 from each class).

Measurement

We had expected fifth graders to be somewhat comfortable measuring lengths in a classroom and were not prepared for the low level of their practical experience with measurement tools and concepts. Early in the unit, many fifth graders exhibited confusion about units and converting units (yards to feet to inches and vice versa). Their classrooms had both metersticks and yardsticks, and some students used these tools interchangeably, considering both to be about one yard long. Some did not know how to measure (lining up the starting mark on the ruler with the beginning of the length to be measured). Students frequently used rulers slightly larger than a foot to represent a foot or cloth tapes slightly longer than five feet to represent five feet. They also had difficulty measuring distances longer than the tool they were using and had to learn to carefully mark the end of one measure as the starting point for the next. Students consistently failed to keep records, which would have allowed the measurements to be reused easily in making a scale drawing.

In all areas, we observed significant progress. The constraint of the software (that measurements be given in feet and inches) motivated students to measure carefully to the nearest inch and to clarify their confusions about units. The desire to make their drawings true to the actual classroom and to eliminate tedious remeasurement led students to become more accurate in their initial measurements and to record measurements on sketches. The need for consistency between the model and the "real" classroom led students to check their first measurements, revising them when necessary.

Geometric Shapes and Transformations

Students were comfortable defining rectangles based on length and width and were able to use the relationship between the diameter of a circle and its radius to draw circular objects (tables) using the software. In the course of their work, students frequently used the Copy Object Tool to create multiple instances of the same shape. They appeared to understand that shapes (and copies of a shape) remain the same after translation and rotation. No emphasis was placed on the measurement of translation or rotation, and students used these capabilities of the software without reflecting on what they meant. Although students could use them well visually, for example, to place chairs evenly around a table, most could not articulate any particular knowledge about rotation or estimate amounts of rotation needed to accomplish a particular move.

Spatial Visualization

Students were generally successful in representing objects in top view and understanding the usefulness and value of doing so. They were able to combine two or more rectangles to represent objects such as a straight-backed chair or an armchair. They were able to use measurements and symmetry considerations to draw nonrectangular objects such as computer monitors, trapezoidal tables, and the like. They did have difficulty representing top views of complex three-dimensional objects such an easel.

Students were also able to approximate shapes with simplified representations: Desks with rounded corners were represented by rectangles, as were chairs with rounded backs. Unfortunately, the prototype software did not support simple representations of noncircular objects with substantial curvature. Representing a semi-circular table was difficult and complex for students—not because of any deficiency in their visualization but because approximating such a table using KidCAD was difficult to accomplish.

Scale and Proportional Reasoning

During the unit, students became comfortable with the idea of a scale drawing as a way to represent a space and the objects in it. They were able to use and understand the idea that one box-length (on grid paper) or one dot-space (on the computer screen) represented one foot of length. They also had no difficulty recognizing that an object drawn on the Furniture Making Screen retained its identify and dimensions when reproduced in a smaller size on the Main Drawing Screen. Students regularly made use of the idea that measurements made of a particular space or object on the computer screen, on the printouts, and in the real world all represented measurements of the same object and could be expressed by the same value in feet and inches. The software (and printouts) were, however, designed to make the proportionality involved in the concept of scale transparent. Students did not directly encounter the idea that the sides of an object remain in proportion to the original when it is represented in a scale drawing or when the same object is drawn at two different scales.

However, students continued to experience difficulty with the use of scale rulers, even by the end of the unit. The use of two different scales on the same triangular ruler may have contributed to their confusion. In their interviews, students had difficulty selecting the appropriate scale ruler to use, even when the grid size was clearly visible on a drawing. Students needed either reminding ("I think you should use the 1/2-inch-scale ruler for this measurement") or questioning ("How far apart are the lines on the grid paper? Which scale ruler should you use?"). On the other hand, students were usually able to estimate scale measurements on a drawing or on the computer screen without using the ruler: They simply counted boxes (on grid paper) or dots (on the computer screen).

By the end of the unit, most students were still not fully comfortable with using the one-box-equals-one-foot convention for scale in all contexts. When given graph paper and asked to draw (by hand) an object they had measured, students, unless prompted to do so, did not consistently use the boxes as guidelines for their drawing or use the same scale to draw different objects in the same floor plan.

THE NEXT STEPS

Overall, the prototype software developed for this research served its purpose well. In particular, students learned to create and work with scale drawings, on a computer or on paper, as a practical representations of real or imagined spaces. They moved comfortably back and forth from the real-world objects to their on-line and printed models. Because the software had limited ability to show curves and no facility for measuring angles, students learned to make practical adjustments and approximations in order to create simplified models of more complex objects. In fact, as noted earlier, the very simplicity of the software turned out to be one of its greatest strengths.

We do, however, need to devote more time to introducing and practicing both full-size and scale measurements and the use of measurement tools. The difficulties students had with rulers, metersticks, and yardsticks surprised us, and more scaffolding here would have been beneficial. Converting real-world measurements to on-screen measurements could have been facilitated by such changes in software as the use of the length of diameter (easier to measure on a real-world object than the radius) to define a circular object; other software changes that could make the tool more flexible and powerful include user access to the setting of screen size and accuracy of measurement (for a wider range of measurement, map-making, and scale drawing purposes), the capacity to store measurements (and area) of an object so that such measurements can also be retrieved for use with a spreadsheet or online calculator, and the capacity to switch between British and metric units.

More time should also be spent discussing the shapes students are using and their properties, area, perimeter, proportions, and so forth. More shape-drawing tools should be available, including those for triangle, trapezoid, semicircle, and quarter circle. More attention should also be paid to having students describe and compare the approximated shapes that they use to make top-view drawings of classroom objects.

CONCLUSIONS

The simplicity of the KidCAD software stands in stark contrast to the excitement of dynamic geometric construction, the measuring and conjecturing tools now widely used in secondary schools. We argue, however, that the secret of the success of any educational software as a vehicle for promoting mathematical power lies not merely within the software itself, but in the way that the software supports students' experiences with mathematical concepts and allows students to make mathematical models. The success of the experiences described in this chapter lies in the way the curriculum and software supported students in making mathematical models of their real-world classroom space and in checking those models against reality by physically measuring distances and objects on their printouts and by comparing what they had created to the creations of other students. This connection between online and off-line activities—this negotiation between model (the scale drawing) and reality (the actual space and the real objects inside it)—stands in strong contrast to many computer learning experiences and is what made these activities a powerful learning experience.

One last thought, a caution. Looking at the differences between students' freehand drawings of their classrooms and the scale representations produced by KidCAD, we cannot fail to notice that the gain in mathematical visualization, seen in the latter, is accompanied by a loss of the individual perspective, imagination, and creativity seen in the former. As educators we must be careful not to stress "mathematical" as superior to "artistic." It would be a shame if through this type of experience students come to feel that their own personalized, idiosyncratic representations of the world around them are somehow less important, or less useful, than the uniformity and consistency of the computer-assisted scale drawing. We need both mathematics and art to enrich and give meaning to our lives and to our work.

REFERENCES

National Council of Teachers of Mathematics. (1989). *Curriculum and evaluation standards for school mathematics*. Reston, VA: Author.

Papert, S. (1980). *Mindstorms: Children, computers, and powerful ideas*. New York: Basic Books.

18

An Introduction to Geometry Through Anchored Instruction

Linda Zech, Nancy J. Vye, John D. Bransford, Susan R. Goldman, Brigid J. Barron, Dan L. Schwartz, Rachelle Kisst-Hackett, Cynthia Mayfield-Stewart, and the Cognition and Technology Group
Vanderbilt University

In this chapter we describe preliminary research on three geometry adventures that are part of *The Adventures of Jasper Woodbury* problem-solving series (Cognition and Technology Group at Vanderbilt University [CTGV], 1990). The overall goal of the series (12 adventures) is to anchor or situate mathematics instruction in meaningful, authentic, real-life problem-solving environments that provide a basis for understanding the value and excitement of mathematical thinking. After solving a *Jasper* adventure, students are encouraged to reflect on their problem-solving strategies and deepen their understanding by continuing to explore important mathematical concepts that were involved in the adventure (e.g., CTGV, 1992, 1993, 1994).

The three *Jasper* adventures that focus on geometry are consistent with Geddes's (1992) recommendations for teaching geometry in the middle school. She emphasized the need for students to explore their world and enjoy new applications of mathematics in their environment. Much of the current instruction in geometry fails to help students appreciate its usefulness for solving real-world problems (Zech et al., 1994).

Our experiences with students' lack of knowledge of the usefulness and applications of geometry had a major impact on our design of the three *Jasper* geometry adventures that we discuss in this chapter. Our goal was to create situations that would help students (and often their teachers) understand the power of geometrical thinking in everyday problem solving. Like all the other *Jasper* adventures, the three that focus on geometry include a 15- to 20-min video story involving one or two main characters. At the end of the story, the characters are faced with a challenge. When the challenge is posed, students in the classroom move from the passive role of watching the videodisc to the active role of trying to "meet" the challenge. All the necessary information to solve the problem is presented as the story unfolds, but students must generate and solve subproblems in order to find the solution information in the story. Because the story line enables students to remember when certain events occurred, students can easily search the videodisc for relevant data. The stories also include models of particular kinds of mathematical thinking, which students can return to when they attempt to solve the *Jasper* challenges. These models provide "just-in-time" information that can help students learn what they need to know.

What follows is a discussion of the theoretical framework for the *Jasper* series, a description of the three adventures based in geometry, a discussion of teaching practices crucial for this series, an analysis of the data collected, and a discussion of new tools we are developing.

THEORETICAL FRAMEWORK

For the past several years, we have been testing an approach to instruction called "anchored instruction" (e.g. CTGV, 1990, 1991), instruction situated in the context of meaningful problem-solving environments and consistent with constructivist theories of learning (e.g., Bransford & Vye, 1989; Clement, 1982; Duffy & Bednar, 1991; Minstrell, 1989; Perkins, 1991; Resnick & Klopfer, 1989; Scardamalia & Bereiter, 1991; Schoenfeld, 1989; Spiro, Feltovich, Jacobson, & Coulson, 1991). Theorists who emphasize the constructive nature of learning argue for the need to change the nature of the teaching and learning process that occurs in many classrooms. Instead of teachers transmitting information that students receive, they should provide opportunities for students to become actively involved in the construction of knowledge. Efforts in mathematics reform also stress that mathematics classrooms need to shift from an emphasis on teacher transmission of knowledge to an emphasis on student attempts to use their current skills and knowledge to solve problems (e.g., Charles & Silver, 1988; National Council of Teachers of Mathematics [NCTM], 1989; Schoenfeld, 1985, 1989; Yackel, Cobb, Wood, Wheatley, & Merkel, 1990). In the words of the National Council of Teachers of Mathematics, "activities should grow out of problem situations" (NCTM, 1989, p. 9).

Centering learning around problems or problem situations helps students engage in activities that promote deep understanding. By solving problems students have opportunities to engage in argumentation, reflect on their own understanding, and refine their existing knowledge. As Resnick and Resnick (1991) noted, in order for concepts and principles to be learned effectively, "they must be used generatively—that is, they have to be called upon over and over again as ways to link, interpret and explain new information" (p. 41). Findings from a number of studies suggest that knowledge that is not acquired and used generatively tends to become what Whitehead (1929) called "inert knowledge"—knowledge that is not used spontaneously even though it is relevant (e.g., Bransford, Franks, Vye, & Sherwood, 1989; Gick & Holyoak, 1980, 1983; Scardamalia

& Bereiter, 1985). Data indicate that knowledge is less likely to remain inert when it is acquired in a problem-solving mode rather than in a factual-knowledge mode (Adams et al., 1988; Lockhart, Lamon, & Gick, 1988).

In order to support generative learning, problem-solving situations must provide a meaningful context to engage students' reasoning and thinking. Anchoring instruction in meaningful contexts simulates the advantages of apprenticeship training (Brown, Collins, & Duguid, 1989; CTGV, 1990, 1991). We refer to our anchored environments as *macrocontexts* because they involve complex situations that require students to formulate and solve a set of interconnected subproblems (Bransford, Sherwood, & Hasselbring, 1988). In contrast, problem sets such as the ones that occur at the end of chapters in textbooks typically involve a series of disconnected *microcontexts*—one for each problem in the problem set. Our anchored approach shares a strong family resemblance to many instructional programs that are case based and problem based (e.g., Barrows, 1985; Gragg, 1940; Spiro et al., 1991; Williams, 1991).

This use of anchors allows students who are relative novices in an area to experience some of the advantages available to experts when they are trying to learn new information about their area (e.g., Bransford, Sherwood, Hasselbring, Kinzer, & Williams, 1990; CTGV, 1990). Theorists such as Dewey (1933), Schwab (1960), and Hanson (1970) emphasized that experts in an area have been immersed in phenomena and are familiar with them. When introduced to new theories, concepts, and principles relevant to their areas of interest, experts can experience the changes in their own thinking that these ideas afford. For novices, however, new concepts and theories often seem merely like new facts or mechanical procedures to be memorized. By "anchoring," novices use their available knowledge to understand the phenomena and activities depicted in the anchor while experiencing the changes in their own noticing and understanding. We especially emphasize the importance of helping novices move from general understanding of a complex problem to one in which they learn to generate and define the distinct subgoals necessary to achieve in-depth understanding.

Another essential element of apprenticeship training that we include in our anchors is modeling by experts (Brown et al., 1989). Modeling helps students in two ways. First, by watching an expert perform in a particular environment, students develop a better understanding of the appropriate performance in that environment. Modeling can also provide coaching and scaffolding for students as they develop their own skills. Without a certain level of preparation, however, students are often overwhelmed with information and do not know what to attend to and what to ignore (e.g. Bransford, Goldman, & Vye, 1991; Bransford et al., 1990). The models in the *Jasper* series are presented as part of the story line of the adventures, but students are not expected to learn effectively through their initial exposure to the models. Videodisc technology makes it easy for students to return to the models as needed while attempting to solve the *Jasper* challenge. This "just-in-time" exposure to information allows students to re-explore models after relevant questions, which define a "need to know," have been generated.

The contexts in the *Jasper* series are in visual rather than textual formats and are on videodisc rather than on videotape (see also Miller & Gildea, 1987; Spiro, Vispoel, Schmitz, Samarapungavan, & Boerger, 1987). Video allows a more veridical representation of events than does text. It is dynamic, visual, and spatial, and students can more easily form rich mental models of the problem situations (e.g., Johnson-Laird, 1985; McNamara, Miller, & Bransford, 1991). This characteristic is particularly important for nontraditional students and for students with little knowledge in the domain of interest (Bransford, Kinzer, Risko, Rowe, & Vye, 1989; Johnson, 1987). Videodisc technology also has random-access capabilities, which allow teachers almost instant access to information for discussion (see Sherwood, Kinzer, Bransford, & Franks, 1987). These random-access capabilities are particularly useful in allowing students to explore the same domain from multiple perspectives.

THE *JASPER* GEOMETRY ADVENTURES

The geometry units in *The Adventures of Jasper Woodbury* highlight the ubiquitous nature of geometry in architecture, wayfinding, and measurement and are designed to provide an introduction to geometry rather than a complete geometry curriculum. Different geometry adventures can be used at different grades, or all three adventures can be used simultaneously in one grade. What follows is a description of the three adventures from the simplest to most the complex.

"Blueprint for Success"

"Blueprint for Success" introduces measurement concepts such as perimeter, area, and volume in the context of architectural design and provides opportunities for students to draw geometric representations of things in their own world, primarily scale drawings (see Fig. 18.1).

To solve the challenge, students generate the subproblems they must consider in order to design the playground (e.g., What is a realistic height for a swingset? What angle should a slide make with the ground? What are the dimensions of a sandbox that holds 32 cubic feet of sand?). All of the information students need to design the playground is in the video. For example, to determine a realistic height for a swingset, students can revisit a scene that shows a man who is 6 ft tall standing next to a swingset. By realizing that the swingset is approximately twice the man's height, students can determine a realistic measure for the height of the swingset. Alternatively, teachers can have students explore their school's playground and use the equipment there to estimate desired dimensions.

The story helps students see what architects do, and the challenge of designing a playground gives students an authentic task. The story also provides modeling by experts who help students construct representations, knowledge, and strategies. For example, students can use the blue-

Christina and Marcus are friends who live in a neighborhood where there is very little space for children to play. Several children have recently been hurt by traffic, including a friend of Christina's brother. Christina and Marcus try to console her brother as his friend is taken to the hospital.

The next day Christina and Marcus participate in career day by visiting an architectural firm. They meet the firm's partners, Gloria and Barry. While working with Gloria, Christina views architectural drawings and learns about the measurements they must include.

To give Christina a better idea of how architects use geometry, Gloria asks Christina to find the length of a ladder's legs if the ladder is to be 4 feet tall. Christina uses graph paper to answer Gloria's question. She also explores the question about what shape provides the most area.

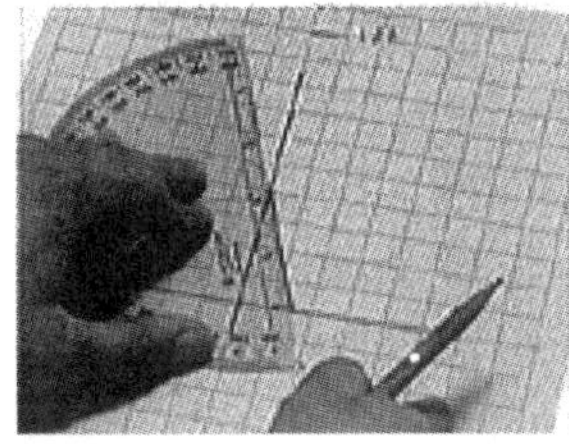

Christina and Marcus meet a developer who decides to donate a lot in their neighborhood for a playground. They want to be involved in helping transform the lot into a playground and volunteer to design the playground equipment. The material to build the playground equipment will be donated.

The Challenge: Create a design or model of the playground for the builders.

FIG. 18.1 Story summary for video anchor "Blueprint for Success."

prints that Gloria shows Christina to generate important information and methods that enable them to draw their own blueprints for the playground equipment. They can revisit the scene in which Christina makes a scale drawing of a ladder to determine the length of its legs. They can also learn strategies for reasoning about measurement such as using graph paper and string to determine the shape with the greatest area, just as Christina does.

In addition to the adventure, each videodisc contains analog and extension problems. For example, one of the extension problems for "Blueprint for Success" asks students to determine the perimeter of various rectangular objects that appear in the story. Students are encouraged to find the perimeter of each object in as many ways as they can. They are asked to share their strategies with other students in their class. Eventually, they are asked to identify a symbolic representation, or formula, for finding the perimeter of a rectangle. The purpose of the analog and extension problems is to help students develop flexible knowledge representations, to better understand key mathematical principles embedded in the *Jasper* adventures, or to practice reasoning and thinking in the context of the problem.

Another extension problem for "Blueprint for Success" asks students to explore the problem that Christina confronts in the story: Given a fixed perimeter, what shape has the greatest area? Using graph paper and string as Christina does, students are encouraged to explore the problem in order to identify the variables and how they are related. An additional question in the extension problems asks them to use a different representational system (square tiles) in reasoning about why a square has more area than a rectangle with the same perimeter. Eventually the students are asked to justify their reasoning using symbolic expressions. These types of extension problems help students continue to develop their knowledge and understanding of geometry using the context of the story to connect their understanding with a number of different representations.

"The Right Angle"

"The Right Angle" introduces concepts involved in wayfinding and the invariant properties of certain triangles, such as isosceles right triangles (see Fig. 18.2).

To help Ryan meet the challenge, students in the classroom interpret Paige's grandfather's directions to trace his route on a topographical map. In so doing, the students revisit relevant scenes in the story that contain information about geometry and that help students learn how to read a topographical map. Both types of information are included in embedded teaching scenes throughout the story. For example, one of the grandfather's directions states that a particular leg of the journey involves a rock spire the top of which was at a 45° angle from the ground when the grandfather stood 300 ft from its base. Students can use information about isosceles right triangles contained in three different scenes in the videodisc story to help them decipher this clue.

"The Great Circle Race"

In this story, students learn about geometric properties involved in orienteering methods such as triangulation. They also learn about properties of triangles and explore invariant properties of circles, such as the relationship between the diameter and the circumference of a circle. The two main characters in "The Great Circle Race," Donna and Tommy, enter a race in order to win a $5000 contribution to the charity of their choice. They plan to donate the prize to their community center to make much needed improvements to the playground (see Fig. 18.3).

The challenge for students in the classroom is to determine who wins the race. In order to do this, they must locate the starting point of

Paige and Ryan each have a problem to solve. Paige must follow a set of clues to find a gift left by her grandfather. Ryan must pass his geometry test in order to play in the baseball championships. Paige explains her difficulty in deciphering her grandfather's clues.

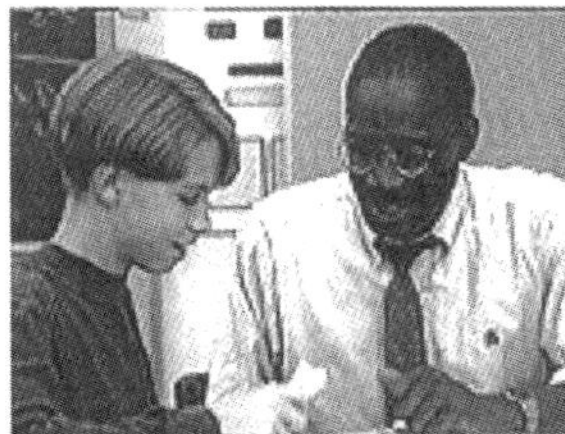

As Ryan meets with his geometry teacher, Mr. Wendell, he discovers some properties of right isosceles triangles. He later tells Paige that he sees no use in knowing these properties. Paige explains how her grandfather taught her how she could determine the height of objects by using the properties of right isosceles triangles.

Paige goes to Larry's barn to care for her horse. While there, Larry explains how to estimate heights of objects by using shadow lengths and also by measuring angles of elevation. During their conversation, Larry gives Paige an important piece of information that she needs to decipher her grandfather's clues.

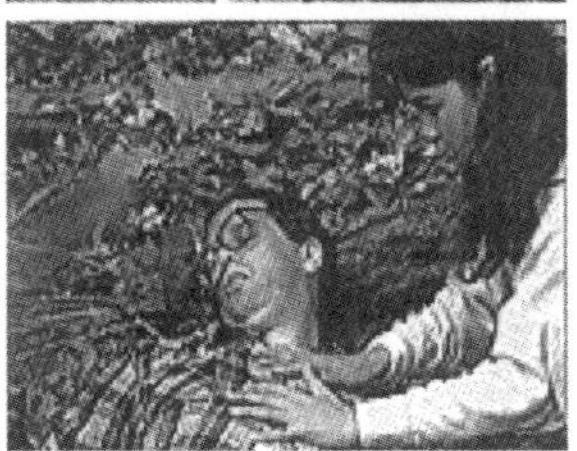

Paige asks Ryan to accompany her on horseback as she retrieves her grandfather's gift. Ryan is late so Larry accompanies Paige to the cave. At the cave, Larry is injured when he is thrown from his horse. Paige sends her horse back with a message to Ryan to send a rescue squad.

The Challenge: Locate the cave and write directions for the rescue squad so that they can find the cave.

FIG. 18.2 Story summary for video anchor "The Right Angle."

each competitor, estimate speeds based on the terrain each competitor has to travel (a topographical map is provided), and calculate the time each competitor needs to complete the race.

RECOMMENDED CLASSROOM PRACTICES

The effective use of the *Jasper* adventures comes not from their being shown once or twice as linear television programs, but from students' generation of goals and subgoals and their revisiting of the adventures to find relevant information about concepts, constraints, and data. In this section, we describe classroom practices that have proven to be effective in

Jasper explains the rules for The Great Circle Race. Competitors will be told where the race ends. Then they must find a place to start that is 5 miles from the finish. Racers can use any nonmotorized mode of transportation during the race. The first place team will select a charity to receive the $5,000 prize.

Tommy and Donna decide to recruit a team to enter the race and meet with Larry in order to learn the technique for locating a starting point on a map. Larry tells how he marked a favorite fishing spot on a map by taking cross bearings of two landmarks with a compass. Donna and Tommy also learn the "great circle secret."

In order to determine the fastest mode of transportation to use in the race, Donna and Tommy collect information about the speed of vehicles like bicycles and roller skates. They even measure how fast Donna can travel in her wheelchair. Based on these time trials they decide to have Tommy ride his bicycle in the race.

On race day, Donna and Tommy suffer a setback when someone slashes both tires on Tommy's bike. Donna decides that their only chance to win rests on her entering the race. The competitors find their starting locations and the race begins. Times of competitors for each of the first three miles of the race are recorded on a chart at race headquarters.

The Challenge: Determine who wins the race and in what time.

FIG. 18.3 Story summary for video anchor "The Great Circle Race."

helping students develop their problem-solving skills and understanding of geometry concepts as they work with each adventure.

We recommend that the instructional sequence start with students watching the adventure. Though some teachers who use *Jasper* for the first time feel the need to teach all of the relevant concepts and vocabulary to students before they watch the adventure, teachers who use *Jasper* usually report that preteaching is unnecessary. Students learn geometry and its vocabulary as they attempt to solve the "challenge."

After students have watched the video, teachers usually engage the class in a discussion where students suggest, and the teacher or a student records, the problems they must solve. Some teachers leave it to the groups or the individuals to identify and solve their own subproblems. In either approach, students are active participants in identifying and solving problems. Because the subproblems are not specified in the challenge, problem generation is an important part of the problem-solving process. Traditional word problems often tell students what they need to find, and students rarely have opportunities to generate the problems themselves. Our data indicate that fifth-grade students can become very good at complex problem formulation on tasks similar to *Jasper* after working with *Jasper* in cooperative learning groups for four to five class sessions (Van Haneghan et al., 1992).

We also recommend that students work in groups as they attempt to solve the challenge. The complexity of the challenges make them difficult to solve alone. Data indicate that students enjoy group work and perform better when they work together (Barron, 1991; Rewey & CTGV, 1992). Working in groups also allows students to share their geometric representations of real-life objects with their peers and receive feedback that helps them to refine their drawings.

Often when students solve the geometry adventures they have difficulty generating appropriate geometric representations. Their initial drawings of real-world objects lack relevant properties or measurements. For example, in drawing a side view of a swingset for the challenge in "Blueprint for Success," students often omit measurements or make errors in representing measurements, such as the length of the legs or the measure of the angle formed by the legs (see Fig. 18.4). We suggest that teachers remind students to think about the information needed to build the swingset. With this type of feedback, students can refine their representations to convey important information accurately.

Opportunities for students to explore geometry in the world are modeled in the story. For example, in one scene in "Blueprint for Success" a safety engineer is shown taking a number of different measurements on a playground in order to determine if the playground meets safety guidelines. Students can use the safety checklist shown in the video to take similar measurements of their school playground to determine its safety. In "The Right Angle," the scenes that show an expert measuring angles of elevation in order to estimate heights of objects have prompted many students to perform similar activities in order to estimate the height of objects in their neighborhoods. In "The Great Circle Race," the scene in which

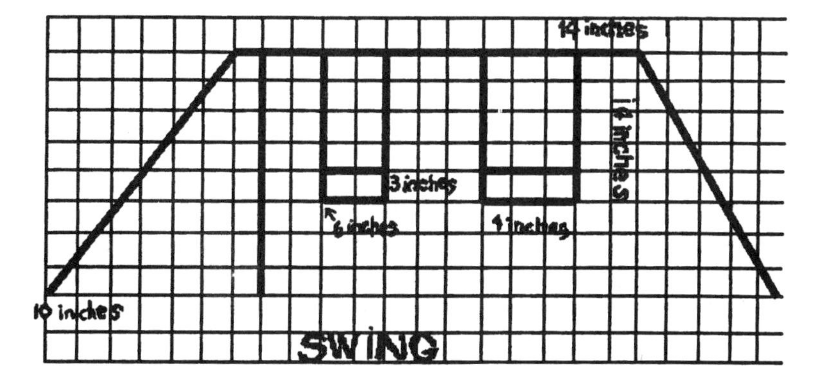

FIG. 18.4 Student drawing of swingset for "Blueprint for Success."

Larry identifies a location on a map by taking two bearings has helped students learn a number of orienteering skills. In completing these activities, students rely heavily on modeling of the experts shown in the adventure. We have observed a similar phenomenon among adults who solve the adventures for the first time.

Solving the challenge can be a base for many other learning activities. We recommend that groups present their solutions to the class and discuss the strengths and weaknesses of different solution strategies. This process helps students consider and analyze alternative solutions, interpret and understand another group's geometric representations, and communicate in the context of solving authentic problems.

Another postchallenge activity that we think is quite useful for students involves projects in students' own communities for which the problem-based activities might serve as a model. In several classes last year, students solved "Blueprint for Success," then designed blueprints for a playhouse, had volunteers use those blueprints to build the playhouses, and donated the playhouses to kindergarten classes in their schools. After an eighth-grade class had solved "The Right Angle," the students drew a map of their school campus. In completing this project, students determined the measurements that they needed to make, made the required measurements by taking bearings and measuring lengths, and correctly represented this information in a map they drew.

There are several advantages to first organizing curricula around anchors and then progressing to hands-on projects. Anchors provide a common ground of shared knowledge that facilitates active participation by students. Organizing instruction around anchors is usually more manageable for teachers than finding all the resources necessary to accomplish actual community-based projects. We find that students' projects are usually of a higher quality if they have first had the opportunity to engage in a related problem-based activity (Moore, Bateman, Bransford, & Goldman, 1995).

RESEARCH ON STUDENT ACHIEVEMENT USING THE *JASPER* SERIES

One of the challenges of our *Jasper* geometry adventures is developing appropriate assessments for the kinds of skills and understandings developed through solving them. Traditional objective tests do not capture such gains. Data from several studies, in which we have assessed multiple aspects of students' understanding of the usefulness of geometry, students' application of geometry in solving real problems, and students' skill at using geometry to represent common objects, show learning gains.

All of the *Jasper* geometry adventures have been tested in a number of middle-school classrooms. Students in Grades 6, 7, and 8 viewed and solved "The Right Angle" and "The Great Circle Race," and students in Grade 5 worked with "Blueprint for Success." One of the major goals of our research on the adventures "The Right Angle" and "The Great Circle Race" was to assess the degree to which they helped students and their teachers understand the usefulness of geometry. Comparing the posttest data with the pretest data of students who solved both adventures shows that students improved dramatically in their ability to identify uses of geometry (see Fig. 18.5). Most of the uses of geometry cited on the posttest were more sophisticated than the simple measurement uses identified on the pretest (e.g., how to find the area of a rectangle or square).

Students' increased knowledge of applications of geometry was also evident in their solutions to problems that required them to recognize and apply geometry principles to real-world situations. For example, one of the application items asked them to find the height of a tree, given that its angle of elevation at a distance of 40 ft from its base is 45°. Another item asked them to find the height of a tree if the shadow of the tree is 70 ft long when the shadow of a person 6 ft tall and standing beside the tree is 2 ft long. Other application items were further removed from the geometry emphasized in the two *Jasper* adventures and were increasingly more difficult. Fig. 18.6 compares the percentage of students who could solve these two application questions on the posttest with the percentage who could solve these problems on the pretest. The percentage of students who could solve each of the other items on the applications posttest also increased,

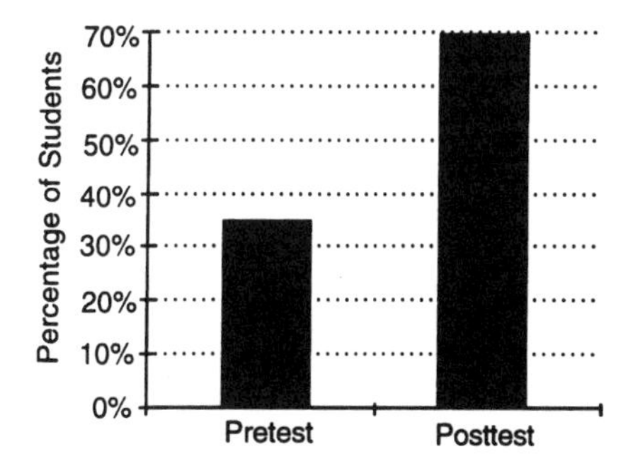

FIG. 18.5 Percentage of middle-school students who identified uses of geometry before and after solving "The Right Angle" and "The Great Circle Race."

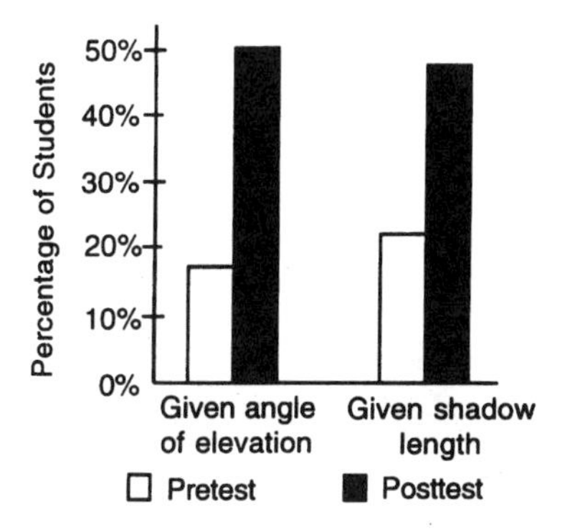

FIG. 18.6 Comparison of percentages of middle-school students finding height of a tree given either angle of elevation or shadow length of nearby object.

although, not surprisingly, the increase was less as problem difficulty increased.

A major goal of our initial research on "Blueprint for Success" was to assess its effects on students' abilities to create accurate representations of real-world objects. For example, on both a pretest and posttest, we asked 106 fifth-grade students to make drawings of a chair that could be used to build it. The pre- and posttest drawings for each student were shown to reviewers with no designation as to whether a particular drawing was from the pretest or posttest. Reviewers compared the drawings on the dimensions of scale, realistic measurements, and types of drawings included to convey accurate measurements (i.e., front view, side view, top view). Based on this type of analysis, reviewers were able to identify the posttest drawings for 97% of the students, which would imply that students' ability to render accurate drawings had improved.

Another performance item on the pre/post assessments asked these students to make a scale drawing of the top view of a swimming pool that measured 12 ft by 18 ft. A three-level coding scheme was developed to capture the degree of accuracy in the drawings. The first level referred to drawings that did not have a scale (or measurements) or sides that measured 3:2. The second level referred to drawings of the pool that had sides in a 3:2 proportional relationship, but that did not include scale or measurements. This category also included drawings that had measurements inconsistent with the scale or that did not have the correct dimensions. The third level referred to drawings that had either a scale or measurements, had sides drawn in a 3:2 proportional relationship, and met the constraints of the problem. Fig. 18.7 shows the percentage of students who stayed in the same ategory, moved to a higher level category (i.e., their drawings improved), or moved to a lower level category (i.e., their draw-

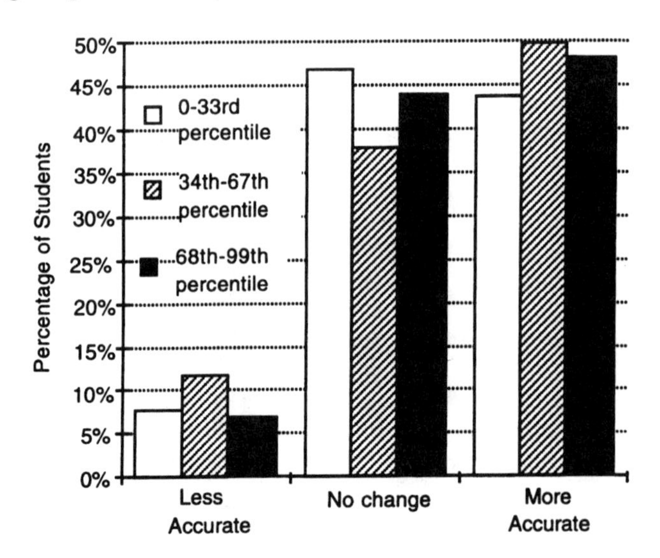

FIG. 18.7 Percentage of students' change in ability to make accurate scale diagrams after solving "Blueprint for Success."

ings became less accurate) from pretest to posttest. The percentages were analyzed as a function of prior mathematics achievement as measured by a standardized test. About half of the students in each achievement group improved their drawings. Out of the group of students who became more accurate in their drawings, 65%, 53%, and 69% of the students in the bottom, middle, and top mathematics achievement groups, respectively, attained top-level scores at posttest. Of the group of students who did not improve their drawings, about half showed some evidence of understanding scale on the pretest. Only 22 of the 106 students in the study did not improve their drawings and still showed no evidence of understanding scale on the posttests.

A second type of performance item asked students to determine the height of a ladder from a scale drawing of that ladder. Figure 18.8 shows that gains on this item were more modest than on the draw-to-scale items, especially for the middle and lower achievement groups. This is not surprising because we know from our classroom observations that students spent relatively little instructional time on interpreting scale drawings. The solution for "Blueprint for Success" required construction of their own drawings to a far greater degree than interpretation of the drawings of others.

The data from our studies of the *Jasper* geometry adventures have been very encouraging. Clearly, however, there is still room for improvement. Most students were far from ceiling on our posttest tasks. Some of the reasons for less-than-perfect scores on our posttests derive from the

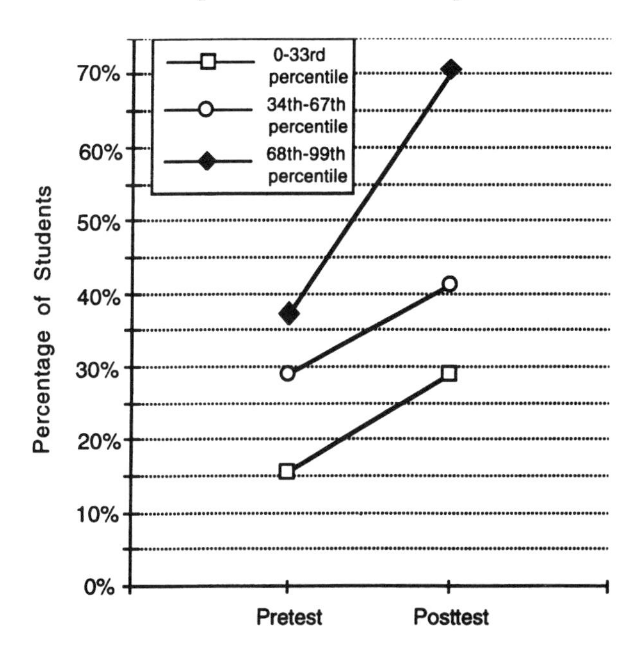

FIG. 18.8 Percentage of students in each of three mathematics achievement groups who drew an accurate scale diagram of a ladder before and after solving "Blueprint for Success."

fact that many of our items required students to put reasons and explanations in writing, and many of the students were very poor at expressing themselves in writing. However, there were other reasons for less-than-perfect scores as well. For example, in talking with teachers after they had taught the adventures, it was clear that many felt they could do a better job the second time around (Zech et al., 1994). In addition, a number of teachers asked for help in managing the group learning that accompanies *Jasper,* in order to ensure that each individual learned effectively. To better understand the challenges faced by teachers, we moved beyond pre/post assessments and began to analyze the teaching and learning processes that occur in the classrooms.

THE TEACHING AND LEARNING PROCESS: INITIAL FINDINGS AND PROTOTYPE TOOLS

In order to learn more about the teaching and learning processes involved in studying geometry through anchored instruction, we analyzed classroom interactions and interviewed teachers and students in six Grade 5 classes who solved "Blueprint for Success." The interactions were captured on videotape and in field notes written during daily observations. Interviews of all teachers and of a randomly selected subset of students were conducted at the completion of the solution process. What follows is a discussion of our findings.

Analysis of Instruction

While watching classroom interactions, we noticed that many students had misconceptions about measurement and geometric relationships and that these misconceptions were often difficult to change. For example, we

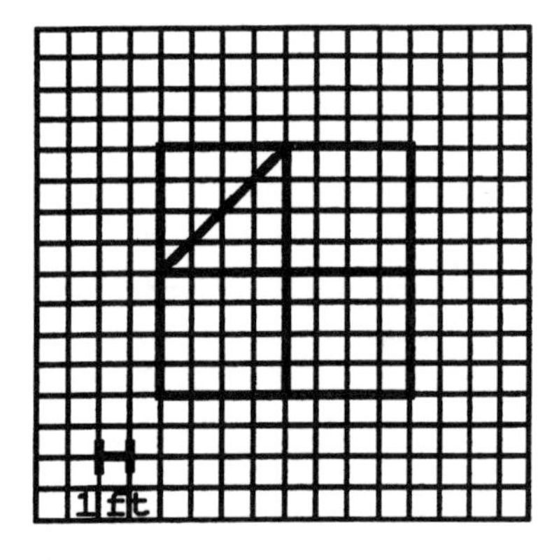

FIG. 18.9 Student scale drawing of four-square court.

observed students in one class who made scale drawings of a "four-square court" that they had on their playground (see Fig. 18.9).

The teacher asked them to determine the length of the diagonal in the serving square:

Teacher: Make one of your boxes the server's square, and draw the diagonal. Look at the drawing where one side of a square is equal to 1 foot.

Teacher: Raise your hand if you know how long the diagonal line is. [Nine students say 4 feet; three say 3 feet, one says 3.5 feet, and the rest do not respond.]

Student 1: Four feet 'cause one square is equal to 1 foot.

Student 2: One square is equal to 1 foot, and the line is going through four squares.

Teacher: Does one square equal to 1 foot?

Student 2: A line in a square equals 1 foot. [Later in discussion]

Teacher: Look at the scale on your drawing, does it show a side of a square or a square? It shows the side of a square. What if I were to tell you that the diagonal line is not 4 feet?

Student 2: We'd be in trouble!

Teacher: The diagonal line is not 4 feet. In your groups, decide whether you think that the line is shorter or longer than 4 feet. Everyone in your group needs to agree and needs to be able to tell me how you determined whether it was shorter or longer.

None of the students in the classroom recognized that the length of the diagonal would be longer than the sides of the square. Many students persisted in this misconception, even after several discussions about the length of different diagonal distances when compared to horizontal or vertical distances on squares.

Students had difficulty interpreting scale drawings of real-world objects. For example, a teacher showed the class a scale drawing of the side view of a swingset (see Fig. 18.10) and asked the students how tall the

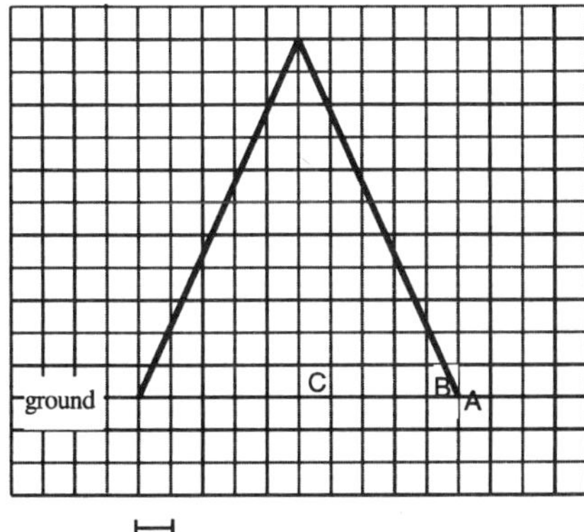

FIG. 18.10 Scale drawing of side view of swingset.

swingset was. All students agreed that the height of the swingset was determined by the distance from the ground to the highest point of the swingset. Students used several different methods to determine the height from the scale drawing. One student started at the base of a swingset leg (A in Fig. 18.10) and counted grid lines as she followed the leg of the swingset. Another student started at the square next to the swingset leg (B in Fig. 18.10) and counted the squares as he moved in a line parallel to the swingset leg, omitting the square at the top of the swingset. A third student started at the square in the middle of the swingset legs (C in Fig. 18.10), numbering the first square zero and omitting the square at the top of the swingset.

During the discussion, the teacher tried to help students correctly interpret the scale drawing of the swingset by clarifying what (in the scale drawing) represented a foot (i.e., a side of a square on the grid paper rather than a square) and where the height of the swingset could be found (i.e., along a vertical line in the middle of the swingset legs rather than along its legs). Some students struggled with the interpretation of their scale drawings of the swingset for quite some time, as evidenced by the incorrect measurements they displayed on their drawings. For example, many students had the lengths of the legs of their swingsets equal to the height of their swingsets. Others found measurements in their drawings with a ruler. This resulted in measurements like 3 in. for the height of their swingset and 3 1/2 in. for the length of the swingset legs.

Teachers also often reported that they had difficulty assessing the reasons for the misconceptions and as a result did not know what instruction might be beneficial to the students. In the postinterview, one teacher said, "When I looked at their drawings... I felt overwhelmed because you see so much wrong, especially at first, that you don't know what to do." This statement was consistent with classroom observations: Teachers would do informal assessment by observing students' work in class; however, they frequently did not provide feedback that would have helped students identify errors in order to revise their drawings. As a result, students' drawings remained full of errors, and misconceptions persisted.

The *Jasper* Toolkit

To help teachers and students assess errors, reflect on and change their thinking, and revise their work, we designed and tested a set of prototype tools, the "*Jasper* Toolkit" (for both teachers and students), to be used with the adventure "Blueprint For Success." The toolkit, like the adventure, is videobased. Students and teachers can watch the tools a number of times for just-in-time information or instruction. We designed the tools based on students' common errors and misconceptions.

The toolkit contains two types of tools for students: visual tools, which help students "see" different objects in the world and think about geometric relationships in and between these objects, and reflective tools, which help students recognize errors. The tools can be used by teachers when they work with a group of students, or by individual students as they try to design their playground equipment or revise drawings they have made.

The visual and reflective tools were designed to engage students in conversations that help them identify and correct their errors. For example, one visual tool, "Draw Like an Architect," shows different views of playground equipment. Students and teachers can use the tool to help them visualize what a swingset, slide, and sandbox look like when viewed from the top, front, and side (see Fig. 18.11). We developed this tool to engage students in conversations about what measurements can be accurately shown in each view. Through these conversations, students began to realize the purpose of drawing difficult views and the measurements that should be included in each view.

A second visual tool, which has proven to be especially helpful, promotes students' understanding of the use of graph paper to draw to scale. In our prototype, a person is represented at several different levels of scale (see Fig. 18.12). This progression helps students talk about realistic measurements and how these measurements can be accurately shown on a scale drawing.

A third visual tool helps students see how to use graph paper as a ruler (see Fig. 18.13) in order to measure diagonal distances. The tool is not presented as if it were a worksheet assignment but is introduced as something that students may use if they wish. It provides a way for students to interpret diagonal distances drawn to scale and to confront misconceptions about their measurements.

The reflection tools are intended to help students recognize and discuss errors that they have made. For example, one reflection tool is a graph that presents students with data about the swing designs of 60 stu-

FIG. 18.11 Prototype visual tool to help students imagine what a swingset, slide, and sandbox look like when viewed from the top, front, and side.

FIG. 18.12 Prototype visual tool to promote understanding of the use of graph paper to draw to scale.

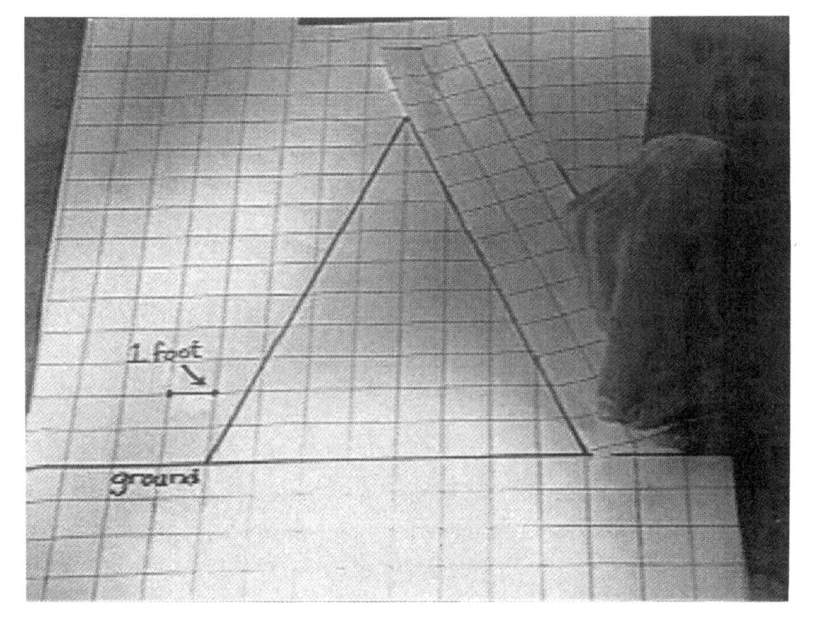

FIG. 18.13 Prototype visual tool to help students make a graph paper ruler to measure lengths in scale drawings.

dents. The data show relationships between the heights and leg lengths of A-frame swingsets. A scatterplot of such data is illustrated in Fig. 18.14a. In Fig. 18.14b, a line connects selected points to show all of the swingset designs that have the same height and leg length, a common error made by students in designing their swingsets. The scatterplot also includes a dynamic visual tool to help students see what a swingset with the same height and leg length would look like (see Fig. 18.15). Teachers have been extremely enthusiastic about the effectiveness of these graphs in getting students to talk about different swingset designs and motivating them to revise their own plans.

In another reflection tool we show different students presenting their drawings of playground equipment. The students on the video make some of the same mistakes that we noticed students commonly make when drawing their playground equipment (e.g., the height of the swingset is the same as the length of the swingset legs). Students in the classroom are to give the students on the video feedback about what is good and what needs to be changed in the drawings. By trying to give the video

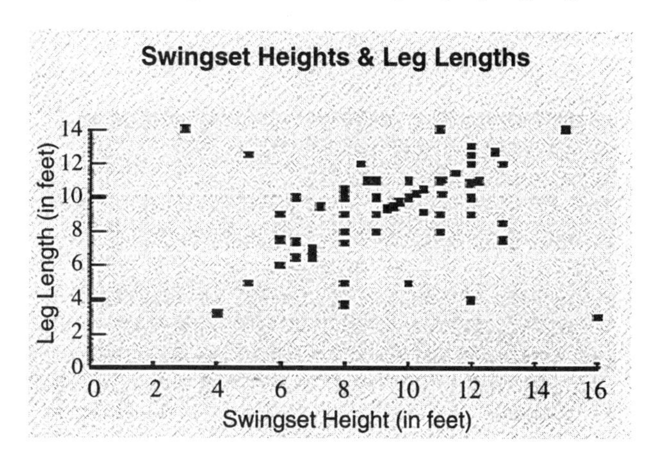

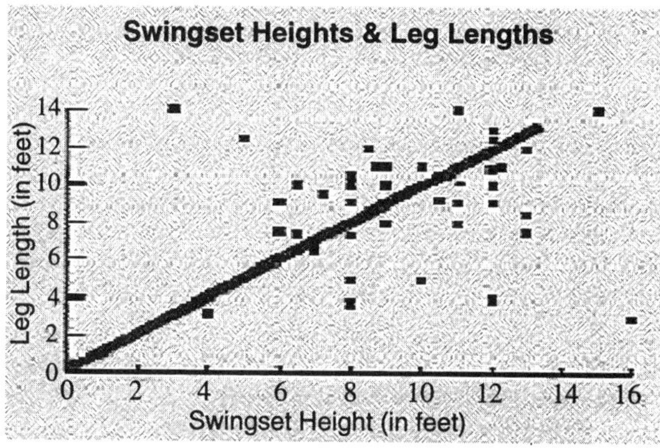

FIG. 18.14 Prototype reflection tool presenting information about swingset designs created by students solving "Blueprint for Success."

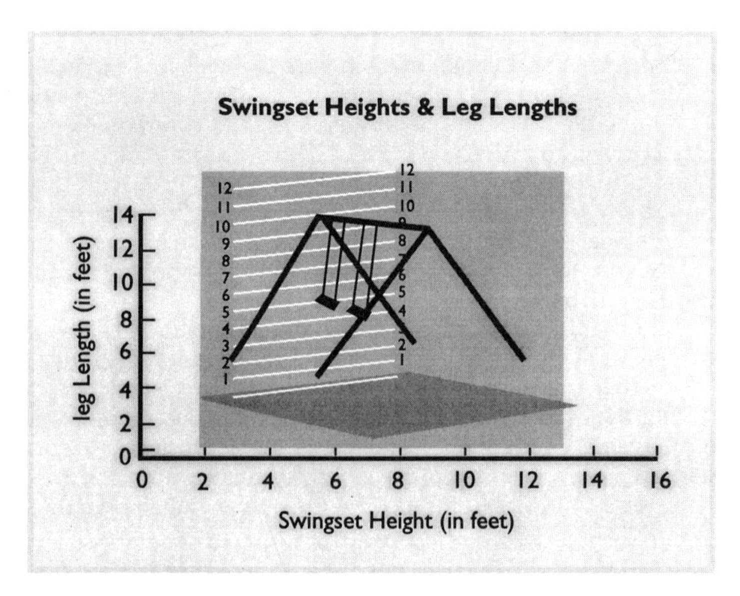

FIG. 18.15 Prototype visual tool showing that a swingset would have to float in the air if its height and leg length were equal.

students feedback, the students engage in conversations that help them reason not only about the drawings in the presentation but also about their own drawings.

One tool, designed for teachers, helps them assess students' drawings. The computer-based tool helps teachers focus on important features of students' drawings, assess these features, and provide appropriate feedback to help students revise their work. Fig. 18.16 shows one of the rubrics used to assess students' use of scale in their drawings.

The suggested feedback contained in the assessment tool is relatively general, but it alerts students to key concepts and provides suggestions for resources that can be used to help them think about their designs. Providing this type of feedback encourages students to reflect on their drawings, have conversations about them in relation to other students' drawings, and try to correct their errors.

Our work with the prototype tools created for "Blueprint for Success" demonstrates the value of creating *Jasper* Toolkits for each adventure to help teachers engage students in meaningful conversations about their thinking and their products. We find that students and teachers make extensive use of the prototype tools, and students often create designs considerably more advanced than they would otherwise have done without the tools. When we interviewed students about their work in designing a playground for the challenge in "Blueprint for Success," students often referred to these tools as they explained the revisions they had made in their work (see Fig. 18.17). In addition, in testing the use of visual tools to help students solve a *Jasper* statistics adventure, we have seen increases in student learning and positive attitudes toward mathematics (Barron et al.,

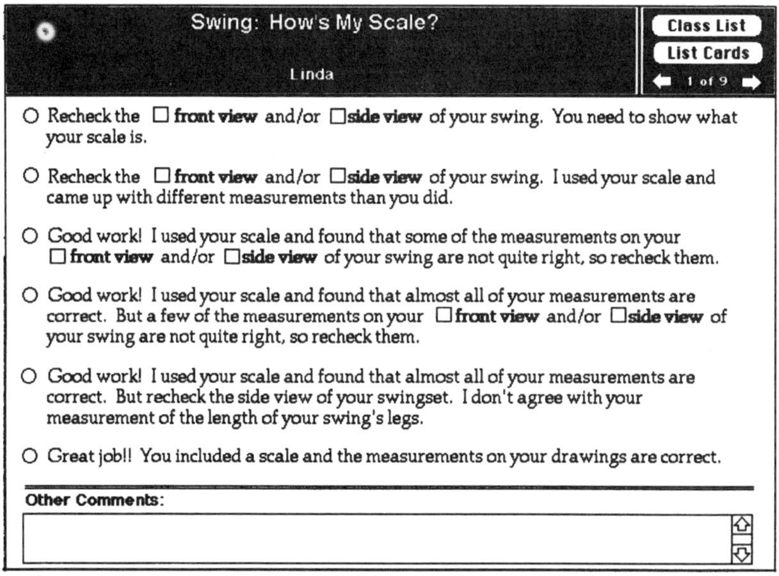

Swing: How's My Scale?

Linda

Class List

List Cards

1 of 9

- ○ Recheck the ☐ **front view** and/or ☐**side view** of your swing. You need to show what your scale is.
- ○ Recheck the ☐ **front view** and/or ☐**side view** of your swing. I used your scale and came up with different measurements than you did.
- ○ Good work! I used your scale and found that some of the measurements on your ☐ **front view** and/or ☐**side view** of your swing are not quite right, so recheck them.
- ○ Good work! I used your scale and found that almost all of your measurements are correct. But a few of the measurements on your ☐ **front view** and/or ☐**side view** of your swing are not quite right, so recheck them.
- ○ Good work! I used your scale and found that almost all of your measurements are correct. But recheck the side view of your swingset. I don't agree with your measurement of the length of your swing's legs.
- ○ Great job!! You included a scale and the measurements on your drawings are correct.

Other Comments:

FIG. 18.16 Percentage of students who reported using particular resources to revise their swingset designs.

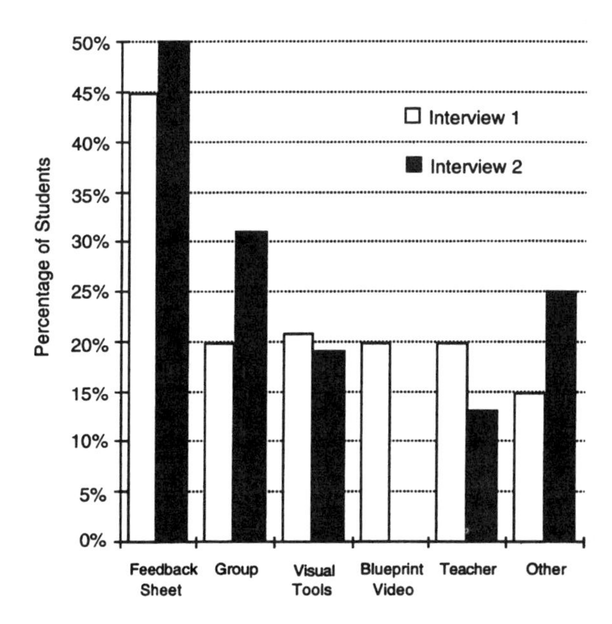

FIG. 18.17 Computer-based scoring rubric to assess students' designs of a swingset for "Blueprint for Success."

1995). We expect similar results when introducing other visual tools to the *Jasper* geometry adventures.

SUMMARY

Each adventure in *The Adventures of Jasper Woodbury* is a video story that creates a visual learning environment to support students' problem generation as well as complex problem solving. All of the information needed to solve the problem is contained in the story, and videodisc technology allows students to easily revisit parts of the story in order to find just-in-time information needed to solve the problem. Modeling by experts occurs as a natural part of the story and helps students see how experts perform in the environment. The modeling also serves to scaffold students' knowledge of geometry.

Results of a number of studies involving middle-school students who have solved the adventures show that students improve in their knowledge of the applications of geometry, in the use of geometry to solve problems, and in the creation of geometric representations of real-world objects. Because an analysis of the teaching and learning process showed that both students and teachers needed support in assessing and revising their work, we designed prototype tools for the adventure "Blueprint for Success." These tools added elements to the learning environment that support reflection and visualization. The initial success of these tools in helping students learn geometry has prompted us to design similar tools for the other *Jasper* adventures.

ACKNOWLEDGMENTS

The preparation of this chapter and the research reported herein were supported by grants from the National Science Foundation (Nos. MDR-9252908, MDR-9050191, and MDR-9252990) and the Dwight D. Eisenhower Act P. L. 100-297, Title II. Any opinions expressed herein are those of the authors and do not necessarily reflect those of the granting agencies. Funding of *The Jasper Woodbury Series* was provided in part by grants from the James S. McDonnell Foundation (No. 91-6) and the National Science Foundation (No. MDR-9050191 and No. MDR-9252990).

REFERENCES

Adams, L., Kasserman, J., Yearwood, A., Perfetto, G., Bransford, J., & Franks, J. (1988). The effects of facts versus problem-oriented acquisition. *Memory & Cognition, 16,* 176–175.

Barron, B. (1991). *Collaborative problem solving: Is team performance greater than what is expected from the most competent member?* Unpublished doctoral dissertation, Vanderbilt University, Nashville, TN.

Barron, B., Vye, N. J., Zech, L., Schwartz, D. Bransford, J. D., Goldman, S. R., Pellegrino, J., Morris, J., Garrison, S., & Kantor, R. (1995). Creating contexts for community based problem solving: The *Jasper* challenge series. In C. Hedley, P. Antonacci, & M. Rabinowitz (Eds.), *Thinking and literacy: The mind at work* (pp. 47–71). Hillsdale, NJ: Lawrence Erlbaum Associates.

Barrows, H. S. (1985). *How to design a problem-based curriculum for the preclinical years.* New York: Springer.

Bransford, J., Franks, J. J., Vye, N. J., & Sherwood, R. D. (1989). New approaches to instruction: Because wisdom can't be told. In S. Vosniadou & A. Ortony (Eds.), *Similarity and analogical reasoning* (pp. 470–497). New York: Cambridge University Press.

Bransford, J. D., Goldman, S. R., & Vye, N. J. (1991). Making a difference in people's abilities to think: Reflections on a decade of work and some hopes for the future. In L. Okagaki & R. J. Sternberg (Eds.), *Directors of development: Influences on children* (pp. 147–180). Hillsdale, NJ: Lawrence Erlbaum Associates.

Bransford, J., Kinzer, C., Risko, V., Rowe, D., & Vye, N. (1989). Designing invitations to thinking: Some initial thoughts. In S. McCormick & J. Zutell (Eds.), *Cognitive and social perspectives for literacy research and instruction* (pp. 35–54). Chicago: National Reading Conference.

Bransford, J., Sherwood, R., & Hasselbring, T. (1988). The video revolution and its effects on development: Some initial thoughts. In G. Foreman & P. Pufall (Eds.), *Constructivism in the computer age* (pp. 173–201). Hillsdale, NJ: Lawrence Erlbaum Associates.

Bransford, J., Sherwood, R. S., Hasselbring, T. S., Kinzer, C. K., & Williams, S. M. (1990). Anchored instruction: Why we need it and how technology can help. In D. Nix and R. Spiro (Eds.), *Cognition, education, and multi–media: Exploring ideas in high technology* (pp. 115–141). Hillsdale, NJ: Lawrence Erlbaum Associates.

Bransford, J., & Vye, N. J. (1989). A perspective on cognitive research and its implications for instruction. In L. Resnick & L. E. Klopfer (Eds.), *Toward the thinking curriculum: Current cognitive research* (pp. 173–205). Alexandria, VA: Association for Supervision and Curriculum Development

Brown, J. S., Collins, A., & Duguid, P. (1989). Situated cognition and the culture of learning. *Educational Researcher, 18,* 32–41.

Charles, R., & Silver, E. A. (Eds.). (1988). *The teaching and assessing of mathematical problem solving.* Hillsdale, NJ: Lawrence Erlbaum Associates and the National Council of Teachers of Mathematics.

Clement, J. (1982). Algebra word problem solutions: Thought processes underlying a common misconception. *Journal of Research in Mathematics Education, 13,* 16–30.

Cognition and Technology Group at Vanderbilt University. (1990). Anchored instruction and its relationship to situated cognition. *Educational Researcher, 19,* 2–10.

Cognition and Technology Group at Vanderbilt University. (1991). Technology and the design of generative learning environments. *Educational Technology, 31*(5), 34–40.

Cognition and Technology Group at Vanderbilt University. (1992). The *Jasper* experiment: An exploration of issues in learning and instructional design. *Educational Technology Research and Development, 40,* 65–80.

Cognition and Technology Group at Vanderbilt University. (1993). Anchored instruction and situated cognition revisited. *Educational Technology, 33*(3), 52–70.

Cognition and Technology Group at Vanderbilt University. (1994). From visual world problems to learning communities: Changing conceptions of cognitive research. In K. McGilly (Ed.), *Classroom lessons: Integrating cognitive theory and classroom practice* (pp. 157–200). Cambridge, MA: MIT Press/Bradford Books.

Dewey, S. (1933). *How we think: Restatement of the relation of reflective thinking to the educative process.* Boston: Heath.

Duffy, T. M., & Bednar, A. K. (1991). Attempting to come to grips with alternative perspectives. *Educational Technology, 31*(9), 12–15.

Geddes, D. (1992). *Geometry in the middle grades* (Addenda series, Grades 5–8). Reston, VA: National Council of Teachers of Mathematics.

Gick, M. L., & Holyoak, K. J. (1980). Analogical problem solving. *Cognitive Psychology, 12,* 306–365.

Gick, M. L., & Holyoak, K. J. (1983). Schema induction and analogical transfer. *Cognitive Psychology, 15,* 1–38.

Gragg, C. I. (1940, October 19). Because wisdom can't be told. *Harvard Alumni Bulletin,* 78–84.

Hanson, N. R. (1970). A picture theory of theory meaning. In R. G. Colodny (Ed.), *The nature and function of scientific theories* (pp. 233–274). Pittsburgh, PA: University of Pittsburgh Press.

Johnson, R. (1987). *The ability to retell a story: Effects of adult mediation in a videodisc context on children's story recall and comprehension.* Unpublished doctoral thesis, Vanderbilt University, Nashville, TN.

Johnson-Laird, P. N. (1985). Deductive reasoning ability. In R. J. Sternberg (Ed.), *Human abilities: An information-processing approach.* (pp. 203–239). New York: Freeman.

Lockhart, R. S., Lamon, M., & Gick, M. L. (1988). Conceptual transfer in simple insight problems. *Memory & Cognition, 16,* 36–44.

McNamara, T. P., Miller, D. L., & Bransford, J. D. (1991). Mental models and reading comprehension. In R. Barr, M. Kamil, P. Mosenthal, T. D. Pearson (Eds.), *Handbook of reading research* (Vol. 2, pp. 490–511). New York: Longman.

Miller, G. A., & Gildea, P. M. (1987). How children learn words. *Scientific American, 257*(3), 94–99.

Minstrell, J. A. (1989). Teaching science for understanding. In L. B. Resnick & L. E. Klopfer (Eds.), *Toward the thinking curriculum: Current cognitive research* (pp. 129–149). Alexandria, VA: Association for Supervision and Curriculum Development.

Moore, A., Sherwood, R., Bateman, H., Bransford, J. D., & Goldman, S. R. (1996, April). Using problem-based learning to prepare for project-based learning. In J. D. Bransford (Chair), *Enhancing project-based learning: Lessons from research and development.* Symposium conducted at the 1996 annual meeting of the American Educational Research Association, New York.

National Council of Teachers of Mathematics. (1989). *Curriculum and evaluation standards for school mathematics.* Reston, VA: Author.

Perkins, D. N. (1991). What constructivism demands of the learner. *Educational Technology, 31*(9), 19–21.

Resnick, L. B., & Klopfer, L. E. (Eds.) (1989). *Toward the thinking curriculum: Current cognitive research.* Alexandria, VA: Association for Supervision and Curriculum Development.

Resnick, L. B., & Resnick, D. P. (1991). Assessing the thinking curriculum: New tools for educational reform. In B. Gifford & C. O'Connor (Eds.), *New approaches to testing: Rethinking aptitude, achievement and assessment.* (pp. 37–76). New York: National Committee on Testing and Public Policy.

Rewey, K., & the Cognition and Technology Group at Vanderbilt University. (1992, April). *Small group problem solving in "The Adventures of Jasper Woodbury" environments: A preliminary examination of dyads.* Paper presented at the annual meeting of the American Education Research Association, San Francisco, CA.

Scardamalia, M., & Bereiter, C. (1985). Fostering the development of self-regulation in children's knowledge processing. In S. F. Chipman, J. W. Segal, & R. Glaser (Eds.), *Thinking and learning skills: Research and open questions* (Vol. 2, pp. 563–578). Hillsdale, NJ: Lawrence Erlbaum Associates.

Scardamalia, M., & Bereiter, C. (1991). Higher levels of agency for children in knowledge building: A challenge for the design of new knowledge media. *Journal of the Learning Sciences, 1,* 37–68.

Schoenfeld, A. (1985). *Mathematical problem solving.* Orlando, FL: Academic Press.

Schoenfeld, A. (1989). Teaching mathematical thinking and problem solving. In L. B. Resnick & L. E. Klopfer (Eds.), *Toward the thinking curriculum: Current cognitive research* (pp. 83–103). Alexandria, VA: Association for Supervision and Curriculum Development.

Schwab, J. J. (1960). What do scientists do? *Behavioral Science, 5*, 1–27.

Sherwood, R. D., Kinzer, C. K., Bransford, J. D., & Franks, J. J. (1987). Some benefits of creating macro-contexts for science instruction: Initial findings. *Journal of Research in Science Teaching, 24*(5), 417–435.

Spiro, R. J., Feltovich, P. L., Jacobson, M. J., & Coulson, R. L. (1991). Cognitive flexibility, constructivism, and hypertext: Random access instruction for advanced knowledge acquisition in ill-structured domains. *Educational Technology, 31*(5), 24–33.

Spiro, R. J., Vispoel, W. L., Schmitz, J., Samarapungavan, A., & Boerger, A. (1987). Knowledge acquisition for application: Cognitive flexibility and transfer in complex content domains. In B. C. Britton & S. Glynn (Eds.), *Executive control processes* (pp. 177–200). Hillsdale, NJ: Lawrence Erlbaum Associates.

Van Haneghan, J. P., Barron, L., Young, M. F., Williams, S. M., Vye, N. J., & Bransford, J. D. (1992). The *Jasper* series: An experiment with new ways to enhance mathematical thinking. In D. F. Halpern (Ed.), *Enhancing thinking skills in the sciences and mathematics* (pp. 15–38). Hillsdale, NJ: Lawrence Erlbaum Associates.

Whitehead, A. N. (1929). *The aims of education*. New York: MacMillan.

Williams, S. M. (1991). Putting case-based instruction into context: Examples from legal and medical education. *The Journal of the Learning Sciences, 2*(4), 367–427.

Yackel, E., Cobb, P., Wood, T., Wheatley, G., & Merkel, G. (1990). The importance of social interaction in children's construction of mathematical knowledge. In T. J. Cooney & C. R. Hirsch (Eds.), *Teaching and learning mathematics in the 1990s: 1990 Yearbook of the National Council of Teachers of Mathematics* (pp. 12–21). Reston, VA: National Council of Teachers of Mathematics.

Zech, L., Vye, N. J., Bransford, J. D., Swink, J., Mayfield-Stewart, C., Goldman, S. R., & Cognition and Technology Group at Vanderbilt University. (1994). Bringing geometry into the classroom with videodisc technology. *Mathematics Teaching in the Middle School Journal, 1*(3), 228–233.

19

Teachers and Students Investigating and Communicating About Geometry: The Math Forum

K. Ann Renninger, Stephen A. Weimar, and Eugene A. Klotz
Swarthmore College

> The mere memorizing of a demonstration in geometry has about the same educational value as the memorizing of a page from a city directory. And yet it must be admitted that a very large number of our pupils do study mathematics in just this way. There can be no doubt that the fault lies with the teaching. (Young, 1925, p. 5)

Although it may seem obvious that teachers (and their students) need to appreciate that learning must be much more than memorization, this realization is almost certainly not enough to enable teachers to move beyond a "review the homework, go over new material, do your homework" approach. Those who come to know the elegance, intuition, and sense of discovery that can characterize the doing of geometry may apprehend it regardless of how they are taught, whereas those who mostly memorize will be convinced that geometry (and, more generally, mathematics) is equivalent to getting right answers.

There are several reasons why this state of affairs has continued in geometry. First, there are many individuals for whom construction of correct proofs is an end in itself because they have been assigned proof after proof, have no questions themselves, and are not really engaged in the questions the proofs represent. These individuals stand in contrast to those who enjoy and continue to work with geometry and for whom proofs are an effective means of communicating what they have figured out.

Second, one function of schooling in the industrial model has been about finding efficient ways to transmit knowledge so that the time it takes to learn things can be dramatically shortened and more can be added to the list of topics to be learned. In this model, because there is so much to convey, geometry has come to mean covering the known facts and knowing how to produce the "correct answer" in a standard format. In fact, proofs have come to epitomize math as a right-answer discipline in which the emphasis is on the speed and accuracy with which the student can reproduce answers already known to the teacher.

A related perceived function of schools is the efficient identification of those who have "ability" in a given area. These two functions, covering as much content as possible and sorting people for access to further education or professions, have led to generations of students who could pass tests but developed no intuition about geometric relationships—the end result of which has been the absence of any noticeable strengthening of the general population's ability to use geometry, and a decided weakness in solving geometry-related problems in life outside of school (Dreyfus & Hadas, 1987; Usiskin, 1987).

In the last half decade there has been a recognition of the deficiencies in these approaches, and alternatives have been formulated by the National Council of Teachers of Mathematics (1989, 1991) and the New Standards Project (1995). It remains a challenge, however, for teachers to know how to implement these standards in their classes.

One effort to enhance the resources available to teachers to effect change in their classes has been made by the Math Forum (formerly the Geometry Forum), a thriving and active National Science Foundation (NSF)–funded Internet community of geometers, teachers, students, researchers, software developers, and math education faculty. Like the teachers in the Urban Mathematics Collective (Webb & Romberg, 1994a), those who participate in the Math Forum are interested in learning—in this case in talking, thinking, and exploring geometry problems, puzzles, software, applications, and pedagogy.

In the Math Forum, the math community finds an environment that purposely seeks to bring teachers, students, publishers, and researchers together to influence and support each other's work. Through their interaction and their involvement with Math Forum projects, these participants generate the information and thinking that forms the basis of the Math Forum's resources. Users have access to some of the best mathematical minds in the world and to advanced computing resources, a support network of peers, new software, direct contact with publishers, projects linking their classrooms with others and outside resources, and databases of current information.

The Math Forum staff is concerned about pedagogical issues, not just the indexing of electronic resources. Through its online projects and materials, as well as its workshops, it attempts to model and facilitate the adoption of emerging teaching standards. Its structure is flexible and encourages teachers to choose the most productive channels for their learning.

In this chapter, we describe the Math Forum as a presence on the Internet and the World Wide Web (WWW). In particular, we overview the interactive online projects and discussions developed by the Math Forum staff in response to and with the community of its users, as well as the Forum Archives and the Internet Resource Collection. Finally, drawing on data from a series of preliminary studies, we document the explicit and implicit ways in which investigating and communicating about geometry on the Math Forum appears to be contributing to reforms in geometry teaching and learning.

THE MATH FORUM AS A PRESENCE ON THE INTERNET AND THE WORLD WIDE WEB

The Math Forum embodies what Papert (1993) referred to as the powerful potential of new technology to enhance learning: the creation of more personalized media capable of meeting the needs of a wide range of learning styles. In particular, the Math Forum provides an inquiry-oriented (Bruner, 1966) add-on to the typical geometry/mathematics class (note that it is not a replacement for such a class), as well as a "forum" for teachers to explore topics related to mathematics and its use, with opportunities to select (actively personalize) the direction of their growth as mathematicians and teachers.

The Math Forum bears a structural similarity to the open-ended characteristics of Logo and Lego-logo software, in that it makes tools available and the user decides when and with what to engage. It is a largely user-driven resource consisting of on-line activity, archives, workshops, and software. The "text" of the Math Forum is like the words found in books, but it is the hypertextual quality of the WWW that makes visible the quality of the Math Forum as a text unlike those found in most schools. Like a constructible book, the Math Forum is composed of bits and pieces strung together by technology.

The ultimate goal of the Math Forum is to provide technologies that will enable the imagination of the user to create an intelligent narrative. Embedded in the hypertext form is the necessity for users to consider a variety of different perspectives on a given topic—regardless of whether they are actually participating in a dialogue or simply "lurking."

There are three online components of the Math Forum in which teachers and students can engage: online projects and discussions, archives of online activity and related resources such as software, and an Internet Resource Collection.

THE MATH FORUM ONLINE

The Internet is a tool for communication, and the Math Forum is built around the premise that making it easier for people to communicate about mathematics is an exciting and powerful use of telecommunications technology. The Math Forum sponsors a number of interactive projects and online discussion forums.

Interactive Projects

The Math Forum offers several different online projects. These are often used by teachers as add-ons to their classes. MathMagic is an example of a project run by others but hosted by the Math Forum. Projects such as the Problem of the Week (POW), Project of the Month (POM), and Ask Dr. Math are Math Forum-generated projects, developed by the staff in conjunction with the online community.

The Problem of the Week (Geometry, http://forum.swarthmore. edu/ geopow/; Elementary, http://forum.swarthmore.edu/elem.pow/; Middle, http://forum. swarthmore. edu/midpow/). Each week a high-school level geometry problem, the Problem of the Week (POW), is posted to the geometry.precollege news group and put on the web (http://forum. swarthmore.edu/geopow/). (An Elementary Problem of the Week [http://forum.swarthmore.edu/elempow/] and a Middle School Problem of the Week [http://forum.swarthmore.edu/midpow/], with a broader scope than simply geometry, are also features of the Math Forum). Some problems are "classic" and follow the traditional geometry curriculum; others may lead to new topics such as tessellations or may involve applications of geometry in areas such as science or construction. Some are directive; others are more open-ended. Often the problems are set in a context: They are questions to which someone actually needs an answer. They can be very different from the kinds of problems that are put in books as exercises, given on exams like the Olympiad, or offered as brainteasers.

The problems are retrieved from the Math Forum as text only and do not require any special equipment or software in order for the student to solve them. Some students have their own accounts, or use their parents' accounts, and send their solutions directly to the Math Forum. More often, teachers photocopy and distribute the problems in their classes, gather solutions, and send them to the Math Forum through their own personal accounts.

As solutions are received, each is read and evaluated, and a personal response to each submission is sent. The student's goal is a well-presented, clearly illustrated, and sufficiently defended answer. POW is about more than getting the answer right; it's about exploring, discussing, and communicating math. To this end, students who submit well-thought-out answers receive a note of appreciation and sometimes suggestions for further exploration as well. Students who submit answers that could be more developed or take questionable directions receive a reply containing questions to stimulate more thinking about the problem and encouragement to resubmit an answer (see Appendix for a sample problem, facilitator response, and student responses).

The feedback process supports individualization: The facilitator does an error analysis and responds with questions that challenge students without overwhelming them. The concept of posting a range of responses and the process of obtaining them are intended to provide students with information about alternative strategies and plans for obtaining an answer. They also provide students with the opportunity to expand their repertoires of possible approaches to a problem. The exchanges with the facilitators can lead to wide-ranging discussions about math and school. Such exchanges are comfortable, honest, and often thought-provoking.

All of the answers in the "correct" pile are combined and posted the following week (sometimes later, if resubmissions of answers are outstanding). When the answers are posted, they include students' names, grades, schools, and notes about the nature of the process involved in obtaining the solution.

A contingent of students (boys and girls from both private and public schools) who have worked the POW as an assignment in past years continue to send in answers to the problems each week, even though they no longer work the problems as math class assignments.

The Project of the Month (http://forum.swarthmore.edu/pom2/). The Project of the Month covers a broader range of topics than the POW and involves forays into such areas as trigonometry, pentominoes, and tangrams. Solutions are submitted to the Math Forum by teams of students who are permitted to use anything they wish to reach a conclusion—the only rule is that the answer be submitted electronically. The teams of students are not expected to prove their answer is correct in order to submit a solution, but they are expected to provide convincing arguments for their claims. These arguments may include explanations accompanied by pictures or well-documented sketches from one of the drawing programs. As with the Problem of the Week, however, no special equipment or software is required to work on the project.

Like the POW, the Project of the Month is intended to enhance students' mathematical power. By asking the students (and teachers) to reflect on their learning and encouraging them to work with others, the project helps them learn to value and develop their confidence in their ability to do mathematics. The fact that they are not required to have a complete answer or proof, but instead to submit their thinking, further emphasizes their skills as mathematical problem solvers and their abilities to communicate and share their reasoning about mathematics. The solutions submitted are read carefully; feedback and questions are posed to the students, similar to those for submissions to the Problem of the Week.

Ask Dr. Math (http://forum. swarthmore.edu/dr.math/). Building on the dormant project "Ask Prof Maths," Ask Dr. Math is a project in which mathematics students at Swarthmore and other colleges across the country and a consulting group of mathematicians, some of whom are world famous, respond to questions sent to them via e-mail by K–12 students. Sometimes the questions posed are as simple as "I'm having problems dividing three-digit numbers"; others broach subjects such as "What is zero?" or "Who invented negative numbers?" The math doctor answers the student (or other respondent) as quickly as possible, at least to indicate that someone is working on the question. Central to the effectiveness of the Ask Dr. Math project is that students are individually engaged in a dialogue about their questions. They do not receive a prewritten response, but questions and discussion focused on their inquiry. In fact, "math doctors" are first apprenticed (regardless of their rank in real life) and then "tenured" based on their ability to work with students. The objective is that students be led to answer their own questions rather than being given answers or algorithms.

E-mail exchanges provide an opportunity for interaction. Each question submitted is taken seriously, confusion is acknowledged, and input from others—even eminent research mathematicians—is possible. Responses can help the students refocus on what is important in mathematics—the process of conjecture, not the memorization of formulas. It is the modeling and discussion of these points that provides the student responded to with the necessary scaffold to push his or her thinking. Other messages conveyed to the student (and readers of this exchange) include the humanity and humility of mathematicians:

> I don't remember either of these formulae [for the surface area of a cone], and I don't think anyone else should! I DO remember the pi.R^2 and 2.pi.R formulae, and I do think everyone else should! (I also remember quite a lot of formulae that I DON'T think anyone else should, but that's beside the point.) The method of finding out such things is what we should be teaching (and remembering).
>
> John Conway.
>
> P.S.- perhaps people involved in the manufacture of funnels and ice-cream cones should be allowed to remember the formulae.

These are friendly responses, not responses from people in a system where a student is either right or wrong, smart or stupid. It is likely that the students can become more confident in their ability to think and ask questions about math, and that they will appreciate what they do know and what they can learn here. Students' comments during these exchanges indicate that they are persevering to understand the questions they pose; they often comment that these exchanges "get them thinking."

The exchanges in Ask Dr. Math can get everyone thinking, as in a discussion that took place among Professor John Conway (Princeton), Professor Steve Maurer (Swarthmore), and a group of high-school precalculus students who thought they were figuring out something about the complex roots of polynomials by looking at the "wiggles" that showed up on their graphing calculators. Both Conway and Maurer wrote back to say that they had never quite looked at these as the students were and that it was interesting to them. Nothing particularly new or exciting was being discovered; in fact, Conway and Maurer could explain what the students were seeing, but when do students in the average high school math class get a visit from practicing mathematicians, much less get appreciation for noticing something that interests experienced people in the field? Such possibilities also permit teachers opportunities to learn with and through their students as observations such as these are brought back to the classroom. Finally, all questions and answers sent in to Ask Dr. Math are archived. Thus, it is possible for teachers to access all of the questions previously posed about polynomials, for example, in order to discover the kinds of questions students may have and the kinds of answers that they could be offered.

Discussion Forums

In addition to interactive projects, the Math Forum sponsors a range of online discussions (e.g., the math-teach mailing list), but the main focus of its efforts has been the geometry newsgroups, which yield many spontaneous discussions, and a Math Forum-facilitated discussion session on Learning and Mathematics. During 1995, these discussions averaged about 30 (substantial) exchanges per topic, with a range of 4 to 79 exchanges.

Newsgroup Discussions. The newsgroup discussions are user dependent and spontaneous. Discussions vary widely, from the use of ruler and compass to the advantages of computer drawing programs, from the utility of two-column proofs to whether students need a full year of geometry, from redesigning the secondary math curriculum to the math needed in planning the bend in a conduit pipe.

Discussions can extend from several days to even weeks. People pose their questions, follow the conversation (whether as "lurkers" or as participants), and respond, addressing a question posed and/or moving a question in a slightly different direction. Discussions usually stay focused. It is not uncommon, however, for portions of a question to be considered and others to be overlooked, unless posed again.

The conversation is distinct from normal teacher conversations. First, there is a culture involved in participating in the Math Forum newsgroups that is distinct from that of the school: Asking questions and pursuing answers about both curriculum and instruction are accepted and encouraged. Second, the conversation benefits from the input of a wide range of individuals. Third, the conversation is in written form, and as such it can be reflected on and taken up again at a later time. Fourth, it is archived (and searchable), so that any time a teacher or student wishes to know something more about geometry and the curriculum, a search of the Math Forum will yield this discussion. It never becomes a "lost" conversation.

The newsgroup discussions provide teachers, whether reading, following a conversation, or actively posting, with an opportunity to think with others, an opportunity not always available in schools. The conversations evolve and change, and teachers themselves decide how and when they will engage. One of the interesting aspects of these discussions is that they become reference points in other discussions on the Math Forum, indicating that these discussions are being read and have become part of the participants' knowledge base.

Discussion Sessions on Learning and Mathematics (http://forum. swarthmore.edu/learning.math.html). The Discussion Sessions on Learning and Mathematics are discussions about research on student learning and mathematics. The Learning and Mathematics series was developed to meet a perceived need for teachers to understand why the *Standards* (NCTM, 1989, 1991) mandated changes in instruction and, to some extent, changes in the focus of the content of instruction. Because research on student learning provides support for the suggestions in the *Standards,* it was thought that providing readable summaries of this work and a forum for discussing them would be useful for teachers.

Each session consists of a readable (mostly jargon- and citation-free) summary of work by math researchers or educational psychologists. Summaries are followed by open discussions among readers of the newsgroup geometry.pre-college, including a group of educational psychology students who know the research but are less versed in practice, and who serve as an informal check on misinterpretation of the posted summaries. A new summary is posted every 2 weeks during the school year, although some of the discussions last beyond this time frame.

This series provides teachers with an opportunity to explore concepts, ask questions, develop a language for thinking about student learning, support their hunches about student functioning, reflect on their own assumptions, and consider alternatives to how they currently work with students. In short, like other discussions and projects on the Math Forum, the teacher drives his or her engagement in the series and, therefore, what he or she learns.

The Forum's Archives

All of the online activity of the Math Forum is archived, thus permitting teachers or students to search and assemble materials in different ways. The best point of access to the archives is through the Math Forum WWW site.

The Math Forum archiving is semiautomated in such a way as to enable staff to add value by linking related discussions, weeding out the insubstantial messages, and annotating topics for easy browsing and enhanced keyword searching. Specifically, all of the Math Forum projects and discussions are "threaded" (pulled together by topic) and described. The descriptions are used to build searchable databases and to create annotated links. The archives are then made available for browsing and keyword searches on the World Wide Web. These features of the archive differentiate the Math Forum site from other sites, which often do not archive their activity, much less massage the results to create substantive, easily navigated material.

The Math Forum's Internet Resource Collection (http://forum.swarthmore.edu/~steve/)

The Math Forum's Internet Resource Collection, a cross-indexed annotated catalog, is a particularly useful resource for mathematics educators. At present, there is an overwhelming amount of "stuff" on the Internet, much of it with little direct value to the educator. Part of the Math Forum staff's task is to make it easy for teachers to find resources that they might be able to use. To this end, the Collection is continually revised to increase the ability of users to select what they need and to enable the Math Forum to organize resources by the topics and sequences of the curriculum and the many other lenses that users bring to it.

There are two versions of the Collection: an easily browsed, compact outline form and an annotated version that contains a brief summary of each site for those who would like to know more before choosing to "visit." The annotated version also provides a basis for sophisticated searches.

Like the projects and discussions on the Math Forum, the Collection provides teachers and students with a wealth of information. The materials facilitate interdisciplinary work, enable searches related to personal interests, and provide information about sources of technological support to assist teachers' efforts to link their schools to the Internet and the Web. Furthermore, the materials make it that much easier to individualize and support student-directed study, tailoring instruction and resources to the needs of students, both in terms of content and through different approaches to teaching. For example, Steve Means, a geometry teacher in Seattle, Washington, assigns his high-school students a year-long project in which they first identify something they love (tap dancing, soccer, auto mechanics, etc.) and then describe it mathematically. In this assignment the students are asked to really think through the mathematics of what they might have assumed were nonmathematical activities. The Collection provides such students with rich and constantly developing links to information and sources on which to draw in completing their assignments.

Consistent with inquiry-based models of education, a tool like the Collection provides teachers and students with the opportunity to explore alternatives in a manner that is both facilitated and regulated. As Bruner (1966) pointed out, teachers and students should not simply be left to explore (or be overwhelmed by) possibilities; rather, the kind of thinking that characterizes real learning is facilitated through knowledge of the goal for the task (the development of a lesson, a project that is due) and the relevance of possible resources. The Collection provides teachers with an annotated, filtered, and categorized place to begin using (and enabling their students to use) the resources of the Internet without feeling either overwhelmed or lost.

THE MATH FORUM AS A SITE FOR INVESTIGATION AND COMMUNICATION

The Math Forum was created and continues to be revised by the staff in conjunction with its users. The original concept for the Math Forum developed out of a need for a more effective means of communicating about software and geometry materials among colleagues, teachers, students, and others. At this time, the Math Forum is a unique resource on the Web. It has expanded to become a virtual center for mathematics education, providing sophisticated ways to track down needed math resources. Few if any other discipline-based resources for teachers like it exist. In fact, the Math Forum has more different Web pages than most sites of any kind.

Two questions continue to be central concerns of the Math Forum staff: What does it take for teachers to get involved in using online resources? And what does it take for teachers to make an investment in contributing to the development of this resource base (in terms of facilitation, technological support, feedback, search mechanisms, etc.)? Introductory and advanced-level workshops run by the Math Forum for its users have been instrumental in providing the Math Forum staff (the workshop facilitators) with answers to these questions. In fact, each workshop serves to "update" the staff's knowledge as technology shifts and the users' capacities become increasingly sophisticated.

The workshops provide opportunities for users to learn to use Math Forum resources, to ask questions, and to follow through with extended discussion and demonstration. In the workshop, participants work collaboratively to learn new skills, explore the use of technology in the classroom generally, and develop projects for their classrooms that are shared and modified in discussion with others. The energy and community life developed in workshops allow for increases in resource production (i.e., Web projects) and new leadership development among Math Forum users (i.e., development and facilitation of the elementary and middle-school versions of POW) that are unparalleled as sources of information about what works for teachers. (For more information and descriptions of workshops, see http://forum.swarthmore. edu/workshops/.)

In addition to an extensive base of formative evaluations used to develop, organize, and learn from the workshops being conducted by the Math Forum staff, a series of preliminary investigations was undertaken. These were intended to both refine and generate hypotheses about the role of the Math Forum as a resource to teachers and their students. The investigations summarized here address (a) Math Forum use, (b) Math Forum user profiles, (c) Math Forum user self-sufficiency, (d) change in Math Forum users' participation, (e) Math Forum accessibility, and (f) student performance and preferences.

Forum Use

One of the difficulties involved in monitoring electronic use of the Math Forum is that there is no good way to determine whether people are reading the newsgroups, the discussions, and so on if they do not actually post responses. In order to find out more about how people are using the Math Forum, 26 users (14 male, 12 female) were interviewed over the phone about Math Forum use by a person new to the staff of the Math Forum (and thus in a position to ask questions that those who had already worked with the users might have found more difficult to pose).

In order to obtain an evenly distributed group of participants, teachers were selected for this study based on our knowledge of their level of postings to the Math Forum. Following the interview, the teachers were classified into four types of users: (a) those with access who were nonusers (two male, three female); (b) those with access (three male, four female); (c) those who did lots of reading, infrequent posting, and some research using the Collection, had a strong personal involvement in the Math Forum, and had integrated it into their classroom activities (four male, five female); and (d) those who, like the users in category (c), did lots of reading, had a strong personal involvement in the Math Forum, and had integrated Math Forum resources into their classrooms (four male, zero female). These last were also doing a lot of postings and were using the Collection and archives frequently both for their own interests and in their classrooms.

The teachers interviewed had taught for an average of 18 years. More specifically, teachers who had access to the Math Forum, but either did not use it or used it passively, could be described as feeling overwhelmed by the Internet generally or as experiencing computer difficulties of one type or another (screen freezing, improper set-up, etc.). They said they needed to come to another workshop because they had understood everything when they had had support, but that without such support they felt lost.

Teachers whose involvement with the Math Forum involved reading, surfing, use of e-mail, and exploring student use of the Internet Resource Collections, and who were excited about using the Web, were technically "passive" users; however, they were on the Net almost every day and found the Math Forum Internet Resource Collection particularly useful for their students. This group (appropriately one third of those interviewed) could be classified as "lurkers" on the Math Forum.

The active users were teachers who had a strong personal involvement in the Math Forum and had integrated Math Forum resources into their classrooms. Typically, they did a lot of posting and used the Collection and archives frequently both for their own interests and in their classroom. They talked more specifically about their use of Math Forum resources and described the Internet as a way to distribute and create information.

Interestingly, each of the user groups was reasonably balanced by gender—in other words, stereotypes about gender, technology, and mathematics, which would suggest higher levels of participation by males, did not appear to apply to Math Forum users, except perhaps to the most frequent users. Perhaps more revealing is the link between individual interest and use suggested by comments from these groups. The most active users were personally involved in the Math Forum (and the Internet) as a resource for themselves as well as for their classes. These two groups had no more training than the others, and as teachers they were not necessarily stronger or weaker, more or less experienced. It is clear, however, that they saw opportunities for themselves and were willing to deal with the corresponding difficulties in order to enrich their classrooms with Internet activity.

Forum User Profiles

In order to learn more about our users and their particular needs, we used an open-ended questionnaire with follow-up e-mail exchanges to "talk" at some length with another group of teachers, people who were either on site (four male, four female) or online (five male, six female) participants in a summer workshop.

The only difference between the groups appears to be that the workshop participants were more likely to be high-school or junior-high-school teachers, whereas approximately half the online participants were high-school teachers and half were mathematics professors.

The similarities between the two groups of users are many. Workshop and online participants had an average of 15–20 years of teaching experience. Almost all participants said they worked in a supportive environment, although it should be noted that "support" sometimes meant they were left alone to make decisions. People in both groups also responded similarly when asked how much control they had over how and what they taught. They overwhelmingly answered that they had control over the presentation of the material. Some in both groups said they collaborated with other teachers on curriculum. A few had to teach from a mandated textbook.

In addition to sharing resources with other teachers and being in leadership positions, all of the teachers in both groups also had outside responsibilities in their school. Almost all were in charge of math or computer-related student activities.

They all began working with computers in the early 1980s, and more than half had computers at home. They noted, however, that to use the Web it was still easier for them to stay at school following classes. Those without equipment at home used school equipment.

People in both groups described their approach to computer problems as largely experimental. In each group, six people specified that they used a trial-and-error approach to the computer. As a second or last resort, they would usually refer to a manual or call the Math Forum's 800 number. Two participants in each group said they went to the manual first.

People in both groups found out about the Math Forum from a wide variety of sources. Many located it by searching for math sites (on the Web, doing gopher or Internet searches, the EDTECH listserv), and another subgroup learned about it through the math-teach mailing list. Other sources of information included colleagues, the Math Forum newsletter, graduate advisors, and meetings.

Many in both groups mentioned that they incorporated the Problem of the Week, Ask Dr. Math, MathMagic, and dynamic geometry software in their classes. Several of those who did not use Math Forum resources in class said that they would be in a position to do so in the near future; their current limitations had to do with computer availability and phone access. There was also a comment that the physical limitation of not having a computer in the classroom did not have to get in the way of participation in something such as the Problem of the Week because teachers could photocopy, distribute, and post responses for their students, but that much more would be possible with classroom computer access.

These users were also asked about their students' responses to the Math Forum. Their responses were generally positive. They viewed the Math Forum as a vehicle for meeting the needs of all types of students—those who need to be challenged, those who need remediation, those who need a change or a different type of instructional setting. They underscored the importance, however, of considering how online activities were actually used in the classroom and school environment. For example, the presence of an active math club in a school changes the function of Ask Dr. Math or POW (the Problem of the Week) as resources. Many teachers cited the popularity of the POW, noting that some students "who wouldn't do anything else dived into the POW problems." Many teachers also noted that their students (and their families) were very enthusiastic:

> They can't wait to get the new problem each week. My students are in the computer room using Sketchpad to work on their solutions before school, during their lunch and study periods, and after school. At parent conferences, many parents have told me that they have gotten hooked on the problems. One mother works the night shift in a hospital, and the whole hospital is usually working on the problem by the end of the night.

In addition to considering the way in which the Math Forum complements the curriculum, however, they also talked about the need to talk with students about the relative successes of their efforts to use the Math Forum and ways in which they might do "x" another time.

What works in some classrooms does not necessarily work in other classrooms. There were teachers who reported that their classes "take off" on the resources available on the Math Forum, and there were others whose students did not find it interactive or searchable enough and instead used it as a backup for class work. The success of the Math Forum also depends on ways in which it can be made responsive to needs for further development and usability. Based on the comments received, it appears that at least two useful enhancements to existing resources might be to include student-generated terms in the categorizing of archived materials and to develop a student newsgroup.

Finally, the teachers were unanimous in their agreement that the Math Forum had contributed to their professional development. What most mentioned, however, was the contact with others that the Math Forum provides:

> I have learned a great deal about geometry and about teaching mathematics from the newsgroups, and I've enjoyed the sense of being able to share ideas with people from other places. I've posted questions on mathematical topics in which I was interested [Penrose tiles and quasicrystals] and followed discussions on others [eversion of the sphere, flexahexagons and so forth]. In addition, I've been involved in the de-

> bates about various teaching techniques especially with regard to congruency proofs. I was also very interested in the discussions of mathematics and language.

Forum User Participation as a Function of Self-Sufficiency

In order to further understand the self-sufficiency of users and the extent to which they needed technological support services in order to use the Math Forum, portraits of independent and dependent users were developed based on interviews conducted with the Math Forum teacher-liaison staff person.

Independent users were characterized as enthusiastic and personally invested in the Math Forum. They typically tried to solve their own computer-related problems. If, however, they were having difficulty, they called or wrote to the Math Forum explaining what they had tried, what they wanted to be able to do, and what they needed to have addressed.

The resources available to them on the Math Forum actually seemed to fuel their energy for teaching. They talked about having found the Math Forum at a good time in their lives. They spent time at home working on their computers; if necessary, they sent files from home when there was no school access. In fact, they perceived the lack of school access as a surmountable problem, even though the topic of "school access" was a major impediment for most other teachers.

They asked for brochures describing the Math Forum and talked to others at school and at conferences about their work with the Math Forum in their classes. These independent users were independent problem solvers. They looked for and helped to create additional resources on the Math Forum.

Dependent users, on the other hand, kept saying "this is great" when something was explained or demonstrated for them, but there appeared to be little follow-through to real understanding without a lot of physical hands-on support. These users did not seem to know which questions to ask, meaning that they either did not ask questions they had, or they appeared to ask questions constantly, regardless of the amount or quality of information provided. They did not appear to be in a position to reflect on or synthesize materials for themselves.

These people often seemed to have such a tenuous understanding of what they were trying to figure out that any difficulty derailed their efforts. (This situation often was further complicated by the social unacceptability of a math teacher or computer resource person without strong math skills being unable to work through an exercise or procedure.) The primary resources that these individuals seemed to need involved intensive one-on-one assistance—keeping their hands "on the mouse" and talking them through their difficulties one step at a time.

Based on these portraits, it appears that the resources needed for independent problem solvers are different from those necessary for dependent problem solvers. In the Math Forum's work to help teachers become more self-sufficient as users, change appears to occur as a direct function of the user's characteristics as a learner, the support provided him or her by the school, and the time he or she invests in practice/usage. Interestingly enough, it does not appear possible to predict the likelihood of movement from dependent-user status to independent-user status, but change does occur and appears to be linked to the user's characteristics as a learner (just not the same characteristics for each user). We have seen dependent users become independent users. We also have worked with dependent users who continue, at least at present, to be dependent users.

Change in Forum Users' Participation: The Specific Case of the Discussion Sessions on Learning and Mathematics

In order to evaluate change across teacher and student participation in the Discussion Sessions on Learning and Mathematics, the contributions of eight teacher participants (four male, four female) and eight educational psychology students (four male, four female) were tracked over a year/term of participation. Each individual's contributions were studied following a complete reading of all of his or her contributions.

Findings from this study suggest that:

1. The personalities of participants did not change over time (e.g., those who were argumentative remained so).
2. Teachers and educational psychology students developed a language for talking about learning over the course of their contributions and used this to develop their ideas, subsequently becoming more developmental in their focus (talking about individual strengths and weaknesses and change in these, rather than categorizing learners only as strong or weak)—they talked about process and product, as well as about access and reflection.
3. There were also shifts in the breadth of both teachers' and students' perspectives. Both groups appeared to appreciate the complexity of student learning in a different way than they had at the outset of their participation.

Several other findings emerged from this study:

1. The depth of the discussion is affected by the topic and whoever else is involved.
2. Teachers have a tendency to rely on and invoke practical experience as evidence, whereas the educational psychology students, despite experience in classrooms and as tutors, were more likely to invoke other research and prior discussions to support their points.
3. Discussion was most effective when participants worked on clarifying perspectives, asked questions, and built on the ideas of others, whereas discussions were less effective when they revolved around opinion based on personal experience or scholarly texts without explanations of the conclusions drawn and/or were divided between practice and research (although it should be noted that constructive discussions also emerge following such dichotomizing).

Findings from this study also suggest that one of the resources afforded by a discussion group is the opportunity to work with ideas and to develop them over time with contributions from those invested in both research and practice as a check on the discussion. Specifically, the opportunity to reflect on one's own ideas and to have those ideas responded to is an important metacognitive tool not typically available to teachers.

Forum Accessibility

In order to consider how an individual who has not participated in a Math Forum-led workshop would view the Math Forum and to obtain insight into the information necessary to enable this type of user to use the Math Forum Internet Resource Collection, groups of prospective student teachers were divided into six focus groups to discuss their understanding, confusion, and recommendations to the Math Forum staff.

Findings from this process indicated recognition that the Collection would be used in lieu of "running around looking for geometry-related projects and activities," that it is constantly updated with new information immediately available, and that it is a useful tool because local tutors and one-on-one help are not always available whereas on the Math Forum a "tutor-like" someone is always around to help.

Recommendations included making it more obvious how much the Math Forum has to offer teachers because many will not spend time independently exploring it to find out what is useful. More specifically, it was suggested that the services needed to be made more clear and that increased emphasis on the visual and manipulable elements of the program would be particularly important to prospective users. Finally, it was suggested that there should perhaps be some sort of introductory program or tour that takes first-time users through the Math Forum resources and enables them to anticipate problems and avoid potential frustration.

In response to this feedback, the home page for the Math Forum was overhauled, and a suggestion box was added to the bottom of every page to facilitate immediate feedback.

Student Performance and Preferences

In order to begin to assess student usage of the Math Forum, problems posted to the high-school Problem of the Week over a 5-month period were monitored for quantity and quality of student response. Approximately 85% of the submitted answers in a given week were accurate; in 1995, the number of submissions in a given week ranged from 42 to 270. (It should be noted, of course, that some teachers only submitted correct answers whereas others relied on the Math Forum to provide students with feedback—needless to say, this affected the figures we have on participation.)

Assessment of the relationship between the type of problem and students' performance was not easy to determine, given that teachers chose to use one particular problem of the week and not another based on such variables as their textbooks, the organization of their classes, and their schools' vacation schedules. Findings from this inquiry do suggest, however, that more students were involved in and submitted correct responses for problems that had an ongoing story line built into them. Furthermore, girls were more likely than boys to send in correct responses.

It appears that there are several dimensions of "situational interest" afforded by the Problem of the Week. First, there is a particular kind of relationship that can be struck up with someone out there on the Internet—someone who does not care what you look like, who gives you a fresh start, and who provides feedback that enables you to keep working on a problem. Second, a problem is more "fun" if it is a continuing story problem or includes, for example, the name of a rock group. Although the context of a problem does not change its difficulty or the skills that students need to complete it, it is possible that context does permit students to persevere longer when a problem is difficult for them (for a related discussion, see Renninger, 1992).

Based on findings from this investigation, it appears that one of the reasons the Problem of the Week is engaging for students (and is a useful resource for teachers) is that it has these components of situational interest. The content of the problems does not deter strong math students from engaging them. It may in fact be the content that enables the so-called less-able students to stay on task at least a little longer.

THE MATH FORUM AS A SITE FOR PROFESSIONAL DEVELOPMENT

Findings from this set of investigations suggest that the Math Forum is being used primarily by experienced teachers who experiment with how to use the Math Forum on their computers and in their classes. They contribute to discussions, develop Web projects, keep online diaries of their classrooms, and so on. In many ways, the Math Forum fits Little's (1993) ideal of professional development in that it moves beyond targeted discussions of practice to enable teachers to rethink their roles as teachers, colleagues, and members of the broader community of educators.

These teachers are people who feel they are able to make choices in their classrooms about curriculum—in fact, they are often in leadership positions and can help others learn to use the Math Forum and the Internet as well. They see the Math Forum as significantly contributing to their own knowledge and teaching of geometry, as well as to their students' learning.

Not all the teachers fit this description, though. In particular, many teachers are constrained by set curricula and/or their perceptions of what is possible in their classrooms. The Math Forum has resources for teachers in all types of settings, although its strength may be that those who use it do so because they want to. Furthermore, because the Math Forum is not labeled as a "reform" effort, it may in fact be in a position to contribute to such efforts in ways that the labeled projects cannot.

In its structure and emphasis on building a community of teachers, the Math Forum has much in common with the Urban Math Collaboratives, whose focus was the intelligent, committed teacher and his or her renewal of knowledge and practice (Webb & Romberg, 1994b). Like the Math Forum, "the collaboratives were to build links to local mathematicians and to the regional and national mathematics education communities so that teachers could improve their mathematical knowledge, feel more at ease with the subject, and develop a better sense of the contemporary uses of mathematics" (Nelson, 1994, p. 20). The efforts and success of the collaboratives were impressive: Teachers did feel less isolated from each other, they did develop broader definitions of possibilities for professional development, they were intellectually stimulated, and teachers' teaching did change based on their experiences (Popkewitz & Myrdal, 1994; Webb & Romberg, 1994a).

The difference between the "forum" of the collaboratives and the Math Forum is that, in addition to building a community that appears to be leading to the development of shared vocabulary and understanding, the content of the Math Forum is archived and searchable. As teachers become ready to consider its usefulness for themselves and their students, it is there.

One of the complications that emerged for the collaboratives (something, in fact, true of most teacher-based work in schools) was that they not only required a tremendous output of effort, but insights arising within them were only available to a select group of teachers and those with whom they associated. Furthermore, as Webb and Romberg (1994b) observed, there were significant pitfalls, including too little time to reflect, too little time to talk, a lack of ideas placed into context, and the problem of new ideas falling prey to old ways. Although these "pitfalls" are not completely overcome by the technology of the Math Forum, the advantage of having both online and archived projects and discussions is that they are available whenever the teacher is ready and that there is a wide variety of users ready to listen and share ideas.

In fact, the juxtaposition of computer literacy and the mathematics basic to the Math Forum may just be discrepant enough for teachers to facilitate changes in the culture of mathematics teaching. Schofield, Evans-Rhodes, and Huber's (1990) evaluation of the Geometry Tutor certainly indicated that the presence of the Geometry Tutor not only led students to get to class early and to be on task during class, but required teachers to alter their roles from transmitters of knowledge to resource persons.

By introducing teachers to a situation that requires them to reflect on what they know of learning and teaching geometry, the Math Forum becomes an adjunct to the geometry class that emphasizes memorization and rote learning. In offering a place for a community of users to cocreate resources, the Math Forum provides the foundation for a conception of mathematics learning that includes teachers and students in an ongoing process of investigation and communication.

ACKNOWLEDGMENTS

The Math Forum, formerly the Geometry Forum, is an NSF-sponsored virtual resource center for math education on the Internet. The research reported in this chapter was supported by NSF Geometry Forum MDR-9155710. All of our servers use forum.swarthmore.edu as their address for ftp, gopher, http (WWW), and news access; e-mail subscriptions to newsgroups are available. We thank Gabriel Quinn Bauridel, Jane Ehrenfeld, Ann E. Fetter, Kristina Lasher, Heather Mateyak, Eric Sasson, and David Weksler for their contributions to the studies reported. Furthermore, we gratefully acknowledge the editorial assistance of Sarah Seastone in the preparation of this manuscript.

REFERENCES

Bruner, J. S. (1966). *Toward a theory of instruction.* Cambridge, MA: Harvard University Press.

Dreyfus, T., & Hadas, N. (1987). Euclid may stay—and even be taught. In M. M. Lindquist & A. P. Shulte (Eds.), *Learning and teaching geometry, K–12* (pp. 47–58). Reston, VA: National Council of Teachers of Mathematics.

Little, J. W. (1993). Teacher's professional development in a climate of educational reform. *Educational Evaluation and Policy Analysis, 15*(2), 129–151.

National Council of Teachers of Mathematics. (1989). *Curriculum and evaluation standards for school mathematics.* Reston, VA: Author.

National Council of Teachers of Mathematics. (1991). *Professional standards for teaching mathematics.* Reston, VA: Author.

Nelson, B. S. (1994). Mathematics and community. In N. L. Webb & T. A. Romberg (Eds.), *Reforming mathematics education in America's cities: The Urban Mathematics Collaborative Project* (pp. 8–23). New York: Teachers College Press.

New Standards Project. (1995). *Performance standards (draft 5.1).* Washington, DC: Author.

Papert, S. (1993). *The children's machine: Rethinking school in the age of the computer.* New York: Harper Collins.

Popkewitz, T. S., & Myrdal, S. (1994). The "urban" in the mathematics collaboratives: Case studies of the eleven projects. In N. L. Webb & T. A. Romberg (Eds.), *Reforming mathematics education in America's cities: The Urban Mathematics Collaborative Project* (pp. 129–150). New York: Teachers College Press.

Renninger, K. A. (1992). Individual interest and development: Implications for theory and practice. In K. A. Renninger, S. Hidi, & A. Krapp (Eds.), *The role of interest in learning and development* (pp. 361–395). Hillsdale, NJ: Lawrence Erlbaum Associates.

Schofield, J. W., Evans-Rhodes, D., & Huber, B. (1990). Artificial intelligence in the classroom: The impact of a computer-based tutor on teachers and students. *Social Science Computer Review, 8*(1), 24–41.

Usiskin, Z. (1987). Resolving the continuing dilemmas in school geometry. In M. M. Lindquist & A. P. Shulte (Eds.), *Learning and teaching geometry, K–12* (pp. 17–31). Reston, VA: National Council of Teachers of Mathematics.

Webb, N. L & Romberg, T. A. (Eds.). (1994a). *Reforming mathematics education in America's cities: The Urban Mathematics Collaborative Project.* (pp. 196–209). New York: Teachers College Press.

Webb, N. L., & Romberg, T. A. (1994b). Collaboration in practice and final thoughts. In N. L. Webb & T. A. Romberg (Eds.), *Reforming mathematics education in America's cities: The*

Urban Mathematics Collaborative Project (pp. 196–209). New York: Teachers College Press.
Young, J. W. (1925). *Lectures on fundamental concepts of algebra and geometry.* New York: Macmillan.

APPENDIX

Sample Problem of the Week

Pythagorean Theorem—March 25-29, 1996

We have done a fair bit with the Pythagorean Theorem this year, so lets take a look at the theorem itself.

One very familiar proof of this theorem is a right triangle with squares on each side. Here's my question: Do we have to use squares? What about hexagons, or other shapes? What would we have to do to use other shapes, if that's at all possible? And why do you suppose squares are usually used for this proof?

Extra: Name a U.S. President who discovered a proof of the Pythagorean Theorem.

Sample Facilitator Response

[Annie, facilitator]

Maybe I am being too hard with my grading this week, but by my count, only two folks got this one right! I received 48 incorrect solutions, some of which got only the extra part right; this may partly be a case of my having an answer in mind and not really asking the right questions, so I'll try to be more explicit next time.

A number of people said no, you don't have to use squares, and suggested that other shapes such as regular polygons would work. A couple of people showed that you could use hexagons, and proved it, but their answers weren't general enough—the key here is that you can use any shape you want, as long as the shapes on the three sides are similar! Thomas and Brian both got that idea in; Thomas actually said similar shapes, and Brian said shapes with proportional lengths. So you could really say the proof says that, given similar shapes on all the edges, the area of the shape on one leg plus the area of the shape on the other leg equals the area of the shape on the hypotenuse. You can check this out in Sketchpad or something—it is pretty cool!

A lot of people said that squares are used because they are easiest, and that's certainly true—the area of a square looks just like the terms we are used to seeing in the Pythagorean Theorem, so this makes it really convenient.

If you come across a problem like this, and you determine that other shapes will work, try to figure out what those other shapes might look like, or what limitations there might be. While you might find a couple that work, see if you can find any that don't, and hence narrow things down a bit.

Sample Student Solutions to the Problem

From: kuo@h3912.chinalake.navy.mil
From: Thomas S. Kuo
Email: ssusd2@owens.ridgecrest.ca.us
School: Murray Junior High School, Ridgecrest, California
Grade: 7th

(1) We don't have to use squares to prove the Pythagorean Theorem.

(2) Hexagons or other shapes could be used too. For a regular hexagon with side a:

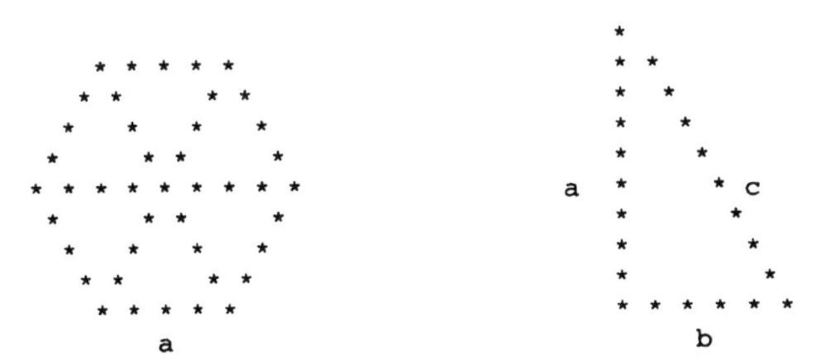

The hexagon is formed by six congruent equilateral triangles. Each triangle has side a and height (sqrt (3)/2) * a. Then the area of a triangle is (1/2) * (a) * (sqrt (3) * a/2) = (a^2) * sqrt * (3)/4. The area of the hexagon is 6 * (a^2) * sqrt (3)/4 = (a^2) * 3 sqrt (3)/2.

Let a, b and c are three side of a right triangle as shown. When we form three hexagons with side a, b, and c, their areas will be (a^2) * 3 * sqrt (3) / 2, (b^2) * 3 * sqrt (3) / 2, and (c^2) * 3 * sqrt (3) / 2, respectively. Now I have to check out if the sum of areas of the first two is equal to that of the third one. After we cancel out the common factor, we have only a^2, b^2 and c^2 left and the relationship a^2 + b^2 = c^2 still holds.

The conclusion is that we can prove the Pythagorean Theorem by using hexagons. It is also true if other shapes are used. The fact is that as long as we use similar shapes with side a, b, and c, their areas will have ratio of a^2, b^2, and c^2. After cancelling out the common factor, it can be simplified to a^2 + b^2 = c^2.

(3) I believe that the reason to use squares in the proof is due to the fact that it is the most direct and easiest method.

* * *

From: bmgordon@ntplx.net
From: Brian Gordon
Grade: 1992
School: Dartmouth

I have an answer for the bonus. It's James Garfield, right? I believe the proof is like this:

```
  |\
b | \  c
  |__\
    a
```

Paste two copies of this triangle together and connect the other acute vertices to make a trapezoid like this:

```
     a
   ____
  |   /*
b |  /c *
  | /    *
  |`-c_  *
a |____\*
     b
```

Now...
the area of the trapezoid = 1/2 * height * sum of bases = 1/2 * (a+b) * (a+b)
The areas of the triangles are:

1. 1/2 * a * b
2. 1/2 * a * b

3. 1/2 * c * c because they meet at a right angle. This can be shown by noting that the straight angle that is the left side of the trapezoid is made up of two complementary angles (from the two congruent right triangles) and the angle of the two sides of length c (which is then 180 - 90 = 90).

Equate the trapezoid formula with the sum of the triangles:

Double everything to eliminate those pesty 1/2's:

(a+b) (a+b) = ab + ab + cc do the FOIL. . .

a^2 + 2ab + b^2 = 2ab + c^2 subtract the 2ab. . . .

a^2 + b^2 = c^2

I like this one much more than the proof I originally learned, which was based on the altitude-on-hypotenuse and similar triangles.

I never actually learned the squares-on-the-sides proof. As for the other shapes, I think you could do it, if all the constructions are done right. That's because the areas of the shapes would be proportional to the square of the lengths of the sides.

I have a formula for a regular n-gon with side s. It's A=.25 * n * s^2 * cotangent (180/n). Could you guys patent that for me? :)

—bri

EPILOGUE

Organization and Freedom in Geometry Learning and Teaching

Rina Hershkowitz
Weizmann Institute of Science

Writing this epilogue was not an easy task for me. I originally intended to reflect on each chapter, as well as on the book as a whole. However, especially in such a rich book, I found that focusing on the tapestry of the whole often led me away from particular, albeit worthy, issues raised in every chapter. Accordingly I foreground what seems to be at the heart of the book and only occasionally focus on particulars raised in any one of the chapters. Nonetheless, I commend each and every chapter because the authors have individually and collectively crafted ground-breaking work in the field.

This volume asserts a focus on "designing learning environments for developing understanding of space and geometry," and indeed almost all the chapters can be viewed and examined on a continuum of a design and development process. The book in its three parts and its 19 different chapters expresses a unique and holistic approach to geometry learning, in which some characteristics seem especially prominent. The authors exercise a refreshing democracy of content, adopting a broad vision of what geometry and visualization could be, rather than what they often are in school. The authors also attend carefully to how students develop knowledge about space and geometry across a wide range of contexts.

FREEDOM IN SELECTING GEOMETRICAL CONTEXT AND CONTENT

The volume is characterized first by the variety of contexts that can serve as "geometrical starting points" or "geometrical springboards," as expressed by Lehrer, Jacobson, et al. in chapter 7, for designing learning environments in geometry. The chapters represent a diverse collection of meaningful situations and problems, a diversity that far exceeds that found in traditional texts or even in recent innovative curriculum development efforts. Diversity in context is accompanied by diversity in content: Students design quilts, model situations, invent units of measure, make conjectures, and establish proofs.

Among the different environments for learning, some are suitable for young children (i.e., chaps., 6, 7, 8, 9) and some for middle or high schools (i.e., chaps. 3, 12, 13, 15, 16). Some environments are based on the power of dynamic tools, by which traditional content is learned in new ways (i.e., chaps. 3, 12, 14, 15, 16). Other environments use tools to address nontraditional content, like mapping a classroom (chap. 17) or designing a quilt (chap. 7). In addition, there are chapters that tell special "stories" of geometry (i.e., Devaney's chap. 4 on chaos and fractals, or Middleton and Corbett, chap. 10, on engineering tasks). Even the traditional high-school course in Euclidean geometry is no longer discussed in terms of "must stay or must go," as if it is the only player on the stage. Beginning with the general view of visualization and geometry described in the first chapter by Goldenberg, Cuoco, and Mark, this volume develops a compelling case for broadening what we consider the content of school geometry. Moreover, the authors describe a variety of novel and productive environments for learning about space and geometry.

LEARNING GEOMETRY

In most of the designed environments, students play active roles in developing geometrical "habits of mind," as Goldenberg et al. (chap. 1) phrase it. Most authors analyze, in theoretical or experimental ways (or both), the processes of geometry learning. Although student roles in these environments range from measuring attributes to designing products, nevertheless the authors' descriptions of student learning share a collective emphasis on internalization, mathematization, and justification.

Internalization

Internalization is used in a Vygotskian spirit, as the "transformation of external activity into internal activity" (Wertsch & Stone, 1989, p. 162). Of special importance are mental actions that follow "real" changes of shape and form under different circumstances. Examples include the change from "what I see" to "how I see" in accordance with the change of the observer's position, described by Gravemeijer in the second chapter, and the continuous change of shapes under similarity and self-similarity in the chapter "Chaos in the Classroom" by Devaney (chap. 4). Further opportunities for transformation of external activity are provided by the "drag mode" in the chapters that deal with dynamic geometry (Chazan & Yerushalmy, Goldenberg & Cuoco , de Villiers, Olive). In all these, the student is put in situations where he or she observes real changes and investigates the invariants under these changes, in order to construct higher mental actions—like better understanding and calculating dimensions in the case of fractals, or generalizing and conjecturing in dynamic geometry environments.

A different sense of internalization is suggested by the Clements et al. chapter on the development of geometric and measurement ideas in the Geo-Logo environment. The authors describe how students develop a "conceptual ruler" composed of "interiorized units of length," and how children have to work quite hard in order to internalize the relationships of turns in order to impose the dynamic transformation (by the tool) onto static figures. In both tasks students connect spatial schemes with numeric schemes (the line and its "length" or the turn and its "angle"). In a somewhat different vein, Lehrer et al. (chap. 7) trace the classroom history by which a unit of construction, a "core square," in lessons on quilting becomes interiorized as a unit of measure in a subsequent series of lessons on area measure.

Mathematization

All the preceding examples and many others in the book also demonstrate, explicitly or implicitly, progressive mathematization. Progressive mathematization is consistent with Freudenthal's philosophy of mathematics as a human activity in which *mathematizing* is seen as a sort of an organizing process by which elements of a context are transformed into mathematical objects and relations (see Gravemeijer in chap. 2). In this sense, making conjectures, as discussed in the different dynamic geometry chapters, can be seen as a mathematizing activity in which inductive investigations in the dynamic situation lead to an "organized" mathematical conjecture. It seems that the student cannot learn with the dynamic geometry environment unless the student both interacts with the visual environment and attempts some sort of mathematization.

The quilt design task described by Lehrer et al. (chap. 7) exemplifies a related sense of mathematization in that children's informal knowledge of quilts and their esthetics about what makes a quilt "interesting" are progressively transformed and reexpressed as the mathematics of the plane. Similarly, Battista and Clements (chap. 9) describe a process in which children's initial conception of a three-dimensional rectangular array of cubes as an "uncoordinated set of faces" is progressively transformed into a coordinated view that enables the construction of cube buildings. Watt (chap. 17) also describes a particularly compelling example of mathematization: children's perception of their classroom is transformed by their efforts to map it. What children "see" changes as they progressively come to understand their classroom geometrically.

Reasoning: Explanation, Justification, and the "Why" Problem

Explanation processes are central to much of the learning described in this volume. Explanation processes (perhaps "reasoning processes" is a better expression), are taken here in a broad sense, meaning the variety of actions that students take in order to explain to others, as well as to themselves, what they see, what they do, and what they think and conclude.

Here again the uniqueness of the book is expressed in the mathematical as well as the cognitive freedom toward legitimate kinds of reasoning. The different chapters discuss quite different justification processes. We are no longer captured only in the deductive reasoning typical of the traditional Euclidean geometry course, in which justification processes are often governed by the rigid rules of two-column proof.

The different kinds of reasoning and explanations emerging from the need to act geometrically in the learning environments described in the book are part of the similarities and differences among these geometrical environments. In some environments, explanations start in actions like measuring, counting, and designing (chaps. 7, 8, 9, 11) or accompany efforts to build and account for the properties of structure (chap. 10).

Inductive reasoning plays a central role in many of the learning environments. For example, Koedinger (chap. 13) writes about "the recognition of the importance of multiple examples as providing better evidence than a single example," and about proving as an extended failure to find a counterexample. In other words, the problem of verifying a conjecture by inductive means is a trigger for many of the investigations conducted by students (see especially the chapters featuring dynamic geometry and chap. 3 by Chazan & Yerushalmy).

Deductive reasoning has many faces and many roles in this volume. The authors focus on deductive justifications rather than on deductive proofs. An important issue is the extent to which deductive reasoning is a vehicle for verifying geometrical (rediscovered) conjectures versus deductive reasoning as a vehicle for explaining *why* the discovered conjecture is true. The first role is criticized, for example by de Villiers (chap. 15). He argues that convincing someone of the truth of a conjecture should take precedence over the process of proof for verification, even in the traditional geometry teaching, and not vice versa, noting that "personal conviction usually depends on a combination of intuition, quasi empirical verification, and the existence of a logical (but not necessarily rigorous) proof." de Villiers goes on to suggest that dynamic geometry environments can contribute to student experimentation in ways that help them come to see proof as a form of explanation rather than of verification, because verification is obtained quickly and relatively easily (by experimentation) when students use the tools of dynamic geometry.

In several chapters the authors describe situations where students are given opportunities not only to construct their ways of reasoning but also to reflect on them as well. For example, Lehrer, Jacobson, et al. (chap. 7) bring many examples from the classrooms. One of them tells about a discourse between teacher and her students about the nature of reasoning they just went through—the limitations of induction. The authors claim that such a discourse "sets the stage for (informal) proof as a form of argument."

Explanation processes are important not only because of their pedagogical value. Researchers also see them as a vehicle for investigating habits of mind: chapter 5 by Pegg and Davey and chapter 6 by Lehrer, Jenkins, and Osana trace the development of children's arguments about shape and form and of other fundamental aspects of reasoning about space. Geometry educators can use the information and conclusions from these kind of investigations in order to design appropriate learning environments, although the contrast between geometric reasoning in everyday and designed contexts obtained by comparing the sixth and seventh chapters is striking.

Additional Comments

Many authors view internalizing, mathematizing, and justifying as processes that are woven together in order to align the views and interactions of people with the spatial world in which they live. I see the concepts of horizontal and vertical mathematization (see chap. 2) as one aspect of internalization; internalization (as a process) is partially expressed in the ability to go through horizontal and vertical mathematization. In order to be able to explain (in a broad sense) one should go through internalization (mathematization) processes. Even in quite simple tasks (e.g., see Battista & Clements, chap. 9), children base their explanations on the process of internalization while going through processes that are mathematization processes as well.

The relationship between the context of the learning environment and mathematical reasoning is worth further exploration. Looking at the variation in contexts presented in these chapters, it seems that conceptual change proceeds along different trajectories and provides a foreground for different aspects (e.g., inductive vs. deductive arguments) of reasoning about space. This means that broad-scope theories of geometrical thinking, like those of Piaget or van Hiele, fail to account well for context. Hence, we need more context-specific research followed by the evaluation of partial theories.

Many chapters discuss the "frame" of the geometrical learning—that is, the pedagogical and epistemological analysis of why this or that environment is appropriate for geometry learning—but only a few give us information on how the above processes occur, especially in a real classroom situation. Lehrer, Jacobson, et al. (chap. 7) are exceptional again; they write about teacher practices, student learning, and a classroom implementation, in which they speak about the transition in task structure (over 3 years) and about the representational fluency in classroom discourse. Here they describe a dramatic change in the proportion of time spent on exploring the mathematics of space, in the awareness of teachers toward their students' thinking and language, and in how teachers use their knowledge of student thinking to generate classroom practices where, among other things, "talk became intimately connected to justification and argument."

Finally, the work of the authors provides compelling evidence for making geometry part of every child's education throughout schooling. This poses formidable problems for professional development and for the redesign of teacher education. Fortunately, this volume provides some signposts for change. For example, Renninger, Weimar, and Klotz's discussion of creation of the Math Forum (chap. 19) shows how human and technological resources can be effectively deployed to create online learning communities that foster professional development even as they serve to further student learning. Innovations like these provide a departure point for the next series of challenges for geometry education and a fitting conclusion to a volume that sets the pace for innovation.

REFERENCE

Wertsch, J. V., & Stone, C A. (1989). The concept of internalization in Vygotsky's account of the genesis of higher mental functions. In J. V. Wertsch (Ed.), *Culture communication and cognition* (pp. 162–179). Cambridge: Cambridge University Press.

Author Index

Subject Index